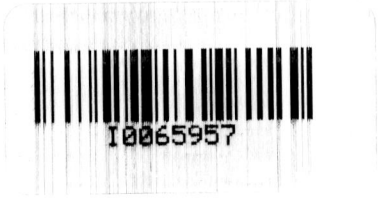

Agricultural Science and Technology

Agricultural Science and Technology

Editor: Adriana Winkler

R CALLISTO REFERENCE

www.callistoreference.com

Callisto Reference,
118-35 Queens Blvd., Suite 400,
Forest Hills, NY 11375, USA

Visit us on the World Wide Web at:
www.callistoreference.com

ISBN: 978-1-64116-091-9 (Hardback)

Cataloging-in-Publication Data

Agricultural science and technology / edited by Adriana Winkler.
    p. cm.
Includes bibliographical references and index.
ISBN 978-1-64116-091-9
1. Agriculture. 2. Agriculture--Research. 3. Agricultural innovations.
4. Agriculture--Technology transfer. I. Winkler, Adriana.
S493 .A37 2019
630--dc23

Table of Contents

Preface

Agricultural science combines the principles and techniques of varied branches of science and engineering. It is concerned with the development and reform of agricultural production, equipment design and machinery. Agricultural technology has implications and relevance in various other scientific fields such as crop production, crop management, livestock production, waste management, etc. This book is compiled to provide detailed information about the theory and practice of agricultural science. It strives to provide a fair idea about this discipline and to help develop a better understanding of the latest advances within this field. The data collated herein presents multiple approaches and perspectives related to this field. Agriculturists, students, researchers and experts will find this book full of unexplored aspects.

The information contained in this book is the result of intensive hard work done by researchers in this field. All due efforts have been made to make this book serve as a complete guiding source for students and researchers. The topics in this book have been comprehensively explained to help readers understand the growing trends in the field.

I would like to thank the entire group of writers who made sincere efforts in this book and my family who supported me in my efforts of working on this book. I take this opportunity to thank all those who have been a guiding force throughout my life.

Editor

Alleviating Effects of Exogenous Glutathione, Glycinebetaine, Brassinosteroids and Salicylic Acid on Cadmium Toxicity in Rice Seedlings (*Oryza Sativa*)

Fangbin Cao[1], Li Liu[1,2], Wasim Ibrahim[1], Yue Cai[1] and Feibo Wu[1]*

[1]*Department of Agronomy, College of Agriculture and Biotechnology, Zhejiang University, Hangzhou, China*
[2]*Hangzhou Wanxiang Vocational and Technical College, Hangzhou, China*

Abstract

A hydroponic experiment was conducted to study the ameliorative effects of 24 h pretreatment with exogenous glutathione (GSH), glycinebetaine (GB), brassinosteroids (BRs) and salicylic acid (SA) upon 50 µM cadmium (Cd) stress to rice seedlings. The results showed that Cd caused a significant reduction in seedling height, chlorophyll content and biomass, the activity of POD in stems, and shoot Mn, shoot/root Zn and root Cu concentration, but improved SOD and POD activities in leaves with elevated MDA accumulation and Fe concentration. Pretreatment with 100 µM GSH, GB or SA greatly alleviated Cd-induced growth inhibition and suppressed Cd-induced MDA accumulation. Compared with Cd alone treatment, pretreatment of GSH, GB and SA markedly increased chlorophyll content; reduced shoot Cd concentration, and GSH also decrease root Cd content. GSH, GB and SA pretreatments counteracted the pattern of alterations in certain antioxidant enzymes induced by Cd, e.g. significantly suppressed Cd-induced increase of leaf SOD activity, GB and BRs also significantly decreased leaf POD activity; GSH significantly elevated the depressed stem POD and SOD activities; and GB elevated stem SOD activity. Compared with Cd alone, GSH pretreatment significantly relieved Cd-induced reduction in Cu or increase in Fe concentration in leaves; GB pretreatment decreased Cd-induced Fe enhancement; SA pretreatment markedly increased shoot Fe, but reduced Mn concentration. Although BRs pretreatment increased plant dry biomass with no effect on chlorophyll content and MDA accumulation, significantly increased Cd concentration both in shoots and roots.

Keywords: Alleviation; Brassinosteroids (BRs); Cadmium (Cd); Glutathione (GSH); Glycinebetaine (GB); Rice (*Oryza sativa* L); Salicylic acid (SA)

Introduction

Heavy metal contamination in soil could resulted in inhibition of plant growth and yield reduction, and even pose a great threat to human health via food chain [1]. Among heavy metals, Cadmium (Cd) in particular causes increasingly international concern [2]. Cd-contaminated soil results in considerable accumulation of Cd in edible parts of crops, and then it enters the food chain through the translocation and accumulation by plants [3-6]. Cereals, especially rice, the staple food in East Asia, is a major source of heavy metal intake [7]. For example, rice, a staple crop for Japanese, was estimated to represent 36-50% of the total oral intake of Cd for Japanese population during 1998-2001 [8]. Correspondingly, it is imperative to find reliable approaches to decrease Cd accumulation in rice aimed for decreasing Cd content in human food.

Previous studies have shown that Cd toxicity causes oxidative stress as accumulation of reactive oxygen species (ROS) and subsequent production of membrane lipid peroxides [9,10]. Reduced glutathione (GSH, γ-Glu-Cys-Gly), an essential metabolite and regulator in plants, plays important role in the cellular defense against abiotic stress [11]. It was reported that GSH can significantly alleviate the Cd induced growth inhibition [12,13]. Zhu et al. [14,15] observed that the over-expression of glutathione synthetase and γ-glutamine homocysteine synthase can markedly improve the tolerance to Cd stress in indian mustard. Salicylic acid (SA) is a cellular signal element and plays an important role against biotic and abiotic stress in plants [16]. Barley seeds were presoaked in SA for 6 h can significantly reduce the Cd content and alleviate the lipid peroxidation in the seedling under Cd stress [17]. Moreover, SA can reduce the damage of photosynthesis system caused by Cd in maize [18]. Glycinebetaine (GB), commonly found in higher plants, is a very important osmoregulation substance. Under abiotic

stress, such as drought and salinity, its concentration would increase [19]. And the foliar spray of GB could markedly alleviate the water stress [20]. Furthermore, the exogenous GB could also improve the tobacco tolerance against Cd stress [21]. Brassino steroids (BRs) are a new class plant steroidal hormone. It plays an important role in a widely spectrum of physiological processes, such as leaf epinasty, pollen tube growth and stem elongation [22]. In addition, BRs has important function against environmental stress. Now, BRs attracted more attention in adaptive response to abiotic stress, particularly in respect to chilled stress [23], water stress [24], and heat stress [22]. Meanwhile, BRs application can decrease the Cd-induced stress through enhancing antioxidant systems in *Brassica juncea* [25]. On the whole, application of GSH, GB, SA and BRs have been widely investigated in alleviating abiotic stresses, including Cd toxicity. However, effects of pretreatment of GSH, GB, SA and BRs against Cd toxicity have not been reported except SA, which could alleviate Cd-induced oxidative damage in rice root. In this study, a hydroponic experiment was conducted to determine: whether GSH, GB, BRs and SA could alleviate Cd toxicity, whether GSH, GB, BRs and SA application altered uptake and translocation of Cd and other elements in rice plants, and whether antioxidant enzymes were involved in the mitigation-measure-mediated protective responses of rice plants exposed to Cd stress.

*Corresponding author: Feibo Wu, Department of Agronomy, College of Agriculture and Biotechnology, Zijingang Campus, Zhejiang University, China
E-mail: wufeibo@zju.edu.cn

Materials and Methods

Culture condition and treatments

A hydroponic experiment was carried out in the greenhouse of Huajiachi campus, Zhejiang University, Hangzhou, China. Rice Xiushui63 (*Oryza sativa* L. *Japonica* unwaxy) seeds were surface sterilized with 2% H_2O_2 for 20 min, and fully rinsed with deionized water, soaked in deionized water at room temperature for 2 days, then germinated for 1 day at 35°C. Germinated seeds were sown in sterilized sand bed and kept in an incubator at 30°C-day/26°C-night under 85% relative humidity for 10 days. At two-leaf stage, the uniform healthy plants were transplanted into plastic containers filled with 5-L basic nutrient solution. The container was covered by a polystyrol plate with 7 evenly spaced holes (2 plants per hole) and placed in greenhouse. After 7 days of transplanting, rice seedlings were pre-treated with 100 μM GSH, 100 μM GB, 10 μM BRs or 100 μM SA for 24 h and then exposed to 50 μM Cd in nutrient solution for 5 d. There were six treatments: (1) Control (basic nutrient solution), (2) Cd (50 μM Cd), and (3) pre-GSH+Cd, (4) pre-GB+Cd, (5) pre-BRs+Cd, (6) pre-SA+Cd, which were pretreated with 100 μM GSH, 100 μM GB, 10 μM BRs, 100 μM SA for 24 h, respectively, then exposed to 50 μM Cd for 5 d. The composition of the basic nutrient solution was (mg l^{-1}): NH_4NO_3 57.1; $NaH_2PO_4\cdot 2H_2O$ 25.2; K_2SO_4 44.7; $CaCl_2$ 55.4; $MgSO_4\cdot 7H_2O$ 202.5; $MnCl_2\cdot 4H_2O$ 0.94; $(NH_4)_6Mo_7O_{24}\cdot 4H_2O$ 0.05; H_3BO_3 0.59; $ZnSO_4\cdot 7H_2O$ 0.03; $CuSO_4\cdot 5H_2O$ 0.02; Fe-citrate 7.44. The experiment was laid in a completely randomized design with 4 replicates. The nutrient solution pH was adjusted to 5.6 ± 0.1 with NaOH or HCl as required. Plant samples were collected after 5 d Cd exposure. The fresh roots and the upper second fully expanded leaves were sampled, and immediately frozen in liquid nitrogen and stored frozen at -80°C for further analyses or directly used for the determination of anti oxidative enzyme activities and MDA content.

Measurements of chlorophyll content and plant growth parameters

The upper second fully leaves were selected to measure SPAD (Soil Plant Analysis Development) value with three replicates. And chlorophyll content of the second uppermost fully expanded leaves was determined according to Cai et al. [12]. After measuring plant height and root length, roots were soaked in 20 mM Na$_2$-EDTA for 3 h and rinsed thoroughly with deionized water. Plants were separated into roots and tops (shoots and leaves), and then dried at 80°C and weighed. Dried shoots and roots were powdered and weighted, then ashed at 550°C for 12 h. The ash was digested with 5 ml 30% HNO_3, and then diluted with deionized water. Metal concentrations were determined using a flame atomic absorption spectrometry (SHIMADZU AA-6300, Japan) [26].

Assay of enzyme activities and MDA content

For the determination of enzyme activities, plant tissue was homogenized in 6 ml 50 mM sodium phosphate buffer (PBS, pH 7.8,) using a pre chilled mortar and pestle, then centrifuged at 10,000×g for 20 min at 4°C. The supernatant was used for enzyme assay. Superoxide dismutase (SOD), peroxidase (POD) and MDA content were measured according to Zeng et al. [27].

Statistic analysis

Statistical assay were performed with SPSS version 17.0. One-way ANOVA was carried out by the Duncan's Multiple Range Test (SSR) to analysis the difference of significance among treatments.

Results

Effect of GSH, GB, BRs and SA pretreatment on growth of Cd-stressed rice

As shown in table 1, exposure to Cd caused obvious inhibition of growth traits. To evaluate the alleviating effects of glutathione (GSH), glycinebetaine (GB), brassino steroids (BRs), salicylic acid (SA) pretreatment for 24 h, we adopt the following formula-based integrated score: SPAD×0.5+shoot height×0.125+root length×0.125+fresh weight×0.125+dry weight×0.125 (Table 1). There is a positive correlation between alleviating effect of different treatments and the integrated scores. According to the integrated scores, the alleviation effect is in order of GSH>GB>SA>BRs. GSH, GB, SA greatly alleviated the Cd-induced growth inhibition. Compared with Cd treatment, GSH pre-treatment increased SPAD value, root length, shoot dry weight and root dry weight by 57.2%, 26.1%, 15.5% and 9.8%, respectively; GB pre-treatment increased by 41.0%, 19.6%, 30.8% and 27.4%, respectively; SA pre-treatment increased by 42.2%, 25.0%, 38.5% and 31.4%, respectively.

Effect of GSH, GB, BRs and SA pretreatment on chlorophyll content

Exposure of the rice seedlings to 50 μM Cd for 5 d markedly reduced Chl and b contents by 11.8% and 14.0%, respectively, compared with control. GSH, GB and SA pretreatment for 24 h greatly relieved Cd-induced reduction in Chl a and Chl b content and the values returned close to or even higher than control, c.f. increased by 43.1%, 19.5% and 24.5% (Chl a), and 48.7%, 22.4% and 26.7% (Chl b), respectively,

	SPAD value		Shoot height (cm)		Root length (cm)		Fresh weight (g per plant)				Dry weight (g per plant)				Interqrated Score	
							Shoot		Root		Shoot		Root			
Control	20.4	c	17.7	a	12.0	a	121	a	66	a	16	ab	5.7	bc	40.0	a
pre-GSH+Cd	26.1	a	16.3	ab	11.6	a	102	b	63	a	15	b	5.6	c	39.7	a
pre-GB+Cd	23.4	b	16.3	ab	11.0	a	113	ab	66	a	17	ab	6.5	ab	40.4	a
pre-BRs+Cd	16.1	d	16.5	ab	11.8	a	109	ab	68	a	17	ab	6.6	a	36.7	b
pre-SA+Cd	23.6	b	16.1	b	11.5	a	115	ab	66	a	18	a	6.7	a	41.0	a
Cd	16.6	d	15.4	b	9.2	b	98	b	60	ab	13	c	5.1	d	33.4	c

Table 1: Effect of GSH, GB, BRs and SA on SPAD value, and growth traits of rice seedlings exposed to Cd stress.
Rice seedlings were pre-treated with 100 μM GSH, 100 μM GB, 10 μM BRs and 100 μM SA for 24 h and then exposed to 50 μM Cd in nutrient solution for 5 d. Control and Cd correspond to basic nutrient solution and 50 μM Cd; *pre*-GSH+Cd, *pre*-GB+Cd, *pre*-BRs+Cd and *pre*-SA+Cd correspond to 5 d 50 μM Cd exposure after 24 h pretreated with 100 μM GSH, 100 μM GB, 10 μM BRs and 100 μM SA, respectively. The same as below. Different letters indicates significant differences among treatments (P<0.05). Error bars represent SD values.
Integrated score=absolute values of [SPAD value×0.5+shoot height×0.125+root length×0.125+fresh weight×0.125+dry weight×0.125].

compared with Cd alone treatment (Figure 1). No effect was detected in BRs pretreatment on Chla and Chl b content.

Effect of GSH, GB, BRs and SA pretreatment on SOD and POD activity in leaf, stem and root of rice

Cd stress caused 54.7% increase in leaf SOD activity, while no effect on stem and root SOD activity relative to control (Figure 2). Pretreated with GSH, GB, BRs and SA significantly lowered Cd-mediated increase in SOD activity by 17.9%, 22.5%, 15.6% and 24.0% in leaves, but no effect on stems and roots except stems of GSH and GB, compared with Cd alone treatment. Concerning POD activity, Cd stress caused 38.9% increase in leaves, but 37.1% reduction in stems and no effect on roots (Figure 2). In leaves, GB and BRs pretreatment significantly decreased POD activity and dropped it back to control level, c.f. decreased by

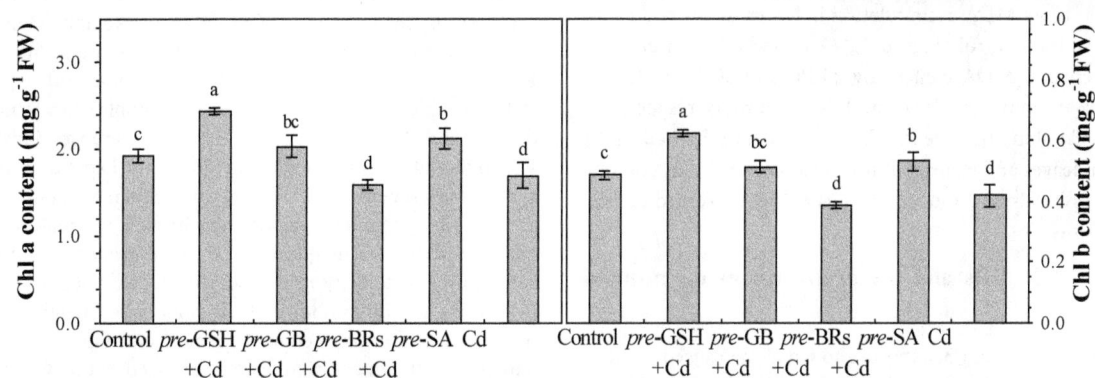

Figure 1: Effect of Cd on chlorophyll content and as affected by pretreatment of GSH, GB, BRs and SA in rice seedlings. Rice seedlings were pre-treated with 100 μM GSH, 100 μM GB, 10 μM BRs and 100 μM SA for 24 h and then exposed to 50 μM Cd in nutrient solution for 5 d. Control and Cd correspond to basic nutrient solution and 50 μM Cd; pre-GSH+Cd, pre-GB+Cd, pre-BRs+Cd and pre-SA+Cd correspond to 5 d 50 μM Cd exposure after 24 h pretreated with 100 μM GSH, 100 μM GB, 10 μM BRs and 100 μM SA, respectively. Error bars refer to SD value. Different letters indicate significant differences (P<0.05) among treatments.

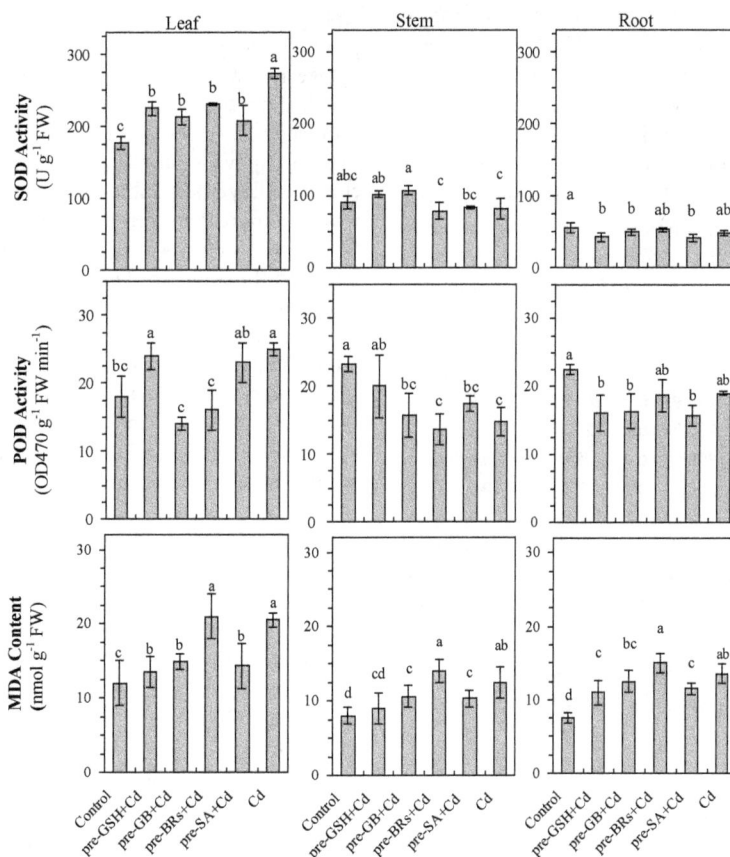

Figure 2: Effect of GSH, GB, BRs and SA on SOD, POD activities and MDA content in leaf, stem and root of rice seedlings exposed to Cd stress. Different letters indicate significant differences (P<0.05) among the 6 treatments.

44.0% and 36.0%, respectively, compared with Cd alone treatment. In stems, only GSH pretreatment markedly elevated Cd-mediated decrease POD activity, increased by 36.4%. Yet, no effect was observed in other 3 mitigation treatments. In roots, GSH, GB, BRs and SA pretreatment kept POD activity similar with Cd alone treatment.

Effect of GSH, GB, BRs and SA pretreatment on MDA content in leaf, stem and root of rice

In comparison with control, Cd alone treatment induced 70.8%, 56.2% and 80.0% more MDA accumulation in leaves, stems and roots, respectively, over the control (Figure 2). GSH and SA pretreatment significantly decreased MDA content by 34.1% and 30.2% in leaves, 28.0% and 17.6% in stems, and 18.5% and 14.8% in roots, respectively, compared with Cd alone treatment. GB pretreatment inhibited MDA accumulation in leaves and stems, but no effect on root MDA content. However, BRs pretreatment did not mitigate MDA overproduction induced by Cd stress.

Effect of GSH, GB, BRs and SA pretreatment on mineral concentration

As shown in tables 2 and 3, exposure to Cd alone caused obvious increase in Fe in shoots/roots, and reduction in shoots/roots Zn, shoots Mn and root Cu, compared with control. GSH, GB and SA pretreatment significantly reduced shoot Cd concentration by 46.1%, 22.0% and 20.5%, respectively, but no effect on root except GSH pretreatment which markedly decreased Cd concentration by 24.3%. On the contrary, BRs pretreatment significantly increase the shoot and root Cd concentration. In leaves, GSH pretreatment significantly decreased Fe and Mn concentration by 23.1% and 13.2%, but increase Cu by 26.9%, compared with Cd alone. GB pretreatment decreased Fe and Mn by 25.6% and 23.4%, respectively. BRs pretreatment markedly elevated Cu by 26.7%, but decreased Fe and Zn by 28.2% and 15.8%, respectively. Moreover, SA pretreatment significantly increased Fe by 30.8%, but decreased Mn by 14.6%. In roots, GSH pretreatment decreased Mn and Cu by 30.3% and 12.5%, compared with Cd alone treatment. GB decreased Fe and Mn by 22.8% and 33.9%, respectively. BRs significantly decreased Fe, Mn and Cu by 40.0%, 34.9% and 18.2%, respectively. SA only decreased Mn by 20.2%.

Treatment	Cd		Fe		Mn		Zn		Cu	
Control	0.0	e	29.6	c	423.9	a	244.6	a	31.9	ab
pre-GSH+Cd	76.3	d	29.8	c	237.3	cd	125.7	bc	33.3	a
pre-GB+Cd	109.6	c	28.7	c	209.0	d	122.9	bc	29.4	ab
pre-BRs+Cd	163.2	a	27.9	c	258.7	bc	117.0	c	33.3	a
pre-SA+Cd	112.5	c	51.0	a	232.7	cd	131.5	bc	31.4	ab
Cd	141.3	b	38.8	b	272.6	b	139.0	b	26.0	b

Table 2: Effect of GSH, GB, BRs and SA on metal concentration (mg kg^{-1} DW) in shoots of rice seedlings exposed to Cd stress. Different letter presents significant difference among treatments (P<0.05).

Treatment	Cd		Fe		Mn		Zn		Cu	
Control	0.0	d	343.5	b	120.7	a	614.5	a	110.8	a
pre-GSH+Cd	473.3	c	433.2	a	75.7	b	361.7	b	76.9	c
pre-GB+Cd	638.8	b	345.5	b	72.2	b	420.1	b	91.7	b
pre-BRs+Cd	676.5	a	269.1	c	70.9	b	335.5	b	71.6	c
pre-SA+Cd	618.4	b	426.9	a	86.6	b	372.1	b	91.6	b
Cd	625.1	b	447.7	a	108.9	a	356.4	b	87.6	b

Table 3: Effect of GSH, GB, BRs and SA on metal concentration (mg kg^{-1} DW) in roots of rice seedlings exposed to Cd stress. Different letter presents significant difference among treatments (P<0.05).

Conclusion

Recent years, Cd contamination has become a worldwide environmental issue. In this work, we analyzed the possible role of exogenous GSH, GB, BRs and SA pretreatment in modulation antioxidant defense system and mineral absorption against Cd stress in rice. Cd stress induced plant growth inhibition has been well described by many researchers [28,29]. In the current study, shoot height, root length, shoot fresh weight and plant dry biomass were severely decreased at 50 μM Cd treatment compared to control. GSH is the direct substrate for the synthesis of phytochelatins (PCs). GSH acts as a first defense line against metal toxicity through complexing metals before induced synthesis of PCs reaches to an effective level. PCs could bind and sequestrate Cd in stable complex, and then transport the complex to vacuolar. The roles of GSH and PCs in heavy metal tolerance were well illustrated in Cd-sensitive mutants of Arabidopsis [30]. Over expression of glutathione synthetase enhances Cd tolerance in Indian mustard with superior heavy-metal phyto remediation capacity [15]. Pretreatment of GSH, GB or SA to Cd stress medium effectively alleviated Cd-induced growth inhibition and toxicity in rice seedlings, described as its capability to preventing the inhibition of SPAD value, plant height, shoot/root dry weight and chlorophyll content (Table 1 and Figure 1). And GSH, GB and SA pretreatment significantly reduced shoot Cd concentration. Under abiotic stress, the concentration of GSH, GB and SA would increase, which play important role in mediating plants responses to the stress [19,31,32]. Moreover, GSH, GB and SA pretreatment could significantly alleviate the Cd-induced growth inhibition and decrease the Cd content in plants [5,33,34]. So, the results suggested a practical potential for exogenous GSH, GB and SA application as an intervention strategy in alleviating Cd stress and reducing Cd translocation in rice plants. Although BRs markedly increased dry weight, it increased Cd uptake and translocation. Chlorophyll content could indicate the plant health and was more accurate and sensitive than shoot dry weight and root length [12]. Our previous study suggested that Cd could suppress the chlorophyll synthesis [12,28]. In the present study, Cd-induced chlorophyll synthesis inhibition was markedly reverted and the content was even more than control when rice seedlings were pre-treated with GSH, GB or SA for 24 h. Contrary to these results, BRs pretreatment did not affect the chlorophyll synthesis compared with Cd alone treatment. This is opposite with Hayat et al. [35], who found BRs could enhance the chlorophyll content under Cd stress by foliar spray. The possible reason might be due to the difference of methods and plant species. MDA content is considered to be an indicator of lipid peroxidation [28]. In the present study, Cd stress induced oxidative stress characterized significant increase in MDA content compared with control. More MDA accumulation could account for presence of the poisoning reactive oxygen species (ROS) [28]. Under environmental stress conditions, such as Cd, plants have evolved antioxidant enzymes systems, including SOD and POD that are involved in cellular elimination of ROS [36]. SOD catalyzes the decomposition of O_2-radicals to H_2O_2 and O_2 [37]. POD can convert the H_2O_2 to H2O [38]. In our study, Cd alone treatment significantly increased leaf SOD and POD activity but decreased the stem POD activity. There is no difference between control and Cd-treated plants in case of stem SOD and root SOD and POD activity. This was in agreement with results of our previous study except leaf SOD activity which was increase after 5 d Cd treatment [5]. The difference might be related to the plant species, interaction between metal and cultivar, and the environment of plant growth. GSH, GB and SA pretreatment effectively alleviate lipid peroxidation as reduced MDA content in shoots, stems and roots (except roots for GB pretreatment). However, BRs pretreatment did not mitigate MDA overproduction.

Meanwhile, application of GSH, GB, BRs and SA did not affect root SOD and POD activity, but reduced leaf SOD activity, compared with Cd alone. GSH and GB pretreatment also increased Cd-induced decrease in stem SOD activity. GB and BRs significantly decreased leaf POD activity. And GSH significantly elevated the depressed stem POD activity. The results showed that the alleviative effects of GSH, GB and SA on Cd phytotoxicity are partly due to the reduction of MDA accumulation and changes of SOD and POD activities. Exposure to 50 μM Cd caused significant reduction in Mn and Zn in shoots, Zn and Cu in roots, and increase in Fe in shoots/roots, which is similar with our previous studies [10,28]. Therefore, excessive Cd accumulation could affect the uptake and distribution of essential mineral elements in crops, and then caused mineral imbalance and inhibition of plant growth. GSH pretreatment decreased Fe in shoots, Mn in shoots/roots, Cu in roots and increased shoot Cu when compared with Cd alone treatment. GB pretreatment decreased the Fe and Mn in shoots/roots. BRs pretreatment decreased the Fe in shoots/roots, Mn and Cu in roots, Zn in shoots, and increased Cu in shoots. And SA pretreatment elevated the Fe in shoots and decreased Mn in shoots/roots. Thus, the varying uptake and distribution of Fe, Mn, Zn and Cu may involve in the plant tolerance against Cd stress. In conclusion, 24 h pretreatment of GSH, GB and SA significant alleviated Cd-induced inhibition on growth and chlorophyll content, reduced shoot Cd concentration and markedly diminished Cd-induced MDA accumulation. GSH, GB and SA pretreatments counteracted the pattern of alterations in certain antioxidant enzymes induced by Cd, e.g. suppressed Cd-induced dramatic increase of leaf SOD activity, GB and BRs also significantly decreased leaf POD activity; GSH significantly elevated the depressed stem POD and SOD activities; and GB elevated stem SOD activity. GSH and GB also counteracted Cd-induced response of element concentration: GSH suppressed Cd-induced dramatic increase of leaf Fe and elevating Cd-depressed leaf Cu; GB decreased Cd-induced increase in root/leaf Fe. However, BRs pretreatment may enhance Cd uptake and translation from root to shoot, accordingly, BRs would be unsuitable for the edible crops grown in Cd contaminated soils to alleviate phyto-toxicity of Cd, although the root length and plant dry weight were increased over Cd alone treatment.

Acknowledgements

This work was supported by the Key Research Foundation of Science and Technology Department of Zhejiang Province of China (2009C12050).

References

1. Lux A, Martinka M, Vaculik M, White PJ (2011) Root responses to cadmium in the rhizosphere: a review. J Exp Bot 62: 21-37.

2. Mulligan CN, Yong RN, Gibbs BF (2001) Remediation technologies for metal-contaminated soils and groundwater: an evaluation. Eng Geol 60: 193-207.

3. Grant CA, Buckley WT, Bailey LD, Selles F (1998) Cadmium accumulation in crops. Can J Plant Sci 78: 1-17.

4. Nan Z, Li J, Zhang J, Cheng G (2002) Cadmium and zinc interactions and their transfer in soil-crop system under actual field conditions. Sci Total Environ 285: 187-195.

5. Cai Y, Cao F, Wei K, Zhang G, Wu F (2011) Genotypic dependent effect of exogenous glutathione on Cd-induced changes in proteins, ultrastructure and antioxidant defense enzymes in rice seedlings. J Hazard Mater 192: 1056-1066.

6. Uraguchi S, Mori S, Kuramata M, Kawasaki A, Arao T, et al. (2009) Root-to-shoot Cd translocation via the xylem is the major process determining shoot and grain cadmium accumulation in rice. J Exp Bot 60: 2677-2688.

7. Liu J, Qian M, Cai G, Yang J, Zhu Q (2007) Uptake and translocation of Cd in different rice cultivars and the relation with Cd accumulation in rice grain. J Hazard Mater 143: 443-447.

8. Kikuchi T, Okazaki M, Kimura SD, Motobayashi T, Baasansuren J, et al. (2008) Suppressive effects of magnesium oxide materials on cadmium uptake and accumulation into rice grains II: Suppression of cadmium uptake and accumulation into rice grains due to application of magnesium oxide materials. J Hazard Mater 154: 294-299.

9. Benavides MP, Gallego SM, Tomaro ML (2005) Cadmium toxicity in plants. Braz J Plant Physiol 17: 21-34.

10. Wu F, Zhang G, Dominy P (2003) Four barley genotypes respond differently to cadmium: lipid peroxidation and activities of antioxidant capacity. Environ Exp Bot 50: 67-78.

11. Ogawa K (2005) Glutathione-associated regulation of plant growth and stress responses. Antioxid Redox Signal 7: 973-981.

12. Cai Y, Lin L, Cheng W, Zhang G, Wu F (2010) Genotypic dependent effect of exogenous glutathioneon Cd-induced changes in cadmium and mineral uptake and accumulation in rice seedlings (Oryza sativa). Plant Soil Environ 56: 516-525.

13. Wang F, Chen F, Cai Y, Zhang G, Wu F (2011) Modulation of exogenous glutathione in ultrastructure and photosynthetic performance against cd stress in the two barley genotypes differing in cd tolerance. Biol Trace Elem Res 144: 1275-1288.

14. Zhu YL, Pilon-Smits EA, Tarun AS, Weber SU, Jouanin L, et al. (1999) Cadmium tolerance and accumulation in indian mustard is enhanced by overexpressing gamma-glutamylcysteine synthetase. Plant Physiol 121: 1169-1178.

15. Liang Zhu Y, Pilon-Smits EA, Jouanin L, Terry N (1999) Overexpression of glutathione synthetase in indian mustard enhances cadmium accumulation and tolerance. Plant Physiol 119: 73-80.

16. Halim VA, Vess A, Scheel D, Rosahl S (2006) The role of salicylic acid and jasmonic acid in pathogen defence. Plant Biol (stuttg) 8: 307-313.

17. Metwally A, Finkemeier I, Georgi M, Dietz KJ (2003) Salicylic acid alleviates the cadmium toxicity in barley seedlings. Plant Physiol 132: 272-281.

18. Krantev A, Yordanova R, Janda T, Szalai G, Popova L (2008) Treatment with salicylic acid decreases the effect of cadmium on photosynthesis in maize plants. J Plant Physiol 165: 920-931.

19. Ashraf M, Foolad MR (2007) Roles of glycine betaine and proline in improving plant abiotic stress resistance. Environ Exp Bot 59: 206-216.

20. Huang J, Hirji R, Adam L, Rozwadowski KL, Hammerlindl JK, et al. (2000) Genetic engineering of glycinebetaine production toward enhancing stress tolerance in plants: metabolic limitations. Plant Physiol 122: 747-756.

21. Islam MM, Hoque MA, Okuma E, Banu MN, Shimoishi Y, et al. (2009) Exogenous proline and glycinebetaine increase antioxidant enzyme activities and confer tolerance to cadmium stress in cultured tobacco cells. J Plant Physiol 166: 1587-1597.

22. Ogweno JO, Song XS, Shi K, Hu WH, Mao WH, et al. (2008) Brassinosteroids Alleviate Heat-Induced Inhibition of Photosynthesis by Increasing Carboxylation Efficiency and Enhancing Antioxidant Systems in Lycopersicon esculentum Plant Growth Regul 27: 49-57.

23. Liu Y, Zhao Z, Si J, Di C, Han J, et al. (2009) Brassinosteroids alleviate chilling-induced oxidative damage by enhancing antioxidant defense system in suspension cultured cells of Chorispora bungeana. Plant Growth Regul 59: 207-214.

24. Yuan GF, Jia CG, Li Z, Sun B, Zhang LP, et al. (2010) Effect of brassinosteroids on drought resistance and abscisic acid concentration in tomato under water stress. Sci Hortic 126: 103-108.

25. Hayat S, Ali B, Hasan SA, Ahmad A (2007) Brassinosteroid enhanced the level of antioxidants under cadmium stress in Brassica juncea. Environ Exp Bot 60: 33-41.

26. Fang Z, Sperling M, Welz B (1991) Flame atomic absorption spectrometric determination of lead in biological samples using a flow injection system with on-line preconcentration by coprecipitation without filtration. J Anal At Spectrom 6: 301-306.

27. Zeng FR, Zhao FS, Qiu BY, Ouyang YN, Wu FB, et al. (2011) Alleviation of chromium toxicity by silicon addition in rice plants. Agr Sci China 10: 1188-1196.

28. Wu F, Zhang G (2002) Alleviation of cadmium-toxicity by application of zinc and ascorbic acid in barley. J Plant Nutr 25: 2745-2761.

29. Van Belleghem F, Cuypers A, Semane B, Smeets K, Vangronsveld J, et al. (2007) Subcellular localization of cadmium in roots and leaves of *Arabidopsis thaliana*. New Phytol 173: 495-508.

30. Howden R, Goldsbrough PB, Andersen CR, Cobbett CS (1995) Cadmium-sensitive, cad1 mutants of Arabidopsis thaliana are phytochelatin deficient. Plant Physiol 107: 1059-1066.

31. Srivalli S, Khanna-Chopra R (2008) Sulfur assimilation and abiotic stress in plants: Role of glutathione in abiotic stress tolerance, Springer, Berlin, Germany.

32. Popova LP, Maslenkova LT, Yordanova RY, Ivanova AP, Krantev AP, et al. (2009) Exogenous treatment with salicylic acid attenuates cadmium toxicity in pea seedlings. Plant Physiol Biochem 47: 224-231.

33. Hossain MA, Hasanuzzaman M, Fujita M (2010) Up-regulation of antioxidant and glyoxalase systems by exogenous glycinebetaine and proline in mung bean confer tolerance to cadmium stress. Physiol Mol Biol Plants 16: 259-272.

34. He J, Ren Y, Pan X, Yan Y, Zhu C, et al. (2010) Salicylic acid alleviates the toxicity effect of cadmium on germination, seedling growth, and amylase activity of rice. J Plant Nutr Soil Sci 173: 300-305.

35. Hayat S, Hasan SA, Hayat Q, Ahmad A (2010) Brassinosteroids protect *Lycopersicon esculentum* from cadmium toxicity applied as shotgun approach. Protoplasma 239: 3-14.

36. Del Rio LA, Corpas FJ, Sandalio LM, Palma JM, Gomez M, et al. (2002) Reactive oxygen species and antioxidant machinery in abiotic stress tolerance in crop plants. J Exp Bot 53: 1255-1272.

37. Gill SS, Tuteja N (2010) Reactive oxygen species and antioxidant machinery in abiotic stress tolerance in crop plants. Plant Physiol Biochem 48: 909-930.

38. Hassan MJ, Zhang G, Wu F, Wei K, Chen Z (2005) Zinc alleviates growth inhibition and oxidative stress caused by cadmium in rice. J Plant Nutr Soil Sci 168: 255-261.

Effects of Organic Selenium and Zinc on the Aging Process of Laying Hens

V.G. Stanley[1]*, P. Shanklyn[1], M. Daley[1], C. Gray[1], V. Vaughan[1], A. Hinton Jr[2] and M. Hume[3]

[1]Prairie View A&M University, Prairie View, Texas 77446, USA
[2]Poultry Processing and Swine Physiology Unit, Agricultural Research Service, United States Department of Agriculture, 950 College Station Road, Russell Research Center, Athens, GA 30604, USA
[3]United States Department of Agriculture, Agricultural Research Service, College Station, Texas 77845, USA

Abstract

The objective of the study was to determine whether supplementing the diets of post-molted hens with organic selenium (Se) (Sel-Plex®) and/or organic Zinc (Zn) (Bio-Plex®)[1] could improve laying hen performance. Prior to molting, 120-78 wk old laying hens were separated into four treatment groups of 30 hens per treatment and were subjected to molting. Molting was induced by reducing photoperiod from 16 h per day to 8 h, and the diet was changed from a standard layer diet (17% CP; 2830 ME/kg) to a straight crushed corn diet. When egg production was reduced to zero, the hens were fed a control diet, or a diet supplemented with 0.3 ppm Se/kg of feed; 20 ppm Zn/kg of feed, or a combination of Se and Zn. Lighting was restored gradually to post-molting period. Changes in daily egg production, egg weight, egg quality (albumen, yolk, and shell weights), feed utilization and hen mortality were recorded. Results indicated that mean egg production was significantly ($P<0.05$) greater and feed utilization was significantly ($P<0.05$) lower for hens fed diet supplement with the combined treatment of Se and Zn compared to the other diets. Single treatment of Zn significantly ($P<0.05$) lowered mortality and increased egg production, but significantly ($P<0.05$) reduced egg, albumen and shell weights.

Keywords: Post-molted hens; Organic selenium; Organic zinc; Layer diet; Egg production

Introduction

Laying hens are typically slaughtered when egg production levels fall below 55%, and in the United States, about 100 million hens fall into this category annually [1]. Before 2009, spent-hens were processed, and the meat from the hens was supplied to the National School Lunch Program [2]. However, this practice has recently come under intense public condemnation. Therefore, the productivity of older hens has become an increased concern to table-egg producers [3], and the layer industry is exploring nutritional strategies to extend the productive life of layers. These strategies include the utilization of high fiber diets [4], skip-a-day feeding [5], high calcium diets [6] and low energy feed [7]. Some of these strategies have raised animal welfare concerns because of the possibility of undesirable effects on the overall well-being of the laying hens [8]. The well-being of the birds during molting and their subsequent performance during post-molting have become a major concern [9], and some fast food industries will no longer purchase eggs produced by laying operations that use forced-molting programs [3].

Previous research has demonstrated that dietary selenium (Se) combined with zinc (Zn) can increase egg production and egg size in commercial laying hens [10]. Selenium is needed in the body as a core element of the enzyme glutathione peroxidase (GSH-Px) [11]. Zinc is also a trace mineral which is essential for animal and human health, and it is the principal component of the hormone insulin which regulates blood glucose level. The addition of Zn to basal diet of layers may significantly increase layer body weight without significantly effecting feed conversion or layer mortality [12]. Therefore, the objective of this study was to determine whether supplementing the diets of post-molted hens with organic selenium (Se) and/or organic zinc (Zn) could improve laying hen performance.

Materials and Methods

Experimental design

One hundred twenty 78 wk old White Leghorn hens were obtained

from the laying house at the Poultry Research Center at Prairie View A&M University, Prairie View, Texas. Before the study, hens were fed a standard corn-soybean based layer ration consisting of 17% CP and 2830 ME/kg of feed and formulated to meet or exceed [13] specifications. All diets were iso-caloric and iso-nitrogenous and were provided ad libitum.

Induced molting

Molting was induced by altering photoperiod and ration of the hens. The standard ration fed previously was replaced with a straight crushed corn diet and provided ad libitum. Water was also made available ad libitum. Photoperiod was reduced to 8 h daylight per 24 h until egg production was reduced to zero. Before the start of the experiment the average daily egg production for the hens was 55% (Table 1). The hens were returned to lay 2 wks after the forced-molting process. Hens that died were replaced before beginning the experiment. Prior to molting the hens, the lighting period was 16 h per day, and the daily feed consumption was 165 g per hen.

Housing

Forced-molted hens were transferred to a naturally ventilated, open-sided layer house in a non-caged laying operation system. The house was cleaned, washed, and disinfected before stocking. Hens were separated into groups of 10 hens each with a stocking density of 0.81 m²/hen on concrete floors covered with fresh pine wood shavings. Water was provided using automatic plastic hanging drinkers with one water

*Corresponding author: V.G. Stanley, Prairie View A&M University, Prairie View, Texas, 77446, USA, E-mail: vgstanley@pvamu.edu

Ingredients and content	Diet 1	Diet 2	Diet 3	Diet 4
	------------------------- Percentage -------------------------			
Yellow Corn	55.93	55.93	55.93	55.93
Soybean Meal (44%CP)	22.10	22.10	22.10	22.10
Alfalfa Meal (17% CP)	5.00	5.00	5.00	5.00
Meat and Bone Meal	3.00	3.00	3.00	3.00
Animal and Vegetable Fat	3.00	3.00	3.00	3.00
Limestone	8.07	8.07	8.07	8.07
Di-calcium Phosphate	1.15	1.15	1.15	1.15
Iodine Salt	0.25	0.25	0.25	0.25
Vitamin Trace Mineral Premix[1]	1.50	1.50	1.50	1.50
Selenium (ppm)	0.00	0.30	0.00	0.30
Zinc (ppm)	0.00	0.00	20.00	20.00
Calculated Values				
Crude Protein (%)	17.00			
ME, kcal/kg	2830			
Phosphorus (available)	0.35			
Calcium	3.20			
Methionine	0.34			
Methionine and Cystine	0.62			
Lysine	0.76			

As-fed basis
Provided the following per kilogram of diet: vitamin A (as vitamin A acetate), 12,000 IU; cholecalciferol (as activated animal sterol), 3,000 IU; vitamin E (as á-tocopheryl acetate), 20 IU; menadione sodium bisulfite, 2.0 mg; thiamine, 1.5 mg; riboflavin, 8.0 mg; niacin, 30.0 mg; pantothenic acid, 15.0 mg; pyridoxine, 4.0 mg; vitamin B$_{12}$, 15 µg; folic acid, 1.0 mg; biotin, 150 µg; cobalt, 0.2 mg; copper, 10 mg; iron, 80 mg; iodine 1.0 mg; manganese, 120 mg; butylated hydroxytoluene (BHT), 150 mg; zinc bacitracin, 20 mg.

Table 1: Composition of diets.

Treatments	Egg production		Feed utilization (g/g)	Mortality (%)
	before[1]	after[2]		
Control	55[a]	60[c]	4.51[a]	13.85[a]
Se	55[a]	62[c]	4.01[a]	11.25[a]
Zn	55[a]	73[b]	4.25[a]	0.00[b]
Se +Zn	55[a]	75[a]	3.15[b]	10.75[a]
SEM	0.96	7.59	0.57	6.13

a-cNumbers within columns with the same letter are not significantly different (P<0.05). n=30
SEM represents standard error of means.
[1]Mean egg production before forced-molted.
[2]Mean egg production after forced-molted.

Table 2: Single and combined effects of organic selenium (Se) and zinc (Zn) on egg production, feed utilization, and mortality of post-molted hens.

per pen, and feed was provided using one tube-type feeder per pen. A 2-tier, 10-hole nest box was placed in each pen for egg collection.

Experimental design, diets and data collection

Molted hens were placed on a standard layer diet supplemented with either organic Se or Zn (Alltech Biotech, Co, Lexington, KY). Diet 1 (control) was not supplemented with Se or Zn [8]; Diet 2 was supplemented with Se only (0.3 ppm/kg of feed); Diet 3 was supplemented with Zn only (20 ppm/kg of feed); and Diet 4 was supplemented with Se (0.3 ppm/kg) and Zn (20 ppm/kg) of feed. Hens were provided water with the diets ad libitum for 5 wks. The experimental design was 2 X 2 factorial arrangements of two levels of Se (0 and 0.3 ppm) and two levels of Zn (0 and 20 ppm).

Eggs laid by hens were collected daily and classified as nest-laid or

Treatments[1]	Egg weight	Yolk weight	Albumen weight	Shell weight
	-- g/g --			
Control	67.61[a]	19.80[a]	40.20[a]	12.5[a]
Se	67.52[a]	20.50[a]	40.50[a]	12.5[a]
Zn	62.20[b]	19.10[a]	37.32[b]	10.0[b]
Se + Zn	67.85[a]	19.25[a]	40.61[a]	12.5[a]
SEM	10.38	0.63	1.55	5.76

a-cNumbers within columns with the same letter are not significantly different (P<0.05). n=30 SEM represents standard error of means.

Table 3: Single and combined effects of selenium and zinc on egg, yolk, and albumen weights of post-molted hens.

floor-laid and as cracked, broken, or soft-shelled. All layed eggs were used to evaluate daily hen-day production and mean egg weight. For each trial, 20 eggs were randomly selected from each treatment group, cracked open, and separated into yolk, albumen and shell. The yolk, albumen and shell weights were recorded. Feed was weighed at the beginning and end of the experiment to determine feed consumption. Mortality was recorded as it occurred. Each experimental treatment was replicated three times.

Statistical analysis

Data were collected on egg production, feed utilization, mortality, egg yolk, albumen and shell weights. Selenium and zinc were the main effects. All statistical analyses were done using ANOVA for a factorial arrangement of treatments. Means were compared using Duncan's Range Test. Main effects and interactions were considered significant at P<0.05.

Results and Discussion

Results indicate that organic Se and Zn, single and combined, have affected the performance of hens after forced molting. Hens fed Se plus Zn supplemented diets after 5 wks, had the highest egg production (75%), which was significantly higher (P<0.05) than (67.61%) the control. Feed utilization was also significantly (P<0.05) affected by Se plus Zn supplementation. Compared to the control (4.51 g/g), feed utilization improved significantly (P<0.05) (3.15 g/g) with the inclusion of Se plus Zn. Except for the single Zn-treatment (0.0%) mortality among the other treatments, including the control, was not significantly different (Table 2).

Table 3 shows that egg, albumen and shell weights (62.20, 37.32, 10.01 g/g, respectively) produced by hens fed Zn-treated diet, were significantly (P<0.05) lower than all the other groups, including the control. Egg, albumen and shell weights from hens on Se-supplemented feed were not significantly (P<0.05) different from the control. When Se was combined with Zn the egg weight, along with the albumen and shell weights were restored to the control levels. Yolk weights from the layers were not affected by the treatments.

Numerous researches have shown that trace minerals are now being used in poultry feed as prebiotics to enhance performance. Most research has examined the effects of supplementing feed with a single mineral, but little research had examined the effect of supplementing feed with 2 or more minerals. Selenium and zinc are trace minerals that play significant roles in the biochemistry of cellular functions in human and animals [14]. Deficiencies in these minerals can lead to immunological and structural problems of body cells. Selenium may play a critical role in the maintenance of optimal health; as it is required to maximize the activity of plasma glutathione peroxidase and other possible health benefit [14]. Suggested dietary levels of Se for poultry are 300 µg/d which are comparable to levels used in the present study. It has also been reported that although Se is an essential nutrient, excessive Se intake can be toxic in its organic form as sodium selenite [15]. To avoid Se toxicity, FDA has approved an organic form of Se, which is less toxic at higher dietary levels. Selenium and Zinc in the organic form also appeared to have higher bioavailability in laying hens performance compared to inorganic Se and Zn [11].

Traditionally, molting has been induced by implementing feed withdrawal ranging from 4 to 14 d accompanied by light restriction and/or the total removal of water for up to 3 d [16]. To reduce stress and induce molting, the hens in this study were subjected only to 7 d of low density feed, instead of feed and water withdrawal. Feeding crushed corn only to induce molting has not been reported to alter post-molting performance [8]. It appears that the method used in our study did not severely affect the physiological requirements of the hens. Furthermore, laying hens are prandial drinkers, and there is a close relationship that exists between feeding and drinking [17]. The free access to drinking water during the pre-molting stage, even though nutrient density was reduced, did not alter the post-molted performance of the hens.

During molt, reproductive tract of laying hens regresses and egg production ceases [4]. The improvement in egg production, egg, yolk, and albumen weights associated with the consumption of feed supplemented with Se and Zn would suggest that activity of these minerals has a synergistic effect on the integrity of the reproductive tract, and the improvement observed in performance could be due to the synergistic or additive effects of these two trace minerals.

As laying hens aged, their productive performance declines [16]. Older laying hens produce fewer and larger eggs, and poor eggshell quality. The results show that the combination of Se and Zn in the organic form as a dietary supplement increased egg production, produced larger eggs, lower feed utilization, and heavier eggshell weight in post-molted hens. The Se dietary supplement level was 0.3 ppm/kg, which did not appear to be toxic as it increased performance of post-molted hens. Also, during the aging process, laying hens increased the production of free oxygen radicals. Hence, older hens need high levels of antioxidant protection to protect cells and enhance tissue development. Older hens under induced forced-molting conditions require an effective antioxidant system which is dependent upon Se status of the cells and tissues. Also, as the hen aged, the need for energy increases. Zinc, which is a critical component of insulin synthesis, could have increased the bioavailability of energy for the transfer of glucose from the blood to the liver to be metabolized into energy. Eggshell quality is important as egg size affects the marketing of eggs. Poor eggshell quality leads to increased breakage and loss of revenues for the producer and the processor. Zinc supplement decreased eggshell weight which was elevated to the control level when it was combined with Se. By supplementing the diet of post-molted hens with organic Se and Zn the productive life of these hens could be extended.

Summary and Conclusion

Molting of older hens can be effectively implemented by reducing the nutritional density of the ration instead of withdrawing feed or water. The combination of Se and Zn as dietary feed supplements appears to have a positive synergistic or additive effect on the performance of post-molted laying hens. Commercial laying operations should consider supplementing feed with Se and Zn to extend the productive life of laying hens.

Acknowledgments

The authors wish to thank the students and colleagues for their assistance during the experiment. We also thank Alltech Biotech Laboratory Corporation, Lexington, Kentucky, USA, for providing the organic selenium and zinc (Sel-Plex and Bio-Plex) and to Dr. Ted Sefton for his assistance.

[1]Sel-Plex and Bio-Plex are trade names for organic selenium and organic zinc produced and distributed by Alltech, Inc., Lexington, Kentucky, USA.

References

1. United Egg Producers (2008) Molting Animal Husbandry Guidelines for U.S. Egg Laying Flocks. United Egg Producers, Alphasetta.

2. Anonymous (2000) McDonald's targets the egg industry. Egg Ind 105: 10-13.

3. United Egg Producers (2010) U.S. Egg Industry Stats. U.S. Department of Agriculture.

4. Dunkley CS (2006) High fiber low energy diet for molt induction in laying hens: The impact of alfalfa on physiology, immunology, and behavior. Dissertation Texas A&M University, TX 1: 1-39.

5. Webster AB (2000) Behavior of White Leghorn laying hens after withdrawal of feed. Poult Sci 79: 192-200.

6. Koelkbeck KW, Anderson KE (2007) Molting layers - alternative methods and their effectiveness. Poult Sci 86: 1260-1264.

7. Dickey ER (2008) Evaluation of a calcium pre-molt and lows-energy molt program: Effects on laying hen behavior, production, and physiology before, during, and after a fasting or non-fasting molt (Thesis) Iowa State University 2: 3-21.

8. Mejia L, Mayer ET, Utterback PL, Ulterback CW, Parsons CM, et al. (2010) Evaluation of limit feeding corn and distillers dried grains with solubles in non-feed withdrawal molt programs for laying hens. Poult Sci 89: 386-392.

9. Appleby MC, Mench JA, Hughes BO (2004) Poultry behavior and welfare. CABI. Publishing Cambridge, MA.

10. Black KS, Stanley VG (2007) Single and combined effects of organic selenium and zinc on the performance of laying hens. Thesis 1-34.

11. Richter GM, Leiterer R, Kermise WL, Ochrimenko, Arnhold W (2006) Comparative investigation of dietary supplements of organic and inorganic board selenium in laying hens. Tierarztl Umsch 61: 155-161.

12. Burrell AL, Dozier WA 3rd, Davis AJ, Freeman MM, Vandrell PF, et al. (2004) Responses of broilers to dietary zinc concentrations and sources in relation to environmental implications. Br Poult Sci 45: 255-263.

13. NRC (1994) Nutrient Requirements of Poultry. (9thedn), National Academy Press, Washington, DC.

14. Fisinin VI, Papzyan TT, Soraie PF (2009) Producing selenium-enriched eggs and meat to improve the selenium status of the general population. Crit Rev Biotechnol 29: 18-28.

15. Bennett DC, Cheng KM (2010) Selenium enrichment of table eggs. Poult Sci 89: 2166-2172.

16. Berry WE (2003) The physiology of induced molting. Poult Sci 82: 971-980.

17. Leeson A, Dummons JD (2005) Commercial Poultry Production. (3rd edn), University Books Guelph, Ontario, Canada.

18. SAS Institute (2003) SAS (User's Guide Statistics. Version 9.0) SAS Institute Inc., Gary, NC.

Combining Ability Analysis to Identify Superior F1 Hybrids for Yield and Quality Improvement in Tomato (*Solanum lycopersicum* L.)

Adhi Shankar[1]*, RVSK Reddy[2], M Sujatha[3] and M Pratap[1]

[1]*Department of Horticulture, College of Horticulture, Rajendranagar, Dr. Y.S.R. Horticultural University, Hyderabad, India*
[2]*Principal Scientist (Hort.), Vegetable Research Station, Rajendranagar, Dr. Y.S.R. Horticultural, Hyderabad, India*
[3]*Department of Genetics and Plant Breeding, College of Agriculture, Hyderabad, India*

Abstract

Combining ability and gene effects for yield and quality traits in tomato were studied by involving twenty four cross combinations obtained from crossing eight diverse lines with three testers in line×tester mating fashion. The analysis of variance revealed that the variance due to line×tester effects were highly significant for all the traits except lycopene content under study. Combining ability analysis revealed that magnitude of Specific Combining Ability (SCA) variance was greater than General Combining Ability (SCA) variance suggesting the predominance of non-additive gene action for yield per plant, pericarp thickness, TSS, titrable acidity, lycopene and shelf life. The degree of dominance revealed that over dominance is the cause of heterosis for these traits. Based on gca effects of parents, the lines LE-62 and LE-53 and the testers Arka Meghali and Arka Vikas were good general combiners for most of the traits under study. The crosses viz., EC-157568×Arka Vikas, EC-163611×Arka Alok, LE-62×Arka Alok and LE-64×Arka Vikas were found to be superior specific combiners for yield per plant. For quality traits, the cross EC-165749×Arka Alok was also superior specific combiner for yield per plant, TSS, ascorbic acid and shelf life and the cross EC-157568×Arka Alok was superior specific combiner for TSS, titrable acidity and lycopene.

Keywords: *Solanum lycopersicum* L.; Combining ability; Gene action; Quality; Tomato; Yield

Introduction

Tomato (*Solanum lycopersicum* L.) is one of the most important vegetable crops grown throughout the world because of its wider adaptability, high yielding potential and suitability for variety of uses in fresh as well as processed food industries. The fruits are available year round and eaten raw or cooked. Tomato in large quantities is used to produce soup, juice, ketchup, puree, paste and powder; it supplies ascorbic acid and adds variety of colours and flavors to the food.

All fruit quality attributes were expressions of genotypic and environmental effect of interactions. Hence quality attributes have to be considered together for future genetic improvement of tomato quality. Total soluble solids (TSS) and ascorbic acid content have been recognized as the most desirable attributes in tomato for processing industry. The increase of 1% TSS in fruits results to increase 20% recovery of processed products [1]. High ascorbic acid content in addition to improving the nutrition also helps in the better retention of natural colour and flavour of the tomato products. The red pigment in tomato (lycopene) is now being considered as the "world's most powerful natural antioxidant" [2]. Therefore, tomato is one of the most important 'protective foods' because of its special nutritive value. The shelf life is an important quality trait for marketing, transportation and domestic use. This trait is controlled by genetic factors as well as environmental factors such as temperature. Characters like whole fruit firmness, number of locules per fruit and pericarp thickness are the important parameters contributing towards shelf life besides biochemical changes [3]. However, pericarp thickness alone accounts for 64% of fruit firmness [4]. Hence, development of firm tomato is the basic need for longer shelf life.

However, the average national productivity is very low (19.5 tonnes/ha as compared to other countries like USA (81 t/ha), Spain (74 t/ha) and Brazil (60.7 t/ha) [5]. This indicates that there is a need to increase the productivity of this crop by developing high yielding varieties

through appropriate breeding work to meet the demand of domestic and export markets. The ultimate objective in any crop improvement programme is to identify the best parent(s) and hybrid(s). Combining ability analysis is a common biometrical tool used in the breeding programme for testing the performance of lines in hybrid combinations and also for characterizing the nature and magnitude of gene action involved in the expression of traits. Line×Tester analysis is a useful tool for preliminary evaluation of genetic stock for use in hybridization programme with a view to identify good combiners, which may be used to build up a population with favorable and fixable genes for effective yield and quality improvement. Thus present investigation aimed to study the genetic architecture of quality traits for the development of high quality varieties/hybrids for processing tomato.

Materials and Methods

The present investigation was conducted during *Rabi* (Oct-Jan) and *Kharif* (June-Sept.) 2010- 2011 at Vegetable Research Station, Dr. Y.S.R. Horticultural University, Hyderabad (A.P.), India, which is situated at 17° 191 North latitude, 78° 231 East longitude at mean altitude of 542.3 m above the Mean Sea Level. The climate of Hyderabad was classified as dry tropical and semi arid. The rain fall mainly received from South West Monsoon approximately 555.1 mm and maximum mean temperature ranged from 27.70C to 41.8 0C, while mean minimum

---

*Corresponding author: Adhi Shankar, Department of Horticulture, College of Horticulture, Rajendranagar, Dr. Y.S.R. Horticultural University, Hyderabad, India
E-mail: shankar1104@gmail.com

temperature varied from 9.7°C to 28.2°C recorded during crop growth period.

Eight genetically divers lines (EC-165749, EC-157568, EC-164838, EC-163611, LE-53, LE-56, LE-62 and LE-64) were crossed with three testers (Arka Alok, Arka Meghali and Arka Vikas) in line×tester mating design during *rabi* 2009. The resultant 24 F1's were evaluated along with their parents and two standard checks (Lakshmi and US-618) during *Kharif* and *Rabi* 2010-11 in randomized block design which was replicated thrice. Each entry was grown in one row with 10 plants in each row by adopting inter row spacing of 60 cm and intra row spacing of 45 cm. The observations were recorded on five randomly selected plants for viz., fruit yield per plant (kg/plant), number of locules per fruit, pericarp thickness (mm), TSS (°Brix), titrable acidity (%), ascorbic acid (mg/100 mg), lycopene (mg/100 mg) and shelf life (days). Data collected during the two growing seasons for above characters were pooled and analysis of variance and combining ability analysis were done as suggested by Panse and Sukhatme [6] and Kempthorne [7], respectively.

Results and Discussion

The pooled results obtained in the present study (*Kharif* and *Rabi* 2010-11) pertaining to combining ability, gene action and ANOVA for yield and quality characters are discussed here under. Analysis of variance (Table 1) for combining ability revealed that the variance due to lines effects were highly significant (@P ≤ 0.01) for yield per plant (kg), ascorbic acid content (mg/100 g) and lycopene (mg/100 g) whereas, mean squares due to testers were non significant for all traits under study except number of locules per fruit, which indicated the existence of substantial genetic diversity in the parents for those traits. While the variance due to line×tester effects were highly significant (@P ≤ 0.01) for all traits under study except lycopene, representing specific combining ability and suggested manifestation of parental genetic variability in their crosses or possibility of better selection of cross combinations among 24 F1 hybrids for these traits (Table 1).

General combining ability

General combining ability refers to the average performance of a line in a series of cross combinations and it is attributable to additive (fixable) gene action. The estimates of *gca* effects provides a measure of general combining ability of each genotype, thus aids in selection of superior ones as parents for breeding programmes.

The line LE-53 registered significant highest *gca* effect for yield per plant, while LE-62 exhibited significant highest *gca* effects for pericarp thickness and lycopene content and number of locules per fruit (desirable negative highest) (Table 2). The highest significant *gca* effects for TSS and shelf life were recorded by EC-163611. The highest

significant *gca* effects were exhibited by EC-165749 for titrable acidity and LE-56 for ascorbic acid (Table 2).

Among testers, Arka Alok registered positively significant highest *gca* effect for pericarp thickness, whereas Arka Meghali exhibited highest significant *gca* effects for yield per plant, ascorbic acid and lycopene. The tester Arka Vikas recorded highest significant *gca* effects for number of locules per fruit, TSS, titrable acidity and shelf life.

Comprehensive assessment of parents by considering *gca* effects for 8 characters studied has resulted into identification of lines LE-62 and LE-53 as good general combiners for overall characters (Table 2). Among testers, Arka Meghali and Arka Vikas were identified as good general combiners for overall characters (Table 2). Hence, these can be utilized in commercial breeding programmes as good donors for yield and quality. The higher *per se* performance of the lines EC-165749 and LE-56 along with high *gca* estimates for titrable acidity and ascorbic acid, respectively suggesting that these two parents may be used in further breeding programme to improve yield and quality by adopting pedigree method.

Specific combining ability

The specific combining ability reveals the best cross combination among the genotypes which can be useful for developing hybrids with high vigour for the traits. Results revealed that was no cross combinations consistently good for all the traits. However, some of the crosses exhibited significant *sca* effects for more than one character.

The cross EC-157568×Arka Vikas was found to exhibit highest *sca* effects in the desirable direction for yield per plant and pericarp thickness (Table 3). The eight cross combinations viz., EL-157568×Arka Vikas (0.85), EC-163611×Arka Alok (0.60), LE-62×Arka Alok (0.56), LE-64×Arka Vikas (0.54), EC-165749×Arka Alok (0.40), EC-164838×Arka Vikas (0.33) and LE-53×Arka Alok (0.26) recorded significant *sca* effects and high per se performance for the character yield per plant (Table 3). The gene action involved in the crosses which recorded significant *sca* effects is dominant or epistatic in nature, which is non-fixable. Hence, heterosis breeding is recommended. The hybrid EC-164838×Arka Vikas was best specific combination for titrable acidity and ascorbic acid. While the crosses EC-157568×Arka Alok, LE-56×Arka Meghali and EC-16364×Arka Vikas were found to be the best specific combiners for TSS, lycopene and shelf life, respectively (Table 3).

Some of the hybrids showed significant *sca* effects in desired direction along with yield per plant such as EC-157568×Arka Vikas for pericarp thickness and TSS, EC-163611×Arka Alok for titrable acidity, ascorbic acid and shelf life, LE-62×Arka Alok for lycopene, LE-64×Arka Vikas for number of locules per fruit and EC-165749×Arka

Source	df	Mean Sum of Squares							
		Yield per plant (kg)	No. of locules per fruit	Pericarp thickness (mm)	TSS (0Brix)	Titrable acidity (%)	Ascorbic acid (mg 100 g⁻¹)	Lycopene (mg 100 g⁻¹)	Shelf life (days)
Replications	2	0.834**	0.384	0.002	0.001	0.026**	395.814**	0.051	0.849
Genotypes	34	1.858**	1.981**	1.813**	0.506**	0.054**	141.109**	17.311**	7.463**
Line Effect	7	2.656*	1.283	2.536	0.749	0.043	340.854**	15.565**	11.980
Tester Effect	2	0.146	2.702*	1.846	0.037	0.046	123.028	20.027	2.169
Line * Tester Eff.	14	0.873**	0.506**	1.006**	0.550**	0.050**	47.418**	9.720	4.917**
Error	68	0.014	0.179	0.065	0.049	0.004	0.042	0.173**	0.678
Total	104	0.633	0.772	0.635	0.197	0.021	53.771	5.773	2.900

*Significant at 5% level, **Significant at 1% level

Table 1: ANOVA for combining ability for yield and quality traits in tomato.

	Yield per plant (kg)	No. of locules per fruit	Pericarp thickness (mm)	TSS (0Brix)	Titrable acidity (%)	Ascorbic Acid (mg 100 g⁻¹)	Lycopene (mg 100 g⁻¹)	Shelf life (days)
Lines								
EC–165749	-0.67**	-0.04**	-0.03**	-0.55**	0.13**	-5.44**	-1.11**	-1.33**
EC–157568	-0.11**	-0.28**	0.38**	0.09**	-0.07**	-0.33**	-1.94**	-0.33**
EC–164838	-0.59**	0.00**	-0.71**	0.10**	0.01**	1.22**	0.34**	1.05**
EC–163611	-0.40**	0.02	-0.63**	0.43**	0.02**	-1.01**	0.02	1.44**
LE–53	0.77**	0.42**	0.52**	0.12**	0.04**	-3.67**	1.55**	1.17**
LE–56	0.24**	-0.18**	-0.34**	-0.12**	-0.09**	14.33**	-1.11**	0.33*
LE–62	0.10**	-0.57**	0.74**	0.12**	-0.02**	-1.89**	1.76**	-1.33**
LE–64	0.67**	0.62**	0.07**	-0.19**	-0.02**	-3.22**	0.49**	-0.99**
SE (i)	0.04	0.14	0.09	0.07	0.02	0.07	0.14	0.28
Testers								
Arka Alok	0.02**	0.23**	0.24**	0.01*	0.00**	-1.22**	-0.78**	-0.02**
Arka Meghali	0.06**	0.15**	0.06**	-0.04**	-0.04**	2.61**	1.00**	-0.29**
Arka Vikas	-0.09**	-0.38**	-0.30**	0.03**	0.04**	-1.39**	-0.22**	0.31**
SE (i)	0.02	0.09	0.05	0.05	0.01	0.04	0.08	0.17

*Significant at 5% level   **Significant at 1% level

Table 2: Estimates of gca effects of lines and testers for fruit yield and quality traits in tomato.

S. No.	Crosses	Yield per plant (kg)	No. of Locules per fruit	Pericarp thickness (mm)	TSS (0Brix)	Titrable acidity (%)	Ascorbic acid (mg 100 g⁻¹)	Lycopene (mg 100 g⁻¹)	Shelf life (days)
1.	EC–165749×Arka Alok	0.40**	0.26	0.70**	0.64**	0.01	5.44**	-0.32	1.02*
2.	EC–165749×Arka Meghali	0.03	-0.70 **	-0.31*	-0.23	0.12**	-3.72**	-0.80**	0.29
3.	EC–165749×Arka Vikas	-0.42**	0.44	-0.39*	-0.41**	-0.13**	-1.72**	1.12**	-1.31**
4.	EC–157568×Arka Alok	-0.86**	-0.21	-0.61**	-0.46**	0.11**	3.67**	2.01**	1.02*
5.	EC–157568×Arka Meghali	0.00	0.10	-0.52**	-0.20	0.02	-2.17**	-1.15**	0.29
6.	EC–157568×Arka Vikas	0.85**	0.11	1.13**	0.66**	-0.13**	-1.50**	-0.86**	-1.31**
7.	EC–164838×Arka Alok	-0.35**	-0.42	-0.26	-0.07	-0.03	-3.89**	-2.54**	-0.87
8.	EC–164838×Arka Meghali	0.02	0.12	0.30*	0.09	-0.12**	-1.72**	1.14**	-0.10
9.	EC–164838×Arka Vikas	0.33**	0.30	-0.04	-0.02	0.16**	5.61**	1.40**	0.97*
10.	EC–163611×Arka Alok	0.60**	0.22	0.20	-0.37**	0.12**	1.68**	0.22	1.24*
11.	EC–163611×Arka Meghali	-0.24**	-0.36	-0.01	-0.08	-0.20**	-0.16	-1.04**	0.51
12.	EC–163611×Arka Vikas	-0.36**	0.14	-0.19	0.45**	0.08*	-1.52**	0.82**	-1.76**
13.	LE–53×Arka Alok	0.26**	-0.18	0.06	0.61**	0.00	1.00**	-1.38**	-0.48
14.	LE–53×Arka Meghali	-0.07	0.14	0.04	-0.20	0.04	-1.50**	-0.30	0.29
15.	LE–53×Arka Vikas	-0.19**	0.04	-0.10	-0.41**	-0.04	0.50**	1.68**	0.19
16.	LE–56×Arka Alok	-0.05	0.09	0.78**	-0.01	0.00	1.00**	-1.08**	-1.64**
17.	LE–56×Arka Meghali	0.18*	-0.10	-0.23	0.04	-0.09*	-0.83**	2.60**	0.62
18.	LE–56×Arka Vikas	-0.13	0.01	-0.54**	-0.03	0.09*	-0.16	-1.52**	1.02*
19.	LE–62×Arka Alok	0.56**	0.11	-0.47**	-0.02	-0.13**	-3.45**	1.71**	0.02
20.	LE–62×Arka Meghali	0.06	0.22	0.19	0.23	0.11**	4.72**	0.45	-1.71**
21.	LE–62×Arka Vikas	-0.62**	-0.34	0.28	-0.21	0.02	-1.27**	-2.16**	1.69**
22.	LE–64×Arka Alok	-0.57**	0.12	-0.40**	-0.31*	-0.07	-5.45**	1.38**	-0.32
23.	LE–64×Arka Meghali	0.03	0.57*	0.55**	0.34**	0.11**	5.39**	-0.90**	-0.19
24.	LE–64×Arka Vikas	0.54**	-0.69**	-0.15	-0.03	-0.04	0.06	-0.48	0.51
	S.Ed	0.07	0.24	0.15	0.13	0.04	0.12	0.24	0.48
	CD @ 5 % level	0.14	0.49	0.30	0.26	0.08	0.24	0.48	0.96
	CD @ 1 % level	0.18	0.66	0.40	0.34	0.10	0.32	0.65	1.28

*Significant at 5% level   **Significant at 1% level

Table 3: Estimates of specific combining ability (sca) effects for yield and quality traits in tomato.

Alok for pericarp thickness, TSS, ascorbic acid and shelf life (Table 3).

Gene action

The variance due to *gca* and *sca* were found to be significant for all the traits except TSS and titrable acidity revealing the presence of both additive and non-additive type of gene action for the inheritance of yield and quality traits. In case of TSS and titrable acidity, *sca* variance alone was significant indicating that only non-additive genetic component is involved in the inheritance of this character (Table 4).

The additive variance was larger than its counterpart non-additive variance for number of locules per fruit, the estimates of average degree of dominance had also confirmed the additive gene action for this trait (Table 4). Similar results were reported by Dhaliwal et al. [8]. Hence, significant advancement could be achieved in the segregating generations using simple selection procedures or conventional breeding methods such as pedigree and bulk selection.

Non additive genetic variance had greater estimates than additive

S.No.	Character	$\sigma^2 gca$	$\sigma^2 sca$	$\sigma^2 gca/\sigma^2 sca$	$\sigma^2 A$	$\sigma^2 D$	$\sigma^2 A/\sigma^2 D$	Degree of Dominance
1	Yield/plant (kg)	0.08*	0.29**	0.28	0.17	0.29	0.59	1.30
2	No. of locules/fruit	0.11**	0.11**	1.00	0.22	0.11	2.02	0.70
3	Pericarp thickness (mm)	0.13**	0.31**	0.42	0.26	0.31	0.82	1.10
4	TSS (0Brix)	0.02	0.17**	0.12	0.04	0.17	0.25	2.00
5	Titrable Acidity (%)	0.00	0.02**	0	0.00	0.02	0.32	1.77
6	Ascorbic acid (mg/100g)	14.05**	15.79**	0.89	28.11	15.79	1.78	0.75
7	Lycopene (mg/100g)	1.07*	3.18**	0.34	2.14	3.18	0.67	1.22
8	Shelf life (days)	0.39*	1.41**	0.28	0.78	1.41	0.55	1.35

*Significant at 5% level   **Significant at 1% level

Table 4: Estimates of variance components and degree of dominance for yield and quality traits in tomato.

genetic variance and the ratio of additive variance and non additive genetic variance is less than unity, establishing the predominance of non additive gene action in the inheritance of these traits viz., TSS, titrable acidity and lycopene [9], pericarp thickness and shelf life [10] and yield per plant [11] (Table 4). The presence of non-additive gene action for these traits requires maintenance of heterozygosity in the population. Hence, it is necessary to follow modified breeding methods such as bi-parental cross or triple test cross design or any other form of recurrent selection method in early generations, which is more useful for exploitation of non-additive gene action in order to recover transgressive segregates by breaking linkages, releasing concealed variability, improving the concentration of favourable genes and changing linkage equilibrium, otherwise heterosis breeding would be a main breeding method in improvement of these traits.

While, the greater values of additive variance than non-additive variance and greater *sca* variance than *gca* variance revealed the involvement of both additive and non additive gene action in the inheritance of ascorbic acid [12]. There is a possibility of deriving high performing pure lines for this trait because longer proportion of non-additive effects in self pollinated crops seems to be due to additive×additive epistatic effect. Hence, reciprocal recurrent selection or bi parental mating is useful for improvement of ascorbic acid content since, it exploits both the components of genetic variance.

In conclusion, the present investigation suggests that heterosis breeding can be used efficiently to improve tomato yield together with good quality for processing purpose.

References

1. Berry SZ, Uddin MR (1991) Breeding tomato for quality and processing attributes. In: Genetic Improvement of Tomato, G. Kalloo (eds.). Springer-Verlag Inc. 196-206

2. Jones JB (1999) Tomato plant culture. In: The field, green house and house garden. CRC Press, LLC 2000, Boca Raton, Florida, USA. 199.

3. Bekov RK (1968) Initial material and breeding tomato varieties for mechanical harvesting. Autoreferate Dess Moscow 21.

4. Ahrens MJ, Huber DJ, Scoot JW (1987) Firmness and mealiness of selected Florida grown tomato cultivars. Proceedings of Florida State Horticultural Society 101: 39-41.

5. NHB 2011. Indian Horticulture Database.

6. Panse VG, Sukhatme PV (1967) Statistical Methods for Agricultural Workers, ICAR, New Delhi.

7. Kempthorne O (1957) An Introduction to genetic Statistic. John Wiley and Sons, Inc. New York, USA. 208-223.

8. Dhaliwal MS, Singh S, Cheema DS (2000) Estimating combining ability effects of the genetic male sterile lines of tomato for their use in hybrid breeding. J of Gen and Plant Breeding 54: 199-205.

9. Mondal C, Sarkar S, Hazra P (2009) Line x Tester analysis of combining ability in tomato (*Lycopersicon esculentum* Mill.). J of Crop and Weed 5: 53-57.

10. Joshi A, Thakur MC, Kohli UK (2005) Heterosis and combining ability for shelf life, whole fruit firmness and related traits in tomato. Indian J. of Hort 62: 33-36.

11. Singh B, Kaul S, Kumar D, Kumar V (2010) Combining ability for yield and its contributing characters in tomato. Indian Journal of Horticulture 67: 50-55.

12. Patil AA, Bojappa KM (1986) Combining ability for certain quality traits in tomato (*Lycopersicon esculentum* Mill.). Prog. Hort. 18: 73-76.

Effect of Nano-Zinc Oxide on the Leaf Physical and Nutritional Quality of Spinach

Kisan B*, Shruthi H, Sharanagouda H, Revanappa SB and Pramod NK

Department of Biotechnology, College of Agriculture, University of Agricultural Sciences, Raichur-584104, India

Abstract

Spinach (*Spinacia oleracea*) belongs to family *Amaranthaceae* and is one of the important and nutritious leafy vegetable consumed in India. The pot culture experiment is carried out during 2014-15 to study the effect of nano-zinc oxide particles on the leaf physical and nutritional traits of spinach. The spinach plants were sprayed with graded concentration of zinc oxide nanoparticles (ZnO NPs) after 14 days of sowing. The leaf physical parameters like leaf length, leaf width and leaf surface area are recorded at the time of maturity (45-50 days). The protein, carbohydrate, fat and dietary fiber content in leaf samples are determined. The plants sprayed with ZnO NPs at the concentration of 500 and 1000 ppm showed the increased leaf length, width, surface area and colour of leaf samples when compared to control leaf samples. Similarly treated plants with ZnO NPs at the concentration of 500 and 1000 ppm showed higher values of protein and dietary fibre content in comparison to control leaf samples of spinach. Hence our study suggests that the nano-zinc oxide sprayed spinach is more nutritious to vegetarian diet by providing, protein, fiber and required amount of vegetarian fat to diet.

Keywords: Nano-zinc oxide; Spinach; Leaf physical properties; Protein; Fibre

Introduction

Spinach (*Spinacia oleracea*) is a green-leafy vegetable belongs to family *Amaranthaceae*. It is often recognized as one of the *functional foods* for its wholesome nutritional, antioxidants and anti-cancer composition. The major micronutrients in spinach are vitamins A (from β-carotene), C, K and folate, and the minerals, calcium, iron and potassium. Spinach also provides fibre and is low in calories. Its tender, crispy, dark-green leaves are one of the favorite ingredients of chefs all around the world. Vegetables are also valuable in maintaining alkaline reserve of the body. They are valued mainly for high carbohydrate, vitamin and mineral contents [1].

Micronutrient fertilizers can increase the tolerance of plants to environmental stresses like drought and salinity [2]. Zinc has been considered as an essential micronutrient for metabolic activities in plants. It regulates the various enzyme activities and required in biochemical reactions leading to formations of chlorophyll and carbohydrates [2,3]. The crop yield and quality of produce can be affected by the deficiency of Zn [4]. Zinc nano-particle is used in various agricultural experiments to understand its effect on growth, germination, and various other properties. Most of the farmers are using either zinc sulfate or EDTA-Zn chelate for soil and foliar applications, however the efficacy is low [5] have demonstrated essentiality and role of zinc in plant growth, reproduction and yield. It has been indicated that the retention time of Zn in the plant system is low and hence, the bioavailability of Zn for long period is not sure with the use of ZnSO$_4$ fertilizer. Under high temperatures conditions ZnSO$_4$ has a large salt index and it may show burning injury if the plants are soft or sensitive [6].

Nano-particles with smaller particle size and large surface area are expected to be the ideal material for use as Zn fertilizer in plants. Application of micronutrient in the form of nano-particles (NPs) is an important route to release required nutrients gradually and in a controlled way, which is essential to mitigate the problems of soil pollution caused by the excess use of chemical fertilizers. A number of researchers have reported the essentiality and role of zinc for plant growth and yield [5,6]. Reynolds [7] demonstrated the use of micronutrients in the form of nano-particles can be used in crop production to increase yield.

It has been postulated that nano-particles are more effective, can be utilized in agriculture for the precision farming and enhance productivity crop yields [7,8]. Several studies are concerned with the synthesis of nanomaterials using biological routes. Only limited studies have been reported on the promotory effects of nanoparticles on plants in low concentrations. Nanoscale titanium dioxide (TiO$_2$) was reported to promote photosynthesis, and growth of spinach. The present study deals with the effect of nano-ZnO, particle suspension as micro-nutrient on the physical and nutritional quality of spinach leaves by foliar spray method. Foliar spray method is more practical from an agronomic standpoint as plant can absorb essential elements through their leaves more efficiently compared to root feeding.

Materials and Methods

Zinc oxide nanoparticles of mean size of 50 nm diameter (Sigma Aldrich) were characterized with zeta potential was used in the study. The stock solution of 10,000 ppm solution prepared and dilutions of 0, 100, 500, 1000 ppm were used for the study. The stock solutions were prepared by directly suspending the nano-particles in deionized water and dispersed by ultrasonic vibration (100 W, 40 KHz) for 1 hour. Magnetic bars were placed in the suspensions for stirring before use to avoid aggregation of the particles. The nano-particle solutions used for the experiment purpose were prepared from the stock solutions.

*Corresponding author: Dr. Kisan BJ, Department of Biotechnology, College of Agriculture, University of Agricultural Sciences, Raichur -584104, Karnataka, India E-mail: kisanb1@gmail.com

Spinach seeds of variety 'All green variety' were used by sowing in pots (20 cm × 40 cm) filled with equal quantity of soil and watered to field capacity. Proper care was taken to use similar soil in all the pots to minimize soil heterogeneity effects. At 14 days after sowing plants were sprayed with the different concentration of ZnO nano-particles. The leaf samples were harvested at 45th day of sowing and were further processed for proximate analysis.

The physical properties of fresh spinach leafs viz., Length and width were measured using venire callipers and surface area of fresh spinach leafs was determined by using digital Planimeter (Make: Placom; model: KP90N roller-type digital Planimeter). The leaf parameters of fresh spinach leafs i.e., color and water activity were determined using Hunter lab colorimeter (Colour Flex EZ, Hunter Associates LAB INC., C04-1005-631, Taiwan) and Rotronic Hygrolab water activity analyzer (Model: a_w-HP23).

The proximate composition of fresh spinach leafs viz. crude protein, crude fat, total ash, crude fiber and carbohydrates were estimated by the recommended methods of the Association of Official Analytical chemists, (2005) in triplicate. Experiments were carried out in triplicate. Data recorded from three replications were subjected to single way analysis of variance (ANOVA) and critical differences were calculated at p=0.05 level.

Results and Discussion

Effect of different concentration of ZnO nanoparticles on leaf physical properties such as leaf length, leaf width and leaf surface area etc., were given in Figure 1. The leaf length was maximum in ZnO-NPs treated leaf samples at the concentration of 500 ppm and minimum in control leaf samples. With respect to leaf width, there is no significant variations were observed between treated and control leaf samples of spinach. However, there is slight improvement in leaf width was observed in ZnO-NPs treated leaf samples. The maximum leaf surface area was observed in ZnO-NPs treated leaf samples at concentration of 1000 ppm followed by at 500 ppm and lowest value observed in control leaf samples. Similarly, Prasad [9] reported that, groundnut seeds treated with nanoscale zinc oxide particles with a concentration of 1000 ppm have shown significant increment in germination, shoot length, root length and vigor index over the control samples. Raskar and Laware [8] studied effect of ZnO NPs on seed germination and seedling growth in onion and observed that seed germination increased in lower concentrations of ZnO NPs but showed decrease in values at higher concentrations.

The color values viz., L', a' and b' of the fresh spinach leaf was found to be is in the range of 37.01-47.11, from -9.79 to -8.97 and 19.68-24.92, respectively and concluded that green in color and water activity of fresh spinach leafs were found to be is in the range of 0.928-0.959. Since the initial moisture content of fresh spinach leaf was more, the presence of available moisture is sufficient to grow micro-organisms also. These values were in confirmation with those reported by Nangula [10] who reported the moisture content of fresh Spinaceaoleraceais found to be 92 g/100 g. Laware and Raskar [6] reported that ZnO-NPs can reduce flowering period by 12-14 days and produce healthy seeds of onion vegetable.

The results of effect of different concentration of ZnO nanoparticles on leaf nutritional traits were given in Table 1.

The Protein content of the raw spinach leaves ranged between 1.54%and 3.99% (Table 1). The plants treated with ZnO-NPs-1000

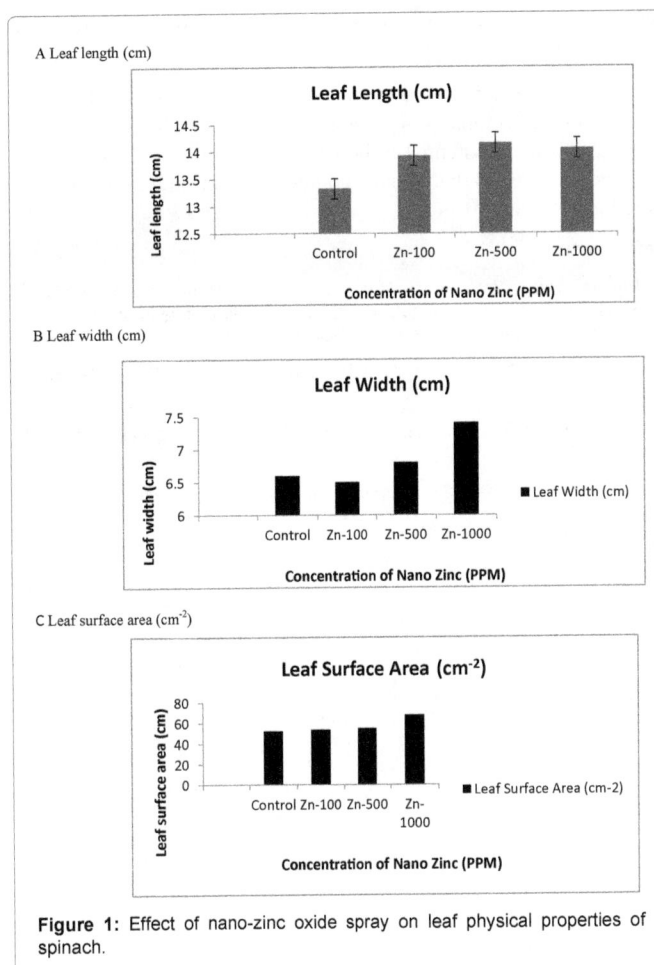

Figure 1: Effect of nano-zinc oxide spray on leaf physical properties of spinach.

S. No.	Traits	Control	Zn-100	Zn-500	Zn-1000
1	Protein (%)	1.54	2.69	2.95	3.87
2	Fat (%)	0.1	0.6	1.06	1.07
3	Fiber (%)	2.01	2.23	3.4	6.97
4	Ash (%)	0.18	2.16	1.90	2.06
5	Carbohydrate (%)	0.20	0.23	0.58	0.52

Table 1: Effect of nano-zinc oxide spray on leaf nutritional quality of spinach.

ppm was found to have the highest protein content (3.99%) and control (1.54%) had the lowest protein content. The protein content to be obtained in the spinach leaf was close to the values previously reported [1] was 2.10 ± 0.15 g. The fat, fiber and carbohydrates content of fresh spinach leaf were found to be is in the range of 0.10-1.57%, 2.01-6.97% and 0.20-0.58%, respectively. Similar results were also concluded by Nangula et al. [10] who reported the fat, fiber and carbohydrates of fresh Spinaceaoleracea was 0.4 g, 3 g and 1 g per 100 g, respectively.

Zinc is an essential micronutrient for normal growth, development, and health of plants and human beings. Zinc enhances cation-exchange capacity of the roots, which in turn enhances absorption of essential nutrients, especially nitrogen which is responsible for higher protein content. Zinc plays vital role in carbohydrate and proteins metabolism as well as it controls plant growth hormone i.e. IAA. Zn is also an essential component of dehydrogenase, proteinase, and peptides enzymes as well as promotes starch formation, seed maturation and production [5]. The application of slow/controlled release fertilizer coated and felted by nano-materials were reported to improve grain yield along with an increase in protein content and a decrease in soluble sugar content in wheat [9].

These results indicated that the nano-zinc oxide enhanced the leaf physical and nutritional properties of spinach leaves. Nano-zinc oxide (1000 ppm) has a potential to be used as a biofortification agent to improve protein and dietary fibre contents of spinach leaves and their by reduce the malnutrition.

References

1. Rumeza Hanif, Zafar Iqbal, Mudassar Iqbal, Shaheena Hanif, Masooma Rasheed (2006) Use of vegetables as nutritional food: role in human health. Journal of Agricultural and Biological Science 1:18-22.

2. Baybordi A (2006) Zinc in soils and crop nutrition. Parivar Press (1stedn) p. 179.

3. Auld DS (2001) Zinc coordination sphere in biochemical zinc sites. Bio metals 14:271-313.

4. Jamali G, Enteshari SH, Hosseini SM (2011) Study effect adjustment drought stress application potassium and zinc in corn. Iranian Journal of crop ecophysiology 3:216-222.

5. Fageria NK, Baligar VC, Clark RB (2002) Micronutrients in crop production. Advances in Agronomy 77: 189-272.

6. Laware SL, Raskar SV (2014) Influence of Zinc Oxide Nanoparticles on Growth, Flowering and Seed Productivity in Onion. Int J Curr Microbiol App Sci 3:874-881.

7. Reynolds GH (2002) Forward to the future nanotechnology and regulatory policy. Pacific Research Institute 24:1-23.

8. Raskar SV, Laware SL (2014) Effect of zinc oxide nanoparticles on cytology and seed germination in onion. Int J Curr Microbiol App Sci 3: 467-473.

9. Prasad TNVKV, Sudhakar P, Sreenivasulu Y, Latha P, Munaswamy V, et al. (2012). Effect of nanoscale zinc oxide particles on the germination, growth and yield of peanut. Journal of Plant Nutrition 35: 905-927.

10. Nangula PU, Oelofse A, Duodu KG, Bester MJ, Faber M (2010) Nutritional value of leafy vegetables of sub-Saharan Africa and their potential contribution to human health: A review, Journal of Food Composition and Analysis 23: 499-509.

Effect of Organic Manures and Amendments on Quality Attributes and Shelf Life of Banana cv. Grand Naine

Vanilarasu K[1]* and Balakrishnamurthy G[2]

[1]Ph. D. Scholar (Horticulture) in Fruit Science, Department of Fruit Crops, Tamil Nadu Agricultural University,

Coimbatore – 641 003, Tamil Nadu, India

[2]Professor (Horticulture), in Fruit Science, Department of Fruit Crops, Tamil Nadu Agricultural University,

Coimbatore – 641 003, Tamil Nadu, India

*Corresponding author: Vanilarasu K, Ph. D. Scholar (Horticulture) in Fruit Science, Department of Fruit Crops, Tamil Nadu Agricultural University, Coimbatore – 641 003, Tamil Nadu, India, E-mail: arasuvani88@gmail.com

Abstract

Of late growing awareness on health makes consumer more concerned for food quality and safety. In banana Total soluble solids, acidity and sugar content mostly determine the degree of acceptability of fruit. An experiment was conducted during 2010-11 with 12 different treatment combinations of Farmyard manure, Vermicompost, Neem cake, Wood ash and green manures (organic sources) along with and without microbial inoculants (Arbuscular mycorrhizae, *Azospirllium*, Phosphate solubilising bacteria and *Trichoderma harzianum*) comparison with inorganic sources alone on quality attributes and shelf life of banana cv. Grand Naine (AAA). Results revealed that the treatment T10 (Farmyard manure @ 10 kg + Neem cake @ 1.25 kg + Vermicompost @ 5 kg and Wood ash @ 1.75 kg /plant + Triple green manuring with sunhemp + Double intercropping of Cow pea + biofertilizers viz., Vesicular Arbuscular Mycorrhizae @ 25 g, *Azospirillum* @ 50 g, Phosphate solubilizing bacteria – @ 50 g and *Trichoderma harzianum* @ 50 g/plant) registered the maximum quality attributes (TSS – 23.23%, Acidity – 0.82%, Ascorbic acid – 12.92 mg. 100 g -1, Non-reducing and Total sugars - 6.06 and 14.92%) besides enhancing the shelf life of banana (14.03 days) and reduced physiological loss in weight (7.44%).

Keywords: Green manures; Organic manures; Bio-fertilizers; Inorganic fertilizers; Quality and shelf life

Introduction

Banana (Musa sps.) is the most important fruit crops of the world. It has nutritional, medicinal, industrial as well as aesthetic value in Hindu religion. Out of the large number of varieties grown in India, Grand Naine is the popular variety grown mostly in all export oriented countries of Asia, South America and Africa. It is a superior selection of Giant Cavendish which was introduced to India in 1990's. Due to many desirable traits like excellent fruit quality, immunity to *fusarium* wilt etc, it has proved as a better variety [1]. The quality attributes of ripe fruit are mainly influenced by the genotype, the nutritional status of the soil also plays a significant role [2]. Continuous use of inorganic fertilizers as source of nutrient in imbalanced proportion is also a problem, causing inefficiency, damage to the environment and in certain situations, harms the plants themselves and also to human being who consumes them. Some studies have suggested that organic manures gave better quality and post-harvest life of fruits when comparing to inorganic sources of nutrients in banana [3]. Many investigators studied the combined application of organic manures and amendments can enhance the yield, quality and post-harvest attributes of fruit crops Patel et al. [4] in banana and Akash Sharma et al., [5] in Guava. Organic manures contain macro and micronutrients, plant growth promoting substances like auxins, gibberellins, and cytokinins [6]. However, information regarding the type of organic manure, their optimum dose, and their interaction with bio fertilizers on different quality attributes and shelf life of banana is sketchy. With

this background the present study was conducted to determine the impact of different doses of organic manures, amendments and their combinations on important quality attributes and post-harvest life of banana.

Materials and Methods

The present investigation was carried out at Horticultural College and Research Institute, TNAU, Coimbatore, during the year 2010-11 with banana (Musa spp.) cv. Grand Naine (AAA). The experiment was laid out in a Randomized Block Design with twelve treatments and four replications. The treatments comprised of organic manures, amendments and green manures viz., FYM @ 10 kg/plant + Neem Cake @ 1.25 kg/plant + Vermicompost @ 5 kg/plant and Wood ash @ 1.75 kg/plant (T$_1$), FYM @ 10 kg/plant + Neem Cake @ 1.25 kg/plant + Vermicompost @ 5 kg/plant and Wood ash @ 3.75 kg/plant (T$_2$), FYM @ 15 kg/plant + Neem Cake @ 1.875 kg/plant + Vermicompost @ 7.5 kg/plant and Wood ash @ 625 g/plant (T$_3$), FYM @ 15 kg/plant + Neem Cake @ 1.875 kg/plant + Vermicompost @ 7.5 kg/plant and Wood ash @ 2.625 kg/plant (T$_4$), Control - absence of organic and inorganic sources (T$_5$), Triple green manuring with sunhemp + Cow pea + Cow pea as inter - crop (T$_6$), Arbuscular Mycorrhizae @ 25 g/plant + *Azospirillum* @ 50 g/plant + PSB @ 50 g and *Trichoderma harzianum* @ 50 g/plant (T$_7$), T$_1$ + T$_6$ (T$_8$), T$_1$ + T$_7$ (T$_9$), T$_1$ + T$_6$ + T$_7$ (T$_{10}$) and the absolute control treatments (inorganic) 300 : 100 : 300 g NPK /plant (T$_{11}$), 110 : 35 : 330 g NPK /plant (T$_{12}$).

The recommended spacing of 1.8 m×1.8 m was adopted for planting of banana cv. Grand Naine obtained from organic field. Among the twelve treatments, ten treatments were organic treatments

(Nutrients equal to the recommended dose of inorganic fertilizers) supplied through organic manures and amendments (FYM and Neem Cake were applied as basal dose, Vermicompost, Vesicular arbuscular mycorrhizae, *Azospirlllium,* Phosphate solubilizing bacteria and *Trichoderma harzianum* were applied after three month of planting and Wood ash was applied after five month of planting) and rest of them were inorganic treatments with three levels of inorganic fertilizers were applied at 3rd , 5th and 7th month after planting of suckers. Drip irrigation was provided to the experimental plots depending on soil moisture availability. Recommended cultural practices (except nutrient management) and plant protection measures were carried out regularly.

Estimation of important quality traits

Fully matured representative fingers were allowed for natural and uniform ripening. These fruits were subjected for determining the quality biochemical parameters. The total soluble solids were determined by using Carl-Zeiss hand refractrometer and expressed in per cent. Titrable acidity was estimated by adopting the method of A.O.A.C [7]. (1960) by titrating against N/10 KOH using phenolphthalein indicator and expressed in terms of percentage of citric acid. Ascorbic acid content was estimated using 2,6-dichlorophenol indophenol dye and expressed as milligrams of ascorbic acid 100 g-1 [8]. Total, reducing and non-reducing sugars were estimated as per the method suggested by Somogyi [9].

Estimation of post-harvest characteristics

Shelf life of the fruit was estimated by days taken for the fruits to loose their edible quality as evident by over softening and onset of decay was taken and expressed in number of days. Physiological Loss in Weight (PLW) was assessed by taken Initial weight of the fruits in

different treatments were recorded and the final weight was taken as and when the fruits reached the stage of yellow flecked with brown, in each treatment. Physiological loss in weight of fruits was computed at the end of full ripening stage by weight/weight basis by adopting the following formula and the value expressed in percentage.

$$PLW = \frac{\text{initial weight- weight after storage}}{\text{initial weight}} \times 100$$

Results and Discussion

Effect of different organic manures and amendments on quality of banana

The fruit quality parameters like TSS (23.23%), total sugars and non-reducing sugars (14.92% and 6.06) and ascorbic acid (12.92 mg. 100 g -1) contents were registered highest values in plants treated with organic amendments as compared to the inorganic treatments (Table 1), due to the better role of nutrients which is involved in the carbohydrate synthesis, breakdown and translocation of starch, synthesis of protein and neutralization of physiologically important organic acids. These findings are in concordance with the results of Anon [10] in Sapota, Anon [11] in custard apple; Pereira and Mitra [12] in guava and Athani and Hulamni [13] in banana and reported that the increased fruit quality parameters are due to the addition of different organic manures and amendments to the soil and in turn to plants, which might had enhanced the biosynthesis and translocation of carbohydrates in to fruits. Further, the availability of nitrogen from different sources might have increased leaf area with higher synthesis of assimilates which is due to enhanced rate of photosynthesis. Such effects have been attributed to increase rate of translocation of photosynthetic products from leaves to developing fruits and thereby increasing total sugars [14-17].

Treatments	TSS (%)	Acidity (%)	Ascorbic acid (mg. 100 g -1)	Non-reducing sugars (%)	Reducing sugars (%)	Total sugars (%)	PLW (%)	Shelf life (days)
T1	21.49	0.83	12.04	5.36	9.08	14.44	9.64	13.18
T2	21.83	0.84	11.56	5.51	8.69	14.20	9.32	12.74
T3	21.89	0.82	12.43	5.85	8.90	14.75	9.45	12.97
T4	21.58	0.84	12.37	5.82	8.72	14.54	10.37	13.84
T5	20.56	0.84	11.06	4.45	7.03	11.48	12.05	9.73
T6	21.45	0.84	11.96	5.25	7.63	12.88	13.22	11.07
T7	21.72	0.83	12.65	5.66	8.23	13.89	8.85	13.27
T8	21.63	0.83	12.24	5.73	7.93	13.66	9.36	13.93
T9	22.47	0.84	12.46	5.95	9.10	15.05	7.84	13.87
T10	23.23	0.82	12.92	6.06	8.86	14.92	7.44	14.03
T11	21.93	0.83	11.91	5.15	8.41	13.56	11.84	10.26
T12	22.08	0.84	12.33	5.45	8.37	13.82	11.28	10.87
SEd	0.24	NS	0.14	0.06	0.07	0.19	0.14	0.14

CD (0.05)	0.49	NS	0.29	0.12	0.14	0.40	0.29	0.28

Table 1: Effect of organic manures and amendments on quality and post harvest characteristics of banana cv. Grand Naine

Effect different organic manures and amendments on post-harvest characters of banana

Shelf life of banana is an important parameter and influenced directly by the pre harvest nutritional status of the fruits. The influence of nutrients derived from organic sources had a positive effect on the post-harvest characters of banana (Table 1), the highest shelf life (14.03 days) of fruits and least physiological loss in weight (7.44 per cent) were observed in the treatment T_{10}. On the other hand, the shelf life of the fruits was minimum (9.73 days) and physiological loss in weight was maximum in T_5 (12.05 per cent) next to T_6 (13.23 per cent). The extended shelf life observed in the present study might be due to the consequence of reduced weight loss and other physiological process like reduced respiration and transpiration. This result lends support to the findings of Athani and Hulamni [13] in banana.

Conclusion

Substitution of organic manures and amendments (biofertilizers and biocontrol agents) and green manures combination significantly enhanced the important quality attributes and post-harvest life of banana compared to either organic manure alone or inorganic sources alone.

References

1. Singh HP, Chundawat BS (2002) Improved technology of banana Ministry of Agriculture, Government of India: 1-46.

2. Roy SK, Chakroborty AK, MacRae RM, Robinso RK, Sandler MJ (1993) Vegetables of tropical climate-commercial and dietary importance, Encyclopaedia of Food Science. J. Food Technology and Nutrition, Academic press, London.

3. Patel PS, Kolambe BN, Patel HM, Patel TU (2010) Quality of banana as influence by organic farming. Int. J. Bioscience Reporter 8: 175-176.

4. Patel KM, Patel HC, Patel KA, Chauhan VB, Patel JS (2012) Effect of organic manures or chemical fertilizers on yield and quality of banana fruits. The Asian Journal of Horticulture 7: 420-422.

5. Akash Sharma VK, Wali P, Bakshi A, Jasrotia, Sardar CV (2013) Effect of organic and inorganic fertilizers on quality and shelf life of guava (psidium guajava). The Bioscan 8: 1247-1250.

6. Krishnamoorthy RV, Vajrabhiah SN (1986) Biological activity of earthworm casts: an assessment of plant growth promoter levels in casts. Proceedings of the Indian Academy of Science (Animal Sci) 95: 341–351.

7. AOAC (1960) Official Methods of Analysis. Published by AOAC, Washington, D.C.

8. Freed M (1966) Methods of Vitamin Assay. Inter science Pub. Inc, New York.

9. Somogyi M (1952) Notes on sugar determination. J. Biol. Chem 195: 19-23.

10. Anonymus (2000) 35th Meeting of Horticulture and Forestry Sub-Committee of Agricultural Research Council of GAU, Navsari: 479.

11. Anonymus (2001) 36th Meeting of Horticulture and Forestry Sub-Committee of Agricultural Research Council of GAU, Navsari: 67-68.

12. Pereira LS, Mitra SK (1999) Studies on organic along with inorganic nutrition in guava. Indian J Agric 43: 155-160.

13. Athani SI, Hulamani NC (2000) Effect of vermicompost on fruit yield and quality of banana cv. Rajapuri (Musa AAB), Karnataka. J. Agric. Sci 13: 942-946.

14. Dey P, Rai M, Nath V, Das B, Reddy N (2005) Effect of biofertilizer on physico-chemical characteristics of guava (Psidium guajava L.) fruit. Indian Journal of Agricultural Sciences 75: 95-96.

15. Kaniszewski S, Elkner K (1988) The effect of irrigation and nitrogen fertilization on the yield and quality of direct sown tomatoes. Biuletyn Warzywniczy 32: 29-52.

16. Katyal JC (1989) Fertilizer use and impact on environment. FAI Seminar, New Delhi. Fertilizer, agriculture and national economy 6: 1-8.

17. Kumaran SS, Natarajan S, Thamburaj S (1998) Effect of inorganic and organic fertilizers on growth and yield of tomato. South Indian Horticulture 46: 203-205.

Assessment of Protective Antioxidant Mechanisms in some Ethno-Medicinally Important wild Edible Fruits of Odisha, India

Uday Chand Basak*, Ajay K Mahapatra and Satarupa Mishra

Regional Plant Resource Centre, R and D Institute of Forest and Environment Department, Govt. of Odisha, Bhubaneswar, India

Abstract

Fruits and vegetables have now been documented as nutraceuticals or functional foods useful for health and medical benefits including prevention and treatment of diseases. With the background of ethno medicinal evidences and the view to utilize the wild fruit resources of Odisha, India as functional food enriched with antioxidants, 4 wild edible fruits were studied for *in vitro* radical scavenging activity and antioxidant enzymes such as Peroxidase, Catalase and Superoxide dismutase (SOD) following standard methods. It was found that the fruit with highest DPPH scavenging activity is *Antidesma ghaesembilla* (1020.6 AEAC mg/100 g dwt) and the lowest recorded in *Morinda tinctoria* (235 AEAC mg/100 g dwt). The highest FRAP value was recorded in *Antidesma ghaesembilla* (2114 µM AEAC/g dwt) and the fruit with lowest FRAP value was *Careya arborea* (538 µM AEAC/g dwt) *Antidesma ghaesembilla* showed the highest Peroxidase value of 1.12 OD/min/g tissue wt while the lowest was found in *Morinda tinctoria* (0.054 OD/min/g tissue wt). Catalase was found in high amounts in *Antidesma ghaesembilla* (5.4×10^4 IEU/g fresh tissue), the lowest value was observed in *Dillenia pentagyna* (1.2×10^4 IEU/g fresh tissue). Similarly for superoxide dismutase (SOD), the highest value was recorded in Morinda tinctoria (4.43 Δ OD/min/mg protein) and lowest in *Careya arborea* (1.12 Δ OD/min/mg protein). Current research reveals that these wild edible fruits, especially *Antidesma ghaesembilla* are rich source of antioxidants and can further be subjected to identification of individual compounds responsible for such high antioxidant activity.

Keywords: Antioxidants; Catalase; Enzymes; Fruits; Peroxidase; Superoxide dismutase

Introduction

It is now well accepted fact that fruits and vegetables are nutraceuticals or functional foods [1]. Nutraceuticals are food or food products that are reported to provide health and medical benefits, including the prevention and treatment of disease. Besides the commercially grown and popularly consumed fruits, wild edible fruits can also be considered part of the continuum since they may be potential source of nutrients and antioxidants.

Antioxidant is a substance that has the ability to delay the oxidation of a substrate by inhibiting the initiation or propagation of oxidising chain reactions caused by free radicals. It plays important roles to prevent fats and oils from becoming rancid and protects human body from detrimental effects of free radicals [2]. Each cell in the body has adequate protective mechanisms against any harmful effects of free radicals i.e. superoxide dismutase (SOD), glutathione peroxidase, glutathione reductase, thioredoxin, thiols and disulfide bonding are buffering systems in every cell. α-Tocopherol (vitamin E) is an essential nutrient which functions as a chain-breaking antioxidant which prevents the propagation of free radical reactions in all cell membranes in the human body. Ascorbic acid (vitamin C) is also part of the normal protecting mechanism. Other non-enzymatic antioxidants include carotenoids, flavonoids and related polyphenols, α-lipoic acid, glutathione etc.

There are several methods to investigate the in vitro antioxidant potential of biological samples. These methods differ in the way the free radicals are scavenged by the sample molecules. In this study we have adopted the DPPH (2, 2-diphenyl-1-picrylhydrazyl) and FRAP (Ferric reducing antioxidant capacity) assay to evaluate the total antioxidant potential of the fruit pulp extracts. Similarly for the enzyme assays Peroxidase assay was carried out following Luhova et al. [3], Catalase following Beauchamp et al. [4].

The present research paper explores 4 ethno-medicinally important wild edible fruits namely *Antidesma ghaesembilla*, *Careya arborea*, *Dillenia pentagyna* and *Morinda tinctoria* for their in vitro free radical scavenging potential and antioxidant enzymatic activity.

Materials and Methods

Materials

Four wild edible fruits namely *Antidesma gaesembilla*, *Careya arborea*, *Dillenia pentagyna* and *Morinda tinctoria* were shortlisted and selected owing to their ethno-medicinal importance. Healthy and infection-free ripe wild fruits were collected from Similipal Biosphere Reserve (District Mayurbhanj), Odisha, India. Voucher specimens were identified in the institutional herbarium. A general account of the selected fruit plants and their uses is presented in Table 1.

Methods

Sample preparation for anti-oxidant activity: The fruits were cleaned and dried at 40°C (not exceeding 50°C) following the suggestion by Khamsah et al. [5]. Then, the dried fruits were ground into fine powder using mortar and pestle. 1 g dried powder of each fruit was weighed and transferred into a beaker. 20 mL of solvent (i.e.

---

*Corresponding author: Uday Chand Basak Regional Plant Resource Centre, R andD Institute of Forest and Environment Department, Govt. of Odisha, Nayapalli, Bhubaneswar-751015, Odisha, India
E-mail: uc_basak07@yahoo.co.in

S.No	Fruit Species	Local Name	Ethno medicinal Uses	Referances
1	*Antidesma ghaesembilla*	Nuniari	Fruit contains high amount of vitamin C.	[12]
2	*Morinda tinctoria*	Achhu	There is greater demand for fruit extract of Morinda species in treatment for arthritis, cancer, gastric ulcer and other heart diseases.	[13]
3	*Dillenia pentagyna*	Rai	Tribal folks use various parts of the plant for the treatment of ailments and diseases like delivery (stem), bone fracture (bark), body pain (root), piles (leaf), diabetes (bark), diarrhea and dysentery (bark).	[14]
4	*Careya arborea*	Kumbhi	The fruit extract is used as decoction to promote digestion.	[15]

Table 1: General account of 4 selected wild edible fruits and their ethno medicinal importance.

absolute methanol) was added into the beaker and the mixture was shaken using mechanical shaker for 24 h at room temperature. Each extract was filtered using Whatman No.1 filter paper. The filtrate was collected and the residue was re-extracted twice. The two extracts were then pooled. The solvents (i.e. absolute methanol) in the extract were removed under reduced pressure at 40°C using rotary evaporator. The extracts were collected and stored at 4°C until further uses.

DPPH radical method: The ability of the extract to scavenge the DPPH (2, 2-diphenyl-1-picrylhydrazyl) radical was evaluated as described in the literature [6]. The antioxidant content was determined using a standard curve of ascorbic acid (0-10 μg/mL). The results were expressed as mg of ascorbic acid equivalent antioxidant content (AEAC) per 100 g of fruit weight.

FRAP method: The total antioxidant activity was measured using Ferric reducing antioxidant power (FRAP) assay [7]. FRAP assay was determined based on the reduction of Fe^{3+}-TPTZ to a blue coloured Fe^{2+} TPTZ. The FRAP reagent was prepared by mixing 300 mM acetate buffer (pH 3.6), 10 mM TPTZ and 20 mM $FeCl_3\ 6H_2O$ in a ratio of 10:1:1, at to 37°C. FRAP reagent (3 ml) was pipetted into test tubes. A total of 100 μl of sample and 300 μl of distilled water was then added to the same test tubes, and incubated at 37°C for 4 min. Each sample was run in triplicate. Absorbance was measured at 593 nm. FRAP value was calculated according to the equation:

FRAP (mM)=0-4 min of ΔA593 nm of test sample/0-4 min of ΔA593 nm of standard sample×[standard] mM

Peroxidase activity–Peroxidase activity was measured by a modified method of Angelini et al. [8] as described by Luhova et al. [3] taking O-dinisidine.

Superoxide Dismutase (SOD)-All extracts were assayed for SOD activity photochemically, using the assay system consisting of methionine, riboflavin, and NBT [4].

Catalase-One unit of catalase activity is defined as that amount of enzyme which breaks down I μmol of H_2O_2/min under the defined assay conditions [9] with slight modifications. Five milliliters of the assay mixture for the catalase activity comprised: 0.2 M of phosphate buffer (pH 6.8), 0.4 N of H_2O_2, and I ml of the twice diluted enzyme extracted. After incubation at 25°C for I min, the reaction was stopped

by adding 10 ml of 2% (v/v) H_2SO_4, phosphate buffer and the residual H_2O_2 was titrated against 0.01 N $KMnO_4$ until a faint purple color persisted for at least 15 sec. A control was run at the same time in which the enzyme activity was stopped at "zero" time.

Results and Discussion

Antioxidant capacity

The antioxidant capacity measured by DPPH and FRAP assay demonstrates that the fruit pulp extracts have different free radical scavenging ability for the above methods. This finding corroborates with the findings of Pellegrini et al. [10]. The findings are presented in Table 2 and Figure 1.

DPPH radical scavenging assay

The DPPH test is the oldest indirect method for determining the antioxidant activity which is based on the ability of the stable free radical 2, 2-diphenyl-1-picrylhydrazyl to react with hydrogen donors including phenols [11]. The antioxidant capacity of 4 ethno medicinally important wild edible fruits evaluated by DPPH assay and expressed as AEAC or ascorbic acid equivalent antioxidant capacity, ranged between 235 and 1020.6 AEAC mg/100 g. The DPPH assay depicts that the fruit with highest DPPH scavenging activity is *Antidesma ghaesembilla* (1020.6 AEAC mg/100g) and the lowest recorded in *Morinda tinctoria* (235 AEAC mg/100 g).

FRAP antioxidant assay

The FRAP assay measures the ability of antioxidant to reduce Fe (3+) to Fe (2+). FRAP values were expressed as μM AEAC/g dw. The highest FRAP value was recorded in *Antidesma ghaesembilla* (2114 μM AEAC/g dw) followed by *Dillenia pentagyna* (1099 μM AEAC/g dw). The fruit with lowest FRAP value was *Careya arborea* (538 μM AEAC/g dw).

S.No	Fruit Sample	DPPH AEAC mg/100 g dry weight	FRAP μM AEAC/g dry weight
1	*Antidesma ghaesembilla*	1020.66 ± 3.21	2114 ± 1.00
2	*Careya arborea*	850.66 ± 3.00	538 ± 0.01
3	*Dillenia pentagyna*	550 ± 3.60	1099 ± 4.00
4	*Morinda tinctoria*	235 ± 13.20	724.5 ± 3.70

Values are mean ± SD (n=3).

Table 2: DPPH and FRAP assay of 4 ethno medicinally important wild edible fruits of Odisha.

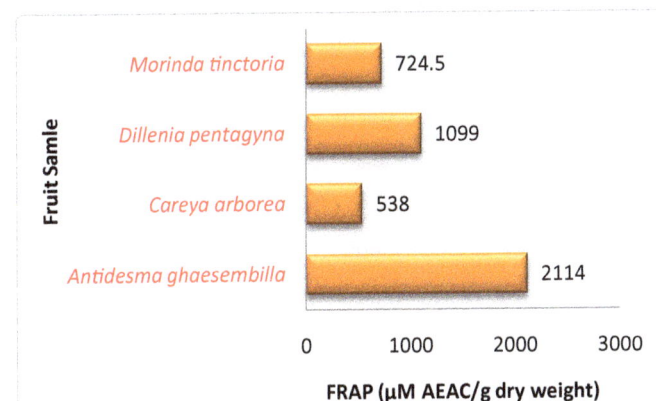

Figure 1: FRAP assay of 4 ethno medicinally important wild edible fruits of Odisha.

Antioxidant enzyme assay

The findings of Peroxidase, Catalase and superoxide dismutase assay were revealed in Table 3 and Figures 2-4.

Peroxidase assay

Antidesma ghaesembilla showed the highest Peroxidase value of 1.12 OD/min/g tissue weight while the lowest was found in *Morinda tinctoria* (0.054 OD/min/g tissue weight). *Careya arborea* recorded 0.09 while *Dillenia pentagyna* showed 0.76 OD/min/g tissue weight of Peroxidase activity.

S.No	Fruit	Peroxidase (OD/min/g tissue wt)	Catalase (IEU)	SOD (Δ OD/min/mg protein)
1	*Antidesma ghaesembilla*	1.12	5.4×10^4	2.66
2	*Careya arborea*	0.09	3.1×10^4	1.12
3	*Dillenia pentagyna*	0.76	1.2×10^4	1.9
4	*Morinda tinctoria*	0.054	2.7×10^4	4.438

Table 3: Antioxidant enzyme assay of 4 ethno medicinally important wild edible fruits of Odisha.

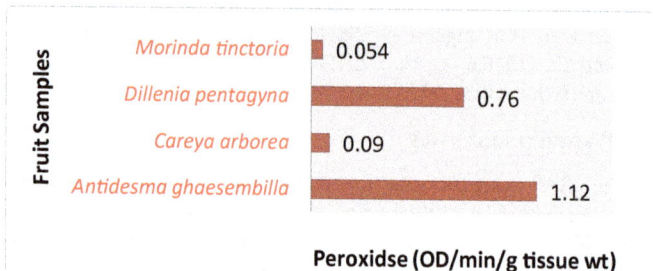

Figure 2: Peroxidase assay of 4 ethno medicinally important wild edible fruits of Odisha.

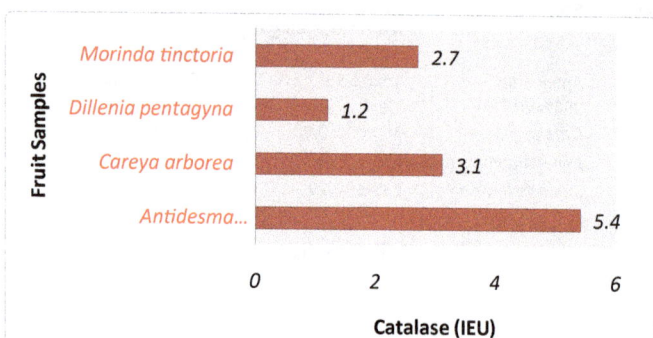

Figure 3: Catalase assay of 4 ethno medicinally important wild edible fruits of Odisha.

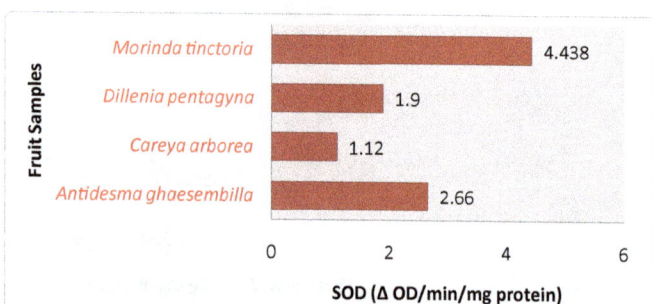

Figure 4: Superoxide dismutase (SOD) assay of 4 ethno medicinally important wild edible fruits of Odisha.

Catalase assay

Catalase helps in breaking down hydrogen peroxide into water and oxygen. Highest Catalase activity was found to be again in *Antidesma ghaesembilla* 5.4×10^4 IEU followed by Careya arborea 3.1×10^4 IEU and subsequently *Morinda tinctoria* 2.7×10^4 IEU. The lowest value was observed in *Dillenia pentagyna* 1.2×10^4 IEU.

Superoxide Dismutase assay (SOD)

SOD ranged between 1.12 and 4.43 Δ OD/min/mg protein. The highest being *Morinda tinctoria* (4.43 Δ OD/min/mg protein) followed by *Antidesma ghaesembilla* 2.66 Δ OD/min/mg protein) and thereby lowest in *Careya arborea* (1.12 Δ OD/min/mg protein).

This study establishes that all the 4 wild edible fruits can serve as supplements of natural antioxidants. But notably *Antidesma ghaesembilla* is the fruit that is not only relishing in taste but is a powerhouse of antioxidants.

Conclusion

Four wild edible fruits namely *Antidesma ghaesembilla, Careya arborea, Dillenia pentagyna* and *Morinda tinctoria* were analyzed for their total antioxidant capacity through in vitro radical scavenging assays such as DPPH and FRAP and antioxidant enzyme content for three enzymes namely Peroxidase, Catalase and superoxide dismutase (SOD). The result so obtained clearly signifies that these fruits are rich source of antioxidants (both enzymatic and non-enzymatic). Since antioxidants play crucial role in prevention of many degenerative diseases, these fruits have the scope to be included in functional foods that provide natural antioxidant supplement.

Acknowledgements

The authors are thankful to the Forest and Environment Department, Govt. of Odisha for supporting this institutional project work under State Plan Grant.

References

1. Shui G, LP Leong (2006) Residue from star fruit as valuable source for functional food ingredients and antioxidant nutraceuticals. Food Chemistry 97: 277-84.

2. Almey A, Khan AA,Ahmed Jalal, Syed Zahir C,Mustapha Suleiman K, et al. (2010) Total phenolic content and primary antioxidant activity of methanolic and ethanolic extracts of aromatic plants' leaves. International Food Research Journal. 17: 1077-84.

3. Luhova L, Lebeda A, Hedererova, Pec P (2003) Activities of amine oxidase, Peroxidase and catalase in seedlings of Pisum sativum L. under different light conditions. Plant Soil Env 49: 151–157.

4. Beauchamp C, Fridovich I (1971) Superoxide dismutase: improved assays and an assay applicable to acrylamide gels. Anal Biochem 44: 276-287.

5. Khamsah SM, Akowah G, Zhari I (2006) Antioxidant activity and phenolic content of Orthosiphon stamineus Benth from different geographical origin. J Sustainability Sci and Management 1: 14-20.

6. Meda, A, Lamien CE, Romito M, Millogo J; Nacoulma OG (2005) Determination of the total phenolic, flavonoid and proline contents in Burkina Fasan honey, as well as their radical radical scavenging activity. Food Chem.91: 571-577.

7. Benzie IFF, Strain JJ (1999) Ferric reducing/antioxidant power assay: direct measure of total antioxidant activity of biological fluids and modified version for simultaneous measurement of total antioxidant power and ascorbic acid concentration. Methods in Enzymology 299: 15-27.

8. Angelini R, Manes F, Federico R (1990) Spatial and functional correlation between diamine-oxidase and peroxidase activities and their dependence upon de-etiolation and wounding in chick-pea stems. Planta 182: 89-96.

9. Chance B, Maehly AC (1955) Assay of catalase and peroxidases. Methods Enzymol, 2: 764-775.

10. Pellegrini N, Serafini M, Colombi B, Del Rio D, Salvatore S, et al. (2003) Total antioxidant capacity of plant foods, beverages and oils consumed in Italy assessed by three different in vitro assays. J Nutr 133: 2812-2819.

11. Roginsky V, Lissi EA (2005) Review of methods to determine chain-breaking antioxidant activity in food. Food Chem, 92: 235-254.

12. Nazarudeen A (2010) Nutritional composition of some lesser-known fruits used by ethnic communities and local folks of Kerala. Ind J Traditional Knowl 9: 398-402.

13. Mathivanan N, Surendiran G, Srinivasan K, Malarvizhi K (2006) Morinda pubescens JE Smith (Morinda tinctoria Roxb) fruit extract accelerates wound healing in rats. J Med Food 9: 591-593.

14. Dubey PC, RLS Sikarwar, KK Khanna, AP Tiwari. Ethnobotany of Dillenia pentagyna Roxb. In Vindhya region of Madhya Pradesh, India. Niscair Online Periodicals Repository. 8: 546-548

15. Gupta PC, Sharma N, Rao ChV (2012) Pharmacognostic studies of the leaves and stem of Careya arborea Roxb. Asian Pac J Trop Biomed 2: 404-408.

Evaluation of Variation and Uncertainty in the Potential Yield of Soybeans in South Korea Using Multi-model Ensemble Climate Change Scenarios

Chung U[1]*, Kim YU[2], Seo BS[2] and Seo MC[3]

[1]Climate Application Department, APEC Climate Center, Busan 48059, Republic of Korea
[2]Department of Plant Science, College of Agriculture and Life Science, Seoul National University, Seoul 08826, Republic of Korea
[3]Crop Production and Physiology Research Division, National Institute of Crop Science, Rural Development Administration, Jeonju 55365, Republic of Korea

Abstract

Recently, information provided by various Global Climate Models (GCMs) has been applied to various research fields. A Multi-model Ensemble (MME) approach, which assesses the impact of climate change on agricultural crop production using one or more climate datasets from GCMs, has been widely used. We estimated the changes in soybean potential yield at 16 sites using the climate change scenarios, and then predicted the relative change in predicted potential yield for each single GCM, producing an observation climate-based simulated potential yield. Lastly, we assessed the degree of uncertainty for changes in potential yield predicted from MME approach.

In the results, although there were differences in the values themselves, the Standard Deviations (SD) of predicted soybean potential yield for each individual GCM were not significantly different from the SD of observation climate-based simulated potential yield, and there were no correlations between the predicted soybean potential yield for each individual GCMs and observation climate-based simulated potential yield in most sites. The estimation error decreased as the number of participating GCMs in the MME increased, but it did not decrease to zero. The means, but not the variance, of the MME of potential yield of soybean was similar to that of the observation climate-based simulated potential yield. The relative changes for predicted soybean potential yield for individual GCMs values of the Representative Concentration Pathways 4.5 and 8.5 scenarios increased in the northern regions of South Korea, such as Chuncheon and Hongcheon. In contrast, differences between them were not significant in most southwestern regions.

Keywords: Climate change; CROPGRO-soybean; Multi-model Ensemble (MME); Rainfed; Uncertainty

Introduction

Korean farmers have recently been encouraged to grow field crops, particularly rainfed crops, by government policies aimed at promoting rice production and at achieving self-sufficiency in grain production. In South Korea, due to the climate, soybeans have been cultivated using a single cropping system in Chuncheon, Gangwon Province, and Yeoncheon, Gyeonggi-do Province. In the southern regions of South Korea, soybeans have been cultivated using a double cropping system, with soybeans planted after winter and spring crops such as winter barley, garlic, and onions. Recently, due to climate change, soybeans have been cultivated as summer crops, alongside potatoes and cabbages. However, concerns for the safe growth and yield of crops have been increasing due to the increased frequency of extreme weather events, such as heat waves and droughts. The optimum average air temperature for the reproductive growth period of the soybean is 29°C, and at higher temperatures (e.g., 37°C) the yield could potentially be greatly reduced [1]. The optimum average temperature for the reproductive growth stages of maize is 32°C [2,3]; maize yields could be reduced by 5–13% even if temperatures increased by only 2°C during the growing season [4]. According to the Korean Meteorological Administration climate change report [5,6], the daily average temperature of the Korean Peninsula will likely rise by between 2°C and 6°C respectively under the RCP (Representative Concentration Pathways) 4.5 and RCP8.5 scenarios by 2100. In addition, the number of days with abnormally high temperatures, such as those caused by heat waves, is expected to increase to 13.1 days under the RCP4.5 scenario, and 30.2 days under the RCP8.5 scenario by 2100, when compared to the present (7.3 days).

Responses to high temperature in the representative growth stages of wheat growth models such as CERES-Wheat, APSIM-N wheat, and Wheat Grow have been improved, and many studies on the growth and yield responses of future climate scenarios that utilize these models have been published recently [7,8]. There have been numerous studies providing genetic information for various rice varieties, as well as on the effect of high temperature on rice production using rice growth models such as CERES-Rice and Oryza 2000 in Korea [9-12]. Research based on paddy growth models is relatively well structured in Korea, but there has been a lack of pure and applied research on field crops and on assessing the impacts of climate change. This is particularly the case for soybeans and maize, due to a lack of awareness regarding the importance of field crops, and because a database has not been established. Although there have been few studies that have used crop growth models, evaluated the growth and yield response of soybeans to the longer growing season and the later planting allowed for by the warming effects of climate change predicted the potential yield of soybeans under a future climate change scenarios for the Korean peninsula using CROPGRO-Soybean, a soybean growth model and suggested fluctuations in soybean production levels were likely assessed changes in the potential grain yield of soybeans due to changes in the plating date caused by higher temperatures during the growing season.

*Corresponding author: Chung U, Climate Application Department, APEC Climate Center, Busan 48059, Republic of Korea, E-mail: uchung@apcc21.org

While various future climate change scenarios have been used in climate change impact assessments in many applications, concerns regarding uncertainty in the future climate scenarios predicted by climate models have increased. Climate change data produced from the Global Climate Model (GCM) may only be suitable for analyzing changes in the average atmospheric characteristics at synoptic scales due to the low spatial resolutions (200 to 400 km) and simplifications of physical processes. To overcome the limitations of the spatial resolution of the GCM, and to produce more detailed climate information, dynamic downscaling methods uses Regional Climate Models (RCM) that can take into account the physical processes of a particular area have been developed [13-15]. As another way to resolve this uncertainty, a few methods have been developed. One such method is a bias correction method that corrects for the systematic bias between GCMs by comparing past climate change scenarios predicted by GCMs with observational data from the same period to evaluate whether GCM that have already been developed can reproduce past climates. Models that apply spatial downscaling to make predictions at the national scale are currently being developed, using the statistical multi-criteria selection method of GCM or RCM data [16,17]. Another method that is used in climate science to reduce uncertainty and produce reliable future climate change data is the multi-model ensemble (MME) approach, a method that combines climate information from more than one GCM [18-23]. MME simulation, that uses more than one crop growth model, is also currently applied to the evaluation of crop productivity in agriculture research [24-26].

The purpose of this study is therefore to examine changes in yield and response of soybeans in crop growth models where the various future climate scenarios have been downscaled to reflect the topography of South Korea. The study will also determine whether an MME approach can contribute to the assessment of the impacts of climatic uncertainty on the potential grain yields of soybeans under various future climate change scenarios.

Materials and Methods

Crop model: CROPGRO-soybean

This study used DSSAT (Decision Support System for Agrotechnology Transfer), a software package that can simulate the growth of various crops by use of the same input/output files, and is one of the most widely used crop growth models world-wide [24,27]. Beans such as peanuts and soybeans are simulated by the DSSAT CROPGRO growth model. CROPGRO can simulate the growth of soybeans and the balance of carbon and nitrogen within agricultural systems (such as uptake, fixation, and formation by the soil system) by inputting daily weather conditions to the model [28-30].

Observation and genetic parameters

The National Institute of Crop Science (NICS) of Korea has regional offices and experimental fields across South Korea, where studies on topics such as crop adaptation are conducted. Chuncheon, Suwom, Jeonju, Miryang, Jinju, and Daegu are representative experimental fields that have previously been used as test sites for studies on crop adaptation (Figure 1). Among the Korean soybean cultivars grown, Taegwang is the main cultivar in the southern part of Korea [31,32] have calibrated and validated the genetic parameters of Taegwang, based on the experiment data of Yeoncheon and Suwon collected in Gyeonggi-do and in Jinju in the southern region of South Korea from 1992 to 2000 (data not show). In this study, the genetic parameters, which were calibrated using the CROPGRO-Soybean by Kim et al.

[32] were tested at six sites (Daegu, Jinju, Miryang, Jeonju, Suwon, and Chuncheon) for 11 years from 2003 to 2013. Weather data from the Automated Surface Observing System (ASOS) run by the Korean Meteorological Administration (KMA) were used because, as detailed earlier, we examined the growth and yield responses of soybeans using ASOS weather data in the crop model, and explored changes in potential grain yield based on future climate scenarios from Global Climate Models (GCMs) that had been downscaled using topographical information from ASOS. To evaluate genetic parameters, the weather data for 2003 to 2013 were collected from 6 ASOS weather stations near where the experimental fields were located. Additionally, the weather data of 10 ASOS weather stations for the same period (2003-2013) were collected and applied to the model (Figure 1 and Table 1).

Soil data

All soil information required for DSSAT soil input parameters was taken from the precision digital soil map produced by the National Institute of Agricultural Science (NAS). This map provides detailed information on the physical and chemical properties of the soil texture. Information from DSSAT soil profiles regarding the locations of the 16 ASOS weather stations was extracted from the digital map for use in this study.

Climate scenario data

Past and future climate change scenarios of global climate models: GCMs included in CMIP5 (Coupled Model Intercomparison Project) were selected for this study. In particular, we collected data from eight GCMs and one regional climate model (RCM) from the Representative Concentration Pathways (RCP) scenarios that were able to provide the minimum input meteorological variables (e.g., daily maximum and minimum temperature, precipitation, and solar radiation) required to run the crop model from 1976 to 2100 (Table 2).

Downscaled climate change scenarios: Spatial downscaling of the climate models to suit the sites was required since the data collected from the nine climate change scenarios did not reflect the geographical features and regional characteristics of the selected weather stations. Therefore, the data from the relevant climate scenarios were downscaled using the nonparametric quantile mapping methods described by Gudmundsson et al. [16] that can simultaneously apply spatial downscaling and bias correction [23].

Application and analysis

Simulation options used for the crop model: In South Korea, soybeans are planted in the middle of June for double cropping systems in the southern region, or at the end of May for single cropping systems in the central and northern regions. The planting dates for Chuncheon, Hongcheon, and Suwon, which are located in the middle and northern regions of the Korean peninsula, were May 25. The planting date for Cheongju, Daejeon, Gunsan, Jeonju, Buan, Jeongeup, Daegu, Jinju, Yeongdeok, Miryang, and Jangheung in the southern regions was June 10 (Table 1). Initial fertilization was provided in the form of N-P-K, with an input of 40 kg/ha N, 30 kg/ha P and 30 kg/ha K. May 25 and June 10 were respectively applied as an estimate of future planting dates according to the location of the sites, and CROPGRO-Soybean was run with an assumption of no abiotic stress (e.g., stress due to lack of water or nitrogen) being present. Two methods are available for setting up projected CO_2 levels for future climate change scenarios. One method is to modify the CO2045. WDA file included with DSSAT, and the other method is to set CO_2 environment parameters manually in the DSSAT experiment file (fileX). For this study, the CO_2 environment for RCP 4.5 and RCP 8.5 was set manually in the experiment fileX.

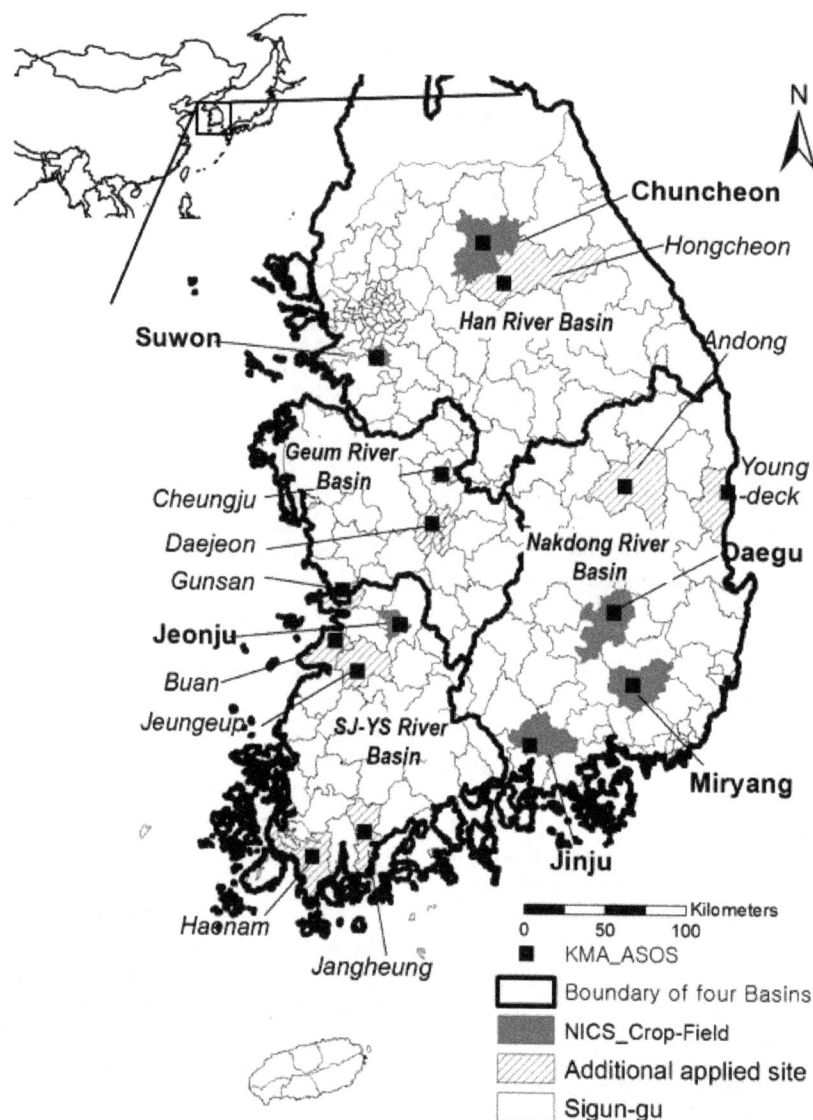

Figure 1: Geographical locations of the six sites (bold letters and gray polygons) of NICS (National Institute of Crop Science) at which the genetic parameters of "Taegwang" were calibrated and validated. In addition, the locations of 10 sites (italic letters and hash polygons) for CROPGRO-Soybean simulation are shown. Bold boundaries delineate the four main basins (Han-River, Nakdong-River, Geum-River, and Seum Jin-Young San (SJ-YS)-River) in South Korea.

Statistical analysis: The simulation period of CROPGRO-Soybean for past climate change scenarios was from 1976 to 2005. This period was chosen because the future climate change scenarios used by CMIP5 in this study had been forecast since 2006, and we then selected the 30 year period from 1976 to 2005 of past climate change scenarios for this analysis. For projections from 2006 to 2100, the near future period used in this study was set from 2021 to 2050. The coefficient of determination (R^2) and root mean squared error (RMSE) were calculated to validate the impact on the potential yields of the genetic parameters of Taegwang.

The potential yield of Taegwang estimated from the observed weather data for the past period (1976-2005) was used as the observation climate-based simulated potential yield (OBS-SIM-PYD) for this analysis. The potential yields of Taegwang predicted by the past and future climate change scenarios of eight individual GCMs and one RCM were also used as the simulated potential yields for individual GCMs (individual-SIM-PYDs). The interquartile range (IQR) of individual-SIM-PYD for the past 30 years from 1976 to 2005 was calculated. The IQR can be obtained by subtracting the first quartile from the 3rd quartile.

The multi-model ensemble (MME) approach was constructed from the averages of the individual-SIM-PYDs. First, the average of two individual-SIM-PYDs that had high correlation with OBS-SIM-PYD was named as the high correlation MME averaged potential yield (MME2C-PYD). Second, the high IQR MME averaged potential yield (MME2H-PYD) was calculated by averaging two individual-SIM-PYDs that had the most similar IQR. Third, the MME4 averaged potential yield (MME4-PYD) was calculated by averaging two individual-SIM-PYDs with high correlation and two individual-SIM-PYDs with high IQR values. Lastly, the average of all nine individual-SIM-PYDs produced the MME9 averaged potential yield (MME9-PYD).

Station ID	Site	Planting day	Basin
ID131	Cheungju	10-Jun	
ID133	Daejeon	10-Jun	
ID140	Gunsan	10-Jun	
ID146	Jeonju	10-Jun	Geum-River Basin
ID159	Buan	10-Jun	
ID245	Jeoungeup	10-Jun	
ID143	Daegu	10-Jun	
ID192	Jinju	10-Jun	
ID277	Youngdeck	10-Jun	Nakdong-River Basin
ID288	Miryang	10-Jun	
ID136	Andong	10-Jun	
ID260	Jangheung	10-Jun	SJ-YS-River Basin
ID261	Haenam	10-Jun	
ID101	Chuncheon	25-May	
ID119	Suwon	25-May	Han-River Basin
ID212	Hongcheon	25-May	

Table 1: Information about basins in which the 16 sites were located and the planting time at each site. Refer to Figure 1 for geographical location of basins.

Model	Origin	Country	Resolution
KMA-12 km	Korea Meteorological Administration	Korea	12.5 × 12.5 km
CanESM2	Canadian Centre for Climate Modeling and Analysis	Canada	2.8° × 2.8°
GFDL-ESM2G	NOAA/GFDL (Geophysical Fluid	USA	2.5° × 2.0°
GFDL-ESM2M	Dynamic Laboratory)		
HadGEM2-CC	Meteorological Office Hadley Center	UK	1.88° × 1.25°
inmcm4	Institute for Numerical Mathematics	Russia	2° × 1.5°
IPSL-CM5A-LR	Institute Pierre Simon Laplace	France	3.75° × 1.8°
MIROC-ESM	Atmosphere and Ocean Research Institute, National Institute for Environmental Studies, and Japan Agency for Marine-Earth Science and Technology	Japan	2.8° × 2.8°
MIROC-ESM-CHEM			

Table 2: List of 8 individual Global Climate Models (GCMs) and one individual Regional Climate Model (RCM) used in this study.

The relative change of the predicted potential yields of each individual climate model (individual-SIM-PYD) for the observation climate-based simulated potential yield (OBS-SIM-PYD) was calculated using Equation 1.

$$Relative\ Change = \frac{(MME2\ C\ of\ PYD\ or\ MME2H\ of\ PYD\ or\ MME9\ of\ PYD) - (OBS\text{-}SIM\text{-}PYD)}{(OBS\text{-}SIM\text{-}PYD)} \times 100 \quad (1)$$

MME2C of PYD, MME2H of PYD, or MME9 of PYD are the average ensemble of individual-SIM-PYDs according to the average ensemble method (i.e., type of GCMs and the number of participants), and *OBS-SIM-PYD* is the potential yield that was estimated from the observed weather data.

Therefore, we first analyzed the changes in potential yield of 16 sites from the climate change scenarios, and then calculated the relative change in the predicted potential yield of each single GCM for the observation climate-based simulated potential yield, and lastly assessed the degree of uncertainty regarding changes in potential yield predicted using the MME approach.

Results

Evaluation of genetic parameters

We used the CROPGRO-Soybean model to simulate soybean cultivation at six National Institute of Crop Science (NICS) sites during 2003 to 2013 to estimate the anthesis day and the potential yield of Taegwang, and to validate the genetic parameters described by Kim et al. [32]. The coefficient of determination (R^2) of the potential yield predicted by the observation were 0.82 in Jinju, 0.43, 0.41, 0.38, and 0.37 for Daegu, Jeonju, Suwon, and Chuncheon, respectively, and 0.37 for Miryang. The spatial average R^2 was 0.40 in all six sites, and the predictability of potential yield was determined to not be high (Figure 2). For anthesis day prediction, the highest R^2 was 0.88 in Miryang, where the R^2 of the potential yield was not useful. The R^2 of the predicted anthesis days in Jinju and Daegu were 0.87 and 0.58, respectively, and 0.29, 0.39, and 0.45 for Jeonju, Suwon, and Chuncheon, respectively. The spatial average R^2 of the predicted anthesis days in six sites was 0.58, which was better than the predictability of the potential yield.

Potential yield of soybeans under different climate change scenarios

Changes of potential yield of soybeans in individual past climate change data: Figure 3 shows the Standard Deviations (SD) and the correlation coefficients of the predicted potential yield for the eight individual Global Climate Models (GCMs) and one Regional Climate Model (RCM) in all six sites during 1976 to 2005. Black circles represent the SDs and correlation coefficients of the observation climate-based simulated potential yield (OBS-SIM-PYD) at all six sites, and circles of red, yellow, orange, pink, blue, green, brown, sky blue, magenta color indicated the SDs and correlation coefficients of the predicted potential yield of each individual climate model (Individual-SIM-PYDs). Each single climate model was indicated in the order CanESM2, MIROC-ESM-CHEM, GFDL-ESM2G, HadGEM2-CC, INMCM4, KMA125 km, IPSL–CM5A-LR, GFDL-ESM2M, and MIROC-ESM. White circles show the SD and correlation coefficient of the nine Multi-model Ensemble (MME) averaged potential yield (MME9-PYD) that was produced from the averaged potential yields of eight individual GCMs and one RCM. Although there were differences in the values themselves, the SD of individual-SIM-PYDs did not deviate from the SD of OBS-SIM-PYD. There were no correlations between individual-SIM-PYDs and OBS-SIM-PYD for most sites. However, especially in Miryang and Jinju, the SD of individual-SIM-PYDs for KMA125 km of RCM were large but showed a trend towards increasing correlation with OBS-SIM-PYD.

The change in potential yield of soybeans predicted by the multi-model ensemble: If the number of models included in an MME increases, the error does not decrease to zero [23,25]. The root mean square errors (RMSEs) were calculated according to the averaged MME ensemble for the combination of type and number of GCMs (Figure 4). In 16 sites, the average of RMSE of individual-SIM-PYD was 508 kg/ha, the averaged RMSE decreased as the number of GCMs increased from 2 to 3. The decline in the averaged RMSE slowed at six GCMs, and stopped at eight GCMs; the averaged RMSE was not significantly reduced even at 7, 8, or 9 GCMs. The estimation error in the individual-SIM-PYDs reduced as the number of models included in the MME increased. The way in which the potential yield averaged by MME represented variations in the observation climate-based predicted potential yields (OBS-SIM-PYD) is also important. Variation (e.g., IQR) s in the individual-SIM-PYDs during the past period (1976-2005) averaged by MME depending on the individual-SIM-PYDs of each single GCM included in the average were compared (Figure 5). Although the variation in individual-SIM-PYDs averaged by MME varied depending on the type and number of included climate models, generally the variance of MME2C-PYD and MME2H-PYD showed

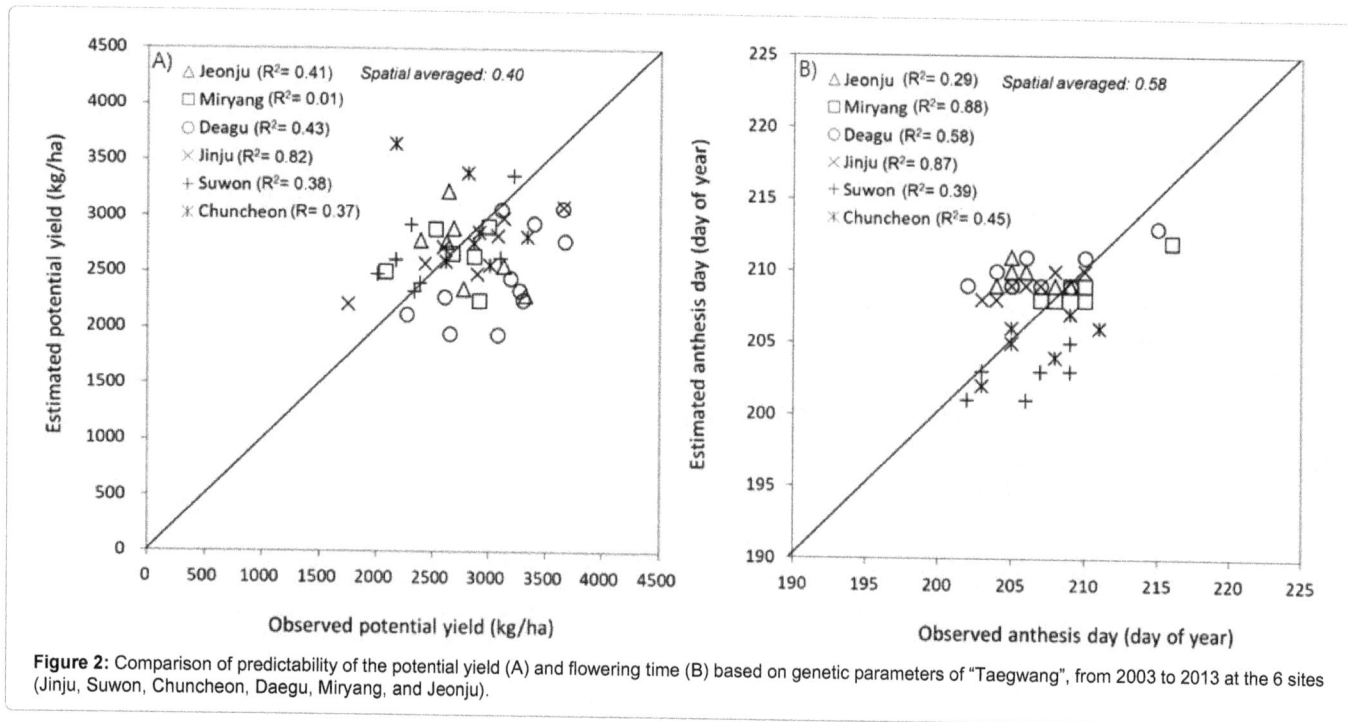

Figure 2: Comparison of predictability of the potential yield (A) and flowering time (B) based on genetic parameters of "Taegwang", from 2003 to 2013 at the 6 sites (Jinju, Suwon, Chuncheon, Deagu, Miryang, and Jeonju).

better reproducibility of the variance in the OBS-SIM-PYD than MME4-PYD or MME9-PYD. In order words, the mean of MME4-PYD or MME9-PYD seemed to be similar to the mean of OBS-SIM-PYD, but they were too averaged to have a small fluctuation range (i.e., IQR) and they could not effectively reproduce the variation of OBS-SIM-PYD.

The relative change in the predicted potential yield of soybeans under future climate change scenarios

The relative change of the predicted potential yield of soybeans under the future climate scenarios of the Representative Concentration Pathways (RCP) 4.5 and 8.5, respectively, for 2021-2050 was calculated using Equation 1, based on the individual-SIM-PYDs of past climates from the GCMs. In 16 sites from 2021 to 2050, the averaged relative change in individual-SIM-PYDs for future climate change scenarios for eight GCMs and one RCM under RCP4.5 and RCP8.5 was: -12.6%/-16.3% in Daegu, -8.6%/-12.2% in Miryang, -8.5%/-10.4% in Jeonju, -8.9%/-10.6% in Jinju, -2.9%/-4.7% in Suwon, 1.2%/-1.2% in Chuncheon, -0.6%/-2.8% in Cheongju, -4.5%/-6.4% in Daejeon, -8.7%/-11.5% in Gunsan, -9.0%/-10.0% in Buan, -11.1%/-11.2% in Jeungeup, -11.8%/-13.7% in Jangheung, -11.6%/-12.7% in Haenam, 5.7%/4.1% in Hongcheon, -3.2%/-3.8% in Andong, and -2.9%/-4.8% in Youngdeck (Figure 6). In particular, the relative changes in individual-SIM-PYDs of other regions, with the exception of Hongcheon and Chuncheon (which are located in the middle north of South Korea), decreased during the near future period. In Hongcheon and Chuncheon, the relative changes in the individual-SIM-PYDs increased, and showed a positive impact on future temperature increases.

Discussion

The predictability of genetic parameters in Taegwang

The prediction for the anthesis day and potential yield in Jinju during 2003-2013 proved reliable since the Taegwang genetic parameters of Kim et al. [32] were constructed using data from Jinju.

Given that soybeans are sensitive to day length, the predictions of the anthesis day for Miryang and Daegu were accurate, as these two sites are geographically close to Jinju and hence have similar day lengths. In contrast with the accurate prediction of the anthesis day in Miryang, the prediction of the potential yield was not accurate. We were not able to determine whether the prediction of the potential yield was not accurate due to insufficient observed yield data from the Mirayng site. However, as mentioned in the Introduction, it could be suggested that, due to the lack of existing case studies of crop model in South Korea, not enough genetic parameters or high quality observation data are available for the calibration and validation of input parameters for crop modeling [33-36].

The predictability of the potential yield of soybeans with climate change

The reproducibility of the potential yield of soybeans under past climate change scenarios: As shown Figure 3, during the past period (1976-2005), the predicted potential yield for individual global climate models (GCMs) (individual-SIM-PYDs) did not reproduce the observation climate-based simulated potential yield (OBS-SIM-PYD) since the correlation between the individual-SIM-PYDs and OBS-SIM-PYD was low. This result was similar to the observations [37], who found that the cherry flowering day predicted based on the past climate change scenarios from individual GCMs did not match the observed cherry flowering day. However, the correlation between the individual-SIM-PYDs of regional climate model (RCM) and the OBS-SIM-PYD is higher than that of the individual-SIM-PYDs of GCMs. This would be expected, because the RCM was dynamically downscaled from the GCM using regional topogeographical factors.

Even if the past climate changes modelled by GCMs were appropriately downscaled and reproduced the observations made during the past period, and this appropriately downscaled climate information was input into the crop model, the crop model may still show different results to what were actually observed. This is because

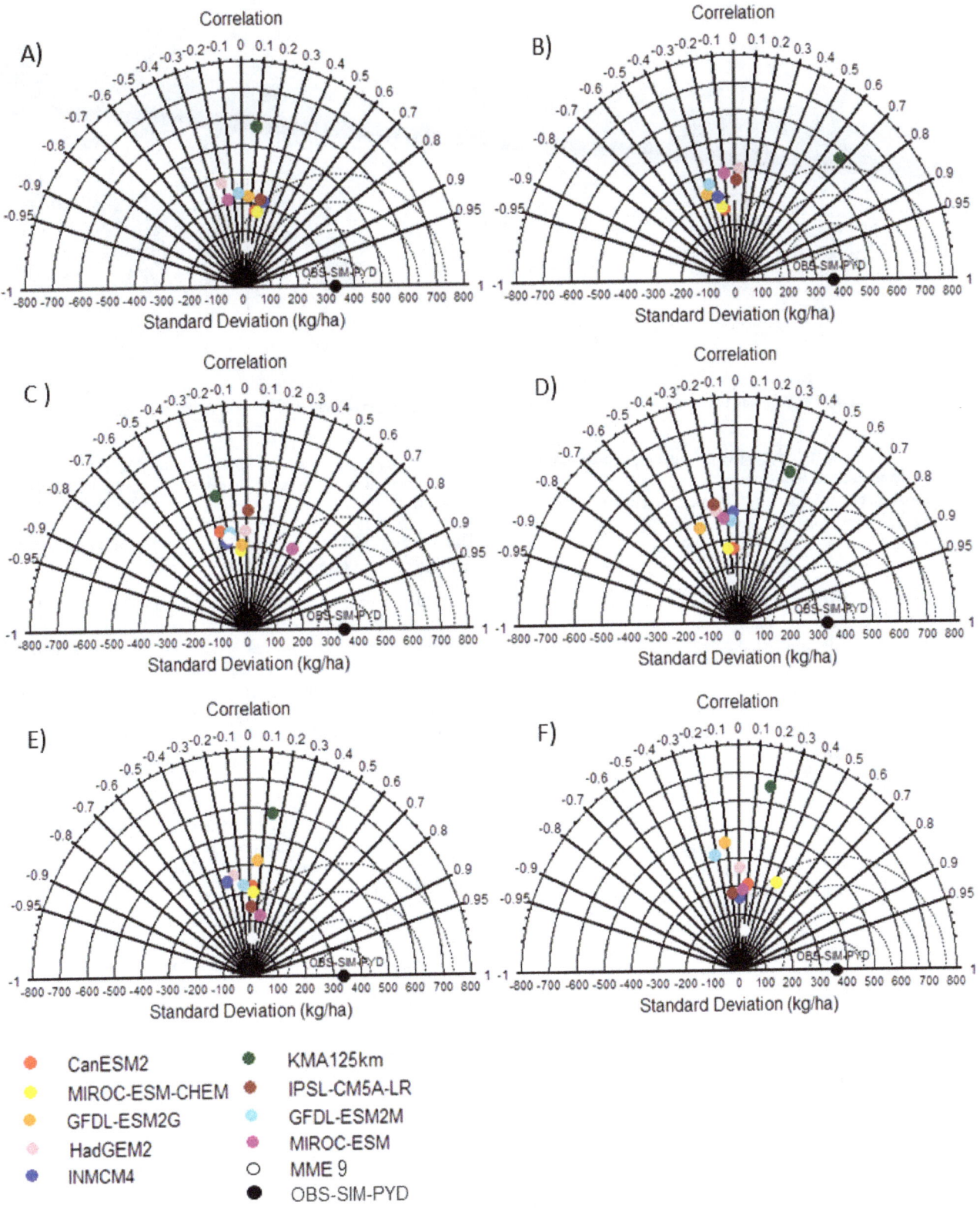

Figure 3: The correlation coefficients versus the standard deviations (SD) for the potential yields (SIM-PYDs) of eight global climate models and one regional climate model (RCM), simulated from CROPGRO-Soybean at six stations (A: Daegu, B: Miryang, C: Jeonju, D: Jinju, E: Suwon, F: Chuncheon). The colored dots indicate the correlation coefficients and the SDs of the 8 individual global climate models (GCMs). The solid black circle represents the correlation coefficient and the SD of the observed predicted potential yield (OBS-SIM PYD) between 1976 and 2005. The solid white circle expresses the correlation coefficient and the SD of the mean averaged potential yield (MME9) simulated from the eight GCMs and one RCM during the same period.

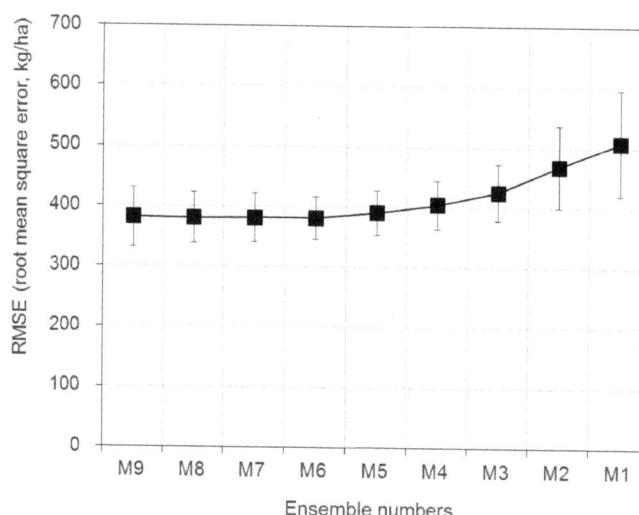

Figure 4: The Root Mean Square Error (RMSE) versus the ensemble type (numbered M1-M8) for each individual climate models.

the crop model is not a simple model (e.g., growing degree day) that predicts plant responses to temperature. It is therefore necessary to analyze in more detail whether the downscaled climate data from GCMs during the major crop growth periods reproduced the observed climate accurately enough, and to evaluate the plant responses in the crop model.

The uncertainty of the potential yield of soybeans by multi-model ensemble: According to Martre et al. [25], as the number of GCMs participating in the Multi-model Ensemble (MME) increased, the Root Mean Squared Error (RMSE) decreased, but the RMSE converged at a certain number instead of continuously decreasing. For the assessment of the agricultural climate index on the future climate change on the Korean peninsula, the RMSE decreased when the number of participation of GCMs reached the maximum, but converged at a certain number rather than continuously increasing [23]. As shown in Figure 4, the estimation error (e.g., RMSE) decreased as the number of GMCs included in the MME increased, but it did not decrease to zero. It is necessary to provide information on the type and number of individual GCMs that can reduce the estimation error by as much as possible, rather than including arbitrarily large numbers of GCMs in the MME.

In addition, the mean of the MME4 averaged potential yields (MME4-PYD) or the MME9 averaged potential yields (MME9-PYD) were similar to the OBS-SIM-PYD, but the range of variations (e.g., interquartile range) of the predicted potential yield was small and showed the typical features of the statistical method, so that the potential yield could not be predicted for any given climate change scenario, such as high temperature events. In a study by Chung et al. [37], which was conducted from 1976 to 2005, the average cherry flowering day predicted by an MME were similar to the observed cherry blossoming day, however, the yearly variations in the cherry flowering day that were caused by annual weather differences were not reproduced. The average predicted potential yield, which is the last information available from the climate change and crop model, is relatively similar to the observed yield; the predicted potential yield was not reproducible due to annual variations in the weather. It could be more important to provide information on the individual GCMs that are able to reproduce changes in the predicted potential yield of

soybeans under conditions of climate change in South Korea, rather than to involve more, or more or many types of individual GCMs, in the MME.

The relative change in predicted potential yield for future climate change scenarios

In most southern regions, the future potential soybean yields will be lower than at present, since in MME the relative change of the potential yields of soybean was expected to decrease in the near future under RCP4.5 and RCP8.5 scenarios at the most sites, with the exception of only two sites among the 16 sites studied. However, the relative change in the potential yield at Chuncheon was increased by 4% for the RCP4.5 scenario and was decreased to -1% for the RCP8.5 scenario. In Hongcheon, the relative change in the potential yield was increased to 6% for the RCP4.5 scenario and 4% for the RCP8.5 scenario, respectively. There is an increased probability that the potential yield of soybeans in Gangwon and the north of Gyeonggi-do in South Korea may itself increase rather than decrease.

Conclusion

In general, a change in the potential yield of a crop will naturally occur in response to local differences. In the northern regions of South Korea, such as Chuncheon and Hongcheon, the temperature increase had a positive effect, and the predicted potential yield of soybeans was increased by future climate change in those areas. The relative changes in predicted potential grain yields under RCP4.5 and RCP8.5 scenarios increased, although the difference between them was not significant in the most southwestern regions. If no experimental data available that can be validated under conditions such as a temperature gradient, such models will remain merely theoretical and without practical applicability. Lately, studies have been actively conducted to reproduce the future climate conditions in South Korea and to directly estimate the growth responses of crops. Improved results for crop modelling will be produced if experimental studies and crop modelling are carried out together.

It could not be concluded that the Multi-model Ensemble (MME) approach reduced the estimation error, but it did reduce the uncertainty of the predicted potential yield of soybeans under future

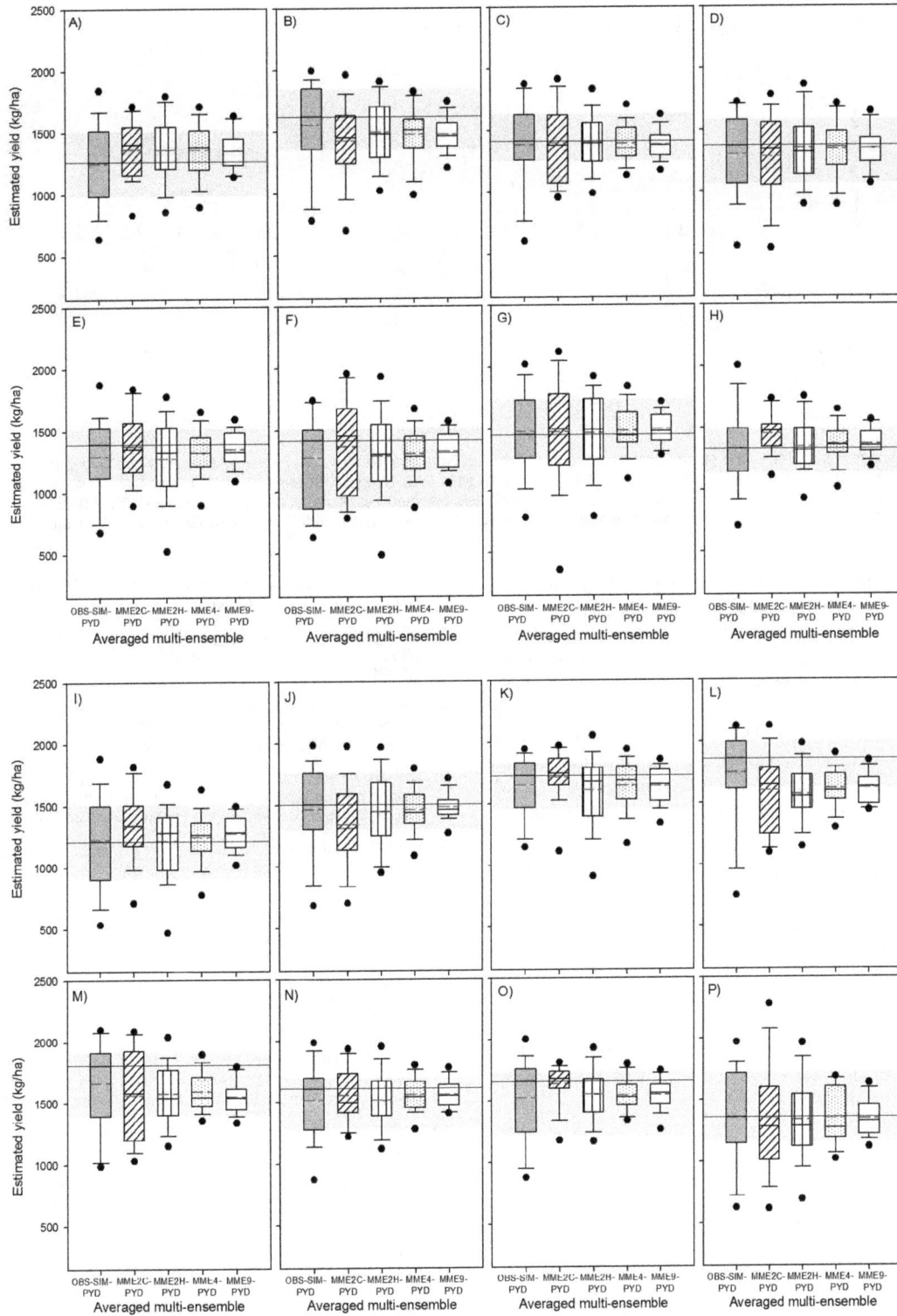

Figure 5: Comparison of the mean potential yields averaged by applying four ensemble methods to each of 16 sites (A: Daegu B: Miryang, C: Jeonju, D: Jinju, E: Suwon, F: Chuncheon, G: Hongcheon, H: Andong, I: Cheungju, J: Daejeon, K: Gunsan, L: Buan, M: Jeungeup, N: Jangheung, O: Haenam, P: Youngdeck). OBS-SIM-PYD estimated the potential yield from CROPGRO-Soybean for the observed weather from 1976 to 2005. MME2C-PYD estimated the mean potential yield averaged from the individual SIM-PYDs of two GCMs with the highest correlation to OBS-SIM-PYD among the individual SIM-PYD of nine GCMs. MME2H-PYD estimated the mean potential yield averaged from the individual SIM-PYDs of two GCMs with the highest correlation to the interquartile range (IQR) of OBS-SIM-PYD. MME4-PYD estimated the mean potential yield averaged from four individual SIM-PYDs, including two individual SIM-PYDs of MME2C-PYD and two individual SIM-PYDs of MME2H-PYD. MME9 represents the mean potential yield averaged from the individual SIM-PYDs of all nine GCMs. Box-and-whisker plots represent means and SDs, red lines and black dots represent medians and ranges.

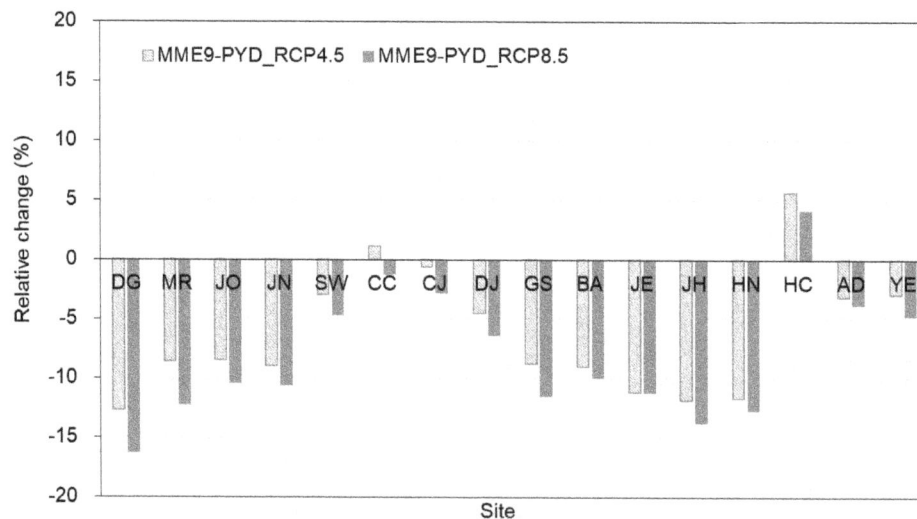

Figure 6: Relative change of the MME9-PYD of the Representative Concentration Pathways (RCP) 4.5 and 8.5 scenarios predicted for a future period (2021-2050), compared with the MME9-PYD during the historical period (1976-2005) at 16 sites (DG: Daegu, MR: Miryang, JO: Jeonju, SW: Suwaon, CC: Chuncheon, CJ: Cheungju, DJ: Daejeon, GS: Gusan, BA: Buan, JE: Jeungeup, JH: Jangheung, HN: Haenam, AD: Andong, and YE: Youngdeck). The MME9-PYD_RCP4.5 and the MME9-PYD_RCP8.5 represented respectively the mean potential yields averaged from the individual SIM-PYD of nine GCMs for RCP4.5 and RCP8.5.

climate change scenarios. That is, the MME approach, according to the type and number of included Global Climate Models (GCMs) can more accurately predict the mean and reduce the estimation error (e.g., RMSE). However, the MME approach is not suitable for the estimation of the potential yield during extreme or abnormal climate events due to the large error in the annual variation of the predicted potential yield. Since climate has too many influencing factors, such as topography and regional land cover, modelling will need to find downscaling methods that can predict local climates from Regional Climate Models (RCMs) or GCMs. Research to improve results must be continuously conducted.

The relative changes in the predicted potential yields of soybeans using individual climate models varied, with the spatial averages of these relative changes for the predicted potential yield at 16 sites during the near future period (2021-2050) estimated as -6.1% and -8.0% for the RCP4.5 and RCP8.5 scenarios, respectively. However, these averages do not represent the whole of South Korea in terms of changes in the potential yield of soybeans under future climate change scenarios. In conclusion, based on the results of this study, it is expected that the importance of research on field crops and modeling of climate change effects will be raised in South Korea.

Acknowledgement

This study was supported by PJ011425032016 of Rural Development Administration (RDA), and conducted by the APEC Climate Center.

References

1. Gibson LR, Mullen RE (1996) Influence of day and night temperature on soybean seed yield. Crop Sci 36: 98-104.

2. Herrero MP, Johnson RR (1980) High temperature stress and pollen viability of maize. Crop Sci 20: 796-800.

3. https://agcrops.osu.edu/newsletters/2012/23

4. Lobell DB, Hammer GL, McLean G, Messina C, Roberts MJ, et al. (2013) The critical role of extreme heat for maize production in the United States. Nat Clim Change 3: 497-501.

5. KMA (2012) Climate Outlook for Korean Peninsula. Korean Meteorological Administration.

6. KMA (2015) Report of Extreme Climate in Korean Peninsula. Korean Meteorological Administration.

7. Liu B, Asseng S, Liu L, Tang L, Cao W (2016) Testing the responses of four wheat crop models to heat stress at anthesis and grain filling. Glob Chang Biol 22: 1890-1903.

8. Maiorano A, Martre P, Asseng S, Ewert F, Muller C, et al. (2017) Crop model improvement reduces the uncertainty of the response to temperature of multi-model ensembles. Field Crops Res 202: 5-20.

9. Lee CK, Kim J, Shon J, Yang WH, Yoon YH, et al. (2012) Impacts of climate change on rice production and adaptation method in Korea as evaluated by simulation study. Korean J Agric For Meteorol 14: 207-221.

10. Hong SY, Hur J, Ahn JB, Lee JM, Min BK, et al. (2012) Estimating rice yield using MODIS NDVI and meteorological data in Korea. Korean J Remote Sens 28: 509-520.

11. Lee KJ, Nguyen DN, Choi DH, Ban HY, Lee BW (2015) Effects of elevated air temperature on yield and yield components of rice. K J Agric Forest Meteorol 17: 156-164.

12. Lee KJ, Kim DJ, Ban HY, Lee BW (2015) Genotypic differences in yield and yield-related elements of rice under elevated air temperature conditions. K J Agric Forest Meteorol 17: 306-316.

13. Lee KM, Baek HJ, Park SH, Kang HS, Cho CH (2012) Future projection of changes in extreme temperatures using high resolution regional climate change scenario in the Republic of Korea. Korean J Geogr Soc 47: 208-225.

14. Becker N, Ulbrich U, Klein R (2015) Systematic large-scale secondary circulations in a regional climate model. Geophysic Res Lett 42: 4142-4149.

15. Giorgi F, Coppola E, Solmon F, Mariotti L, Sylla MB, et al. (2012) RegCM4: model description and preliminary tests over multiple CORDEX domains. Clim Res 52: 7-29.

16. Gudmundsson L, Bremnes JB, Haugen JE, Skaugen TE (2012) Downscaling RCM precipitation to the station scale using quantile mapping–a comparison of methods. Hydrol Earth Sys Sci 16: 3383-3390.

17. Eum HI, Cannon AJ, Murdock TQ (2017) Intercomparison of multiple statistical downscaling methods: multi-criteria model selection for South Korea. Stochastic Env Res Risk Assess 31: 683-703.

18. Murphy JM, Sexton DM, Barnett DN, Jones GS (2004) Quantification of modelling uncertainties in a large ensemble of climate change simulations. Nature 430: 768-772.

19. Semenov MA, Stratonovitch P (2010) Use of multi-model ensembles from global climate models for assessment of climate change impacts. Clim Res 41: 1-14.

Evaluation of Variation and Uncertainty in the Potential Yield of Soybeans in South Korea Using Multi-model Ensemble...

33

20. Diffenbaugh NS, Giorgi F (2012) Climate change hotspots in the CMIP5 global climate model ensemble. Clim Change 114: 813-822.

21. The Intergovernmental Panel on Climate Change (IPCC) (2013) The physical science basis: contribution of working group I to the fifth assessment report of the intergovernmental panel on climate change. Cambridge University Press, Cambridge, United Kingdom, NY, USA.

22. The Intergovernmental Panel on Climate Change (IPCC) (2014) Climate change 2014: synthesis report. Contribution of working groups I, II and III to the fifth assessment report of the intergovernmental panel on climate change, IPCC, Geneva, Switzerland.

23. Chung U, Cho JP, Seo MC, Jung WS (2015) Evaluation of agro-climatic index in Korean Peninsular using multi-model ensemble downscaled climate prediction of CIMP5. Abstract book of International Scientific Conference, Paris, France, pp: 316-317.

24. Bassu S, Brisson N, Durand JL, Boote K, Lizaso J, et al. (2014) How do various maize crop models vary in their responses to climate change factors? Glob Change Biol 20: 2301-2320.

25. Martre P, Wallach D, Asseng S, Ewert F, Jones JW, et al. (2015) Multimodel ensembles of wheat growth: many models are better than one. Glob Chang Biol 21: 911-925.

26. McDermid SP (2015) The AgMIP coordinated climate crop modeling project (C3MP): methods and protocols. Handbook of Climate Change and Agroecosystems: The Agricultural Model Intercomparison and Improvement Project (AgMIP). Imperial College Press, London.

27. Di Paola A, Valentini R, Santini M (2016) An overview of available crop growth and yield models for studies and assessments in agriculture. J Sci Food Agric 96: 709-714.

28. Boote KJ, Jone JW, Hoogenboom G, Wilkerson GG, Jagtap SSS (1989) PNUTGRO V1.02: Peanut crop growth simulation model. User's Guide. Florida Agricultural Experiment Station Journal No. 8420. University of Florida, Gainesville, FL, USA p: 76.

29. Jones JW, Hoogenboom G, Porter CH, Boote K, Batchelor WD, et al. (2003) The DSSAT cropping system model. Eur J Agron 18: 235-265.

30. Hoogenboom G, Jones JW, Wilkens PW, Porter CH, Boote KJ, et al. (2010) Decision Support System for Agrotechnology Transfer (DSSAT) Version 4.5 [CDROM]. University of Hawaii, Honolulu.

31. Kim SD, Hong EH, Lee YH, Moon YH, Kim HS, et al. (1992) Resistant to disease, good in seed quality, high yielding and widely adapted new soybean variety" Taekwangkong". Research Reports of the Rural Development Administration (Korea Republic).

32. Kim SK, Park JS, Lee ES, Jang JH, Chung U, et al. (2004) Development and use of digital climate models in Northern Gyunggi province–I. Derivation of DCMs from historical climate data and local land surface features. Korean J Agric For Meteorol 6: 49-60.

33. Kim DJ, Kim SO, Moon KH, Yun JI (2012) An outlook on cereal grains production in South Korea based on crop growth simulation under the RCP8. 5 climate change scenarios. Korean J Agric For Meteorol 14: 132-141.

34. Park HJ, Han WY, Oh KW, Kim HT, Shin SO, et al. (2014) Growth and yield components responses to delayed planting of soybean in southern region of Korea. Korean J Crop Sci 59: 483-491.

35. Lee D, Min SK, Jin J, Lee JW, Cha DH, et al. (2017) Thermodynamic and dynamic contributions to future changes in summer precipitation over Northeast Asia and Korea: a multi-RCM study. Clim Dyn 1-19.

36. Chung U, Cho HS, Kim JH, Sang WG, Shin P, et al. (2016) Responses of soybean yield to high temperature stress during growing season: A case study of the Korean soybean. Korean J Agric For Meteorol 18:188-198.

37. Chung U, Kim JH, Kim KH (2016) Variation and uncertainty in the predicted flowering dates of cherry blossoms using the CMIP5 climate change scenario. Asia-Pacific J Atmos Sci 52: 509-518.

8

Effect of Conservation Trenches on Plantation Crop in Degraded Watershed in Kandhamal District of Odisha

Subudhi R*

Department of Soil and Water Conservation Engineering, College of Agricultural Engineering and Technology, Orissa University of Agriculture and Technology, Bhubaneswar, Odisha, India

Abstract

Kandhamal district situated in central part of Orissa receives an annual rainfall of 1396 mm, and this region is highly prone to soil and runoff loss due to heavy rainfall during kharif. A trial was conducted during 2001-2004 to study the effect of conservation trenches on plantation crop. This trial was conducted on farmers' field of Sudreju village of Kandhamal district under National Agricultural Technology Project (NATP, RRPS-7), with the following objectives.

1. To conserve moisture for establishment of plantation crop

2. To reduce erosion from upstream area.

3. To increase production of timber, fruit species, fuel wood and fodder .The following treatments were tried.

a. No treatment (Control) b. Continuous V-ditches at 10 m horizontal interval. c. Continuous V-ditches at 20 m horizontal interval. d. V-ditches staggered at 5 m horizontal interval. e. V-ditches staggered at 10 m horizontal interval. Mango varieties Pusa Amrapalli was tried during kharif and during, rabi Black gram (PU-30) was tried in between mango rows. It is observed that in, cont. contour V-ditch at 10 m interval rate of growth was 2.06 cm/month in case of Amrapalli,which is 46% higher compared to control. The grain yield of niger, black gram and mustard are 33.4%, 23.5% and 26.6% higher than control, respectively. Though the cost of construction is little high, it is recommended to practice contour V-ditch at 10 m intervals, to conserve soil and moisture, and to get more grain yield in degraded watershed of Kandhamal district.

Keywords: Conservation trenches; Plantation crop

Introduction

Kandhamal, though receives rainfall around 1396 mm, due to its uneven distribution, heavy downpour of rain at times results in sudden high runoff, which ultimately causes substantial soil loss. The uneven distribution of rainwater and movement of soil within the watershed, results heavy loss to farmers. So conservation trenches for plantation crops helps to conserve the soil and moisture, and ultimately improves grain yield of the farmers. The objectives of the experiment are to conserve moisture for establishment of plantation crop, to reduce soil erosion from upstream area, and to increase production of timber, fruit species, fuel wood and fodder.

Samra JS [1] reported that renovation of terrace and plantation of fruit plants, timber plants improved biomass production, net returns, growth of crop, productivity, reduction of runoff in the range of 1.5-10.8 times, peak flow rate by 20 times and soil loss in the range of 1.2 to 5.2 times, as well as water table rise. Subudhi et al. [2] have reported that effect of vegetative barrier like Vetiver has increased the rice yield, decreased the soil loss and decreased the runoff compared to farmers practice. Arora and Gupta [3] reported that there is a growing need for rain water management, since 96 m ha out of 142 m ha of net cultivated land of the country is rain fed. Scientific use of these resources will definitely increase the productivity and conservation of resources like soil and water. Kumar [4] reported that impact of different soil and water conservation techniques *viz.* contour bunding,terracing, land leveling, smoothening and gully plugging, sowing across the slope, vegetative barrier, increase the Kharif crops by 25-30%. Establishment of vegetative barrier with mechanical measures were more effective in controlling soil erosion (3.8 t ha^{-1}) over conventional method (9.64 t ha^{-1}), and runoff thereby making more moisture available for crop growth. Anonymous [5] reported that V-ditch at 10 m CCVD increased the crop yield significantly compared to no treatment.

Materials and Methods

The study area lies in the Pila-Salki Watershed of Mahanadi Catchment. It falls under Sudreju revenue village of Khajuripada block in Phulbani district. As per Soil Conservation Department Govt. of Orissa, it is a part of watershed ORM 3-9-6-5. As per watershed map classification reported by the Orissa Remote Sensing Application Center (Department of Science and Technology, Govt. Of Orissa), the selected Micro-Watershed falls under Sub-Watershed No 17-07-31-01-01. This sub-watershed consists of parts of Survey of India Topographical Sheet Nos. 73D/2, 73D/6, 73D/3 and 73D/7. However, the Micro-Watershed under study falls only under Topo Sheet No. 73D/6. These Micro-Watersheds are located at a distance of about 10 km from Phulbani district headquarters on Phulbani-Sudrukumpa State Highway. An on farm trial was conducted in the year 2001-2004, at Sudreju under Dry land Agril Research Project, Orissa University of Agriculture and Technology, Phulbani, financed through National Agriculture Technology Project, Rain fed Rice Production System-7. Five following treatments were tested with 4 replication in randomized block design.

***Corresponding author:** Subudhi R, Associate Professor, Department of Soil and Water Conservation Engineering, College of Agricultural Engineering and Technology, Orissa University of Agriculture and Technology, Bhubaneswar, Odisha, India, E-mail: rsubudhi5906@gmail.com

Treatments

T1-No treatment; T2-Continuous V-ditches at 10 m horizontal intervals; T3-Continuous V-ditches at 20 m horizontal intervals; T4-V-ditches staggered at 5 m horizontal interval; T5-V- ditches staggered at 10 m horizontal intervals.

The name of farmer is Kisore Pradhan. Mango variety Amrapalli was tried during Kharif in 5 meter spacing and Niger, Black gram and Mustard were tried during Rabi with 30 cm spacing. Weather was favorable for all crops (Table 1).

Disease and pest

Mango hopper in all Mango varieties. Crop stand: Good. Slope: Field was contour surveyed, and the slope was 4.15%. Soil loss was measured after the rainy season in the V-ditches; the soil was completely filled in 10 m CCVD. So soil conserved was calculated as we know the size of the V-ditch before and after the rainy season.

Results and Discussion

Monthly rainfall is presented in Table 2. It is observed from above table that the year 2002 is a drought years, it received only 74% of rainfall, a deficit of 36% from mean rainfall. But 2001 and 2003 are good years receiving 39.6% and 4% more than the mean annual rainfall, respectively. The mean annual rainfall is 1396.14 mm. The fluctuation shows the rainfall is very erratic in all the three years.

Table 3 shows rate of growth of mango. The rate of growth is highest (3.02 cm/month) in T2-CCVD at 10 meter interval, and lowest (1.22 cm/month) in control from 2001-2003. The grain yield of Niger, black gram and mustard are 33.4%, 23.5% and 26.6% higher than control, respectively (Table 3). This may be due to more soil and water conserved at root zone of the crop as the moisture content in T2 is more compared to all other treatments and lowest in control, as there was no V-ditch (Table 3). The soil conserved in T2 is 6.2 ton/ha, followed by T5 where soil conserved was 5.5 t/ha. Patil et al. [6] has obtained similar result, they got lowest soil loss (1.51 t/ha) and highest survival percentage of cashew nut plantation in Continuous contour trench compared to staggered trench (3.95 t/ha) and control (16.55 t/ha). So it can be concluded that 10 m CCVD should be recommended for uplands of degraded watershed at Kandhamal district of Orissa.

Summary and Conclusion

The present study reveals that grain yield of niger, black gram and mustard are 33.4%, 23.5% and 26.6% higher than control, respectively. Though the cost of construction is little high, it is recommended to practice contour V-ditch at 10 m intervals, to conserve soil and moisture, and to get more grain yield in degraded watershed of Kandhamal district. It is observed that in cont. contour V-ditch at 10 m interval rate of growth was 3.02 cm/month in case of Amrapalli, which is 46% higher, compared to control. Also, we can conserve 6.2 t/ha of soil by 10 sm CCVD, which is highest among all the treatments.

Sl.No.	Name of the farmer	Depth (cm)	Crop	pH (1:2.5)	EC (dsm⁻¹) q	OC (g/kg)	OM (g/kg)
1	Kishore Pradhan	0-30	Mango	5.42	0.0174	5.62	9.67
2	Kishore Pradhan	30-60	Mango	5.98	0.042	3.26	5.61

Table 1: Soil analysis report.

Month	Monthly normal	Actual in 2001	Deviation from normal, %	Actual in 2002	Deviation from normal, %	Actual in 2003	Deviation from normal, %
January	9.18	-	-100	13.0	+41.6	0.0	-100
February	14.07	-	-100	-	-100	23.5	+67.0
March	21.70	56.0	+158.1	20.0	-7.8	12.5	-57.6
April	30.40	-	-100	32.0	+5.2	89.0	+192.7
May	57.48	48.0	-16.5	70.0	+21.8	7.0	-87.8
June	191.62	504.9	+163.5	149.0	-22.2	117.0	-38.9
July	353.62	797.6	+125.6	129.0	-63.5	237.0	-33.0
August	378.65	300.1	-20.7	329.0	-13.1	358.1	-5.4
September	218.57	124.7	-42.9	134.9	-38.3	350.1	+60.2
October	88.93	111.5	+25.4	11.0	-87.6	216.0	+142.9
November	27.48	6.9	-74.9	-	-100	0.0	-100
December	4.45	-	-100	-	-100	42.0	-843.8
Annual	1396.15	1949.7	+39.6	887.9	-36.4	1452.2	+4.0

Soil: The soil data has been presented in Table 2 it reveals that pH is low in top soil (5.42), compared to bottom soil (5.98)

Table 2: Monthly rainfall (mm) during 2001, 2002 and 2003 and their deviation from mean.

Treatments	Niger (q/ha) (2001-02)	Black gram (q/ha) (2002-03)	Mustard(q/ha) (2003-04)	Mean moisture Content (%) At 0-30 cm on weight basis during 2001-03	Mean rate of growth of mango (cm/month) during 2001-2003	Mean Soil conserved in ton/ha
T1-No treatment	2.33	6.12	4.17	3.67	1.22	0
T2-Continuous V-ditches at 10 m horizontal interval.	3.11	8.00	5.28	10.25	3.02	6.2
T3-Continuous V-ditches at 20 m horizontal interval.	2.44	7.12	4.85	5.59	2.47	3.2
T4-V-ditches staggered at 5 m horizontal interval.	2.51	7.37	5.15	8.47	2.42	5.5
T5-V- ditches staggered at 10 m horizontal intervals.	2.49	7.25	5.00	7.02	2.50	3.1
SE (m) +	0.13	0.57	0.05			
CD (0.05)	0.39	NS	0.17			

Table 3: Yield, plant height and moisture content and soil conserved in different treatments.

It can be concluded that 10 m CCVD should be recommended for upland of degraded watershed of Kandhamal district of Orissa.

Acknowledgement

The authors acknowledge the help of Vice Chancellor, O.U.A.T., Director, CRIDA, Hyderabad and Dean of Research. O.U.A.T., Bhubaneswar, for time-to-time guidance and financial help to carry out this project. The authors also acknowledge the help of D.L.A.P. staff of Phulbani and staff of CAET, OUAT, Bhubaneswar, who are helping for the success of the project.

References

1. Samra JS (2002) Watershed management a tool for sustainable production. Proceedings of Indian Association of Soil & Water Conservationists, Dehradun conference held in 20011-10.

2. Subudhi CR, Pradhan PC, Senapati PC (1999) Effect of grass bund on erosion loss and yield of rainfed rice, Orissa, India, T.Vetiver Network 19:32-33.

3. Arora D, Gupta AK (2002) Effect of water conservation measures in a pasture on the productivity of Buffel grass. Proceedings of Indian Association of Soil & Water Conservationists, Dehradun conference held in 2001 65-66.

4. Kumar M (2002) Impact of soil & water conservation on erosion loss and yield of Kharif crops under ravenous watershed. Proceedins of Indian Association of Soil & Water Conservationists, Dehradun conference held in 2001 301-303.

5. Annonymous (2003) Final progress report of NATP, RRPS-7, DLAP, OUAT, Phulbani.

6. Patil PP, Gutal GB, Ganvir BN, Bodake PS (2004) Soil and moisture conservation practices for the hill slopes in Western Ghat of Maharastra. Extended abstracts of National Conferences on Resource Conserving Technologies for Social Upliftment. 122-124.

9

Farmers Utilization of Insecticide Treated Bed Nets for Malaria Prevention in Ahoada East Local Government Area, Rivers State, Nigeria

Franklin E Nlerum*

Department of Agricultural and Applied Economics/Extension, Rivers State University of Science and Technology, Nkpolu-Oroworukwo, Nigeria

Abstract

The study examined the utilization of insecticide treated bed nets for malaria prevention by rural farmers in Ahoada East Local Government Area of Rivers State, Nigeria. The sample size of the study was 100 respondents who were randomly selected from farmers in the area who benefited from the free distribution of the nets by Rivers State Government. The questionnaire was used in eliciting data from the respondents. Data were analyzed with percentage, mean score and multiple regression. Socio-economic result shows that the respondents on the average were 40 years old, earned monthly net income of N24,184.00 ($121.38) and spent 11 years in schooling. More results indicated that while the ownership of the net was as high as 71.73%, the actual utilization was as low as 28.27%. Result of multiple regression analysis indicated a multiple determination (R^2) value of 0.6333. Determinants of the utilization of the net were age, sex, occupation and educational level of the respondents. The two major constraints in the utilization of the net in the area were inadequate information and poor design and inconvenience of hanging. In order to improve the rate of utilization of the net in the area, the study recommends enhanced information which will lead to a better education of the beneficiaries. A better design which will reduce the inconveniences associated with hanging of the nets is also recommended.

Keywords: Farmers; Utilization; Insecticide; Treated bed nets

Introduction

Malaria exerts a significant health and economic burden on Nigerians. "According to the statistics of the Federal Ministry of Health [1]", malaria was responsible for 60% of out-patient visits to health facilities, 30% of childhood deaths, 25% of infant deaths and 11% of maternal death, and an estimated annual loss of 132 billion Naira, in the form of treatment and prevention costs, and loss of man-hours amongst other losses. Rivers State government in the south-south Nigeria distributed more than two million mosquito treated bed nets. The target of the State was to provide two nets to every household in the state of which farmers were among. Ownership as used in this study means all those who received the free nets. Utilization in other hand was used to describe all those who actually put the nets into continuous use. Recently, it has been shown that the use of insecticide-treated bed nets can reduce malaria morbidity by 50% to 60% and malaria mortality by 20% as indicated by Azondelon et al. [2].

Meaning of insecticide treated nets (Itns)

ITN is a form of personal protection that has been shown to reduced malaria illness, severe disease and death caused by malaria endemic regions according to the Centre for Disease Control and Protection [3]. Approximately, 40% of the world's population (mostly those living in the poorest countries are at the risk of malaria as indicated by Adogu and Ijemba [4]. Although bed nets form barrier against mosquitoes around people sleeping under them, bed nets treated with an insecticide are much more protective than untreated nets. ITN was introduced in Nigeria in the year 2000 as an effective means of preventing mosquito bites and malaria transmission as stated by Ukibe et al. [5]. In 2012, an estimated 627,000 people died of malaria in the sub-Sahara Africa according to Centre for Disease Control [3].

Effects of malaria on farmers

Malaria costs Africa more than 12 billion dollars annually and it slowed economic growth in African countries by as much as 1.3% per year as shown by Adebayo et al. [6]. When malaria attacks farmers,

their health is hampered, resulting to the reduction of agricultural production. In the fight to increase food production in the tropics, farming communities suffer most from malaria according to the New Agriculturist [7]. This is because malaria effect on farmers leads to fever which results into incapacitation. The study of Oluwatayo [8] has shown that farmers who were susceptible to malaria infection suffered about 10 days of incapacitation.

Malaria causes reduced farm labour according to Eboh and Okeibunor [9] because its attack leads to the death of farmers. Malaria also leads to financial insecurity in that its attack leads to increased expenditure on malaria treatment. In addition, malaria infection reduces the use of agricultural innovations as shown by GBC Health [10] because premature death of trained farmers from malaria hinders a further use of the acquired agricultural innovations. According to the World Health Organisation [11] there were 214 million new cases of malaria worldwide and 438,000 malaria deaths in 2015 with African as the most affected region. The economic burden of malaria on production efficiency among farmers in Oyo State, Nigeria was as higher as N7, 578.71 according to Adekunle [12].

The rational for the study was that, despite the free distribution of insecticide treated bed nets by the State government of which farmers were among the beneficiaries, malaria problem is still rampant among farmers in the study area. The goal of the study was therefore to know

*Corresponding author: Nlerum FE, Department of Agricultural and Applied Economics/Extension, Rivers State University of Science and Technology, Nkpolu-Oroworukwo, Port Harcourt, Nigeria, E-mail: frankezi@yahoo.com

why farmers still suffer the burden of malaria after receiving the free nets. The study objectives therefore analyzed the socio-economic characteristics of the farmers, determined the level of ownership and utilization and ascertained constraints to the use of the nets by farmers.

Research Methodology

According to the Federal Republic of Nigeria [13], malaria in Nigeria is transmitted by the vector specie called Anopheles mosquito with a very high rate of prevalence in the country. This study was conducted in Ahoada

East Local Government Area of Rivers State. Rivers State is located in the Niger Delta, south-south of Nigeria, and has a topography of flat plains netted in a web of rivers and tributaries as shown by the Rivers State Economic Summit [14]. Ahoada East Local Government Area is one of the 23 Local Government Areas of Rivers. It is in the West Senatorial District. The people are predominantly farmers.

Primary data for the study were obtained with a structured questionnaire which was designed by the researcher. The questions contained in the questionnaire elicited data to cover the objectives of the study. A trained enumerator for this purposed distributed and collected the questionnaire from respondents. The population of the study was made up of all farmers who received the free insecticide treated bed nets in the area. A sample size of 100 respondents was randomly selected from the list of farmers who benefited from the free nets in the Local Government Headquarters. Twenty farmers were drawn from each of the five randomly selected communities of the study area [15]. The communities were Abarikpo, Ihuaje, Odiabide, Ogbo and Ula-Ihuda. Data were analyzed with percentage, means of a Likert-type rating scale and multiple regression analyses using the Statistical Programme for Social Sciences (SPSS) version 15 and Excel, 2007. The scoring for constraints in the use of the bed net was done with a four point Likert- type rating scale. The points in the scale were very severe (4 points), severe (3 points), and averagely severe (2 points) not severe (1 point). A cut-off mean (mid-point) was obtained by adding 4, 3, 2 and 1 and dividing the sum by 4 to have 2.50, as used by Ugwoke [16]. Results showing means which were equal or greater than 2.5 were interpreted as severe constraints, while those with less than 2.50 were interpreted as not severe constraints.

Results and Discussion

The results in Table 1 show that 60.0% of owners of the free insecticide treated bed nets distributed by the state government in the study area were males, while females represented 40%. This finding indicates that more males benefited from this activity of the government than females. Married respondents represented a higher proportion of ownership of the insecticide treated bed nets with 43.0%. The age range of 30-40 years with 71.0% represented the highest age of owners of the treated mosquito nets. The mean age was 40 years. This mean age represented a very active age, showing that many who received the insecticide treated mosquito nets were in their active ages.

The finding however shows that age range of 50-60 was 7.0%, while that of those above 60 years was 5.0%. This result seems to indicate that as the age of respondents increases to 50 and upwards the interests shown in the use of the net decreases. Further results showed that 66.0% of the respondents were full-time farmers, while 34.0% were part-time farmers, showing that farming is the main livelihood source of the people in the study area. The result on the monthly farm net income range of N24,500.00–N32,000.00 with 51.0% was the highest for these respondents. The monthly mean income was N24,158.00

Characteristics	Frequency	Percentage %	Mean	Standard Deviation
Sex	-	-	-	-
Female	40	40	-	-
Male	60	60	-	-
Total	**100**	**100**	-	-
Marital Status				
Single	32	32	-	-
Married	43	43	-	-
Divorced	8	8	-	-
Separated	-	-	-	-
Widowed	17	17	-	-
Total	**100**	**100**	-	-
Age in years				
Below 30	-	-	-	-
30–40	71	71	40 years	0.61
40–50	17	17	-	-
50–60	7	7	-	-
Above 60	5	5	-	-
Total	**100**	**100**	-	-
Occupation				
Full-Time Farming	66	66	-	-
Part-Time Farming	34	34	-	-
Total	**100**	**100**	-	-
Monthly net income (in Naira)				
Less than 7,500.00	6	6	-	-
7,500 – 15,000	11	11	-	-
16,000 – 23,500.00	15	15	-	-
24,500.0032,000.00	51	51	24.158.00	0.78
Above 32,000.00	17	17	-	-
Total	**100**	**100**	-	-
Educational Status (Years of Schooling)				
Primary	14	14	-	-
Secondary	45	45	11.0 years	0.9
Tertiary	25	25	-	-
Non	16	16	-	-
Total	**100**	**100**	-	-

Source: Field survey, 2012. One dollar: N199.25 (Naira) as at october, 2015

Table 1: Socio-economic characteristics of insecticide treated bed nets in Ahoada East Local Government Area.

($121.39). This mean income shows that an average farmer in the area earned more than N18,000.00 which is the monthly minimum wage of civil servants in Rivers State. The mean income also connotes that farmers in the area were out of the poverty line which is below $1.00 per person per day by reason of the fact they earn about $4.05 per day. For educational attainment, secondary level was the highest with 45.0%. The mean number of years spent in schooling was 11 years, while as much as 16.0% of the farmers have not attempted any form of formal education. The low standard deviation for age (0.61), income (0.78) and educational level (0.90) shows the reliability of the socio-economic factors.

Table 2 shows that the highest ownership of insecticide treated bed nets was in Ogbo community with 98.0%. They also occupied the highest position in terms of utilization with 48.0%. This result showed a gap of as much as 50.0% between the ownership of the net and its utilization. Abarikpo community had 96.0% ownership and as low as 5.0% in utilization. This community was the least (5th position) in terms of utilization of the nets. The trend in the utilization of the treated bed nets in this study agrees with the study of Ukibe et al. [5] in Anambra State, Nigeria where out of the 60% of the respondents that owned the treated net, 46% of them actually utilized it. The Anambra State respondents also left some gaps between ownership and utilization as with the findings of this study.

The result of the test of hypothesis in Table 3 shows that the linear function had the highest multiple determination (R^2=0.6333) and a reasonable number of significant variables. The linear function was therefore used for the interpretation of the result. The R^2 of 0.6333 indicates that about 63.33% of the variation in utilization of the net was accounted for by the joint action of the five socio-economic variables investigated. Consequently, the variables with significant coefficients were age (X_1), sex (X_2), occupation (X_5) and educational level (X_6). These four variables were considered as being important determinants in the utilization of insecticide treated mosquito bed nets among these respondents. This result agreed with the one of Sena et al. [17] that age of respondents was significant to utilization of ITNs in Ethiopia. F-ratio which was 0.6052 was significant. This meant that the combined effect of the independent variables played important role in the utilization of the nets.

Table 4 shows that the most severe (3.35) constraint to the utilization of the nets in this study area was inadequate educational information on how to make the best use of the nets. The finding agreed with the study of Teklemariam et al. [18] where lack of enough education was a constraint to the use of ITN by respondents. This constraint tends to imply that the nets were distributed to beneficiaries without adequate education on how to use them. The second severe constraint with the mean of 3.30 was poor design and inconvenience of hanging. This finding agreed with the study of Omotayo and Oyekale [15] which indicated that inconvenience in hanging was a constraint in the use of mosquito treated bed nets in Ido area of Oyo State, Nigeria. The third and fourth severe constraints to the use of the nets in this

study area were distance to health centers for advice and inadequate health extension workers with the mean of 3.22 and 3.20 respectively. The results further shows that out of the eight constraints analyzed in this study, six of them had severe effects in the utilization of the treated bed nets in the area. These constraints would have been responsible for the high disparity between ownership and utilization of the net as identified in Table 2 of this study.

Limitation of the study

The limitation of the study was that poor educational levels of some farmers made it difficult for them to fully respond to the questionnaire on their own. This limitation was however tackled with the use of trained enumerator who explained some parts of the survey instrument to fairly educated respondents.

Conclusion and Recommendations

The study has shown that there was a high disparity between ownership of the insecticide treated bed nets and actual utilization among farmers in this study area. The actual mean utilization of the net was as poor as 24.4%, while the mean ownership on the other hand was as good as 72.4%. Socio-economically, the recipients of the nets on the average were 40 years of age, had monthly net income of N24, 184.00 (Naira), and 11 years of schooling experience. The major constraints which were responsible for the poor utilization of the nets in this study area were inadequate information on how best to use the nets, poor design and inconvenience of hanging. In order to increase the rate of utilization of the nets by farmers in the area, the study recommends improved information delivery on how best to benefit from their use and improvement in design to reduce the inconveniences associated with their hanging.

Community	Ownership Percentage (%)	Ranking	Utilization Percentage (%)	Ranking
Abarikpo	96	2nd	5	5th
Ihuaje	45	5th	38	2nd
Odiabidi	66	3rd	37	3rd
Ogbo	98	1st	48	1st
Ula Ehuda	57	4th	36	4th
Mean	72.4	-	25.4	-

Source: Field survey, 2012. Multiple responses were allowed

Table 2: Percentage distribution of ownership and utilization of insecticide treated bed nets in the area (n=100).

Model Summary & Fitness	Parameters	Linear	Semi-log	Double-log
	Multiple R Square (R^2)	0.6333	1.455	-0.162
	F-ratio	0.6053*	0.1881**	0.725*
	P-value of the f-ratio	0.0044	5.4261	0.7344
Co-efficients estimates	Variables	70.17	4.2509	0.628
BO	Intercept	0.101 (-1.0)NS	-2.22 (0.32)*	-3.36 (-1.53)**
B1	Age (X_1)	7.65 (0.29)*	2.034 (-2.3)*	0.308 (-1.11)NS
B2	Sex (X_2)	0.52 (0.02)*	-0.64 (-0.10)*	-0.188 (-0.07)NS
B3	Marital Status (X_3)	2.91 (-0.11)NS	1.06 (-0.23)NS	(0.12)NS
B4	Income level (X_4)	11.01 (-0.42)NS	2.39 (-1.06)NS	0.38 (0.10)
B5	Occupation (X_5)	21.85 (0.85)**	3.08 (1.65)*	0.48 (2.1)*
B6	Educational Level (X6)	20.70 (0.81)**	2.73 (1.05)*	0.43 (1.5)*

Source: Field survey, 2012. Figures in parentheses are t-ratios: *=Level of significance at 0.05%, **=Level of significance at 0.01%, NS=Non significance

Table 3: Summary of multiple regression analysis of relationship between respondents' socio-economic characteristics and utilization of the treated bed nets.

Constraints	Total Score	Mean	Ranking	Decision
Cultural barrier and superstition	290	2.9	6th	Severe
Religious prohibition	160	1.6	8th	Not Severe
Feeling of suffocation	210	2.1	7th	
Inadequate educational information on use	335	3.35	1st	Not Severe
Inadequate health extension workers	320	3.2	4th	Severe
Distance to health centres for advice	322	3.22	3rd	Severe
Feeling of heat inside net	315	3.15	5th	Severe
Poor design and inconvenience of hanging	330	3.3	2nd	Severe

Source: Field survey, 2012. Cut-off mean=2.50

Table 4: Summary of mean distribution of constraints to utilization of insecticide treated mosquito bed nets (n=100).

Biography of Author

Dr. Franklin Eziho Nlerum, has a Ph.D. in Rural Sociology and Development. His research interests are in Agricultural Extension, Rural Sociology and Rural Development. Currently, he is a Senior Lecturer. He has worked in Rivers State Agricultural Development Programmed as a Block Extension Supervisor, Subject Matter Specialist (Agronomy/Plant Protection) and an Assistant Director of Agriculture. He has published over 40 papers in both local and international journals. He is an Editorial Board member of Pyrex Journals (Journal of Agricultural Extension and Rural Development) and he is also a member of some professional bodies in his field of research interest.

References

1. Federal Ministry of Health (2009) "National Malaria and Vector Control Division". Annual Report 2008. Abuja, Nigeria.

2. Azondelon R, Gnanguenon V, Oke-Agbo F, Housevoessa S, Green M, et al. (2014) "A tracking tool for long-lasting insecticidal (mosquito) net: In 2011 national distribution in Benin", Parasite and vectors.

3. Center for Disease Control and Prevention (2015) "Insecticide treated bed nets".

4. Adogu POU, Ijemba C (2013) "Insecticide treated nets possession and utilization among pregnant women in Enugu, Nigeria: A descriptive cross-sectional study". JNSR 3: 40-47.

5. Ukibe SN, Mbanago JI, Ukibe NR, Ikeakor LC (2013) "Level of awareness and use of insecticide treated bed nets among pregnant women attending ante-natal clinics in Anambra State, South-Eastern Nigeria". JPHE 5: 391-396.

6. Adebayo O, Olagunju K, Adewuyi SA (2015) "The impact of malaria disease on productivity of rural farmers in Osun State, Nigeria. Studia, Mundi-Economics 2: 86-94.

7. New Agriculturist (2015) "Points of View: Malaria and agriculture".

8. Oluwatayo IB (2013) "Socio-economic burden of malaria on productivity of rice farmers in rural south-west, Nigeria". MJSS 5: 175-185.

9. Eboh EC, Okeibunor JC (2005) "Prevalence and impacts of farm household labour uses and productivity in the irrigated rice production system of Omor community, Nigeria". Tanzania Health Research Bulletin 7: 7-15.

10. GBC Health (2012) "Linkages between malaria and agriculture".

11. World Health Organisation (2015) "Fact sheet: World malaria report 2015".

12. Adekunle SW (2013) "Economic burden of malaria and production efficiency among farming households in Oyo State, Nigeria". IJRSR 4: 1467-1650.

13. Federal Republic of Nigeria, Nigeria malaria indicator survey (2010)" Final report, National Population Commission, Abuja, Nigeria, 2012.

14. Rivers State Economic Summit (2001) "Report on First Rivers State Economic Summit Committee, Port Harcourt" April 21st – 3rd May.

15. Omotayo AO, Oyekale AS (2013) "Effect of Malaria on farming households' welfare in Ido Local Government Area of Oyo State, Nigeria". JHE 14: 189-194.

16. Ugwoke FO (2005) "Constraints in implementing the agricultural sector employment programme of National Directorate of Employment in Enugu State, Nigeria. Global Approaches to Extension Practice 14: 55-63.

17. Sena LD, Deressa WA, Ali AA (2013) "Predictors of Long-lasting insecticide-treated bed net ownership and utilization. Evidence from community-based cross sectional comparative study, Southwest, Ethiopia". Malaria J 12: 1-9.

18. Teklemariam Z, Awoke A, Desie Y, Weldegebreal F (2015) "Ownership and utilization of insecticide-treated nets for malaria control in Harari National Regional State, Eastern Ethiopia". Panafrican-Med-J 21: 1-9.

10

Determination of Influencing Factors for Integrated Pest Management Adoption: A Logistic Regression Analysis

Talukder A[1]*, Sakib MS[2] and Islam MA[1]

[1]*Statistics Discipline, Khulna University, Khulna-9208, Bangladesh*
[2]*Department of Statistics, Jagannath University, Dhaka-1100, Bangladesh*

Abstract

   Adoption of an environment friendly agricultural technique, namely Integrated Pest Management (IPM) depends on various socio-economic and demographic factors. This study attempts to determine the factors that influence farmers' decision to receive IPM. For analysis purpose several socio-economic and demographic information were collect from 617 farmers of five division (Dhaka, Chittagong, Rangpur, Khulna, Barisal), Bangladesh by prepared a structured query. To ensure randomness, simple random sampling technique was used for data collection. Farmers' ten background characteristics were analyzed in both bivariate and multivariate setup. In bivariate setup, association between selected factors and adoption status of IPM were investigated by performing a chi-square test. To get the adjusted effect, a binary logistic regression model was estimated in multivariate setup. The results of the model provide evidence that farmers' age, education level, farming experience, training on IPM and membership status of IPM club are the highly significant ($P<0.05$) factors for IPM adoption. Farm ownership status and Barisal division also found significant ($P<0.10$) factors for IPM adoption in Bangladesh.

Keywords: Adoption; IPM; Chi-square test; Logistic regression model

Introduction

   Management of pesticide to control pest has become a major cause for concern in most of the developing countries. It is acute in Bangladesh, a densely (964/sq.km) populated country having increasing growth rate of 1.34% [1]. Every year we have lost a substantial amount of production because of the attack of various pests. A previous study conducted by MOA [2] estimate that on an average 16% of rice, 30-40% of vegetables, 15% of jute, 25% of pules, 11% of wheat were annually lost due to the serious attack of different types of pastes. For controlling pests, most of the farmers still fully depend on the application of chemical pesticide in their cultivated land. The fully dependence on chemical pesticide is not a good sign for environment, since excessive use of chemical has so much negative effects on soil, health and environment [3]. Therefore we need to think an alternative approach that can control not only pest but also helpful for environment. In this context, organic agriculture is a powerful tool having no negative impact on ecology [4-6]. But only one approach may not enough to meet the increasing demand of growing population in Bangladesh. So we have to think an integrated approach that should be helpful for better production and ensure the safety of environment. In this dilemma, there is a simple solution, which is Integrated Pest Management (IPM) [7].

   In literature, IPM has more than 65 definitions [8]. However, the actual meaning of IPM is still a mystery for us [9]. By considering the ecological behavior of IPM, several scholars define IPM according to their own way. For example, IPM is said to be an environmentally friendly agricultural technique in Dasgupta et al. [10], economically sound technique in Prokopy and Kogan [8], and sustainable agriculture in De Souza Filho et al. [11], and clean farming technologies in Veisi [12]. In spite of having different opinions from various scholars about the definition of IPM, the original massage remains almost same: 'Controlling pest in a sustainable manner'.

   Now-a-days development of a sustainable agriculture is the most challenging task in Bangladesh. Moreover, sustainable agriculture remains incomplete without adopting IPM practice. Generally, adoption of any new method depends on several social, economic, demographic and physiological factors. These factors may influence the mentality of human being in any moment. Hence, adoption of IPM is not exception from this setup. In our study, we try to determine the possible influencing factors for the adoption of IPM in the context of Bangladesh by fitting a binary logistic regression model.

Materials and Methods

Data sources

   Since the determination of influencing factors for IPM adoption was the main focus of this study, all Bangladeshi farmers were the major sources of data collection. In the practical point of view, it is totally impossible to gather information from all Bangladeshi farmers (target population). However, we can interview a part (study population) of our target population by determining appropriate sample size.

Determination of sample size

   In this research our determined sample size is 617 (for detailed calculation of sample size determination (Appendix). Therefore we have to choose 617 farmers randomly, and collect desired information from them.

Data collection

   For collecting desired information from study population, a structured questionnaire was prepared. The questionnaire was made as simple as possible and only relevant questions were included. Best

*Corresponding author: Talukder A, Statistics Discipline, Khulna University, Khulna-9208, Bangladesh, E-mail: ashistalukder27@yahoo.com

efforts were made to obtain unbiased answers from the respondents. To collect several socio-economic and demographic information we randomly interviewed 450 conventional farmers (those who use chemical pesticides) and 167 IPM farmers (those who recently adopted IPM) from five divisions (Dhaka, Chittagong, Rangpur, Khulna, Barisal) of Bangladesh. Therefore we have a total of 617 (desired sample size) Bangladeshi farmers for further analysis. To ensure the randomness we consider simple random sampling as a sampling technique for data collection.

Variable selection

The adoption status of IPM (1, if adopt IPM, 0, otherwise) is considered as the main variable of interest. The farmer is considered to be IPM farmer if he recently adopt IPM, otherwise the farmers is treated to be a conventional farmer. Besides the main variable, we also consider respondents' age (categorized into five groups), education level (no/primary education, secondary, higher), division, farming experience (at least 10 years of farming experience, more than 10 years), farm size (categorized into three groups based on existing size of the cultivated land), ownership of the farm (yes, no), training on IPM (yes, no), membership on IPM club (yes, no), attending in farmer field school (FFS) (yes, no) as possible influencing factors of IPM adoption. These variables were found significant in previous studies [7,10,13-19].

Statistical analysis

To assess the influencing factors of IPM adoption, we conduct our analysis into both bivariate and multivariate setup. In bivariate setup, we perform chi-square test to assess the unadjusted effect of the selected explanatory variables on IPM status. Since bivariate analysis fail to explain the adjusted effect of explanatory variables, we consider a statistical model appropriate for binary response namely binary logistic regression model in multivariate setup. For details on binary logistic regression model [20]. The software that we used for data analysis purpose is SPSS (version 20 for windows).

Results

Univariate analysis

The average age of the surveyed farmers is found to be 35 years with a standard deviation of 10 years. It is investigated that among 617 individuals, 48.9% have no or primary education. On the other hand 37.1% and 13.9% of participants have secondary and higher education, respectively. Most of the farmers (71%) have greater than 10 years of farming experience. The distribution of sampled farmers among the divisions Dhaka, Chittagong, Rangpur, Khulna and Barisal are 20.1%, 19.8%, 20.9%, 18.6% and 20.6%, respectively. More than 50% farmers have 1 to 1.5 acres of cultivated land. It is found than 66.1% farmers have a cultivated land of their own and 78.1% receive training on IPM farming. Moreover, 26.9% and 19.8% farmers are the member of IPM club and FFS, respectively. However, less than one-third of the sampled farmers (27.1%) adopt IPM farming.

Bivariate analysis

The results obtained from bivariate analysis are displayed in Table 1. It is very much surprising to observe that, adoption rate of IPM increases as age of the farmer increases. This implies, older farmers are more interested to adopt IPM compared to younger farmers. Among the education level, the higher education group has higher proportion of IPM farmers (44.2%); whereas other two groups has almost similar (around 24%). Considering the divisions, farmers from Barisal are more

likely to adopt IPM (37%); whereas it is least in Khulna (21.7%). Among the IPM farmers, 29.5% have more than 10 years of farming experience, 39.6% have more than 1.5 acres of cultivated land, 28.7% are the owner of his farm, 30.3% receive training on IPM, 28.3% join the IPM club and 29.5% are the member of FFS. Note that age, education level, farm size and training on IPM are reported to be highly significant (P<0.01) factors for IPM adoption. Beside these factors, division and farming experience of the farmers also found significant at 5% level (P<0.05). However, ownership status of farm and membership status of IPM club or FFS show insignificant effect on IPM adoption (P>0.10).

Regression analysis

To get the adjusted effect of selected factors for IPM adoption, we consider a binary logistic regression model, since our main variable IPM adoption status has two category (farmer adopt IPM or not). The necessary results are given in Table 2. From this table we observe that farmers' age has a positive effect on IPM adoption. That is, as age of the farmer increases the chance of adopting IPM also increases in a

Factors	IPM adoption status		P value
	Yes n (%)	No n (%)	
Age group			
<30	29 (17.3)	139 (82.7)	
30-34	33 (18.9)	142 (81.1)	
35-39	36 (30.5)	82 (69.5)	0.000***
40-44	37 (42.5)	50 (57.5)	
45 or more	32 (46.4)	37 (53.6)	
Education level			
No/primary	74 (24.5)	228 (75.5)	
Secondary	55 (24.0)	174 (76.0)	0.001***
Higher	38 (44.2)	48 (55.8)	
Division			
Dhaka	37 (29.8)	87 (70.2)	
Chittagong	28 (23.0)	94 (77.0)	
Rangpur	30 (23.3)	99 (76.7)	0.033**
Khulna	25 (21.7)	90 (78.3)	
Barisal	47 (37.0)	80 (63.0)	
Farming experience			
At least 10 years	38 (21.2)	141 (78.8)	0.022**
More than 10 years	129 (29.5)	309 (70.5)	
Farm size			
<1 acre	34 (22.8)	115 (77.2)	
1-1.5 acre	89 (24.9)	268 (75.1)	0.004**
>1.5 acre	44 (39.6)	67 (60.4)	
Farm ownership status			
Yes	107 (26.2)	301 (73.8)	0.566
No	60 (28.7)	149 (71.3)	
Training on IPM			
Yes	146 (30.3)	336 (69.7)	0.001***
No	21 (15.6)	114 (84.4)	
Member of IPM club			
Yes	47 (28.3)	119 (71.7)	0.372
No	120 (26.6)	331 (73.4)	
Member of FFS			
Yes	36 (29.5)	86 (70.5)	0.284
No	131 (26.5)	364 (73.5)	
***P value <0.01, **P value <0.05, *P value <0.10			

Table 1: Assessing association between IPM adoption status and selected factors with P values obtained from chi-square test.

significant manner (P<0.01). To be specific, the odds of adopting IPM is more than double for the farmers belong to the age group (35-39) [OR=2.338] and (40-44) [OR=2.415], compared to the farmers having age less than 30. This odds is much higher [OR=3.267] for the farmers with age 45 or more. It is interesting to see that, the higher educated farmers [OR=3.407] are more likely to adopt IPM than no or primary educated farmers. Moreover, the farmers with more than 10 years of farming experience [OR=2.425] and receive training on IPM [OR=3.836] have more than double chance for IPM adoption. Farmers belonging to IPM club are 43% [OR=1.426] more likely to adopt IPM. The farm ownership status and Barisal division have significant positive effect at 10% level of significant (P<0.10). However, farm size and membership status of FFS have no significant effect on IPM adoption (P>0.10).

Discussion and Conclusion

The main target of this study is to identify the influencing factors for the adoption of IPM in Bangladesh. To fulfill our target we first performed a bivariate analysis in our collected dataset. From the bivariate analysis we observed that farmer's age, education level, division, farming experience, farm size and training status on IPM are the highly significant factors for IPM adoption in Bangladesh. On the other hand, we estimate the adjusted effect of the suspected factors by fitting a binary logistic regression model.

Based on the fitted logistic model, we get strong evidence that the older farmers are more likely to adopt IPM compared to their counterpart. Higher educated farmers have higher odds of adopting IPM. This may due to the awareness of educated farmers about the bad effects of chemical pesticides on human health and environment. The farmers having more farming experience and receive training on IPM have more chance to practice IPM farming to control pests. Farmers from Barisal division and having a cultivated land of their own, also have higher odds of IPM adoption. One of the reasons for getting significant of the ownership status of a farm is that the farmers having own land are mentally ready to practice new techniques on their cultivated land. We also get evidence that the farmers of an IPM club have more chance to receive IPM farming. However, membership status of FFS has found no significant effect on IPM adoption in Bangladesh.

Any effort taken by government to unfurl IPM farming will be useless if the majority of the farmers reject it. We observed that less than one-third (27.1%) of our sampled farmer adopt this farming. It is very much practical that before unfurling any new techniques at farm level, one have to realize the background characteristics of the farmers. In literature, several studies were conducted to identify farmers' background characteristics that can influence the adoption status of IPM [21]. However, different scholars used different techniques to fulfill their objectives and get different results. Considering this fact, our research attempt to explore the factors for IPM adoption by fitting a binary logistic regression model.

Our findings coincide with several previous studies. For example, farmers' age, education level, farming experience, training on IPM is found to be highly significant factors for increasing the adoption rate of IPM farming. These factors also found significant in previous studies [13-19]. According to our findings, we strongly suggest the policy maker to take initiatives for increasing farmers' education level and facilitate more training programme on IPM. Several motivational seminars on IPM may be organized by establishing IPM club in village level.

Factors	Estimate	Odds ratio (95% CI)	P value
Age group			
<30 (ref)	-	-	-
30-34	-0.186	0.830 (0.461, 1.494)	0.534
35-39	0.849	2.338 (1.288, 4.244)	0.005***
40-44	0.882	2.415 (1.299, 4.488)	0.005***
45 or more	1.184	3.267 (1.705, 6.261)	0.000***
Education level			
No/primary (ref)	-	-	-
Secondary	0.034	1.034 (0.545, 1.962)	0.917
Higher	1.226	3.407 (1.085, 5.698)	0.036**
Division			
Dhaka (ref)	-	-	--
Chittagong	-0.302	0.739 (0.381, 1.433)	0.371
Rangpur	-0.452	0.636 (0.245, 1.653)	0.353
Khulna	-0.764	0.466 (0.693, 2.513)	0.399
Barisal	0.277	1.139 (0.210, 1.033)	0.060*
Farming experience			
At least 10 years (ref)	-	-	-
More than 10 years	0.886	2.425 (1.467, 4.008)	0.001***
Farm size			
<1 acre (ref)	-	-	-
1-1.5 acre	-0.04	0.961 (0.392, 4.990)	0.91
>1.5 acre	-0.336	0.715 (0.464, 3.893)	0.605
Farm ownership status			
No (ref)	-	-	-
Yes	0.721	2.057 (0.988, 4.284)	0.054*
Training on IPM			
No (ref)	-	-	-
Yes	1.345	3.836 (1.985, 5.413)	0.000***
Member of IPM club			
No (ref)	-	-	-
Yes	0.355	1.426 (1.389, 3.326)	0.000***
Member of FFS			
No (ref)	-	-	-
Yes	-0.441	0.643 (0.350, 1.181)	0.154

***P value <0.01, **P value <0.05, *P value <0.10

Table 2: Logistic regression model based adjusted effects of selected factor for IPM adoption.

This research has several limitations. There is lot of socio-economic and demographic factors for farmers' adoption decision of IPM. Because of time and money we consider only nine factors for our analysis. Moreover, for policy development, it is important to consider all significant factors that can influence IPM adoption. A recent study conducted by Kabir et al. [22], report that FFS has significant effect on IPM farming. However, we do not get any evidence of significance of FFS. This may due to the data pattern that was used for analysis. So, more research should conduct by considering more factors that can influence this environmentally friendly farming technique.

Appendix

We used following formula for estimating the sample size.

$$n = \frac{Z_{\alpha/2}^2 pq}{d^2}$$

Where,

$Z_{\alpha/2}$: Standard normal deviate usually set at 1.96, which corresponds to the 95% confidence level.

p: Assumed proportion in the target population estimated to have a particular characteristic.

q=1-p

d: Allowable maximum error in estimating population proportion.

Here we consider: p=0.5; q=0.5; d=0.0394.

Therefore,

$$n = \frac{(1.96)^2 \times 0.5 \times 0.5}{(0.0394)^2} = 617.617 \cong 617$$

Here we have taken 617 samples for data analysis.

Acknowledgement

We would like to thank Dr. Md. Sawar Jahan, Professor, Khulna University for his kind help to understand IPM practice in Bangladesh. The author would also like to thank Mr. Rassel Kabir, Assistant professor, Khulna University for his advice at different stages of this study.

References

1. BBS (2011) Preliminary report on population and housing census. Bangladesh Bureau of Statistics, Ministry of Agriculture, Dhaka, Bangladesh.

2. MOA (2002) National integrated pest management policy. Ministry of Agriculture, Government of the People's Republic of Bangladesh.

3. Kabir MH, Alam MM, Roy R (2010) Farmers' perception on the harmful effects of Agro-chemicals on sustainable environment. J Sustain Agr Tech 6: 1-5.

4. Henning J, Baker L, Thomassin P (1991) Economics issues in organic agriculture. Can J Agr Econ 39: 877-889.

5. Kilcher L (2006) How can organic agriculture contribute to sustainable development? Conference of prosperity and poverty in a globalized world: challenges for agriculture research.

6. Lampkin NH, Padel S (1994) The economics of organic farming: An international perspective. Biddles Ltd., UK.

7. Bonabana-Wabbi J, Taylor DB, Kasenge V (2006) A limited dependent variable Analysis of integrated pest management adoption in Uganda. American Agricultural Economics Association Annual Meeting, Long Beach, California.

8. Prokopy R, Kogan M (2003) Integrated pest management, in Encyclopedia of insects. In: Resh VH and Carde RT (eds.) Academic Press, San Diego, pp: 589-595.

9. World Bank (2003) Agriculture and rural development working paper 5. Integrated pest management in development: review of trends and implementation strategies.

10. Dasgupta S, Meisner C, Wheeler D (2007) Is environmentally friendly agriculture Less profitable for farmers? Evidence on integrated pest management in Bangladesh. Appl Econ Perspect Policy 29: 103-118.

11. De Souza Filho HM, Young T, Burton MP (1999) Factors influencing the adoption of sustainable agricultural technologies: evidence from the State of Espírito Santo, Brazil. Technol Forecast Soc Change 60: 97-112.

12. Veisi H (2012) Exploring the determinants of adoption behaviour of clean technologies in agriculture: a case of integrated pest management, Asian J Technol Innov 20: 67-82.

13. Amir HM, Shamsudin MN, Mohamed ZA, Hussein MA, Radam A (2012) Economic evaluation of Rice IPM practices in MADA, Malaysia. J Econ Sustain Dev 3: 47-55.

14. Ofuoku AU, Egho EO, Enujeke EC (2008) Integrated pest management (IPM) Adoption among farmers in central agro-ecological zone of Delta state, Nigeria. Afr J Agr Res 3: 852-856.

15. Chaves B, Riley J (2001) Determination of factors influencing integrated pest management adoption in coffee berry borer in Colombian farms. Agr Environ 87: 159-177.

16. Fernandez-Cornejo J, Jans S (1996) The economic impact of IPM adoption for orange producers in California and Florida. In: Proceedings of the XIII International Symposium on Hort Econ Acta Hort 429: 325-334.

17. Ricker-Gilbert J (2005) Cost-effectiveness evaluation of integrated pest management (IPM) extension methods and programs: The case of Bangladesh. MS thesis, Virginia Polytechnic Institute and State University.

18. Hristovska T (2009) economic impacts of integrated pest management in developing countries: evidence from the IPM CRSP. MS thesis, Virginia Polytechnic Institute and State University.

19. Bonabana-Wabbi J (2002) Assessing factors affecting adoption of agricultural technologies: The case of integrated pest management (IPM) in Kumi district, Eastern Uganda. Virginia Polytechnic Institute and State University.

20. Hosmer DW, Lemeshow S (2000) Applied logistic regression. John Wiley & Sons.

21. Kabir MH, Rainis R (2013) Determinants and methods of integrated pest managementadoption in bangladesh: an environment friendly approach. J Am Eur J Sustain Agr 7: 99-107.

22. Kabir MH, Rainis R, Azad MJ (2017) Are spatial factors important in the adoption of eco-friendly agricultural technologies? Evidaence on Integrated Pest Management (IPM). J Geograph Info Sys 9: 98-113.

11

Analysis of Morphological and Physiological Responses to Drought and Salinity in Four Rice (*Oryza sativa* L.) Varieties

Chowdery RA[1,3], Shashidhar HE[2] and Mathew MK[1]*

[1]National Centre for Biological Sciences (TIFR), Bellary Road, GKVK Campus, Bangalore, Karnataka 560065, India
[2]Department of Plant Biotechnology, University of Agricultural Sciences, Bellary Road, GKVK Campus, Bangalore, Karnataka 560065, India
[3]Manipal University, Madhav Nagar, Manipal, Karnataka 576104, India

Abstract

Salinity and drought adversely affect rice production globally. Here we have examined physiological responses to drought and salinity across four rice cultivars with varying sensitivity to these stresses. The salt tolerant Pokkali restricts fluid entry to limit Na^+ uptake under saline stress, while the drought-tolerant ARB6 needs to enhance fluid uptake under drought. Surprisingly, Pokkali does reasonably well when subjected to drought as does ARB6 under saline stress-in contrast to the stress-sensitive but high yielding varieties IR-20 and Jaya. Both tolerant varieties use long roots to mine water under deficit conditions, increasing aerenchyma and suberization of the exodermis to provide oxygen to deep-reaching roots. Major alterations in patterns of suberization in both exodermis and endodermis are undertaken, the patterns being dramatically different under the two stresses. Genes implicated in suberin biosynthesis also showed variation in transcript levels under stress, corresponding with the observed suberization patterns. Osmolyte accumulation drives uptake of water under deficit conditions, while restricting fluid flow to symplastic routes minimizes Na^+ entry. Overall, the morphological and physiological responses of the tolerant varieties ensure adequate fluid flow through the transpiration stream without excessive salt uptake, thereby promoting growth under both drought and salinity.

Keywords: Drought; Endodermis; Passage Cells (PCs); Salinity; Suberin; Xylem sap

Introduction

Rice (*Oryza sativa* L.) a dietary staple of more than half of the world, is generally cultivated in semi-aquatic conditions and is very salt sensitive [1-3]. Short term salinity leads to physiological drought and its persistence causes ionic stress [4,5] leading to altered K^+/Na^+ ratios, which affects all aspects of plant growth. Salinization of irrigated land leads to over a million hectares of land being lost to production annually. Salinization of arable land may have contributed to the demise of the early Mesopotamian civilization, making this a 6000 year old problem [6].

When grown in paddies with submersion, rice consumes an estimated 3000 to 5000 liters of water to produce a kilogram of grain [7]. Consequently, scientists have bred "aerobic rice" [8-10], which does not require water logging [11]. Coarse, long roots, capable of branching and penetrating into the deep layers of soil have been selected for in breeding for drought tolerance [12-17]. Recent studies have shown aerobic rice to perform excellently in terms of high grain yield under severe drought stress [18,19]. However, the mechanism(s) employed to withstand drought have not been delineated. Nor is it clear as to whether traits that promote drought tolerance would adversely affect the ability of the plant to combat salt stress.

Some traditional varieties like Pokkali and Nona-Bokra are highly salt tolerant [20-22] whereas IR-20 is highly sensitive to salt [23]. Na^+ compartmentalization in the shoot influences survival with high Na^+ in the apoplast being correlated with poor survival [24-27]. Pokkali restricts Na^+ loading into the xylem stream by building substantial apoplastic barriers in roots [28] and maintains low cytosolic Na^+ [27,29] by reducing plasma membrane permeability to Na^+ as well as sequestering Na^+ in the vacuole [30]. It is not clear as to whether these mechanisms would be of use in combating drought stress.

The common apoplastic barriers in roots are casparian bands and suberin lamellae [31,32]. While Casparian Bands are deposited on radial and anticlinal walls of cells, suberin lamellae are deposited as secondary wall thickenings of the primary cell wall [32,33]. Elongases, hydroxylases and peroxidases are the important enzymes involved in suberin biosynthesis [34-36]. The role of suberin barriers in the response of higher plants to salinity, flooding, mechanical impedance, iron-deficiency and biotic stresses has been well studied [37-39]. Changes in suberin deposition following drought stress have not been investigated nor have the underlying molecular mechanisms been elucidated.

In the light of changing environmental conditions, salinity and drought are expected to create increasingly more severe challenges to rice production in the future [40]. Roots are directly and constantly in contact with soil and are responsible for water and nutrient uptake. Both anatomical and physiological strategies could foster growth and survival of the plant under stress. It would be of interest to investigate whether the mechanisms conferring tolerance to one abiotic stress (for example salinity) are beneficial or antagonistic to countering another abiotic stress such as drought. In the present study, we investigate the responses of four rice varieties to drought and salinity. We have chosen one specialist cultivar that has been shown to do well under drought-ARB6; another under salinity-Pokkali; one cultivar known to be very sensitive to both stresses-IR-20; and finally one that is moderately

*Corresponding author: Mathew MK, National Centre for Biological Sciences (TIFR), Bellary Road, GKVK Campus, Bangalore, Karnataka, India
E-mail: mathew@ncbs.res.in

sensitive to both stresses-Jaya. All four were subjected to drought and salt in separate experiments. The expectation was that the anatomical and physiological responses would be distinct for each stress; that specialists would, in consequence, do well under "their" respective stress, but would be sensitive to the other stress; and that the responses would be graded according to the degree of tolerance/sensitivity among the four cultivars studied. Given the range of sensitivity among the cultivars tested, such findings would set up hypotheses that could be subsequently tested.

Materials and Methods

Experimental setup and growth condition

Four cultivars of rice (Pokkali, ARB6, IR-20 and Jaya) were used based on their differences in tolerance and sensitivity to drought and salinity. Plants were grown and subjected to drought or salinity in PVC pipes of diameter 8 cm and length 60 cm. Field soil with pH of 5.7 was mixed with organic vermicompost (Organic carbon 9-17%, Nitrogen 0.5-1.5%, Phosphorous 0.1-0.3%, Potassium 0.15-0.56%, Sodium 0.06-0.3% and micronutrients) in the ratio 2:1 (soil: manure), and used to fill the pipes. Further compaction of the soil was achieved by watering and pressing the soil at intervals in order to mimic the compaction of soil in field conditions as described in Shashidhar et al. [41]. Seeds were sown in three batches with 72 pipes in each batch. Plants were thinned to one per pipe and grown for 45 days with average temperature of 31.2°C, relative humidity of 67.3%. In total, 216 pipes were used with 18 replications per variety and using 6 replicates for each analysis. Pipes arranged in a randomized complete block design were regularly watered till the 38th day. From the 39th day onwards, one of the batches was exposed to well-watered condition referred to as "Control" and other two batches were exposed to either drought (no irrigation till the soil reached about 30-40% of field capacity) or salinity (150 mM NaCl) stress for one week. Soil moisture content was monitored using moisture meter (Procheck, Decagon Devices Inc., USA) (Supplementary Information 1A-1D). Electrical conductivity of soil under salinity condition was also measured (Supplementary Information 1E).

Analysis of plant morphology and biomass

After one week of stress, the soil column along with the whole plant was taken out of the pipe. Roots were carefully washed free of soil. The number of lateral roots, tillers and leaves was counted manually and total shoot and root lengths measured. Fresh weight of the plants was recorded at this stage and dry weight after oven-drying at 70°C for 48 h.

Measurement of xylem sap exudation

Sap measurements were recorded using the method described [42,43] with some modifications. Rice varieties were grown in six replicates in PVC pipes and exposed to drought and salinity stress as described above. At 06:00 pm on the 45th day, shoots were cut ~5 cm above the soil surface. Cut stems attached to the root system were covered with pre-weighed blotting paper, then covered with polyethylene wrapper, which was in turn tightly sealed at the base with a rubber band (to avoid evaporation of the sap). The setup was left for 12 h, followed by removing the blotting papers which were immediately weighed to quantify the amount of xylem sap sent up by the root system. The values were normalized to the root mass of the plant from which sap was collected. The blotting papers were then used to estimate the Na+ content of the sap.

Measurement of root exudation in the absence (osmotic exudation) and in the presence of hydrostatic pressure gradients

Root L_{pr} measurements were carried out as described by Miyamoto et al. [44], with some modifications. After excising the shoot at 5 cm above the soil surface, the soil was carefully washed away from the roots and the root system submerged in a container of nutrient solution was placed in a pressure chamber. All tillers were closed using clamps except the main tiller which was threaded through a rubber stopper sealed with silicone sealant. Pre-weighed blotting paper was used to cover the cut surface of the main tiller to absorb the exuded xylem sap and was covered with a polyethylene wrapper to avoid evaporation. The blotting paper was weighed to determine the amount of xylem sap exuded by osmotic pressure. The osmotic pressure of the nutrient solution measured using a freezing point Osmometer (Osmomat 030, GONOTEC GmbH, Germany) was 0.0075 MPa. The measured reflection coefficient, σ_{sr} of the nutrient salts was 0.4. Plants used for measuring osmotic water flows through the root were also used to measure the hydrostatic water flow of the roots. The air in the chamber was pressurized and monitored with a pressure transducer. Pressure was slowly raised in steps of 0.05 MPa from 0 MPa to 0.3 MPa (±0.001 MPa). Sap flow was stable at pressures above 0.05 MPa. At a given applied gas pressure (P in MPa), the volume exuded from the root system (V in m³) was plotted against time (Supplementary Information 2A). The slope of the V vs. time curve, normalized to the surface area of the root system, yields the volume flow, J_{vr} in m s⁻¹. Plots of J_{vr} against applied pressure were linear above 0.15 MPa. Root hydraulic conductivity (L_{pr}) was estimated as the slope of the curve in this linear regime, illustrated with red line (Supplementary Information 2B).

Estimation of Crop Canopy Air Temperature Difference (CCATD)

Small plots were prepared in the field and seeds sown in five lines 15cm apart with ten replicates in each line. Control plots were watered regularly till 45th day. Plants in the other set were well-watered for 38 days, following which the drought protocol was followed. At mid-day on the 45th day, ambient and canopy temperature of both control and stressed plants was recorded by using AGRI-THERM III™ infra-red sensing instrument (Everest Interscience Inc., USA). The difference in the two readings gives the Crop Canopy Air Temperature Difference (CCATD).

Measurement of photosynthetic rate

Photosynthetic rate was monitored using an infrared gas analyzer (IRGA) based LI-6400 photosynthesis system (LI-CoR Biosciences) on Day 0 (one day before subjecting the plants to drought and salinity stress) and during 7 days of stress period between 9.00 to 11.00 am, measurements were recorded on intact leaves 6 per plant using 6 plants per cultivar.

Estimation of xylem sap Na+ content

Blotting papers with and without xylem sap was kept in deionized distilled water for 1 h. The blotting paper was then discarded and the supernatant analysed for Na+ ions using a flame photometer (Systronics Flame Photometer 128, Ahmedabad, India). The difference in the two values gives an estimate of Na+ in the xylem stream.

Estimation of apoplastic Na+ and intracellular Na+ and K+ in shoot and root

The method described by Anil et al. [27] was used to release the total shoot Na⁺, apoplastic and intracellular fluid from shoot and root. The Na⁺ and K⁺ in the solution were estimated by flame photometry.

Estimation of shoot and root osmotic adjustment

For the estimation of osmotic adjustment, shoot and root tissue was collected on the 45th day. The tissue was immediately frozen in liquid nitrogen and brought to lab, where it was thawed and chopped into fine pieces with a razor blade. 500 µL of deionized double distilled water was added to chopped tissue and centrifuged at 12,000 rpm to collect the sap. The osmotic strength of the sap was estimated by using Osmometer (Osmomat 030, GONOTEC GmbH, and Germany).

Root microscopy

Thin sections (200 µm) of Base (25 cm), Middle (15 cm) and Tip (5 cm) zones of roots were cut using a McIlwain tissue chopper (The Mickle Laboratory Engineering Co. Ltd, UK). Sections were stained with 0.1% (w/v) berberin hemisulphate for 1 h followed by 0.5% (w/v) aniline blue for 30 min as described by Brundrett et al. [45]. Stained sections were observed under Olympus FV1000 confocal microscope and imaged using 488 nm excitation and 510-540 nm emission. Suberin was seen as a fluorescent light yellow band shaped structure deposited around the endodermal and exodermal cells of the root, whereas Passage Cells (PCs) showed no tangential suberin deposition. For counting and estimation of number of cells with or without suberin deposits, stained sections were observed using Nikon fluorescent microscope under UV light.

Semi-quantitative RT-PCR analysis

Total RNA was isolated from roots of 45 day old plants using TRI-reagent (Sigma–Aldrich, St Louis, MO, USA) following the manufacturer's instructions. RT-PCR was performed with 1 µg of RNA using 100 units of Moloney murine leukemia virus reverse transcriptase (Invitrogen) according to instruction of the manufacturer. The cDNA thus obtained was subjected to a 25-cycle PCR reaction with the following conditions: 3 min at 94°C followed by 25 cycles of 60 s at 94°C, 60 s at 58°C and 80 s at 70°C, and finally 5 min at 70°C. The constitutively expressed *Actin* was used as a control. The primer sequences and predicted amplicon sizes were 5'-CCTCTTCCAGCCTTCCTTCAT-3'(forward) and 5'-ACGGCGATAACAGCTCC TCT T-3'(reverse) for *Actin-1* (Os03g50890) (400 bp), 5'-CGCCTCACCTTCGATAACAT-3' (forward) and 5'-CACTCGCAGTCCATTCTTCA-3' (reverse) for *P450* (Os01g63540) (960 bp), 5'-TCGTAATCTTCTCCGCCATC-3' (forward) and 5'-GATGTAGGCGAGCTCGTACC-3' (reverse) for *Elongase* (Os03g12030) (821 bp), 5'-CCGTCTACCTCGTCGACTTC-3' (forward) and 5'-ATCCATCCACGGGTTCTTCT-3'(reverse) for *Elongase*(Os02g11070)(1190bp),5'-TCGTCAATTGCAGCTTGTTC-3' (forward) and 5'-TCTCCTTCGCTGGATTCACT-3' (reverse) for *Elongase* (Os06g39750) (867 bp).

Statistical analysis

The data in the figures have been presented as the mean values ± SE; n=6. Student's t-test was used to estimate the significant differences at P<0.01 and P<0.05, between control and stress treatments. For root hydraulic conductance experiment, analysis of variance (ANOVA) was used to detect the significant differences and Least Significant Difference (LSD) test was used to determine genotypic differences in each treatment and group them into letter classes.

Results

We selected four cultivars of rice with differing sensitivities to drought and salinity for this study. All four varieties were grown in soil under well-watered conditions for 38 days, following which they were subjected to either drought or saline stress for a week. Plants were harvested on the 45th day for analysis.

Responses to drought stress

Morphophysiological changes: At the end of the one week drought regime, all of the experimental plants were alive. Soil moisture content, measured at a point 20 cm below the surface, declined for all varieties in a similar manner (Supplementary Information 1A-1D). IR-20 and Jaya plants showed clear signs of distress: leaf wilting and chlorosis; whereas ARB6 plants appeared quite healthy and Pokkali plants were not severely affected (Figures 1A and 1B). Photosynthetic rates for all 4 varieties were the same on Day 38-i.e., the day before stress was imposed. The rates did not change substantially over the next 7 days for plants that were well watered (Figure 1C). The plants subjected to drought, however, showed a decline in photosynthetic rate that was extreme for IR-20, large for Jaya and moderate for ARB6 and Pokkali. ARB6 plants were stimulated to grow under drought and exhibited a substantial increase in biomass compared to control plants (Supplementary Information 3A). Pokkali plants showed no significant growth arrest, while IR-20 and Jaya displayed significant decline in growth (Supplementary Information 3A) with final biomass about 30% lower than in control plants.

In the shoot, leaf number increased dramatically in ARB6 and marginally in Pokkali with declines observed in IR-20 and Jaya (Supplementary Information 3B). No significant change in plant height was displayed by any genotype (Supplementary Information 3C). Pokkali exhibited no change in tiller number, whereas the other three varieties exhibited increases under drought (Supplementary Information 3D). Unlike the shoots, roots of all four varieties appeared healthy with enhancement of root number (Figures 1D-1F) the most prominent being Pokkali where the root number almost doubled. Root lengths showed modest increases (Figure 1G) of 20% in ARB6, 10% in Pokkali and none in IR-20 and Jaya. Aerenchymas were extensive in the basal region of control roots of all four varieties. However, under drought stress, these took over a large fraction of root area and extended to the middle and tip regions of ARB6 and Pokkali

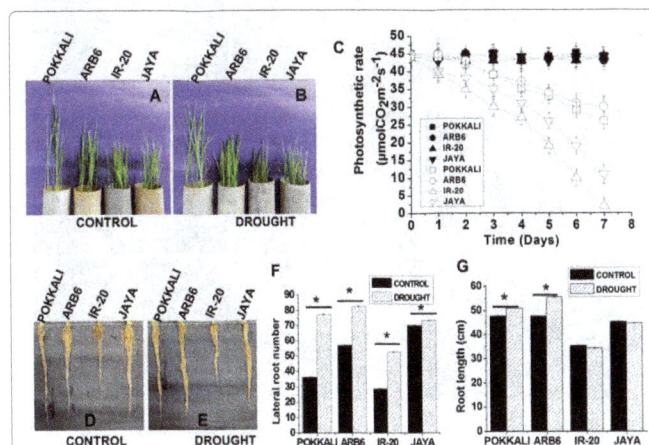

Figure 1: Rice varieties and their roots after being subjected to drought: Plants were grown in PVC pipes for 38days and then stressed with drought (no irrigation) for a week. (A, D) well-watered control; (B, E) drought stressed plants; (c) Photosynthetic rates were monitored throughout the stress period (closed symbols-control and open symbols-drought); (F) Root number; and (G) Root length. Data represents mean (SE; n=6), Asterisk indicates differences between control and drought with P<0.05.

(Supplementary Information 4A). The data for the middle region of the root is presented in Figure 2A. Data for the base and tip regions are presented in Supplementary Information 4B-4D.

Overall, the analysis of gross morphological features demonstrates that ARB6 plants do significantly better over a one week of drought

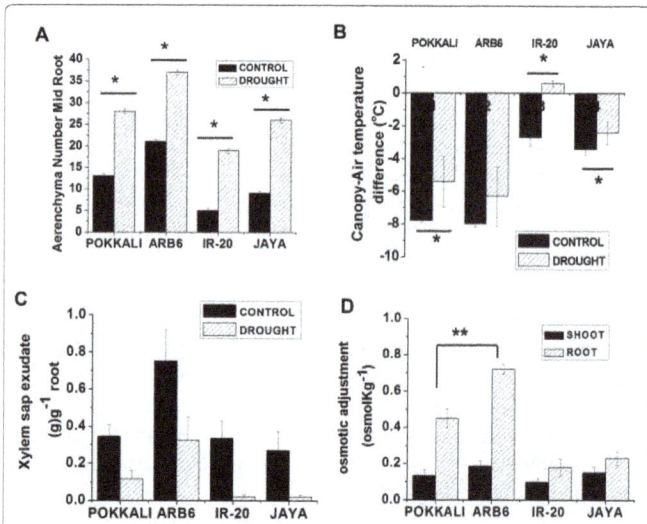

Figure 2: Plants were grown in PVC pipes for 38 days and subjected to control (well watered), or drought (no irrigation) for a week. A) Aerenchyma number of plants at the end of treatment period; B) Crop canopy air temperature difference (CCATD). Ambient and canopy temperature was measured by using infra-red sensing instrument. C) Xylem sap exudation. D) Osmotic adjustment in shoot and root under drought condition. Data represents mean (±SE; n=6), Asterisks indicates differences between control and stressed at *P<0.05 and **P<0.01.

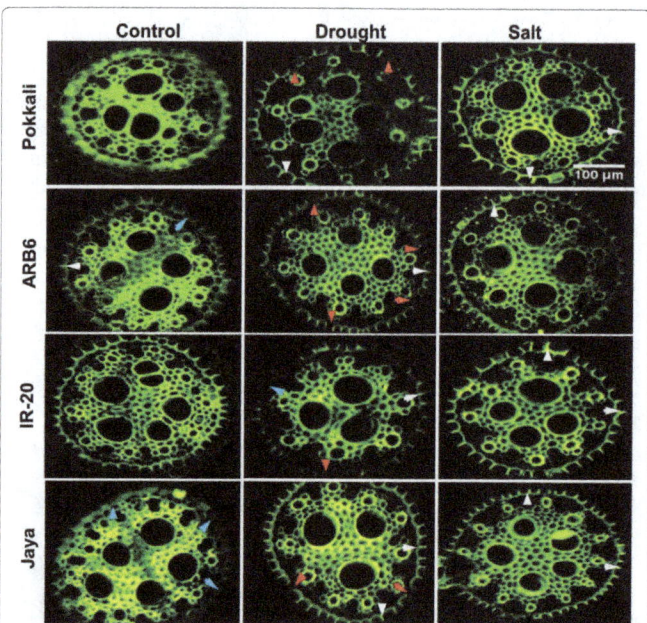

Figure 3: Images of root sections showing unsuberized cells, passage cells and suberin deposits in root endodermis of plants grown in PVC pipes for 38 days and subjected to control (well watered), drought (no irrigation) or salt (150 mM NaCl) for A week. Roots were washed, divided into three zones (Tip–5 cm, Mid–15 cm and Base–25 cm) from the root tip and cut into thin sections (200 µm), then stained with berberine-anilline blue and imaged using Olympus FV1000 confocal microscope, 488 nm laser was used for excitation. Red, white and cyan arrows show passage cells (cells with no tangential suberin deposition), suberin deposits and in suberized cells respectively, scale bar 100 µm.

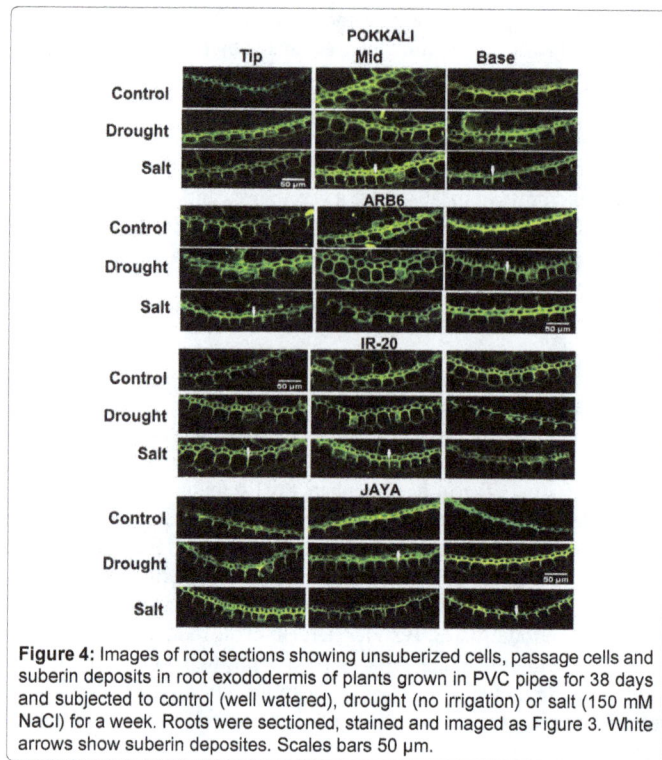

Figure 4: Images of root sections showing unsuberized cells, passage cells and suberin deposits in root exododermis of plants grown in PVC pipes for 38 days and subjected to control (well watered), drought (no irrigation) or salt (150 mM NaCl) for a week. Roots were sectioned, stained and imaged as Figure 3. White arrows show suberin deposites. Scales bars 50 µm.

stress protocol than control plants, whereas IR-20 coped poorly. Jaya exhibited moderate sensitivity to drought. Interestingly, Pokkali which was selected as a salt tolerant variety and reported to display characters that could sensitize it to drought, actually does well under this stress.

Canopy temperature: Increase in biomass can be achieved either by increasing water use efficiency or by accessing additional water sources under drought. Evaporative cooling of leaf tissue utilizes transpired water and reduces canopy temperature below that of the ambient air. The magnitude of this reduction in temperature is thus a good indicator of the extent of transpiration. Canopy temperatures were around 8°C below ambient in ARB6 and Pokkali under control conditions and about 4°C in Jaya and IR-20 (Figure 2B). Canopy temperature differences declined under drought by 1-3°C in ARB6, Pokkali and Jaya, though the decline in ARB6 was not significant. The canopy of IR-20 was at ambient temperature under drought indicating that transpirational cooling was no longer active. Maintenance of large canopy temperature differences in ARB6 and Pokkali is indicative of high transpiration rates, suggesting that the plants were able to access adequate water sources even under drought conditions. Jaya maintained a small but significant canopy temperature difference under drought.

Xylem sap exudation: Measurement of the rate of xylem sap exudation confirms the hypothesis that ARB6 and Pokkali were able to access water under the experimental drought conditions. Exudation rates for Pokkali, Jaya and IR-20 were comparable under control conditions (Figure 2C). ARB6 had almost double this rate of sap exudation under control conditions. Drought dramatically reduced exudation rates in all four varieties, but the final exudation rate in ARB6 was comparable to that exhibited by the other three varieties under well-watered conditions (Figure 2C). Pokkali sap exudation was reduced considerably from well-watered conditions but is still significant. Very little sap was exuded by IR-20 and Jaya using root pressure alone.

Osmotic adjustment: Osmolyte accumulation under stress was estimated in terms of osmolarity differences with control sap. Very low osmotic adjustment was seen in the shoots of all varieties (Figure 2D). Both Pokkali and ARB6 exhibited large osmotic adjustment in the roots with the latter being significantly larger. Jaya exhibited moderate osmotic adjustment, while that of IR-20 was least.

Suberin deposition: Root hydraulics would be critical for delivering fluid to the growing shoot. We examined hydrophobic barriers on the endodermis and exodermis in all four varieties (Figure 3 and 4 and Supplementary Information 5). The percentage of fully suberized cells in the exodermis appeared comparable across all four varieties in control conditions (Supplementary Information 6A). IR-20 had a larger proportion of unsuberized cells than the other three varieties in all three regions analyzed-base, middle (mid) and tip of the root. After a week of drought, there was a significant decrease in the percentage of fully suberized cells in ARB6 and Pokkali with a concomitant increase in PCs (PCs) (Supplementary Information 6B). The shift was most dramatic in ARB6. Taking the middle region of exodermis for example, completely suberized cells dropped from ~30% to 20% while PCs increase from ~3% to 15% with little change in unsuberized cells (Supplementary Information 6B). A similar trend of decrease in suberized cells was observed in the base and tip regions as well, with

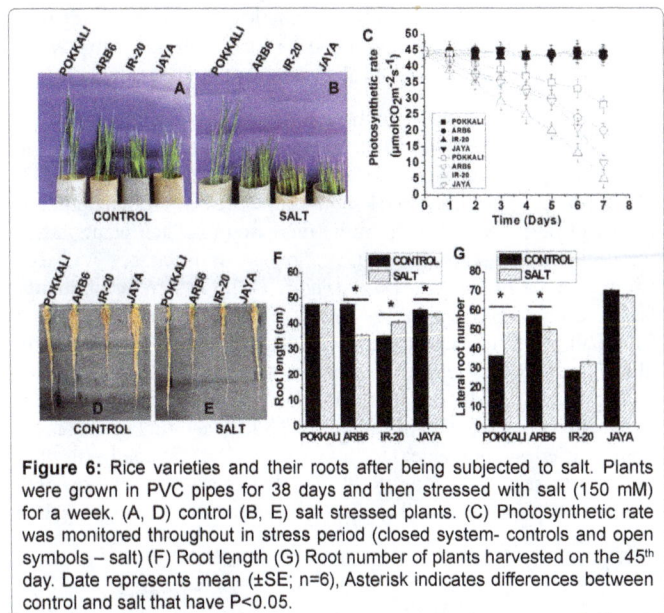

Figure 6: Rice varieties and their roots after being subjected to salt. Plants were grown in PVC pipes for 38 days and then stressed with salt (150 mM) for a week. (A, D) control (B, E) salt stressed plants. (C) Photosynthetic rate was monitored throughout in stress period (closed system- controls and open symbols – salt) (F) Root length (G) Root number of plants harvested on the 45th day. Date represents mean (±SE; n=6), Asterisk indicates differences between control and salt that have P<0.05.

a corresponding increase in PCs (Supplementary Information 6B). No such decrease in completely suberized cells was seen in Jaya or IR-20 (Supplementary Information 6B and 6C).

The change in the pattern of suberization follows the same trend in the endodermis, only more so. Following drought, there was a significant decline in the percentage of completely suberized cells in both Pokkali and ARB6 but not in IR-20 and Jaya (Figures 5A, 5D, and 5G). On the other hand, the percentage of PCs in the Pokkali and ARB6 endodermis was comparable but greater than IR-20 and Jaya (Figures 5B, 5E, and 5H). The changes in IR-20 and Jaya were relatively small and statistically insignificant in most cases (Figures 5B, 5E, and 5H). No change was seen in percentage of unsuberized cells in Pokkali, ARB6 and Jaya. However, a decreasing trend was seen in IR-20 (Figures 5C, 5F, and 5I). The tolerant plants thus increase the number of cells potentially available for uptake of fluid into the symplastic stream under drought stress.

Root hydraulic conductivity (L_{pr}): Osmotic conductivity (L_0) of Pokkali and ARB6 exceeded that of Jaya and IR-20 under control conditions, with little difference between the latter two (Table 1). While there was a trend towards increasing conductivity in all four varieties under drought, the changes were not significant in any case.

Under control conditions, hydrostatic conductivity or L_{pr} was significantly higher in ARB6 than in the others, with Pokkali L_{pr} being greater than Jaya, which exceeded IR-20 (Table 1). Drought stress increased L_{pr} by 10–15% in the first three varieties, while the change in IR-20 was insignificant.

Responses to salt stress

Morphophysiological analysis: At the end of a week of saline stress, all the experimental plants were alive. Pokkali plants looked healthy whereas all the other varieties appeared highly chlorotic (Figures 6A, and 6B). Electrical conductivity of treated samples, measured at various depths, was significantly higher than in control pipes but similar across varieties (Supplementary Information 1E). Photosynthetic rates were the same for all four varieties on Day 38, and did not change over the next 7 days for plants that were well watered (Figure 6C). Rates fell sharply in all varieties when subjected to saline stress (Figure 6C). Both

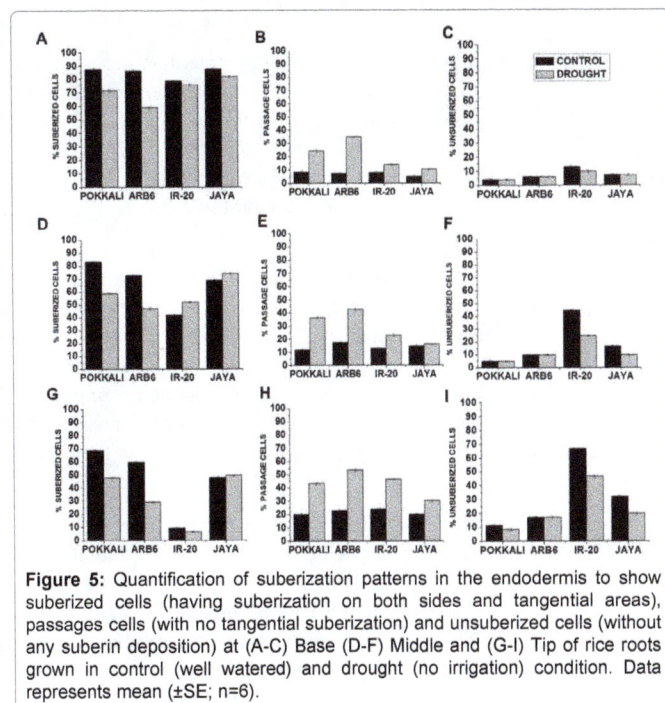

Figure 5: Quantification of suberization patterns in the endodermis to show suberized cells (having suberization on both sides and tangential areas), passages cells (with no tangential suberization) and unsuberized cells (without any suberin deposition) at (A-C) Base (D-F) Middle and (G-I) Tip of rice roots grown in control (well watered) and drought (no irrigation) condition. Data represents mean (±SE; n=6).

	Variety	Control	Drought	Salinity
Hydrostatic L_{pr} (10-8 m MPa^{-1} s^{-1})	Pokkali	41 ± 1.11[a]	47 ± 1.03[ab]	18 ± 1.06[a-c]
	ARB6	49 ± 1.16[b]	56 ± 1.05[bc]	23 ± 1.08[a-d]
	IR-20	31 ± 1.05[c]	34 ± 1.14[cd]	27 ± 1.11[c-e]
	Jaya	37 ± 1.08[d]	42 ± 1.12[de]	22 ± 1.1[a,d-f]
Osmotic L_0 (ms^{-1})	Pokkali	5.2 ± 0.1[ab]	5.6 ± 0.3[c-e]	1.7 ± 0.4[f]
	ARB6	5.4 ± 0.6[cd]	6.0 ± 0.2[d]	2.2 ± 0.3[a-c]
	IR-20	4.1 ± 0.1[d-f]	4.3 ± 0.3[ef]	2.8 ± 0.5[e]
	Jaya	4.4 ± 0.3[b-d]	5.1 ± 0.1[c]	2.2 ± 0.4[a-c]

Table 1: Root hydraulic conductivity driven by external hydrostatic pressure (L_{pr}) or by osmotic pressure alone (L_0). For L_{pr}, Pneumatic pressures were applied to the root medium to vary the driving force, while, no pressure was applied during L_0 measurements (mean ± SE, n=6). Values within a column and rows followed by the different letters are significantly different; and a: indicates no significant difference within a column at P<0.05 by LSD test.

IR-20 and Jaya had very low photosynthetic rates on the 7th days of stress, when the plants were harvested for further experiments. ARB6 decline was moderate while Pokkali decline was modest (Figure 6C).

The biomass of all varieties showed a significant decline of 30-45% (Supplementary Information 7A). No change in root length was seen in Pokkali as against significant declines in ARB6 and Jaya (Figures 6D-6F). Surprisingly, IR-20 showed a significant increase in root length. The number of roots showed a significant increase in Pokkali plants, while they declined in ARB6 (Figure 6G). No significant change was seen in root number of Jaya and IR-20 (Figure 6G). No significant change in leaf number was seen in any variety (Supplementary Information 7B). Small but significant decreases in plant height were observed in Pokkali and ARB6, whereas no change was observed in Jaya and IR-20 (Supplementary Information 7C). Despite the decrease in plant height observed in both Pokkali and ARB6, tiller number increased in both varieties (Supplementary Information 7D). No change in tiller number was seen in Jaya or IR-20 (Supplementary Information 7D). Aerenchyma increased in all regions of all four varieties under saline

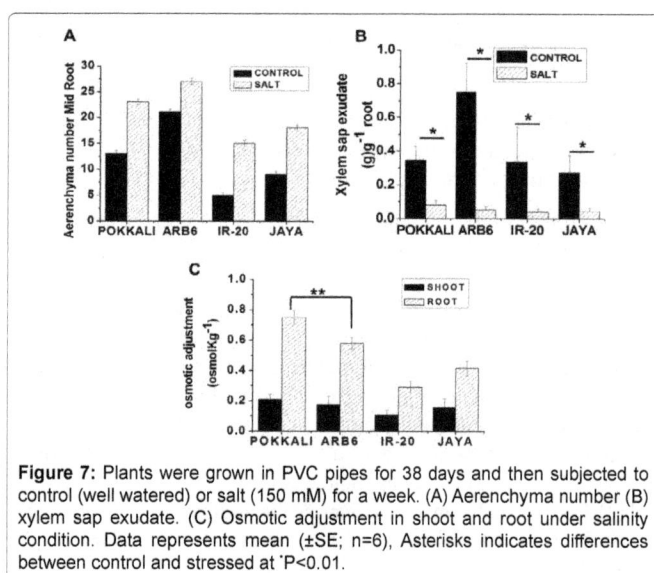

Figure 7: Plants were grown in PVC pipes for 38 days and then subjected to control (well watered) or salt (150 mM) for a week. (A) Aerenchyma number (B) xylem sap exudate. (C) Osmotic adjustment in shoot and root under salinity condition. Data represents mean (±SE; n=6), Asterisks indicates differences between control and stressed at *P<0.01.

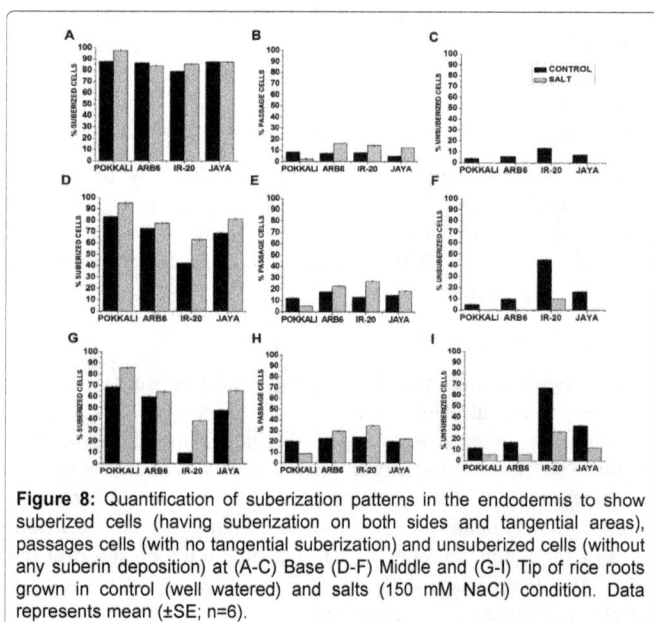

Figure 8: Quantification of suberization patterns in the endodermis to show suberized cells (having suberization on both sides and tangential areas), passages cells (with no tangential suberization) and unsuberized cells (without any suberin deposition) at (A-C) Base (D-F) Middle and (G-I) Tip of rice roots grown in control (well watered) and salts (150 mM NaCl) condition. Data represents mean (±SE; n=6).

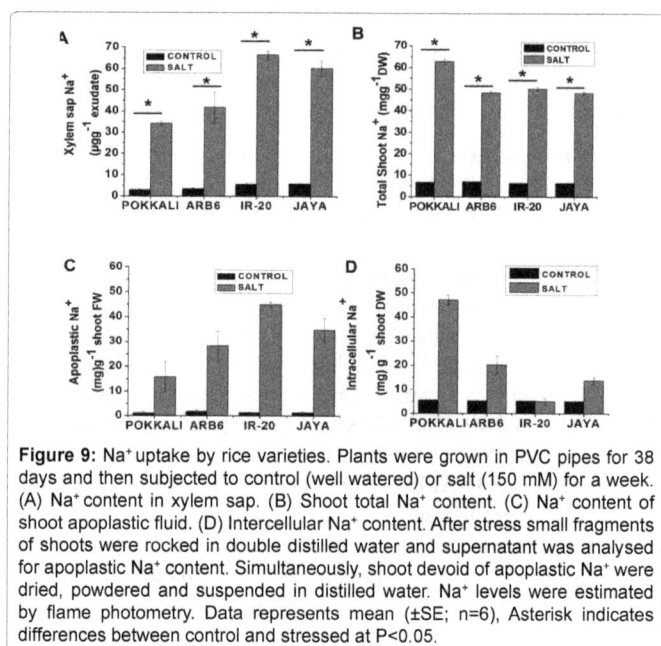

Figure 9: Na+ uptake by rice varieties. Plants were grown in PVC pipes for 38 days and then subjected to control (well watered) or salt (150 mM) for a week. (A) Na+ content in xylem sap. (B) Shoot total Na+ content. (C) Na+ content of shoot apoplastic fluid. (D) Intercellular Na+ content. After stress small fragments of shoots were rocked in double distilled water and supernatant was analysed for apoplastic Na+ content. Simultaneously, shoot devoid of apoplastic Na+ were dried, powdered and suspended in distilled water. Na+ levels were estimated by flame photometry. Data represents mean (±SE; n=6), Asterisk indicates differences between control and stressed at P<0.05.

stress, except in the tips of IR-20 roots (Supplementary Information 4D). Aerenchyma in the middle region of root is presented in Figure 7A. Overall, the morphological analysis confirms that only Pokkali performed well under salinity conditions. Interestingly, ARB6, which was selected as a drought tolerant variety, performed less poorly than Jaya, which is reported to be moderately tolerant to salt [27]. IR-20 did very poorly under salt stress.

Xylem sap exudation and osmotic adjustment: The extent of xylem sap exudation reduced dramatically in all four varieties, with only Pokkali showing significant exudation following saline stress (Figure 7B). Very low osmolyte accumulation was seen in the shoots of all varieties (Figure 7C). Osmolyte accumulation was significant in the roots of all the varieties with Pokkali accumulating significantly more osmolytes than ARB6, while moderate adjustments were seen in Jaya and IR-20 (Figure 7C).

Suberin deposition: The extent of suberization increased in all four varieties and in all regions of both exodermis (Supplementary Information 6C) and endodermis (Figures 8A, 8D, and 8G). The changes were most dramatic in the endodermis of Pokkali. A few endodermal cells lacked suberin deposits in their tangential regions. These cells were adjacent to xylem elements and have been taken to be PCs. An increase in PCs was seen in all varieties except Pokkali where there was a decrease (Figures 8B, 8E, and 8H). No unsuberized cells were seen in the endodermis in either the base or mid regions of any variety except IR-20 (Figures 8C, 8F, and 8I), whereas unsuberized cells were observed in the tip region in all four varieties, where over 30% of cells in IR-20 are unsuberized (Figures 8C, 8F, and 8I). It may be noted that even in the tip region, Pokkali and ARB6 had very few unsuberized cells and Pokkali had few PCs in this region. The other three varieties had comparable fraction of PCs in the tip region.

Na+ content in rice varieties under salt stress: The Na+ content was much higher under saline stress than in control samples in all four varieties, but was highest in IR-20 followed by Jaya and ARB6 (Figure 9A). It was least in Pokkali. Despite maintaining low Na+ content in the xylem sap, Pokkali had somewhat higher total shoot Na+ levels than the other three, which were comparable (Figure 9B). Apoplastic Na+, which we have previously shown to correlate well with survival under salt stress [27], showed a different trend. Pokkali had by far the lowest apoplastic Na^+ and IR-20 had the highest Na^+ content with ARB6 and Jaya being intermediate (Figure 9C). Much of the shoot Na^+ in Pokkali appears to be present in the intracellular compartment with a negligible amount being sequestered in this compartment by IR-20 (Figure 9D). Our earlier studies would suggest that the bulk of the intracellular Na^+ in Pokkali is sequestered in vacuoles [30]. Intracellular K^+ levels were well controlled in all the four varieties under salt stress (Supplementary Information 8)

Root hydraulic conductivity: Hydraulic conductivity decreased significantly under saline stress in all varieties except IR-20 (Table 1). Osmotically driven conductivity (L_0) dropped by a factor of 2 or more in the cases of Pokkali, ARB6 and Jaya, ending at values of around 2 (ms^{-1}) (Table 1). The decline was maximum for Pokkali (67%) and less for ARB6 (60%) and Jaya (50%). IR-20 finished the salt stress period with an osmotically driven hydraulic conductivity that was 65% greater than that of Pokkali. Similar declines were observed for external pressure driven hydraulic conductivity (L_{pr}) with declines exceeding 50% for Pokkali and ARB6 while Jaya exhibited a decline of 40%. These three varieties had final conductivity values around 20×10^{-8} m MPa^{-1} s^{-1}. The decline in IR-20 was marginal.

Expression of suberin synthesis genes in roots under drought and salt stress condition: To check the expression of suberin synthesis genes, total RNA from roots of control and stressed four rice varieties was analyzed by RT-PCR. Figure 10 shows expression of three *Elongases* (Os03g12030, Os02g11070 and Os06g39750) and one *cytochrome P450* (Os01g63540) after one week of drought and salt stress. *Elongase* (Os03g12030) was found to express equally in all four rice varieties under control, drought and salt stress. The expression of *Elongase* (Os02g11070) had dramatically reduced in all four varieties under drought stress compared to salt and control condition. While *Elongase* (Os06g39750) was found to express in all four varieties under salt stress, there was a dramatic decline in expression in Pokkali and ARB6 under drought stress. The expression of *cytochrome P450* (Os01g63540) was essentially the same in all four varieties under both drought and salt stress.

Discussion

Salinity and drought are among the major factors limiting crop productivity worldwide. Both are essentially water deficit problems, although salinity has the added feature of Na^+ toxicity. While some responses are common across several abiotic stresses, others are stress specific and may well prove counterproductive in dealing with others. It was to explore this possibility that we have subjected "specialist" varieties reported to be tolerant to either drought or salinity to both stresses and contrasted their responses. We have used conditions where all the plants survived the stress protocol. Measurement of soil moisture content and electrical conductivity indicated similar and significant amounts of stress being experienced by all varieties in each protocol (Supplementary Information 1A-1E). Photosynthetic rates declined significantly over the period of stress in all cases (Figures 1C and 6C), the extent of decline being moderate for the tolerant varieties and large for the sensitive varieties. It was very surprising to find that both Pokkali and ARB6 maintained good photosynthetic rates when subjected to the stress they had not been selected for.

Root architecture and physiology may be expected to play key roles in surviving drought and salinity. Indeed long and prolific root systems are traits selected for while breeding for drought tolerance [14,46,47].

Minimization of apoplastic flow that bypasses hydrophobic barriers in the endodermis is a valuable trait in selecting for salt tolerant lines [48,49]. This trait could have deleterious effects on root hydraulics and may not be conducive to survival under drought. Thus varieties chosen for their stellar performance in one stress regime may not necessarily do well under a contrasting stress. Features such as the activity and localization of ion transporters or aquaporin's may be critical in one regime while playing minor roles in the other. Our current study has focused on features affecting fluid uptake and delivery.

Pokkali is known to develop extensive hydrophobic barriers around both the endodermis and the exodermis constitutively. The unexpected finding that Pokkali does well under drought stress prompted the question as to whether the two successful varieties deployed different strategies, one enhancing water use efficiency while the other resorts to water mining. The observation of a large differential between canopy and ambient temperature in both ARB6 and Pokkali under drought stress strongly suggests evaporative cooling and hence a substantial transpirational stream even in Pokkali (Figure 2B), which is consistent with water mining. High rates of photosynthesis and xylem fluid exudation by both varieties under drought (Figures 1C and 2C) support this hypothesis. Vigorous transpiration in the face of drought requires access to subsurface water. Both Pokkali and ARB6 have long roots that elongate further under drought stress ending up over 50% longer than IR-20 roots (Figure 1G), facilitating access to deeper layers.

Uptake is driven by accumulation of osmolytes in both ARB6 and Pokkali. Accumulation of osmolytes is much higher in the face of drought stress (Figure 2D) than in salinity in case of ARB6 (Figure 7C), while the opposite is true for Pokkali. Jaya, which was earlier shown to adapt well to salinity, also accumulates a modest amount of osmolytes under saline stress, but not under drought. IR-20, which also adapted to saline stress by building hydrophobic barriers and restricting bypass flow under hydroponic culture [23], did not accumulate any osmolytes under either drought or salinity stress.

The water mining strategy is likely to result in roots growing deep into hypoxic environments. Transporting oxygen down from the shoots requires extensive aerenchyma formation, while suberization of the exodermis would minimize radial oxygen loss to the surrounding soil. Extension of the size and extent of aerenchyma is seen in both ARB6 and Pokkali in response to stress (Supplementary Information 4), along with substantial suberization of the exodermis (Supplementary Information 6). This adaptation would serve both to reduce the number of cells drawing on scarce resources under stress as well as supplying oxygen to distal portions of the root.

The hydraulic conductivity we report is that measured with roots immersed in standard growth medium in all cases. The osmotically-driven conductivity (L_0) is measured in the absence of any external pressure and reflects the hydraulics of the symplastic cell-to-cell pathway which is strongly influenced by aquaporin conductivity [50]. In addition, fluid uptake may be expected to reflect differences in water potential across the plasma membrane. The large increase in osmolyte content under drought would lead to an expectation of enhanced driving force for water with the roots in growth medium and hence enhanced L_0. The underwhelming increase in L_0 is puzzling. Sap exudation rate, on the other hand, is measured with plants in soil and reflects a combination of L_0 and the difference of water potential between root cells and the soil surrounding the root. The fact that ARB6 and Pokkali maintain moderate rates of xylem sap exudation even in drought or salinity must then be ascribed to the osmotic adjustment made in the two cases, together with the reduction in completely suberized cells in these two varieties.

Pressure-driven hydraulic conductivity L_{pr} reflects the hydraulics of the apoplastic pathway, primarily due to suberization patterns of the endodermis [51,52]. Exodermal suberization has been shown not to influence L_{pr} significantly in maize [53] and rice [54]. Small, but significant increases in L_{pr} in ARB6, Pokkali and Jaya contribute to enhanced fluid flow through the apoplastic pathway under drought (Table 1). This corresponds well with the decrease on overall suberization and increase in passage cell number under drought in these varieties (Figure 5 and Supplementary Information 6). Conversely, the lack of such change in suberization patterns in IR-20 is echoed in the finding that L_{pr} is also unaffected under drought (Table 1).

Salinity induces a substantial decrease in osmotic L_0 suggesting either a decrease in aquaporin expression or activity or both. A concomitant decrease in hydrostatic L_{pr} correlates well with the increase in completely suberized cells at the cost of unsuberized cells in the endodermis of Pokkali, the most successful variety under saline stress (Figure 8). ARB6 and Jaya also increase suberization of the endodermis but to a smaller extent. Again, maintenance of xylem sap exudation despite a reduction in L_0 is strongly correlated with osmolyte accumulation in both Pokkali and ARB6.

It is worth noting differences between the behavior of the plants grown in soil and hydroponics. Roots of all varieties are extensively suberized when grown in well-watered soil, as opposed to hydroponics [23]. Further, the barriers formed in response to a week of saline stress do not completely eliminate osmotically driven xylem fluid exudation in soil (Figure 7B) as it does in hydroponics [23]. A week of conditioning stress under hydroponic culture rendered IR-20 almost as tolerant to salt as Pokkali, whereas the extent of adaptation when grown in soil appears to be less effective. It would thus appear that not only are modest barriers already in place under control conditions when grown in soil, but that the extent to which they are fortified in response to salinity stress is much less than under hydroponics. Extension of results from hydroponics to field conditions may need to be taken with a pinch of salt.

The impressive rates of exudation seen in ARB6 and Pokkali, especially under drought, could be attributed to the presence of PCs. Indeed, the dramatic increase in PCs is one of the most striking features of the drought tolerance response of ARB6. It may be noted that while passage cell numbers increase under drought stress, cells with unsuberized radial walls are completely eliminated in Pokkali and largely so in ARB6 under salt stress. Thus the pattern of suberization is dramatically different in the face of different stresses. This is illustrated in Figure 11, which shows a near linear variation of xylem sap exudation rates and xylem sap Na$^+$ content with the number of PCs under drought and salinity (Figures 11A and 11B). The only variety to do well under salt stress is Pokkali, which completely eliminates unsuberized cells in the base and middle regions of the endodermis and also significantly reduces PCs. ARB6 also eliminates unsuberized cells, but slightly increases PCs

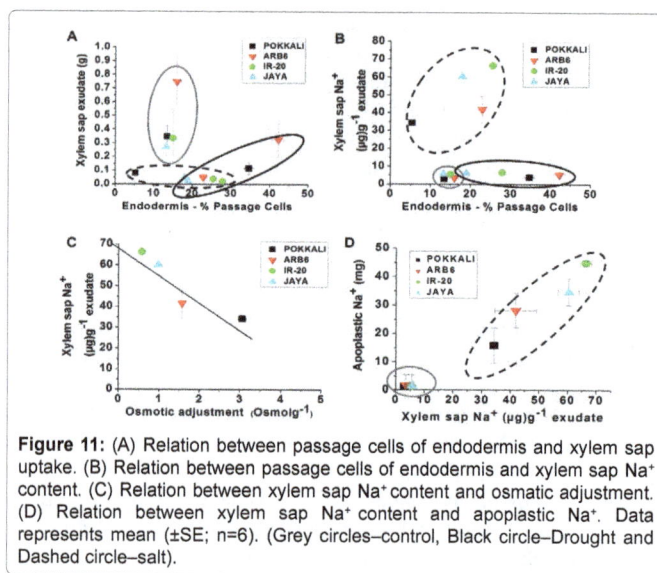

Figure 11: (A) Relation between passage cells of endodermis and xylem sap uptake. (B) Relation between passage cells of endodermis and xylem sap Na$^+$ content. (C) Relation between xylem sap Na$^+$ content and osmatic adjustment. (D) Relation between xylem sap Na$^+$ content and apoplastic Na$^+$. Data represents mean (±SE; n=6). (Grey circles–control, Black circle–Drought and Dashed circle–salt).

in this condition. The dramatic reduction in the number of completely suberized cells in the base region of both endodermis and exodermis of ARB6 and Pokkali in response to drought is puzzling, as it implies the removal of suberin deposits. No mechanism for depolymerization of suberin has thus far been reported. This may well prove to be a suitable system to explore such mechanisms. These findings demonstrate that suberin deposits are laid down in very different patterns in response to the two stresses. Intriguingly, both ARB6 and Pokkali are capable of remodeling their hydrophobic barriers appropriately in both cases while IR-20 is not.

The elimination of unsuberized cells in the endodermis of Pokkali, ARB6 and Jaya under salt means that fluid entry over much of the length of the root occurs through PCs. This ensures that solutes have to cross at least two sets of plasma membranes before entering the xylem stream. Solute movement across cell membranes should be subject to selectivity of the relevant transporters. The large build-up of osmolytes in Pokkali compensates for the restriction in fluid entry points. It also appears to contribute to minimizing Na$^+$ loading of the xylem sap as we observe a nearly linear decline in the Na$^+$ content of sap with osmolyte accumulation (Figure 11C). This reduction in xylem sap Na$^+$ is in turn correlated with limiting Na$^+$ in the shoot apoplast (Figure 11D), which is tightly correlated to survival [28]. A significant part of the total shoot Na$^+$ in Pokkali is located in the intracellular fraction. Our earlier studies on cellular mechanisms of combating salinity would suggest that this reflects the sequestering the Na$^+$ in vacuoles, thereby minimizing the amount present in the apoplastic fraction. This sequestration mechanism appears not to function effectively in IR-20.

In RT-PCR data, the enhanced mRNA expression levels of *P450* and *elongases* (39750, 11070 & 12030) under salinity stress in all four varieties (Figure 10) was consistent with the observations of [23,55]. Whereas, there was significantly low expression of *elongase* (11070) and *elongase* (39750) under drought stress (Figure 10). The suberization pattern and RT-PCR data appears to suggest that the suberin synthesis pathway and deposition in tolerant and sensitive rice varieties under salinity and drought stress could be different.

This study comparing the physiological responses of four cultivars varying widely in tolerance to drought and salinity clearly identifies several features of roots that contribute to survival under these stresses. An extensive root system facilitates water mining in drought.

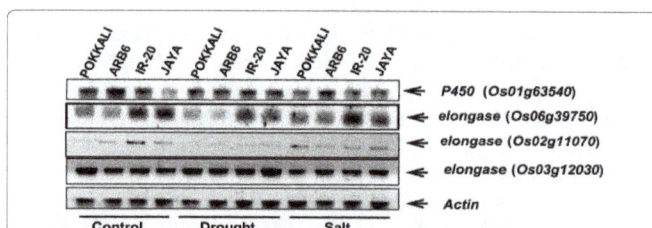

Figure 10: Expression of elongases and P450 in roots of four rice varieties after one week of drought and salt stress monitored by RT-PCR.

Aerenchyma formation reduces the number of resource consuming cells as well as providing a conduit for oxygen to distal portions of the root; suberization of the exodermis minimizes loss of the transported oxygen. The pattern of suberization differs greatly between drought and salinity, but in both cases PCs facilitate fluid uptake. In the latter case, the only endodermal cells lacking tangential suberin are PCs-thereby ensuring solute selectivity. Hydrostatic conductivity changes are prominent under salinity, restricting fluid flow to the symplastic pathway and where solutes can be sifted by selective transporters. Uptake is driven by osmolyte loading, which also contributes to reduction of Na^+ uptake. Our study is purely correlative. However, the use of four varieties with a wide span of sensitivity to each stress provides significant spread in the stress–driven parameters monitored, thereby strengthening the conclusions drawn. We had hypothesized that specialist varieties would do well under "their" respective stress, but would be sensitive to the other stress; and that the responses would be graded according to the degree of tolerance/sensitivity among the four cultivars studied. The latter expectation was indeed upheld with photosynthetic rates spanning a wide range at the end of each stress protocol. The former expectation was based on an assumption that the "specialists" are restricted to one response to all stresses. This is clearly not the case.

In conclusion, the set of adaptations we describe appears to make a significant contribution to the ability of rice plants to survive water deficit conditions in both drought and salinity stress. The study has provided experimental evidence to validate the large quantity of empirical field data related to drought tolerance of ARB6 and salt tolerance of Pokkali. Testing the effectiveness of remodeling suberization patterns will require tools to perturb such patterns. Mechanisms for the detection of stress and the regulation of the responses deployed are areas for further work as is the search for additional mechanism underlying drought and stress tolerance at the root level.

Conflict of Interest

It is declared that the authors have no conflict of interests.

Acknowledgements

We are grateful to Dr. Shivaprasad and Dr. Kalika Prasad for critical reviews of this manuscript. We also thank Ashvini Kumar Dubey, Anirban Baral and Raveendra GM for their timely advice and help. The Central Imaging & Flow Cytometry Facility (CIFF) at National Centre for Biological Sciences (NCBS) Bangalore is gratefully acknowledged. The work was performed with internal NCBS funding. The funding from Department of Biotechnology (DBT), India is also highly acknowledged.

References

1. Akbar M, Ponnamperuma FN (1982) Saline soil of south and Southeast asia as potential rice lands. In Rice research strategies for the future. International Rice Research Institute Los Banos, Laguna, Manila, Philippines pp: 256-281.

2. Khatun S, Rizzo CA, Flowers TJ (1995) Genotypic variation in the effect of salinity on fertility in rice. Plant Soil 173: 239-250.

3. Munns R, Tester M (2008) Mechanisms of salinity tolerance. Annu Rev Plant Biol 59: 651-81.

4. Rai AK, Hagemann M, Erdmann N (1997) Environmental stresses. In: Cyanobacterial Nitrogen Metabolism and Environmental Biotechnology Springer, Heidelberg; Narosa Publishing House, New Delhi.

5. Zhu JK (2002) Salt and drought stress signal transduction in plants. Annu Rev Plant Biol 53: 247-73.

6. Jacobson T, Adams RM (1958) Salt and silt in Ancient Mesopotamia Agriculture. JSTOR 128: 1251-1258.

7. Bouman B, Humphreys E, Tuong TP, Barker R (2007) Rice and water. Adv Agron 92: 187-237.

8. Vijayakumar C (2006) Breeding for high yielding rice (Oryza sativa L.) varieties and hybrids adapted to aerobic (non-flooded, irrigated) conditions - II. Evaluation of released varieties. Indian J Genet 66: 182-186.

9. Shashidhar HE (2007) Aerobic rice - an efficient water rice production. In: Aswathanarayana U (ed.) Food and Water Security. Taylor and Francis, London, UK.

10. Verulkar SB (2010) Breeding resilient and productive genotypes adapted to drought-prone rainfed ecosystem of India. Field Crop Res 117: 197-208.

11. Atlin GN, Tao D, Laza M, Amante M, Brigitte C (2006) Developing rice cultivars for high-fertility upland systems in the Asian tropics. Field Crop Res 97: 43-52.

12. Blum A, Mayer J, Golan G (1989) Agronomic and physiological assessments of genotypic variation for drought resistance in sorghum. Aust J Agric Res 40: 49-61.

13. Samson BK, Hasan M, Wade LJ (2002) Penetration of hardpans by rice lines in the rainfed lowlands. Field Crop Res 76: 175-188.

14. Wang H, Yamauchi A (2006) Growth and function of roots under abiotic stress soils. In: Huang B (ed.) Plant-Environment Interactions (3rd edn.) CRC Press, Taylor and Francis Group, LLC, New York pp: 271-320.

15. Henry A (2011) Variation in root system architecture and drought response in rice (Oryza sativa): Phenotyping of the Oryza SNP panel in rainfed low land fields. Field Crop Res 120: 205-214.

16. Gowda VRP, Henry A, Yamauchi A, Shashidhar HE, Serraj R (2011) Root biology and genetic improvement for drought avoidance in rice. Field Crop Res 122: 1-13.

17. Wasson AP, Richards RA, Chatrath R, Misra SC, Sai Prasad SV, et al. (2012) Traits and selection strategies to improve root systems and water uptake in water-limited wheat crops. J Exp Bot 63: 3485-3498.

18. Kumar A, Verulka SB, Mandal NP, Variar M, Dwivedi JL, et al. (2012) High-yielding, drought-tolerant, stable rice genotypes for the shallow rainfed lowland drought-prone ecosystem. Field Crop Res 133: 37-47.

19. Utharasu S, Anandakumar CR (2013) Heterosis and combining ability analysis for grain yield and its component traits in aerobic rice (Oryza sativa L.) cultivars. Electron. J Plant Breed 4: 1271-1279.

20. Akbar M (1985) Genetics of salt tolerance in rice in IRRI. Rice Genetics. Proceeding of International Rice Genetics Symposium, IRRI, Manila, Philippines pp: 399-409.

21. Gregorio G, Senadhira D (1993) Genetic analysis of salinity tolerance in rice (Oryza sativa L.). Theor Appl Genet 86: 333-338.

22. Ali Y, Aslam Z, Awan AR, Hussain F, Cheema AA (2004) Screening Rice (Oryza sativa L.) Lines/Cultivars against Salinity In Relation To Morphological and Physiological Traits and Yield Components. Int J Agric Biol 6: 572-575.

23. Krishnamurthy P, Ranathunge K, Franke R, Prakash HS, Schreiber L, et al. (2009) The role of root apoplastic transport barriers in salt tolerance of rice (Oryza sativa L.). Planta 230: 119-134.

24. Oertli JJ (1968) Extracellular salt accumulation, a possible mechanism of salt injury in plants. Agrochimica 12: 461-469.

25. Yeo A, Flowers T (1982) Accumulation and localisation of sodium ions within the shoots of rice (Oryza sativa) varieties differing in salinity resistance. Physiol Plant 56: 343-348.

26. Yeo AR, Flowers TJ (1983) Varietal differences in the toxicity of sodium ions in rice leaves. Physiol Plant 59: 189-195.

27. Anil VS, Krishnamurthy P, Kuruvilla S, Sucharitha K, Thomas G, et al. (2005) Regulation of the uptake and distribution of Na^+ in shoots of rice (Oryza sativa) variety Pokkali: role of Ca^{2+} in salt tolerance response. Physiol Plant 124: 451-464.

28. Krishnamurthy P, Ranathunge K, Nayak S, Schreiber L, Mathew MK (2011) Root apoplastic barriers block Na^+ transport to shoots in rice (Oryza sativa L.). J Exp Bot 62: 4215-4228.

29. Kader MA, Lindberg S (2005) Uptake of sodium in protoplasts of salt-sensitive and salt-tolerant cultivars of rice, Oryza sativa L. determined by the fluorescent dye SBFI. J Exp Bot 56: 3149-58.

30. Anil VS, Krishnamurthy H, Mathew MK (2007) Limiting cytosolic Na^+ confers salt tolerance to rice cells in culture: a two-photon microscopy study of SBFI-loaded cells. Physiol Plant 129: 607-621.

31. Perumalla CJ, Peterson CA (1986) Deposition of Casparian bands and suberin lamellae in the exodermis and endodermis of young corn and onion roots. Can J Bot 64: 1873-1878.

32. Schreiber L, Hartmann K, Skrabs M, Zeier J (1999) Apoplastic barriers in roots: chemical composition of endodermal and hypodermal cell walls. J Exp Bot 50: 1267-1280.

33. Naseer S, Leea Y, Lapierreb C, Frankec R, Nawratha C, et al. (2012) Casparian strip diffusion barrier in Arabidopsis is made of a lignin polymer without suberin. Proc Natl Acad Sci USA 109: 10101-10106.

34. Bernards MA, Summerhurst DK, Razem FA (2004) Oxidases, peroxidases and hydrogen peroxide: The suberin connection. Phytochem Rev 3: 113-126.

35. Franke R, Briesen I, Wojciechowski T, Faust A, Yephremov A (2005) Apoplastic polyesters in Arabidopsis surface tissues - A typical suberin and a particular cutin. Phytochemistry 66: 2643-2658.

36. Höfer R, Briesen I, Beck M, Pinot F, Schreiber L, et al. (2008) The Arabidopsis cytochrome P450 CYP86A1 encodes a fatty acid omega-hydroxylase involved in suberin monomer biosynthesis. J Exp Bot 59: 2347-2360.

37. Enstone DE, Enstone, Peterson CA, Fengshan MA (2003) Root Endodermis and Exodermis: Structure, Function, and Responses to the Environment. J Plant Growth Regul 21: 335-351.

38. Schreiber L, Franke RB (2011) Endodermis and exodermis in roots. In Encyclopedia of Life Sciences pp: 1-7.

39. Chen T, Cai X, Wu X, Karahara I, Schreiber L, et al. (2011) Casparian strip development and its potential function in salt tolerance. Plant Signal Behav 6: 1499-502.

40. Wassmann R (2009) Regional vulnerability of climate change impacts on Asian rice production and scope for adaptation. Adv Agron 102: 91-133.

41. Shashidhar H, Vimarsh HSG, Raveedra GM, Pavan JK, Kumar GN (2012) PVC tubes to characterize roots and shoots to complement field plant productivity studies. In Methodologies for Root Drought Studies in Rice. AGKB.

42. Morita S, Abe J (2002) Diurnal and phenological changes of bleeding rate in lowland rice plants, Japanese journal of crop science. Crop Science Society of Japan pp: 383-388.

43. Henry A, Andrew J. Cal, Batoto TC, Torres RO, Serraj R (2012) Root attributes affecting water uptake of rice (Oryza sativa) under drought. J Exp Bot 63: 4751-4763.

44. Miyamoto N, Steudle E, Hirasawa T, Lafitte R (2001) Hydraulic conductivity of rice roots. J Exp Bot 52: 1835-1846.

45. Brundrett MC, Enstone DE, Peterson CA (1988) A berberine-aniline blue fluorescent staining procedure for suberin, lignin, and callose in plant tissue. Protoplasma 146: 133-142.

46. Matsui T, Singh BB (2003) Root characteristics in cowpea related to drought tolerance at the seedling stage. Exp Agric 39: 29-38.

47. Blum A (2005) Drought resistance, water-use efficiency, and yield potential are they compatible, dissonant, or mutually exclusive? Aust J Agric Res 56: 1159-1168.

48. Yeo AR, Yeo ME, Flowers TJ (1987) The contribution of an apoplastic pathway to sodium uptake by rice roots in saline conditions. J Exp Bot 38: 1141-1153.

49. Faiyue B, AL-Azzawi M, Flowers TJ (2012) A new screening technique for salinity resistance in rice (Oryza sativa L.) seedlings using bypass flow. Plant, Cell Environ 35: 1099-1108.

50. Steudle E (1994) The regulation of plant water at the cell, tissue, and organ level: role of active processes and of compartmentation. In: Schulze ED (ed.) Flux control in biological systems: from enzymes to populations and ecosystems pp 237-299.

51. Clarkson D, Robards A (1975) The endodermis, its structural development and physiological role. In: Torrey (ed.) The development and function of roots pp: 415-436.

52. Melchior W, Steudle E (1993) Water Transport in Onion (Allium cepa L.) Roots 1 Changes of Axial and Radial Hydraulic Conductivities during Root Development.

53. Zimmermann HM, Steudle E (1998) Apoplastic transport across young maize roots: effect of the exodermis. Planta.

54. Ranathunge K, Steudle E, Lafitte R (2003) Control of water uptake by rice (Oryza sativa L.): role of the outer part of the root. Planta 217: 193-205.

55. Ranathunge K, LIN J, Steudle E, Schreiber L (2011) Stagnant deoxygenated growth enhances root suberization and lignifications, but differentially affects water and NaCl permeabilities in rice (Oryza sativa L.) roots. Plant Cell Env 34: 1223-1240.

12

Gene Expression Studies in Lignin Synthesis Pathway of Sorghum [*Sorghum Bicolor*] (L. Moench)

Tanmay K[1]*, Umakanth AV[2], Madhu P[2] and Bhat V[2]

[1]*Jawaharlal Nehru Technological University, Hyderabad India*
[2]*Indian Institute of Millets Research, Hyderabad India*

Abstract

Gene expression play significant role in lignin synthesis pathway in sorghum. Expression level of brown-midrib sorghum was studied in brown midrib sorghum *bmr* 6 and *bmr* 12 mutant of Atlas, Kansas collier, Early hagari Sart, Rox Orange. Gene expression levels for *bmr* 6, CAD 4, SBCAD2, *bmr* 12, COMT3 COMT were compared for wild sorghum genotypes with their *bmr* 6 and *bmr* 12 counterparts. *bmr* 6 has negative non-significant correlation with lignin content (-0.075).

Keywords: Brown-midrib sorghum; Gene expression; Lignocellulosic; Kansas collier

Introduction

Gene expression is the process by which genetic information is converted into protein or a functional product. This process uses an intermediate molecule, m-RNA, which is transcribed from DNA and then used as a template to translate the message into a protein product. Studies of gene expression provide a window into how an organism's genetic makeup enables it to function and respond to its environment. Real-Time PCR can be used to quantify gene expression by two methods: relative and absolute quantification. The relative quantification method compares the gene expression of one sample to that of another sample: drug-treated samples to an untreated control, for example, using a reference gene for normalization. Absolute quantification is based on a standard curve, which is prepared from samples of known template concentration. The concentration of any unknown sample can then be determined by simple interpolation of its PCR signal (Cq) into this standard curve [1-6]. As RT-q PCR performance is affected by the RNA integrity, Fleige and Pfaffl recommend an RNA quality score (RIN or RQI) higher than five as good total RNA quality and higher than eight as perfect total RNA for downstream applications [4-7]. A study on the impact of RNA quality on the expression stability of reference genes indicated that it is inappropriate to compare degraded and intact samples, necessitating sample quality control prior to RT-q PCR measurements [8]. (Vermeulen *et al.*, submitted for publication) data indicate that RNA quality has a profound impact on the results, in terms of the significance of differential expression, variability of reference genes and classification performance of a multi-gene signature. In addition or as an alternative to the use of capillary gel electrophoresis methods that assess the integrity of the ribosomal RNA molecules as discussed in [4,8] PCR based tests are also frequently used to determine mRNA integrity. In one such a test, the ratio between the 5' and 3' end of a universally expressed gene is measured upon anchored oligo-dT cDNA synthesis, reflecting integrity of that particular poly-adenylated transcript [7] finally, another PCR based assay is often used in clinical diagnostics to determine sample purity. By comparing the Cq value of a known concentration of a spiked DNA or RNA molecule in both a negative water control and in the sample of unknown quality, enzymatic inhibition can be determined [7] (Figure 1).

Bmr 6: The *bmr* 6 mutation in sorghum encodes cinnamyl alcohol dehydrogenase 2 (CAD2). In the final step of monolignol biosynthesis, CAD catalyzes the reduction of cinnamyl aldehydes (Coniferyl, Coumaryl and Sinapyl aldehyde) to their corresponding cinnamyl alcohols, using NADPH as a cofactor, prior to their incorporation into the lignin polymer (Figure 2). ZmCAD2 is an ortholog to both the sorghum bmr6 and rice Gh2, mutations in either gene resulted in reduced CAD activity and altered lignin composition similar to the bm1 phenotype.

The bmr 6 allelic group consists of *bmr* 6-3, *bmr*6-4, *bmr*6-20, *bmr*6-22, *bmr*6-23, *bmr*6-24, *bmr*6-27, *bmr*6-28, *bmr*6-39, *bmr*6-40 and *bmr*6-41 including *bmr* 6-refference allele. Alleles *bmr* 6-39 and *bmr* 6-40 both are resulted due to G to C 3699bp and *bmr* 6-41 consist of C-to-T transition at 3619bp and G-to-T transversion at 3620bp [9-11]. Although CAD2 protein was absent from *bmr* 6 tissues, CAD activity was still detectable in the tissues, though activity was reduced

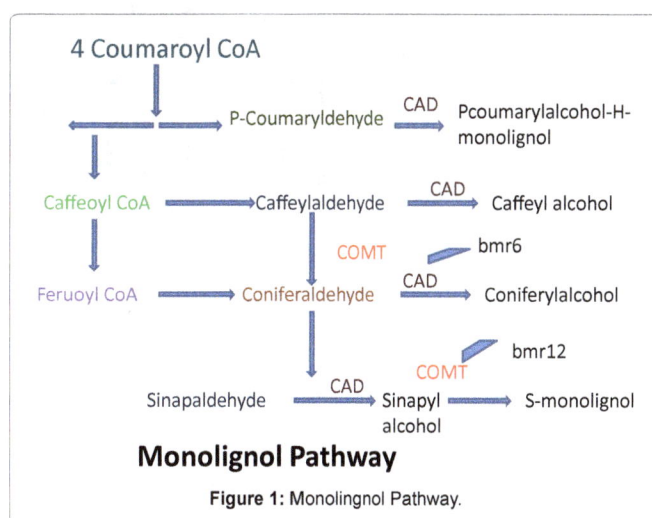

Figure 1: Monolignol Pathway.

*Corresponding author: Tanmay K, Jawaharlal Nehru Technological University, Hyderabad, India, E-mail: vilol.tanmay@gmail.com

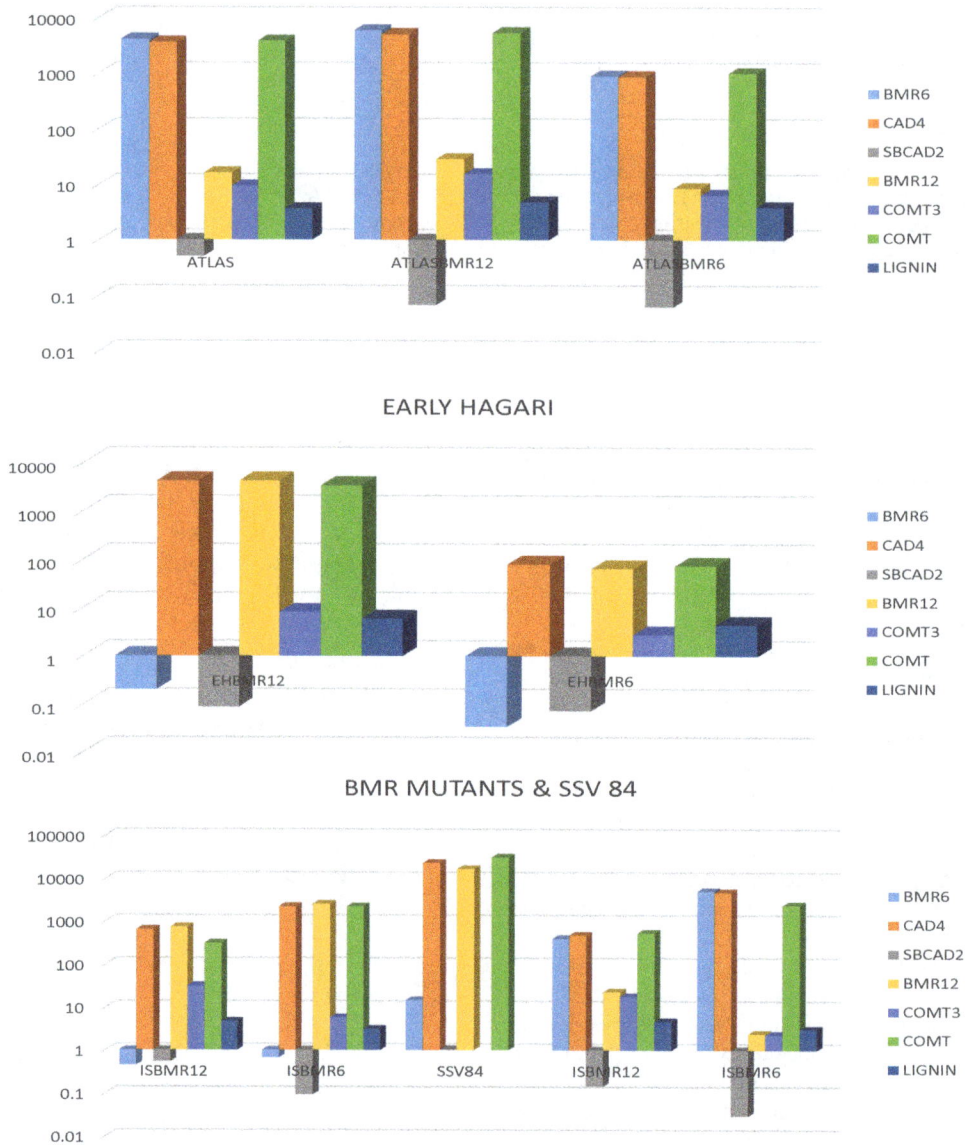

Figure 2: Expression Profile of *Bmr* Genes In Mutant Derivatives.

to 15-50% of wild type activity. This indicate that there are other CAD proteins present in sorghum that can utilize cinnamyl substrates, but the brown midrib phenotype reveals that bmr6 encode the main CAD protein in the monolignol biosynthetic pathway in sorghum (Figure 3).

Mutations in brown mid-rib lines

Bmr **12:** Sorghum bmr12 locus encodes orthologous caffeic O-methyl transferase (COMT). Caffeic O methyl transferase (COMT) is members of an evolutionary conserved O-methyl transferase family, whose function in lignin biosynthesis has been documented in both monocots and dicots. The lignin monomeric compositions of bmr 12 plants has shown that syringyl-lignin was greatly reduced, while p-hydroxyphenyl and guaiacyl-lignin were slightly reduced.

About ten distinct alleles of sorghum *bmr* 12 have been isolated and the mutated sites identified i.e *bmr* 12-ref, *bmr* 12-7, *bmr* 12-15, *bmr* 12-18, *bmr* 12-25, *bmr* 12-26, *bmr* 12-30, *bmr* 12-34, *bmr* 12-35 and *bmr* 12-820 [1,2,9]. The mutations in the *bmr* 12and *bmr* 18 alleles are located 27 (C-to-T 486) and 80nt (G-to-A4 36) upstream of the exon-exon junction respectively, whereas the *bmr* 26 mutation is (G-to-A2292) 388nt downstream of this boundary [2]. The *bmr* 25 consist of same mutation as that of *bmr* 18 Nonsense mutations are responsible for four of the characterized alleles *i.e.*, *bmr* 12-30 consist of G-to-A transition at 2364nt leading to Gly225Asp, bmr12-34 is due to C-to-T at 518 and 2139nt causing Ala71Val and Pro 150 Leu respectively, *bmr* 12-34 and *bmr* 12-820 contain two mutations in *bmr*12, which are identical, *bmr*12-35 is due to G-to-A transition at

Expression profile of *bmr* genes in mutant derivatives
X axis - Genotypes
Y axis- Fold induction in gene expression (log scale)

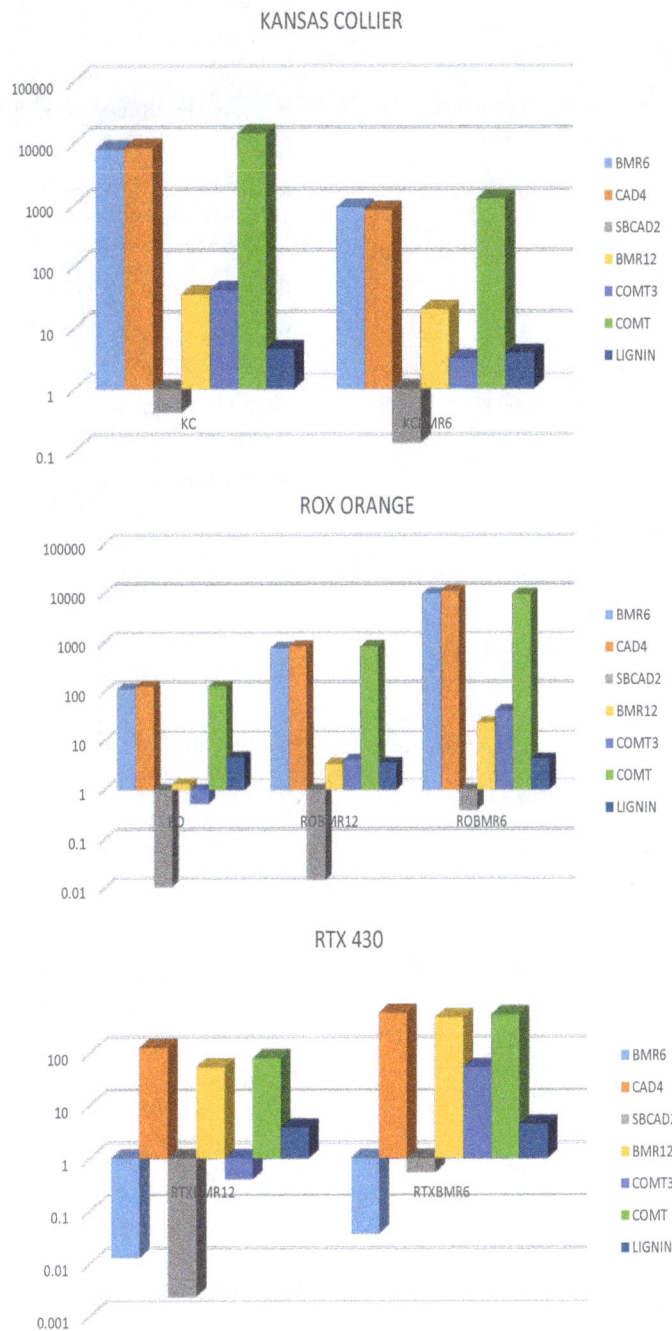

Figure 3: Expression profile of *bmr* genes in mutant derivatives.

2663 leading to Gly325Ser [9,10]. These four nonsense mutations are all presumably null alleles, because the premature stop codons would truncate the polypeptide prior to the SAM binding site of the enzyme.

Lignin is one of the most important biomolecules in vascular plants and is uniquely involved in the structure support, water transport, and other functions [3,5]. Lignin biosynthesis has been subject to intensive study during the past two decades, mainly driven by the significant needs in forage and biofuel industries [12].

Material and Methods

Sorghum DNA isolation

Breaking the cell wall and cell lysis: Plant tissue very well ground up, caution taken to prevent DNA from degrading during the procedure. The cells are lysed in the presence of chaotropic agent CTAB.

1. 9 ml of warm (65° C) CTAB extraction buffer added to 1g freeze dried, ground tissue (young leaves) in a 30 ml centrifuge tube.

2. Incubated for 3h at 65° C in a water bath with occasional mixing.

Separate DNA from other cell components: Chloroform helps bind up the complex proteins and polysaccharides. Chloroform is denser than water solutions and thus after spinning this solution. Chloroform and water will separate into two distinct phases. The lower phase will be chloroform. This is the phase that proteins and polysaccharides find most chemically attractive. The upper aqueous phase phase will contain DNA. Iso-propanol is used to precipitate DNA present in aqueous phase. Precipitation with iso-propanol has the advantage that the volume of liquid to be centrifuged is smaller. However, iso-propanol is less volatile than ethanol and it is more difficult to remove the last traces; moreover, solutes such as sucrose or sodium chloride are more easily co-precipitated with DNA when iso-propanol is used, especially at -70° C.

1. Tubes were removed from water bath, wait for 4-5 min and added 10ml chloroform/iso-amyl alcohol (24:1). Mixed gently for several times.

2. Centrifuged at 6000 rpm for 10 min at RT.

3. Transfer aqueous phase to a clean 30 ml (or 15ml) tube. 6 ml chloroform/isoamyl alcohol (24:1) added and mixed gently several times.

4. Centrifuged at 6000 rpm for 10 min at RT.

5. Transferred aqueous phase to clean 30 ml (or 15 ml) tube. 6 ml (2/3 volume) isopropanol added, mixed gently by inversion several times. Removing DNA by hook minimizes the contamination by salt precipitations, DNA washed to remove impurities.

6. Precipitated DNA removed with glass hook, or centrifuge at 6000 rpm for 10 min and pellets DNA.

7. Place hook with DNA (or DNA pellet) in a 5 ml plastic tube containing 2 ml of washing buffer 1(76 % ethanol, 0.2 M sodium acetate). Leave DNA on hook in tube for at least 20 min.

8. Rinse DNA on hook briefly in 1-2 ml of washing buffer 2 (76% ethanol, 10Mm ammonium acetate) and air dry at 37° C.

9. Transfer DNA to 1.5 ml micro-centrifuge tube containing 0.4 ml of TE buffer and place at 4° C overnight to disperse DNA. Next day, treat with RNase for 3 h at 37° C.

Phenol extraction and ethanol precipitation of DNA:

1. Equal volume of phenol/chloroform/isoamyl alcohol added (24:24:1) to the DNA solution in a 1.5 ml micro centrifuge tube.

2. Vortex vigorously 10 sec and micro centrifuge for 10 min at maximum speed.

3. The top (aqueous) phase removed carefully containing the DNA using a pipette and transfer to a new tube. If a white

precipitate is present at the aqueous organic interface, re-extract the organic phase and pool aqueous phases.

4. An equal volume of chloroform /isoamyl alcohol (24:1) added to the DNA solution in a micro centrifuge tube.

5. Vortex vigorously 10 sec and micro centrifuge for 10min maximum speed.

6. 1/10 volume of 3 M sodium acetate Ph 5.2 added. Mix by vortexing briefly or by flicking the tube several times with a finger.

7. Add 2 to 2.5 volume (calculated after salt addition) of ice-cold 100% ethanol. Mix by vortexing and place in crushed dry ice for 5min or longer.

8. Spin 10 min at high speed and remove supernatant.

9. 1 ml of RT 70% ethanol added. Inverted the tube several times and Micro-centrifuged as in step 6.

10. The supernatant removed, dry the pellet and dissolve in appropriate volume of water or TE buffer, Ph 8.0.

RNA isolation: It is essential to use correct amount of starting material to obtain optimal RNA Yield and purity. A maximum of 100 mg plant material or 1x107 cells can generally be processed.

1. β-Mercaptoethanol (β-ME) added to Buffer RLT or Buffer RLC before use. 10 μl β-ME per 1 ml Buffer RLT or Buffer RLC added. Buffer RLT or Buffer RLC containing β-ME can be stored at room temperature (15-25° C) for up to 1 month.

2. Buffer RPE is supplied as a concentrate. Before using for the first time, add 4 volumes of ethanol (96-100%) as indicated on the bottle to obtain a working solution.

3. 100 mg of plant material weighed. Weighing tissue is the most accurate way to determine the amount.

4. Immediately placed the weighed tissue in liquid nitrogen, and grounded thoroughly with a mortar and pestle. Tissue powder decanted and liquid nitrogen into an RNase-free, liquid-nitrogen cooled, 2 ml micro centrifuge tube. The liquid nitrogen allowed to evaporate, but do not allow the tissue to thaw. Proceed immediately to step3.

5. RNA in plant tissues is not protected until the tissues are flash-frozen in liquid nitrogen. Frozen tissues not allowed to threw during handling.

Discussion

Total RNA was isolated from wild type and *bmr* 6 and *bmr* 12 leaves and q RT-PCR was used to measure *bmr* 6 and *bmr* 12 leaves and q RT-PCR was used to measure *bmr* 6 and *bmr* 12 RNA expressions. Expression levels were determined using the ΔCT method and *bmr* 6, 12 gene expression was relativized against wild trait level. The presence of lignin reduces the quality of lignocellulosic biomass for biofuels (Figure 4). The reduced lignin content characteristic of brown midrib (*bmr*) mutants improves the efficiency of bioethanol conversion from biomass. bmr 6 gene encode cinnamyl alcohol dehydrogenase the final step of monolignol pathway. bmr 6 has overall negative correlation with lignin content for all the genotype studied (Table 1).

Atlas, Rox orange, Kansas collier, Tx 430, Early Hagaris sart, was studied for gene expression. Gene expression levels for *bmr*6, CAD 4,

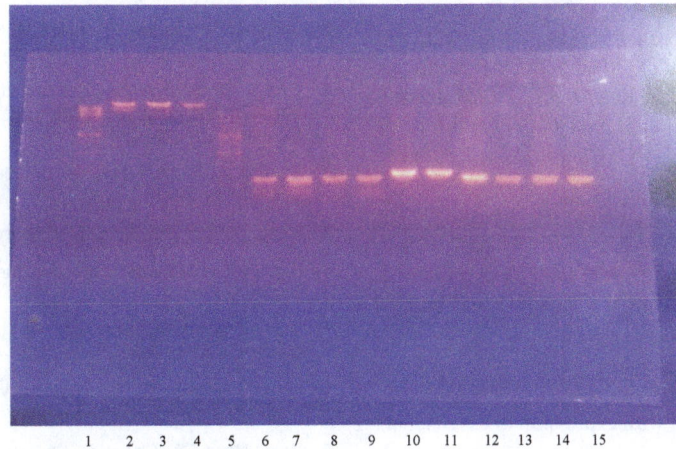

Figure 4: PCR Product at 52° C Showing Bands of Different Brown Mid rib Genotypes Atlas (234), Kansas Collier(67), Early Hagari Sart(89), Rox Orange(10,11) Tx 430 (12,13), Tx 631 (14,15).

	Wild trait	**Bmr 6**	**Bmr 12**
Atlas	√	√	√
Rox Orange	√	√	√
Kansas collier	√	√	√
Early Hagari		√	√
RTx 430		√	√
IS 1		√	√
IS 2		√	√
SSV 84	Sweet sorghum variety		

Table 1: Sorghum genotypes for gene expression studies.

	ATLAS	ATLASBMR12	ATLASBMR6	EHBMR12	EHBMR6	KC	KCBMR6	RO	ROBMR12	ROBMR6	Correlation with lignin content
BMR12	16.111289	28.640802	8.5741877	4211.1542	64.893407	34.059846	19.427118	1.3013419	3.3403517	23.425371	-0.169
BMR6	4039.6092	5873.4807	879.17101	0.2030631	0.0349152	7858.2917	885.28612	114.56321	770.68633	9877.9777	0.170
SbC3H4	0.0587202	1.7776854	0.0308198	0.0356489	0.9592641	0.2624292	0.0448111	0.0100268	0.0064343	0.0211969	-0.184
SbC3HF	3147.5204	4067.7069	709.17605	2957.167	116.97043	8364.1313	497.99933	130.68955	471.13608	8306.3561	-0.064
SbCAD4	3565.7751	5007.9346	861.07793	4182.0657	79.893155	8192	797.86453	128	820.29555	11113.303	-0.093
SbCCR1	43.713288	75.061437	17.267652	16	3.8370565	103.96831	7.8353624	1.4339552	10.556063	112.20553	-0.228
SbCOMT	3769.0886	5293.4772	995.99867	3373.4288	74.028044	14164.578	1243.3356	128	826.00116	9741.9847	-0.192
SbCOMT3	9.3826796	15.454981	6.4980192	8.1116758	2.7510836	40.224428	3.0525184	0.528509	4.0840485	41.069629	-0.149
SbHCT	1024	2368.8974	342.50945	2740.0756	37.530718	3125.7789	436.54906	51.625073	310.83389	2574.3634	-0.287
SbCAD2	0.5069797	0.0674518	0.0629347	0.0877778	0.0743254	0.4117955	0.1330463	0.0107464	0.0147822	0.3895823	-0.195
Lignin content	3.61	4.78	3.9	5.83	4.4	4.42	3.78	4.63	3.54	4.27	

Table 2: Correlation of *bmr* mutants with lignin content.

SBCAD2, *bmr*12, COMT3 COMT were compared for wild sorghum genotypes with their *bmr* 6 and *bmr* 12 counterparts. *bmr* 6 mutation in sorghum encodes cinnamyl alcohol dehydrogenase 2 (CAD2). bmr 6 has negative non-significant correlation with lignin content (-0.075) (Table 2).

Atlas

Gene expression level were compared with lignin content in atlas (WT), atlas *bmr* 6 and bmr 12. It was found that expression level not

much differs in gene *bmr*6, COMT and CAD 4. SBCAD2 gene activity suppressed. Lignin content of Atlas *bmr* 6 (3.9), *bmr* 12(4.78) and Atlas (WT) (3.61) was measured.

Kansas collier

Bmr 6 and CAD 4 were equally expressed in Kansas collier (WT) and bmr 6 genotypes. Lignin content for Kansas collier (WT) (4.42) and Kansas collier *bmr* 6 (3.78) was measured. Gene expression level was less in *bmr* 6 mutant for CAD 4, *bmr* 6 and COMT genes than wild version.

Rox orange

Bmr 6, COMT and CAD 4 were equally expressed for wild and mutant genotypes in Rox orange (WT). Maximum expression for *bmr*6, COMT and CAD was observed in Rox Orange *bmr* 6 mutant. Lignin content for Rox orange (WT) (4.63), *bmr* 6(4.27) and *bmr* 12 (3.54) was measured.

Early hagari

Gene SBCAD 2 and *bmr* 6 supressed in Early Hagari. Lignin content in Early Hagari *bmr* 6 (4.4) and *bmr* 12 (5.83) was measured. *Bmr* 6, COMT and CAD 4 were equally expressed *bmr* 6 and bmr 12 mutants.

RTx 430

Gene *bmr* 6 is negatively correlated with expression of CAD 4, *bmr* 12 and COMT3.

Conclusion

Brown midrib mutants *bmr* 6 and *bmr* 12 are having low lignin content could be used introgression breeding program. Expression level of *bmr* 6 gene was negatively correlated with lignin content in all *bmr* 6 mutants. Sorghum cultivars with reduced lignin can have a better way to increase second generation cellulosic ethanol production as compared with other crop residues and also improve process economics targeting higher conversion efficiency. Reduced lignin content will be highly beneficial for improving biomass conversion yield.

References

1. Bittinger TS, Cantrell RP, Axtell JD (1981) Allelism tests of the brown midrib mutants of sorghum. J Hered 172: 147-148.

2. Bout S, Vermerris W (2003) A candidate-gene approach to clone the sorghum brown midrib gene encoding caffeic acid O-methyltransferase. Mol Gen Genomics 269: 205-214.

3. Boerjan W, Ralph J, Baucher M (2003) Lignin biosynthesis. Annu Rev Plant Biol 54: 519-546.

4. Fleige S, Pfaffl MW (2006) Mol. Aspects Med 27: 126-139.

5. Higuchi T (2006) Look back over the studies of lignin biochemistry. J Wood Sci 52: 2-8.

6. Illumina (2010) Technical Note Real time PCR 9885 Towne Centre Drive, San Diego, CA 92121USA.

7. Nolan T, Hands RE, Ogunkolade W, Bustin SA (2006) Anal. Biochem 351: 308-310.

8. Perez-Novo CA, Claeys C, Speleman F, Van CP, Bachert C, et al. (2005) Biotechniques 39 p. 52.

9. Saballos A, Vermerris W, Rivera L, Ejeta G (2009) Allelic association, chemical characterization and saccharification properties of brown midrib mutants of sorghum (Sorghum bicolor (L.) Moench). BioEnergy Res 1: 193-204.

10. Sattler SE, Palmer NA, Saballos A, Greene AM, Xin Z, et al. (2012) Identification and characterization of four missense mutations in Brown midrib 12 (Bmr12), the caffeic O-methyltranferase (COMT) of sorghum. BioEnergy Res.

11. Gorthy S, Mayandi K, Faldu D, Dalal M (2012) Molecular characterization of allelic variation in spontaneous brown midrib mutants of sorghum (Sorghum bicolor (L.) Moench). Mol Breeding 31: 795-803.

12. Yuan JS, Tiller KH, Al-Ahmad H, Stewart NR, Stewart CN (2008) Plants to power: bioenergy to fuel the future. Trend Plant Sci 13: 421-429.

13

Differential Expression of Two Different *EeSTM* Genes of Leafy Spurge (*Euphorbia esula*) and Root-directed Expression from EeSTM Promoter in Leafy Spurge and *Arabidopsis*

Vijaya K. Varanasi[1], Wun S. Chao[1,2], James V. Anderson[1,2] and David P. Horvath[1,2]*

[1]Department of Plant Sciences, North Dakota State University, Fargo, ND 58102, USA
[2]United States Department of Agriculture, Agricultural Research Service, Biosciences Research Laboratory, P.O. Box 5674, State University Station, Fargo, ND 58105-5674, USA

Abstract

SHOOTMERISTEMLESS (*STM*) encodes a member of the class I *KNOX* homeodomain protein family that is required for meristem development and maintenance. Leafy spurge is a model perennial weed that produces adventitious meristems on its roots and hypocotyl. These buds are capable of displaying para-, endo- and ecodormancy. We have cloned two different full length cDNAs of *STM* (*EeSTM1* and *EeSTM4*) from leafy spurge that displays different tissue specific expression patterns. *EeSTM1* gene appears to be expressed only in young root and hypocotyl tissue. *EeSTM4* is co-expressed with *EeSTM1*, but it is also expressed highly in the shoot apical meristem and in mature roots. An *EeSTM* promoter was able to drive *GUS* expression in the roots, hypocotyl, and shoot apical meristem of *Arabidopsis*, whereas the *AtSTM* promoter only produced *GUS* expression in the shoot apical meristem. Expression analysis indicates that *EeSTM* expression is detectable in dormant adventitious buds but is up-regulated within three days following defoliation, a treatment that initiates bud regrowth. Seasonal changes in expression of *EeSTM* correlated to dormancy status in underground leafy spurge buds were observed.

Keywords: *STM*; Leafy spurge; Promoter; Gene Expression; GUS; Bud Dormancy; *Arabidopsis*

Introduction

Many plants propagate through the development and regulated growth of adventitious shoot buds. Additionally, many plants maintain a pool of viable but dormant shoot buds as a means to survive periods of growth inhibiting environmental conditions. In the perennial weed leafy spurge, such buds initiate and develop on the underground roots and/or crown region and are essentially fully formed shoot meristems. Recent studies on model organisms such as *Arabidopsis* (*Arabidopsis thaliana* Hyne) have provided a greater understanding of the genes required for proper shoot meristem initiation and development. One of the genes required for shoot development is *SHOOTMERSITEMLESS* (*STM*). *STM*, recognized as a developmentally important gene, was first cloned from maize [1] and then later from cauliflower (*Brassica oleracea* var. *botrytis*) and *Arabidopsis* [2]. *STM* is a member of the Class I *KNOX* homeobox gene family. *KNOX* encoded transcription factors are important in regulating meristem development and maintenance [3-5]. *STM* in particular is required for maintaining a supply of non-differentiated cells in the central zone of the shoot apical meristem (SAM).

STM is primarily expressed in the central zone of the SAM in *Arabidopsis*. STM also has a very important role in maintaining meristematic identity in cells of the peripheral zone that are not recruited to form organ primordia. *ROUGH SHEATH2* (*ZmRS2*) in maize and *ASYMMETRIC LEAVES1* (*AS1*) negatively regulate *KNOX* (*KNOTTED1*-like homeobox) genes such as *STM* in Arabidopsis [6-8]. *RS2* and *AS1*encode MYB-domain proteins that down-regulate *KNOX* genes in developing leaves. *ASYMMETRIC LEAVES2* (*AS2*), a *LATERAL ORGAN BOUNDARIES* (*LOB*) protein, along with *AS1* repress *KNOX* gene expression to promote leaf differentiation [9]. *CUP-SHAPED COTYLEDON* (*CUC*) initially induces *AtSTM* expression in the developing SAM [10]. There is evidence of both chromatin modification and transcription factor binding sites playing a role in regulation of *STM* in *Arabidopsis* and other species [11,12].

STM has been extensively studied in annual plant species such as *Arabidopsis* and likely orthologues have been identified in many plant species. However, very little is known about how *STM* is expressed in perennials that produce adventitious buds capable of various states of dormancy. It is unknown if *STM* is expressed in dormant buds, or if it is differentially regulated between growth and dormancy transitions. We have chosen to study *STM* expression in leafy spurge (*Euphorbia esula* L.). Leafy spurge is an herbaceous perennial weed, native to the central Europe, but which has become invasive in the Northern Great Plains of the US and Canada [13]. Genetically, leafy spurge is an auto-allo hexaploid [14].

Leafy spurge is a model perennial that is ideal for studying bud dormancy [15]. Leafy spurge produces large numbers of easily identifiable adventitious buds along the lateral roots (often referred to in the literature as root buds) and on the underground stem (referred to as crown buds) [16]. Leafy spurge root and crown buds display all three types of dormancy (para-, endo- and ecodormancy) [17,18]. After formation in early summer, root and crown buds enter a state of paradormancy. This paradormant state is maintained throughout the growing season by auxin and sugar signals produced in the growing

*Corresponding author: David Horvath, Ph.D, United States Department of Agriculture, Agricultural Research Service, Biosciences Research Laboratory, P.O. Box 5674, State University Station, Fargo, North Dakota 58105-5674, USA E-mail: david.horvath@ars.usda.gov

shoot [19,20]. Root and crown bud growth is reinitiated by separation of buds from the growing shoot. In the fall, root and crown buds of leafy spurge enter an endodormant state and will not grow even if plants are returned to growth-conducive conditions [18,20]. Prolonged periods of cold temperatures release the endodormant state [18]. The buds remain in an ecodormant state and will not develop into growing shoots until growth-conducive conditions return in the spring [18].

We previously isolated and characterized several genomic and cDNA clones of *STM* from the leafy spurge [21]. A comparison to other class I *KNOX* genes indicates that these EeSTM genes represented orthologues of *KN2* from poplar and likely were related to *ARBORKNOX1* of poplar and *STM* of *Arabidopsis*. 5′ Race indicated that the transcription initiation site is close to the start of translation and is conserved between Arabidopsis and leafy spurge. Putative *cis*-acting elements were identified in an *EeSTM* promoter, including several long elements that were conserved between leafy spurge and poplar that were independently identified as part of a larger conserved region in other species [12]. Another identified putative *cis*-acting element resembled a tuber-specific sucrose-responsive element which was hypothesized to play a role in the expression of *EeSTM* in root tissue [21].

We report here the differential expression of *EeSTM* during various dormancy transitions, and tissue specific expression of two different *EeSTM* genes (*EeSTM1* and *EeSTM4*). Additionally, reporter constructs indicate that EeSTM can drive tissue specific expression of a reporter gene in a heterologous system (*Arabidopsis*). We also test the activity of the tuber-specific sucrose-responsive element in driving *STM* expression outside the SAM in *Arabidopsis*. The results should prove beneficial in identifying molecular targets for regulating the development and growth of underground adventitious buds in leafy spurge and could result in better control strategies for leafy spurge and other perennial weeds with similar growth habits.

Materials and Methods

Plant material

Plants of *Euphorbia esula* used for the experiments were grown in cones (5 by 20 cm) with Sunshine mix (SUN GRO Horticulture, WA, USA) under greenhouse growth conditions as single stems under 16 h of natural and artificial lighting. All experiments were replicated at least twice for confirmation of results.

In order to study differential expression of *EeSTM* after paradormancy release, leafy spurge plants growing in the green house were defoliated by excising the aerial portions at the soil surface. Adventitious buds from 21-28 plants were then harvested, pooled, and stored in liquid nitrogen at various times following excision (0, 2, 4, 8, 16, 24, 48, 72 h). All the buds were harvested at the same time of the day to avoid artifacts caused by circadian effects. Real-Time PCR was done to quantify *EeSTM* expression (see below). Fold differences are shown relative to the 0 h time point. The experiment was repeated twice (two independent time course collections done several months apart).

To follow seasonal changes in *EeSTM* expression, adventitious root buds were harvested from leafy spurge from August 2001 through February 2005. Total RNA was extracted from these buds and cDNA synthesized. Semi-quantitative RT-PCR was done on seasonal cDNAs from 2001-02, 02-03 and 03-04. Serial dilutions of *EeSTM* containing plasmid was amplified in parallel as a concentration standard, and resulting gel was blotted and hybridized to ^{32}P labeled EeSTM probe. Label hybridizing to each band was quantified using a Packard Instant Imager. Real-Time PCR was performed on 2004-05 samples. The data

obtained was converted to fold differences in expression for each year using August cDNA as the calibrator.

For tissue specificity studies, leafy spurge seeds were germinated on moist filter paper in Petri dishes. The Petri dishes containing leafy spurge seeds were kept in an incubator at 37°C. The hypocotyl and radicle tissues were collected 8-10 d after emergence and elongation from the seed. Mature root tissue was collected from 4 month old leafy spurge plants. Adventitious buds were removed from the roots before harvest. Leafy spurge leaves were collected from middle 1/3rd of the stalk. Meristems were collected by excising the tip of the leafy spurge plant and removing most of the immature leaves from the shoot tip. All the leafy spurge tissues (hypocotyl, radicle, old root, leaf and meristem) were frozen in liquid nitrogen and stored at -80°C.

Real-Time PCR

Expression analysis of *EeSTM* was characterized by Real-Time PCR. For all analyses except the time course following paradormancy release, TaqMan chemistry was used in a 7300 Real-Time PCR system. A two-step RT-PCR assay was used for relative quantitation of *EeSTM* cDNA synthesized from time-course and seasonal RNA's. Total RNA was extracted from underground adventitious buds of leafy spurge at various times using the pine tree extraction method [22]. cDNA was synthesized from samples and trace amounts of ^{32}P-dCTP were included to help assess the quality and quantity of the resulting cDNA via gel electrophoresis, blotting, and quantification of the incorporated label. All samples were amplified in triplicate using ATG ATT GCT TTT GGA GAC AAC A (5′) and AGT CTG CAT TAT GAT GAT GAT G (3′) primer sequences. The FAM dye-labeled TaqMan probe sequence used for this study was TGG AGG AGG AGG AGG AGG AG. T_m for primers and probe was 60°C and 70°C. Various microarray experiments suggest that a large number of endogenous genes are differentially regulated during dormancy transitions, thus we chose to use an exogenous spike control (total Human cDNA) to normalize quantification of the target cDNA. Multiplex PCR was performed in each reaction tube containing 0.02 µg of total leafy spurge cDNA and 0.01 µg of total human cDNA with primers and probes corresponding to *EeSTM* and human *GAPDH*. A comparative CT (Threshold cycle) method was used in which the *EeSTM* cDNA was normalized to an exogenous (spiked control) reference (Human GAPDH). Serial dilutions (10 fold) of the *EeSTM* cDNA clone were included as external quantification controls. A negative control without *EeSTM* cDNA and a *GAPDH* control were also included in this expression study. All reactions were run in triplicate. For the paradormancy release time course, equal amounts of cDNA were used in each reaction and reactions were run in triplicate. SYBR green and the endogenous ROX reference dye were used to determine relative CT values. Numerous genes with expression patterns confirmed by northern hybridization (*HISTONE H3* being only one such gene shown) were run concurrently to ensure appropriate amplification and quantification of each sample.

Allelic discrimination assay

An allelic discrimination assay was conducted on cDNA synthesized from various leafy spurge tissues. All samples (hypocotyl, radicle, old root, leaf, and meristem) were amplified in triplicate using GCACTACTGGTGGTTCTTCTTCT (5′) and GCAGTGTTTGAGTTTGAGTTTGTGT (3′) primer sequences. Two probes were used, one specific to *EeSTM1* (VIC labeled – TGATCGTCAATAATCAT) allele and the other specific to *EeSTM4* allele (FAM labeled – ATCGTCAAGAATCAT). A standard 96-well reaction plate was used for loading samples and for the Pre-read,

Amplification, and Post-read steps of the assay. Both negative and positive controls were included. Analysis was done using the SDS software (v 1.4.1).

Transgene construction

Primers sequences CCGCTCTAGAAGTACTCTCG (5′) and CTCTCCTCAACAAATCCTAC (3′) were used to amplify 1,918 base pairs (bp) of promoter sequence (including 5′ UTR) from the *EeSTM* genomic clone (Genbank accession #EF636204). Likewise, primers sequences ACGAATCACTGTCCTTAACC (5′) and CTTCTCTTTCTCTCACTAG (3′) were used to amplify up to 1,934 bp of the *STM* promoter sequence (including 5′ UTR) from genomic DNA of Arabidopsis. For functional analysis of the putative sucrose-responsive element - *STK* (*STOREKEEPER*) binding site, primers were developed to replace the *STK* binding site with a sequence containing a BamHI restriction site. Two primer sets, CCGCTCTAGAAGTACTCTCG (5′), ATGGATCCAGTGGTCTCCCG (3′) and GTGGATCCCTTTGTGGCTAT (5′), CTCTCCTCAACAAATCCTAC (3′) were used to amplify two fragments from the *EeSTM* promoter sequences immediately flanking the *STK* element. The PCR amplified promoter fragments were digested with *Bam*HI and ligated to form a single contiguous promoter without the sucrose element. The amplified DNAs were gel purified and ligated into pBI121, replacing the cauliflower mosaic virus 35S promoter (CaMV 35S). Constructs were confirmed by PCR and used to transform *Arabidopsis*.

Transformation of *Arabidopsis*

Seeds of the wild-type *Arabidopsis* (Columbia) were sown in green house pots and grown until they begin flowering and ready for transformation. These flowering *Arabidopsis* plants were transformed by *Agrobacterium tumefaciens* (strain GV3101) harboring the binary vector pBI 121 containing *EeSTM* promoter constructs using the floral dip method [23]. Putative transgenic seeds (T_1) from 7-10 independent lines for each construct were collected and re-screened for kanamycin resistance (Kan^r). Several plants each from at least three lines from each construct that produced Kan^r seeds (T_2) were transplanted in potting mix and allowed to mature. Transformation of T_2 plants was confirmed by PCR using a GUS primer (5′ TTCATGACG ACCAAAGCCAGTAAAGT 3′) and one of the 5′ *STM* promoter primers from leafy spurge or *Arabidopsis*. The amplified DNA band confirmed for each construct by sequencing. Transgenic plants were allowed to grow and produce seeds (T_3). 20-40 T_3 seedlings each from at least three lines for each construct were GUS stained to characterize the activity of the leafy spurge and *Arabidopsis STM* promoters.

Results

EeSTM expression following release from paradormancy

Two full length cDNA (EF636205, EF636206) of *EeSTM* were obtained (*EeSTM1* and *EeSTM4*) [21]. Since both clones were similar, the initial analysis of expression was done using a primer and probe set designed to detect both transcripts. *AtSTM* is required for development and maintenance of the shoot apical meristem in Arabidopsis [3]. However, it is not clear if *EeSTM* was expressed in dormant meristems or if growth induction had any impact on *EeSTM* expression. We investigated *EeSTM* expression in paradormant and growing buds of leafy spurge using Real-Time PCR with primers that would amplify any known *EeSTM* genes (Figure 1). *EeSTM* is initially down-regulated shortly after release of buds from paradormancy by defoliation and then within 4 h, *EeSTM* expression steadily increases above initial levels for 2-3 d. Interestingly, increased *EeSTM* expression clearly precedes

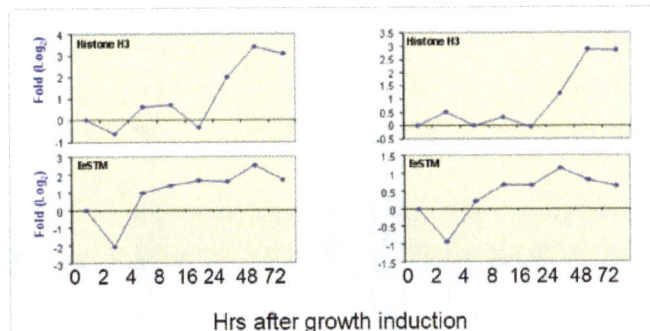

Figure 1: Real-Time PCR analysis of *EeSTM* and *HISTONE H3* expression in underground buds of leafy spurge. Crown buds were collected at various times following release of buds from paradormancy. Real-Time PCR analyses from two independently isolated time course series are shown. Expression pattern was relative to non-growing crown buds (time 0).

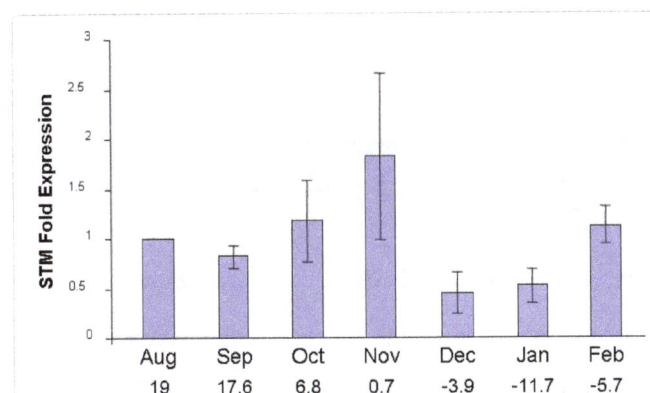

Figure 2: Seasonal expression of *EeSTM*. *EeSTM* expression was examined in crown buds collected monthly over 4 years. August cDNA was used as a calibrator for relative quantification of *EeSTM*. Error bars represent standard error. The average temperature during the week prior to bud harvest is noted below each sample point.

the induction of HISTONE H3 which was up-regulated only after 16 h following growth induction.

Seasonal expression of EeSTM

Leafy spurge buds can be maintained in a paradormant, endodormant or ecodormant state, and the transition from paradormancy to growth resulted in a change in *EeSTM* levels. Thus, it was of interest to determine if *EeSTM* expression was differentially regulated during the seasonal transition from paradormancy to endo- or ecodormancy. *EeSTM* expression was examined in buds collected monthly over several different years. Although there was a significant trend for *EeSTM* expression to drop in Dec. and Jan. and then increased in late winter, *EeSTM* expression fluctuated substantially between monthly samples depending on the year, particularly during Oct. and Nov. when buds were endodormant (Figure 2).

Tissue specific expression of *EeSTM*

The presence of possible root-specific *cis*-acting elements in the promoter of *EeSTM*, and the ability of leafy spurge to develop adventitious buds on its root system suggested the possibility that *EeSTM* could be expressed in root tissue. To determine which tissues *EeSTM* was expressed in leafy spurge, Real-Time PCR was used to analyze RNA from leaf, meristem, young hypocotyl, young root, and old root tissue (devoid of visible adventitious buds) (Figure 3). Surprisingly, these experiments indicated that maximum expression of *EeSTM* was

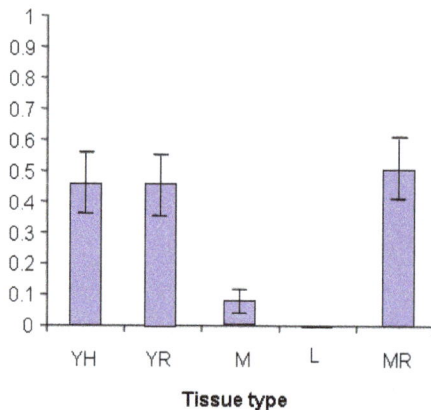

Figure 3: Relative tissue specific expression of *EeSTM* from three independent samples as quantified by Real-Time PCR. Error bars standard error from three different samples for each tissue. Tissues from young hypocotyl (YH), young root (YR), shoot meristem (M), leaves (L), and mature root (MR), were used for this study.

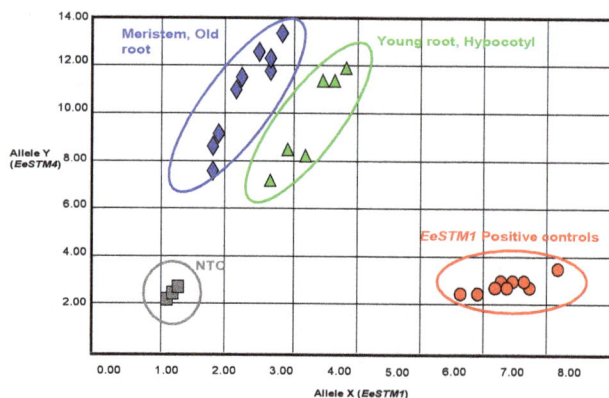

Figure 4: Real-Time Allelic Discrimination Assay of *EeSTM1* and *EeSTM4* alleles. Each dot represents a well in the 96-well reaction plate of Real-Time assay. Tissues containing only *EeSTM4* (diamonds) are aligned more towards the upper left corner of the plot, whereas control samples containing only *EeSTM1* (circles) are aligned more towards the lower right corner of the plot. Tissues containing both alleles (triangles) are approximately midway between the above two groupings. Negative controls (squares) grouped in the bottom left as expected. Data were plotted by using the absolute fluorescence of each reporter dye probe.

in old roots followed by young roots, young hypocotyl, meristem, in that order, but not in the leaf tissue as would be expected for an *STM* orthologue. Expression of *EeSTM* in old and young roots and hypocotyl of leafy spurge seedlings is consistent with the ability of leafy spurge to form adventitious meristems in these tissues.

Differential tissue specific expression of *EeSTM1* and *EeSTM4*

The differences in the coding sequences of *EeSTM1* and *EeSTM4* suggested they might be homeologous or paralogous genes rather than alleles of the same gene. Thus, it was possible that these genes might have different expression profiles. To test this hypothesis, an allelic discrimination assay was conducted to detect variation in expression patterns of *EeSTM1* and *EeSTM4* in various tissues of leafy spurge (Figure 4). *EeSTM4* expression was detected in all tissues tested, however surprisingly *EeSTM1* was only called as "present" in the young root and hypocotyl.

Heterologous activity of the *EeSTM* promoter in *Arabidopsis*

Screening of a leafy spurge genomic library only identified a single clone with defined promoter sequence [21]. The coding sequence from this clone was most similar, but not identical, to *EeSTM4*. The presence of conserved sequences in this *EeSTM* promoter suggested that it might be able to accurately direct expression of a reporter gene in a heterologous system such as Arabidopsis [21]. To test this hypothesis, the 2000 bp fragment of the *EeSTM* genomic clone including the 5'UTR sequence was used to drive *GUS* expression in transgenic Arabidopsis. The results indicated that this *EeSTM* promoter could direct *GUS* expression in the SAM of *Arabidopsis* (Figure 5). Interestingly, *GUS* expression was also observed in the hypocotyl and root pericycle of the transgenic plants carrying the *EeSTM:GUS* construct, whereas the construct containing a similar region of the *AtSTM* promoter only drove *GUS* expression in the meristem (Figure 5). This observation is consistent with expression of *EeSTM* in the root and hypocotyl observed in leafy spurge, and supports the hypothesis that elements within this *EeSTM* promoter might be needed for development of underground adventitious shoots in leafy spurge.

Root and hypocotyl expression is not dependent on *STK* element

Initial characterization of the *EeSTM* promoter identified a putative root-specific sucrose-responsive element (*STOREKEEPER* binding site- *STK*) that was unique to leafy spurge [21]. Consequently we hypothesized that this element might be involved in regulating expression of *EeSTM* in root tissue. We therefore deleted this element in our reporter gene and analyzed *GUS* expression in transgenic *Arabidopsis*. The results clearly indicated that *GUS* was still expressed in roots and hypocotyls of transgenic *Arabidopsis* (Figure 5). Thus, it appeared that deletion of this element had no effect on expression from the *EeSTM* promoter.

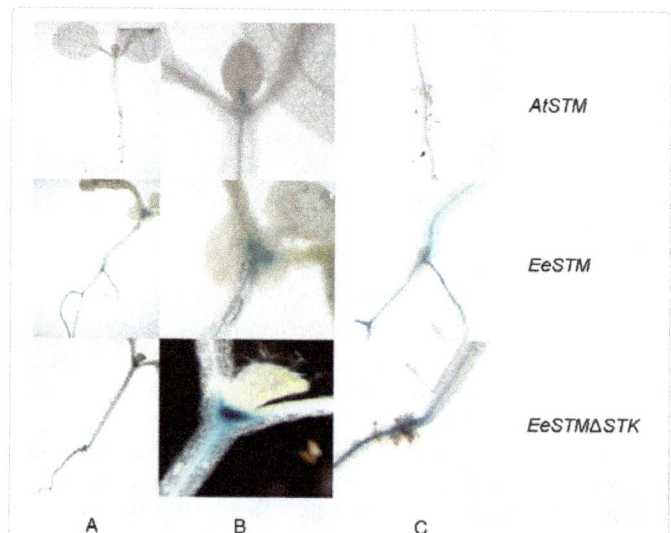

Figure 5: *GUS* expression in transgenic *Arabidopsis* on whole plant (A), close up of meristem region (B), or of hypoctyl-root boundry (C). Representative *GUS*-stained plants transformed with a construct in which the *AtSTM* promoter and 5'UTR are driving *GUS* expression (*AtSTM*), *EeSTM* promoter and 5'UTR driving *GUS* (*EeSTM*), or the *EeSTM* promoter and 5'UTR in which the STK binding site was replaced with a *Bam*HI restriction site (*EeSTMΔSTK*).

Discussion

Characterization of the *EeSTM* genes

The minor differences observed between cDNA and genomic clones of *EeSTM* suggest there may be several alleles of *EeSTM* in leafy spurge [21]. This was expected since leafy spurge is an auto-allo hexaploid [14]. Given that poplar has two *STM* orthologues and that poplar is in the same order as leafy spurge, there could be as many as 12 or more different copies (alleles, paralogues, and homeologues) of *STM* in leafy spurge. *EeSTM* genes form a separate clade with orthologous genes from other perennials such as poplar (*Populus alba*), and snapdragon (*Antirrhinum majus*). The separate clade also includes *AtSTM* from *Arabidopsis* but excludes *KNAT1* from Arabidopsis and the *STM-like* genes from monocots [21]. These results suggest that our *EeSTM* genes represent orthologues or paralogues of *AtSTM* rather than a different member of the Class I *KNOX* gene family.

Regulation of *EeSTM* in leafy spurge and *Arabidopsis*

The observation that the *EeSTM* promoter can direct expression in the roots of leafy spurge and *Arabidopsis* indicates that some conserved element directs root specific expression. Thus, the presence of a putative *STK* binding site in the promoter of the *EeSTM* gene was intriguing. *STOREKEEPER* (*STK*) is a conserved DNA-binding protein shown to recognize the 10 bp motif (GCTAAACAAT) in potato and regulate the expression of patatin [24] in potato tubers. It is interesting to note that this 10 bp motif is not found in the promoter of STM gene in *Arabidopsis* or poplar. *AtSTM* is not normally expressed in the roots of Arabidopsis, and poplar plants do not normally maintain a population of preformed adventitious shoot buds on their root system, although poplar can and does form adventitious shoots from its roots following damage to the root system or other shoot inducing treatments. Leafy spurge however forms viable shoot meristems on both the lateral roots and hypocotyl as early as 8 days following germination [16]. Thus it was hypothesized that this *STK* binding site might be important for bud formation on underground organs of leafy spurge. However, the results of these experiments (Figure 5) suggest that the putative *STK* binding site is not required for *EeSTM* expression in roots of leafy spurge. It will be interesting to determine what other elements are required for root specific expression of *EeSTM* genes. There are several related spurges that only form buds on hypocotyls or that do not form underground buds at all [25]. A comparison between *STM* gene regulations between these related species could identify elements required for root specific expression of *STM*.

In *Arabidopsis*, it appears that *AtSTM* is regulated by mechanisms involving chromatin remodeling [11], however there is also evidence that Myb-type transcription factors such as *AS1* and *CUC* also regulate *STM* expression [8,10]. Although some critical biochemical components of this chromatin remodeling are well characterized, there is insufficient understanding of the targeting and action of such proteins and protein complexes to allow specific manipulation of chromatin regulated genes. Consequently, coordinated cross-species comparisons of regulatory mechanisms could shed much needed light on such processes. Since the *EeSTM* promoter can accurately direct meristem specific expression in a heterologous system such as *Arabidopsis*, and since there appear to be both conserved and species specific elements within the *EeSTM* promoter, chimeric constructs can be produced with elements from leafy spurge and *Arabidopsis* to dissect the sequences responsible for regulating this critical developmental gene.

Tissue-specific expression of *EeSTM* and its alleles in leafy spurge

The Real-Time expression of *EeSTM* in hypocotyl, old root, young root (radicle), and meristem and not in the leaf tissue is consistent with the *EeSTM* promoter driven *GUS* expression in *Arabidopsis*. We detected expression of both *EeSTM1* and *EeSTM4* genes in the juvenile tissues (radicle and hypocotyl) but only *EeSTM4* in older tissues (meristem, and old root). This suggests that *EeSTM1* is responding to developmental specific signals that either turn this gene on in juvenile tissue or turn it off in older tissue. Other perennials such as poplar also have paralogous copies of *STM*. It would be interesting to determine if there is also tissue specific regulation of paralogous in these other species. It was also surprising to note that there was relatively higher expression of *EeSTM* in the root and hypocotyl tissues than in the meristem. The reason for this is unclear, and needs further investigation. However, the fact that a relatively limited number of cells in the SAM express *STM* in *Arabidopsis* and that our meristem samples contained a number of immature leaves and a small portion of the stem could account for the relatively lower concentration of *STM* mRNA in these samples. In situ hybridizations with *EeSTM* specific probes and antibodies on root and meristem sections will be needed to determine if indeed *EeSTM* was more highly expressed on a cell by cell basis between the various root cell types and cells in the central zone of the meristem, and to determine if the *EeSTM* gene expression is indicative of *EeSTM* protein accumulation.

EeSTM expression in relation to growth induction

There is differential regulation of other genes responsive to cell cycle in growing versus non-growing tissues (dormant buds) of leafy spurge [19,26]. Genes like *HISTONE H3*, *TUBULIN*, and *CYCLIN D3* are up-regulated in underground buds after induction of growth by defoliation [19]. The Real-Time PCR results indicate that induction of *EeSTM* follows a roughly similar pattern of expression observed for *CYCLIN D* or *HISTONE H3* but that the subsequent induction of *EeSTM* expression occurs earlier than the rise of *HISTONE H3* or *CYCLIN D3* expression (Figure 1) [26]. *AtSTM* up-regulates the cytokinin biosynthesis gene *isopentenyl transferase 7* (*AtIPT7*) and activates cytokinin response factor, resulting in increased synthesis of cytokinin and up-regulation of cytokinin-responsive *CYCLIN-D* [27]. *AtSTM* and *CYCLIN D3* are both up-regulated in *abnormal meristem mutant 1* (*amp 1*), indicating a link between growth and development [28]. Thus, our observations are consistent with studies that suggest *AtSTM* expression may impact cell division processes and *CYCLIN* expression. Our results also suggest that induction of developmental responsive signaling systems precede induction of growth *per se*. Additionally, the up-regulation of *EeSTM* following induction of active growth suggests that *EeSTM* may be needed to sustain shoot growth and development in adventitious buds following re-initiation of growth.

The initial drop in *EeSTM* expression is also intriguing. The leafy spurge plants could be responding to wounding stress following decapitation. Down-regulation of genes for a brief period might be one of the ways to conserve energy under stress and to readjust physiology before putting forth new shoots. Elevated levels of jasmonic acid and oxophytodienoic acid produced during wounding stress [29] have an antagonistic effect on the endogenous cytokinin levels [30]. *AtSTM* is positively regulated by cytokinins [28]. A decreased cytokinin level could therefore down-regulate *EeSTM* immediately after decapitation.

Expression of *EeSTM* during dormancy transitions

The relative down regulation of *EeSTM* in early winter as the buds transition into ecodormancy, and subsequent up-regulation as the buds move out of ecodormancy in early spring is consistent with similar expression patterns observed for various cell cycle genes such as *HISTONE H3* and *CYCLIN D3* [26]. Also, there was substantial variation in *EeSTM* expression during Oct. and Nov. when the buds are in endodormancy. This variation did not appear to correlate either with temperatures near the time of harvest or the depth of the dormant state of the plants as noted in parallel experiments (data not shown). This indicates that *EeSTM* expression during endodormancy may be regulated by other unknown factors.

Acknowledgements

We thank Laura Kelley for technical assistance and Cheryl Kimberlin for growing leafy spurge.

References

1. Hake S, Vollbrecht E, Freeling M (1989) Cloning *Knotted*, the dominant morphological mutant in maize using *Ds2* as a transposon tag. EMBO J 8: 15-22.

2. Granger CL, Callos JD, Medford JI (1996) Isolation of an *Arabidopsis* homologue of the maize homeobox *Knotted-1* gene. Plant Mol Biol 31: 373-378.

3. Barton MK, Poethig RS (1993) Formation of the shoot apical meristem in *Arabidopsis thaliana*: an analysis of development in the wild type and in the shoot meristemless mutant. Development 119: 823-831.

4. Long J, McConnell J, Fernandez A, Grbic V, Barton MK (1996) Developmental Genetics Of Shoot Apical MERISTEM FORMATION IN *ARABIDOPSIS*. Plant Physiol 111: 40002.

5. Groover AT, Mansfield SD, DiFazio SP, Dupper G, Fontana JR, et al. (2006) The *Populus* homeobox gene *ARBORKNOX1* reveals overlapping mechanisms regulating the shoot apical meristem and the vascular cambium. Plant Mol Biol 61: 917-932.

6. Schneeberger R, Tsiantis M, Freeling M, Langdale JA (1998) The *rough sheath2* gene negatively regulates homeobox gene expression during maize leaf development. Development 125: 2857-2865.

7. Tsiantis M, Schneeberger R, Golz JF, Freeling M, Langdale JA (1999) The maize rough sheath2 gene and leaf development programs in monocot and dicot plants. Science 284: 154-156.

8. Byrne ME, Simorowski J, Martienssen RA (2002) ASYMMETRIC LEAVES1 reveals knox gene redundancy in *Arabidopsis*. Development 129: 1957-1965.

9. Xu L, Xu Y, Dong A, Sun Y, Pi L, et al. (2003) Novel as1 and as2 defects in leaf adaxial-abaxial polarity reveal the requirement for ASYMMETRIC LEAVES1 and 2 and *ERECTA* functions in specifying leaf adaxial identity. Development 130: 4097-4107.

10. Kidner CA, Martienssen RA (2005) The role of ARGONAUTE1 (*AGO1*) in meristem formation and identity. Dev Biol 280: 504-517.

11. Phelps-Durr TL, Thomas J, Vahab P, Timmermans MC (2005) Maize *rough sheath2* and its *Arabidopsis* orthologue ASYMMETRIC LEAVES1 interact with HIRA, a predicted histone chaperone, to maintain *knox* gene silencing and determinacy during organogenesis. Plant Cell 17: 2886-2898.

12. Uchida N, Townsley B, Chung KH, Sinha N (2007) Regulation of SHOOT MERISTEMLESS genes via an upstream-conserved noncoding sequence coordinates leaf development. Proc Natl Acad Sci U S A 104: 15953-15958.

13. Chao W, Anderson J (2004) Euphorbia esula. In:Crop Protection Compendium. CAB International, Wallingford.

14. Schulzschaeffer J, Gerhardt S (1987) Cytotaxonomic Analysis of the *Euphorbia* Spp (Leafy Spurge) Complex. Biol Zent Bl 106: 429-438.

15. Chao WS, Horvath DP, Anderson JV, Foley ME (2005) Potential model weeds to study genomics, ecology, and physiology in the 21st century. Weed Science 53: 929-937.

16. Raju MVS (1975) EXPERIMENTAL STUDIES ON LEAFY SPURGE (EUPHORBIA ESULA L) .1. ONTOGENY AND DISTRIBUTION OF BUDS AND SHOOTS ON HYPOCOTYL. Bot Gaz 136: 254-261.

17. Lang GA, Early JD, Martin GC, Darnell RL (1987) Endodormancy, Paradormancy, and Ecodormancy - Physiological Terminology and Classification for Dormancy Research. HortScience 22: 371-377.

18. Anderson JV, Gesch RW, Jia Y, Chao WS, Horvath DP (2005) Seasonal shifts in dormancy status, carbohydrate metabolism, and related gene expression in crown buds of leafy spurge. Plant Cell Environ 28: 1567-1578.

19. Horvath DP, Chao WS, Anderson JV (2002) Molecular analysis of signals controlling dormancy and growth in underground adventitious buds of leafy spurge. Plant Physiol 128: 1439-1446.

20. Chao WS, Serpe MD, Anderson JV, Gesch RW, Horvath DP (2006) Sugars, hormones, and environment affect the dormancy status in underground adventitious buds of leafy spurge (*Euphorbia esula*). Weed Science 54: 59-68.

21. Varanasi V, Slotta T, Horvath D (2008) Cloning and characterization of a critical meristem developmental gene (*EeSTM*) from leafy spurge (*Euphorbia esula*). Weed Science 56: 490-495.

22. Chang S, Puryear J, Cairney J (1993) A simple and efficient method for isolating RNA from pine trees. Plant Mol Biol Rep 11: 113-116.

23. Clough SJ, Bent AF (1998) Floral dip: a simplified method for *Agrobacterium*-mediated transformation of *Arabidopsis thaliana*. Plant J 16: 735-743.

24. Zourelidou M, de Torres-Zabala M, Smith C, Bevan MW (2002) *STOREKEEPER* defines a new class of plant-specific DNA-binding proteins and is a putative regulator of patatin expression. Plant J 30: 489-497.

25. Klimesova J, Martinkova J (2004) Intermediate growth forms as a model for the study of plant clonality functioning: an example with root sprouters. Evolutionary Ecology 18: 669-681.

26. Horvath DP, Anderson JV, Jia Y, Chao WS (2005) Cloning, characterization, and expression of growth regulator *CYCLIN D3-2* in leafy spurge (*Euphorbia esula*). Weed Science 53: 431-437.

27. Jasinski S, Piazza P, Craft J, Hay A, Woolley L, et al. (2005) *KNOX* action in Arabidopsis is mediated by coordinate regulation of cytokinin and gibberellin activities. Curr Biol 15: 1560-1565.

28. Rupp HM, Frank M, Werner T, Strnad M, Schmulling T (1999) Increased steady state mRNA levels of the *STM* and *KNAT1* homeobox genes in cytokinin overproducing *Arabidopsis thaliana* indicate a role for cytokinins in the shoot apical meristem. Plant J 18: 557-563.

29. Reymond P, Farmer EE (1998) Jasmonate and salicylate as global signals for defense gene expression. Curr Opin Plant Biol 1: 404-411.

30. Ananieva K, Malbeck J, Kaminek M, Van Staden J (2004) Methyl jasmonate down-regulates endogenous cytokinin levels in cotyledons of *Cucurbita pepo* (zucchini) seedlings. Physiol Plant 122: 496-503.

14

Bioactivity Effect of *Piper nigrum* L. and J*atropha curcas* L. Extracts Against *Corcyra cephalonica* [Stainton]

Mousa Khani[1,2]*, Rita Muhamad Awang[2], Dzolkhifli Omar[2] and Mawardi Rahmani[3]

[1]*Department of Cultivation and Development of medicinal plants, Iranian Institute of Medicinal Plants, [ACECR], P.O. Box: 13145-1446, Tehran, Iran*
[2]*Department of Plant Protection, Faculty of Agriculture, University Putra Malaysia, 43400 UPM Serdang, Selangor, Malaysia*
[3]*Department of Chemistry, Faculty of Science, University Putra Malaysia, 43400 UPM Serdang, Selangor, Malaysia*

Abstract

Petroleum ether extract of black pepper [*Piper nigrum*] and physic nut [*Jatropha curcas*] were shown to have insecticidal efficacies against rice moth [*Corcyra cephalonic*) [Stainton]. The *C. cephalonica* larvae [16 day old] were shown to have similarities susceptibility to petroleum ether extract of *P. nigrum* and *J. curcas* with LC_{50} values of 12.52 and 13.22 µL/mL, respectively. In a bioassay using no-choice tests, the parameters used to evaluate antifeedant activity were Relative Growth Rate [RGR]; Relative Consumption Rate [RCR], efficiency on conversion of ingested food [ECI] and Feeding Deterrence Indices [FDI]. Both extracts showed high bioactivity at all doses against *C. cephalonica* larvae and antifeedant action was increased with increasing plant extract concentrations. The petroleum ether extract of *P. nigrum* and *J. curcas* showed strong inhibition on egg hatchability and adult emergence of *C. cephalonica* at the lowest concentration. Based on the results of this study petroleum ether extracts of *P. nigrum* and *J. curcas* could be used in IPM program for rice moth.

Keywords: Antifeedant; Feeding deterrence; *Jatropha curcas*; *Piper nigrum*; *Corcyra cephalonica*; Egg hatchability; Adult emergence

Introduction

The rice moth, *Corcyra cephalonica* [St.] is the major and important pests of stored commodities in the tropics [1], Asia, South America and Africa [2,3] . The larvae feed on rice, corn, cocoa, chocolate, dried fruit, biscuits, coffee and other seeds. The rice moth is a worldwide pest of stored foodstuffs. Control of these insects generally requires the use of chemical insecticides that are toxic to humans and domestic animals and harmful to the environment [4]. In addition the larvae while feeding, leaving silken threads and contaminate the grain by producing dense webbing containing their fecal material and cast skins. The webbing formed is noticeably dense and tough adding the damage caused [5,2]. For the control of stored produced insects, it is frequently more safe to use plant materials with insecticidal, antifeedant or repellent properties than the use synthetic insecticides [6,2] and several plant species have been noted for these purposes in pest management. However, no similar work has been carried out with rice moth, *C. cephalonica*. So black pepper [*Piper nigrum*] and physic nut [*Jatropha curcas*] are some of these plants that may possess insecticidal or antifeedant properties [7,8,9]. In this study we evaluated the efficacy of petroleum ether extract of *P. nigrum* and *J. curcas* against rice moth [*C. cephalonica*] larvae.

Materials and Methods

Insects rearing

The rice moth [*C. cephalonica*] larvae were obtained from an entomology laboratory stock culture of University Putra Malaysia [UPM] and reared on medium including finely ground rice and maize flour in the ratio 1:1 [w/w] at 27 ± 1°C, 75 ± 5% RH with a 12:12 h light : dark cycle as method of [2] with some modifications. The food media were sterilized in an autoclave before experimentation. The subcultures and the tests were set up under the same conditions. *C. cephalonica* larvae [16 ± 1 days old] was used to the experiments. All the cultures in plastic containers [28×18×18 cm] were held in trays with guards submerged in water to prevent insects from crawling into them [2].

Plant materials

Fruits of *P. nigrum* were supplied from Sarawak State in the North-East of Malaysia and seeds of *J. curcas* were prepared from botanical garden in University campus. Plant extracts of *P. nigrum* and *J. curcas* were prepared by the percolation method described by [10]. Fruits of *P. nigrum* and seeds of *J. curcas* were ground to powder using a grinder prior to oil extraction. The powders [300 g] were soaked in methanol [95%] for 48 h, filtered and the residues were extracted afterwards. An equal volume of water added to the crude extract and extraction were done by petroleum ether [11]. The prepared extracts were concentrated by rotary evaporator [40°C] and stored at 4°C for further use.

Toxicity of plant extracts

Laboratory bioassays were conducted to evaluate toxicity of petroleum ether extracts of *P. nigrum* and *J. curcas* against *C. cephalonica*. To prepare stock solutions [w/v] of each extract, 10 gram of crude extract was dissolved in 100 ml of respective solvent. Solutions were diluted using the formula $C_1V_1 = C_2V_2$ [12], where C_1 and C_2 are concentration of 1st and 2nd solution, V_1 and V_2 are volume of 1st and 2nd solution, respectively. For evaluating efficacy of plant extracts, rice kernels were treated with 2, 4, 6, 8, 10% of prepared dilutions with n-Hexane, then were shaken to ensure uniform coverage of extracts on rice kernels. After shaking, treated rice in conical flask was placed on filter paper to evaporate the solvent. Then the rice kernels were divided five parts by electric balance [each part 5 g]. After that each part was placed in the Petri dishes and 20 larvae [16 ± 1 days old] of *C.*

*Corresponding author: Mousa Khani, Department of Plant Protection, University Putra Malaysia, Selangor, Malaysia, E-mail: khanimousa@yahoo.com

cephalonica were introduced into each Petri dish. Infested Petri dishes were incubated at 27 ± 1°C, 75 ± 5% RH with a 12:12 h light: dark cycle. Petri dishes were checked out after 72 h to count the number of dead larvae for evaluating toxicity and followed 7 days to evaluate total mortality of *C. cephalonica* larvae. Using Polo-Plus Software [13] to evaluate LC_{50} values and percentage larvae mortality was calculated by probit analysis [14].

Evaluation antifeedant

No-choice test as described by [15] and [16] was carried out to determine antifeedant activity of plant extracts with some modifications. One mL of prepared concentrations of 1, 2, 3, 4, 5% [or 2, 4, 6, 8, 10 µL of plant extracts] or 1 mL solvent alone as control were applied on to 5 gram rice kernels against larvae [16 ± 1 days old] of *C. cephalonica*. After evaporating the solvent, the rice kernels were placed back in Petri dishes [5 cm dia.]. Then ten group-weighted larvae of *C. cephalonica* [starved for 24 hours] were transferred to each pre-weighed rice kernels in Petri dishes. After 3 days of feeding under laboratory conditions [27 ± 1 °C, 75 ± 5% RH with a 12:12 h light: dark cycle], the rice kernels and live insects were re-weighed, and mortality of insects, if any, was recorded. Five replicates of each treatment were prepared including the control. Weight loss and nutritional indices were calculated as described by [17,15].

The following parameters were calculated: Weight Loss [%WL] = [IW-FW]×100/IW, where the IW is the initial weight and FW is the final weight; Relative Growth Rate [RGR] = [A – B] / [B × day], where A is weight of live larvae on the third day [mg] / no. of live larvae on the third day, B is original weight of live larvae [mg] / original no. of larvae; Relative Consumption Rate [RCR] = D / [B × day], where D is biomass ingested [mg] / no. of live larvae on the third day; Efficiency of Conversion of Ingested food [ECI] [%] = [RGR / RCR] × 100. The antifeedant action was evaluated by calculating the Feeding Deterrence Index [FDI] by formula, FDI [%] = [C – T] / C × 100, where C is the consumption of control rice kernels, and T is the consumption of treated rice kernels [18,19]. The mortality data were adjusted for mortality in the control using Abbott's formula and expressed as percentages [20]. Results of nutritional studies were expressed as means ± SEM and the significance of mean difference between treatments and control was assessed using the analysis of variance procedure at 5% probability level with individual pair wise comparisons with Tukey's test using the SAS v. 9.1.3 software package in Microsoft Windows 7 [21].

Effect of plant extracts on egg hatchability

Five grams treated rice with the petroleum ether extracts of *P. nigrum* and *J. curcas* at the 1, 2, 3, 4 and 5% dose level were placed in plastic Petri dishes [5 cm dia.] and twenty uncollapsed eggs [0-24 h old] of *C. cephalonica* were introduced into each Petri dish using a fine brush. Five replicates of each treatment and untreated rice kernel were set up. After one week, Petri dishes were checked for eggs that failed to hatch [3] and percentage egg hatch was calculated. Results were analyzed by ANOVA and mean values were adjusted by Tukey's comparison test.

Evaluation adult emergence

Rice kernels [5 g] were treated with plant extracts at 1, 2, 3, 4, 5% and solvent only as a control. Then impregnated rice were placed in 25 ml Petri dishes and 20 larvae [3rd instar] of *C. cephalonica* were allowed to feed on to produce adults. Number of adults emerged [2] was recorded at the end of the experiment. Five replicates of each treatment and untreated rice kernel were set up. Data were analyzed by using

Treatment	Slope ± SEM	(x²)	Df[1]	LC₅₀ (µL/ml) (Min-Max)[2]
P. nigrum (fruits)	2.26 ± 0.35	2.19	3	12.52 (11.23-16.08)
J. curcas (seeds)	2.26 ± 0.35	2.19	3	13.22 (11.23-16.08)

1- Degree of freedom
2- 95% lower and upper fiducial limits are shown in parenthesis
3- No mortality

Table 1: LC_{50} value of petroleum ether extracts of Piper *nigrum* and *Jatropha curcas* against Corcyra cephalonica larvae at 72 hours after commencement of exposure.

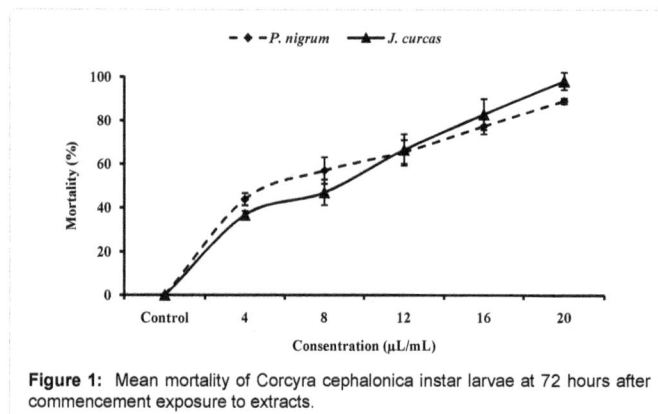

Figure 1: Mean mortality of Corcyra cephalonica instar larvae at 72 hours after commencement exposure to extracts.

analysis of variance [ANOVA] and Tukey's Multiple Comparison Test was used for means and comparison of means.

Results

Efficacy of plant materials against *Corcyra cephalonica*

Experiments were carried out to evaluate insecticidal activities of plant extracts from fruits of *P. nigrum* and seeds of *J. curcas* against *C. cephalonica* larvae. The medium lethal concentration [LC_{50}] of plant extracts on the *C. cephalonica* larvae 72 hours from commencement of exposure are presented in Table 1. Results also showed that the petroleum ether extracts from fruits of *P. nigrum* and seeds of *J. curcas* had LC_{50} values of 12.5 and 13.2 µL/mL against *C. cephalonica* larvae, respectively. Overlapping of the 95% fiducial limits showed that differences between the applied extracts against *C. cephalonica* larvae were non-significant. Petroleum ether extracts of *P. nigrum* and *J. curcas* at 12 µL/mL dose level caused mortality against *C. cephalonica* larvae with values of 65.6 and 66.5%, respectively. While, high mortality was observed at 20 µL/ml dose level with values of 88.9 and 98.0%, respectively (Figure 1).

Evaluation antifeedant efficacy of plant materials against *C. cephalonica*

The results showed reduction in RGR, RCR and ECI of *C. cephalonica* larvae when the rice kernels treated with *P. nigrum* and *J. curcas* extracts and significantly reduced the RGR of *C. cephalonica* larvae at the lowest concentration dose level of 2 µL/g rice kernels (Table 2). Feeding Deterrence Indices [FDI] showed that the plant extracts had antifeedant action against *C. cephalonica* larvae at all concentrations. Also, petroleum ether extracts of *P. nigrum* and *J. curcas* showed antifeedant action against *C. cephalonica* larvae at a concentration of 6 µL/g rice kernels, with 55.87 and 48.08% reduction in feeding, respectively (Table 2). All the treatments showed significant

Extract	Concentration (µL/g rice kernels)	RGR (mean ± SEM) (mg/mg/day)	RCR (mean ± SEM) (mg/mg/day)	ECI (mean ± SEM) (%)	Mortality (%)	FDI (mean ± SEM) (%)	Weight loss (mean ± SEM) (%)
Petroleum ether extract of P. nigrum	0	0.071 ± 0.004 a	0.448 ± 0.037 a	16.41 ± 2.01 a	0	-	1.65 ± 0.13 a
	2	0.041 ± 0.003 bc	0.429 ± 0.057 a	10.45 ± 1.92 b	2	13.90 ± 9.98 c	1.42 ± 0.22 ab
	4	0.030 ± 0.008 c	0.318 ± 0.103 ab	11.36 ± 1.61 ab	4	39.30 ± 20.44 b	0.99 ± 0.32 ab
	6	0.028 ± 0.003 c	0.256 ± 0.043 b	11.74 ± 1.26 ab	14	55.87 ± 5.61 ab	0.72 ± 0.10 ab
	8	0.016 ± 0.007 cd	0.196 ± 0.106 bc	9.30 ± 3.95 b	16	67.70 ± 16.82 a	0.58 ± 0.32 b
	10	0.013 ± 0.014 cd	0.192 ± 0.045 bc	6.72 ± 6.10 bc	18	60.21 ± 32.53 a	0.52 ± 0.39 b
Petroleum ether extract of J. curcas	0	0.078 ± 0.007 a	0.482 ± 0.019 a	16.24 ± 1.48 a	0	-	1.70 ± 0.08 a
	2	0.043 ± 0.009 bc	0.367 ± 0.073 ab	12.49 ± 2.14 ab	4	24.78 ± 12.23 bc	1.36 ± 0.29 ab
	4	0.034 ± 0.001 c	0.302 ± 0.036 ab	11.83 ± 1.24 ab	8	39.77 ± 11.94 b	0.99 ± 0.15 ab
	6	0.032 ± 0.011 c	0.290 ± 0.058 b	10.66 ± 2.26 ab	8	48.08 ± 9.14 ab	0.89 ± 0.16 ab
	8	0.007 ± 0.002 d	0.183 ± 0.037 bc	3.95 ± 1.33 bc	8	64.70 ± 7.71 a	0.59 ± 0.13 b
	10	0.002 ± 0.002 d	0.148 ± 0.054 bc	0.88 ± 1.32 c	14	74.96 ± 7.63 a	0.44 ± 0.14 b

Each datum represents the mean of five replicates
RGR, Relative Growth Rate; RCR, Relative Consumption Rate; ECI, Efficiency of Conversion of Ingested food; FDI, Feeding Deterrence Index (Huang and Ho, 1998)
Means within columns followed by the same letters are not significantly different (P<0.05; Tukey's Comparison Test)
Table 2: Nutritional and feeding deterrence indices of Corcyra cephalonica larvae fed on rice kernels treated with petroleum ether extracts of P. nigrum and J. curcas at sublethal concentrations.

weight loss in rice kernels because of feeding by C. cephalonica larvae. Also the weight loss [%] due to C. cephalonica larvae feeding on treated rice kernels with petroleum ether extracts of P. nigrum and J. curcas were significantly different at dose level of 8 µL/g compared with the control (Table 2). Petroleum ether extract of P. nigrum and J. curcas were the most effective treatment. This agreed with [22] who showed that neem leaf powder, nochi leaf powder and neem oil are effective in controlling the rice moth [C. cephalonica] in groundnut kernels and pods. [23] Showed that Acarus calamus essential oils are effective on C. cephalonica larvae. In another study, [24] evaluated the effect of neem oil volatiles by confining the adults and larvae of C. cephalonica in a chamber containing neem oil. They recorded a marked decline in the reproductive potential and egg hatchability.

Effect of plant materials on egg hatchability

Results in Table 3 showed that egg hatchability was reduced by petroleum ether extract of P. nigrum and J. curcas at the lowest dose level [2 µL/mL] with values of 59 and 58%, respectively [Table 3]. Significant reductions in egg hatchability revealed the harmful effects of petroleum ether extracts of P. nigrum and J. curcas towards C. cephalonica eggs. This observation is in agreement with that of [25] who reported that food treated with Jatropha gossypifolia seed extract strongly inhibit the fecundity of Tribolium castaneum compared with Tribolium confusum at doses of 8000 and 16000 ppm. Bunker and Bhargava determined the effect of vegetable oils on the eggs of C. cephalonica. The treatments were castor bean [Ricinus communis], coconut [Cocos nucifera], groundnut [Arachis hypogaea], Indian mustard [Brassica juncea], sesame [Sesamum indicum], and sunflower [Helianthus annuus] oils at 0.5, 1.0, 2.0, 3.0 and 5.0%. All the vegetable oil concentrations were significantly superior to the control in reducing egg hatchability. In this study the percentage of egg hatch inhibition in all the treatments increased with an increase in concentration.

Effect of plant materials on adult produce

The number of adults of C. cephalonica that emerged from the treated rice kernels decreased with increasing concentration of plant extracts. All the treatments strongly suppressed adult emergence of C. cephalonica larvae at the lowest concentrations with 2 µL/mL dosage (Table 4). The mean percent of F1 adults of C. cephalonica that emerged from the treated rice kernels strongly decreased even at the lowest concentrations of petroleum ether extracts from P. nigrum and J. curcas.

These observations are in agreement with [26] who showed that the J. curcas oil at the lowest concentration of 0.5% suppressed adult emergence in Callosobruchus maculatus, also indicating that the oil had ovicidal activity. The adults of C. cephalonica that emerged from rice kernels that treated with petroleum ether extract of P. nigrum and J. curcas at the lowest concentration [2 µL/mL] were 3 and 8% compare with untreated rice kernels with 86 and 85%, respectively. In this regards, [22] reported on the efficacy of a range of plant products, including neem leaf powder and edible oil in protecting stored groundnuts against the rice moth [C. cephalonica] and noted that even though all the plant products and edible oils afforded protection, neem leaf powder and neem oil were most effective Allotey [2] studied on groundnut kernels treated with Citrus sinensis at dosage of 0.5 g and 2.0 g per 40 g of legume seeds, but differences between botanicals were not significant when all dose levels were considered. They reported that Eichhornia crassipes suppressed the emergence of C. cephalonica to a greater extent at dosage of 0.5, 1.0 and 2.0 g than Citrus sinensis and Chromolaena odorata.

Discussion

The Piperaceae family has been reported to have insecticidal activities due to presence of many potential phyto-chemicals. The P. nigrum extracts offer a unique and beneficial source of bio-pesticide material for the control of insect pests on a small scale [9,27]. The major components of P. nigrum fruit extracts such as piperine, caryophyllene and limonene are reported as having insecticidal properties. Many insecticidal components of plant extracts are mainly monoterpenes such as limonene which have been shown to be toxic to Tribolium castaneum [28]. Early studies on extracts of P. nigrum seeds had indicated that piperine and other active piperamides were responsible for the toxicity of the extracts to the adzuki bean weevil, Callosobruchus chinensis L. [28,29]. Piper nigrum seed oil formulations were found to be effective for protecting stored wheat from both stored grain pests, Sitophilus oryzae [L.] and Rhyzopertha dominica [F.], for more than 30 days, at concentrations of 100 mg/L and higher [30]. Ethyl acetate extracts of P. nigrum seeds were also reported to be toxic to Lepidopteran and Hymenopteran herbivorous insects such as eastern tent caterpillar Malacosoma americanum F., forest tent caterpillar Malacosoma disstria Hubner, pine sawfly Diprion similis Hartig, gypsy moth Lymantria dispar [L.], spruce budworm Choristoneura fumiferana [Clemens], European pine sawfly Neodiprion sertifer Geoffroy and spindle ermine moth larvae Yponomeuta cagnagella [Hubner] [31,32].

Extract	Dosage (μL/mL)	Egg hatching % (Mean ± SEM)*
Petroleum ether extract of P. nigrum	Control	93 ± 2.55 a
	2	59 ± 4.00 b
	4	49 ± 1.87 bc
	6	27 ± 2.55 e
	8	22 ± 7.00 ef
	10	9 ± 2.92 f
Petroleum ether extract of J. curcas	Control	91 ± 2.92 a
	2	58 ± 2.55 bc
	4	48 ± 3.39 bc
	6	32 ± 2.55 de
	8	20 ± 2.24 ef
	10	8 ± 1.22 f

*Data are average of 5 replicates. Means within columns followed by the same letters are not significantly different (P<0.05; Tukey's multiple comparison test)

Table 3: Effect of petroleum ether extracts of Piper nigrum and Jatropha curcas on egg hatchability in Corcyra cephalonica.

Extract	Dosage (μL/mL)	Adult emergence % (Mean ± SEM)*
Petroleum ether extract of P. nigrum	Control	86 ± 1.87 a
	2	3 ± 1.22 bc
	4	1 ± 1.00 bc
	6	0 ± 0.00 c
	8	0 ± 0.00 c
	10	0 ± 0.00 c
Petroleum ether extract of J. curcas	Control	85 ± 2.74 a
	2	8 ± 1.22 b
	4	5 ± 1.58 bc
	6	3 ± 2.00 bc
	8	2 ± 1.22 bc
	10	0 ± 0.00 c

* Data are average of 5 replicates. Means within columns followed by the same letters are not significantly different (P<0.05; Tukey's multiple comparison test)

Table 4: *Corcyra cephalonica* adult emergence from rice kernels treated with petroleum ether extracts of *Piper nigrum and Jatropha curcas*.

The toxic effect of *P. nigrum* was also reported against some test insects. *Piper nigrum* was shown to be most toxic to *Callosobruchus chinensis, Acanthoscelides obtectus, C. cephalonica, Ephestia cautella* Hubn., followed by *Oryzaephilus surinamensis* [L.], *Sitophilus zeamais Mosteh, Rhyzopertha dominica* [Fab.] and *Tribolium castaneum* Herbst [Ponce de Leon, 1983]. The high toxicity effects of *P. nigrum* extracts against *C. cephalonica* larvae are attributed to the presence of high concentrations of well-known toxic components such as caryophyllene and piperine.

In the case of *Jatropha curcas*, it is a multipurpose plant with many properties and considerable insecticidal potential [33]. Different parts of *J. curcas* contain the curcin and phorbol ester which are toxic alkaloids that inhibit animals from feeding on it [34]. The insecticidal and inhibition of progeny emergence activities of oil extracted from seeds of *J. curcas* has been reported by earlier researchers against several insect pests [26,35-46]. This study is agreement with earlier studies, the petroleum ether extract of *J. curcas* seeds showed insecticidal activity against *C. cephalonica* larvae. These effects are attributed to the presence of oleic and linoleic acids which are well known toxic components.

The LC_{50} values of petroleum ether extract of *J. curcas* seeds at 72 hours after exposure against *C. cephalonica* larvae were 13.2μL/mL. [37] Investigated toxicity of petroleum ether extracts of three different sources of *J. curcas* seeds, and noted LC_{50} values of 8.0, 3.1 and 24.4 g/L against *S. oryzae*, respectively. [46] Studied larvicidal activity of

crude methanol leaf extracts from *J. curcas* and noted LC_{50} of 92.1 ppm against 3rd instar larvae of *Anopheles arabiensis*. In similar study, [45] reported high toxic activity of methanol leaf extracts of *J. curcas* against *Culex quinquefasciatus* Say from first to fourth instar larvae at dose levels of 1.2, 1.3, 1.4 and 1.5%. [47] Reported toxic effects of petroleum ether extracts of *J. curcas* against larvae of *Culex quinquefasciatus* Say [LC_{50}=11.3 ppm] and *Aedes aegypti* [LC_{50}=8.8 ppm]. The petroleum ether extracts of *J. curcas* were reported to be more efficient than other tested plant extracts. Mortality percent also was highly significant for petroleum ether extract of *P. nigrum* and *J. curcas*. According to Chauhan the extracts of *Croton sparsiflorus* [LC_{50}=0.073], *Anona squamosa* [LC_{50}=0.278] and *Acorus calamus* [LC_{50}=1.072] showed potential as safe insecticide. Pathak and Tiwari reported at 0.25% dose level of neem leaf larval mortality 17 ± 1.78%, while 100% mortality they reported at 3.5% dose level of neem leaf. Jadhav reported LC_{50} values of *Annona squamosa* [14.36], *Tephrosia purpurea* [38.05] and *Acorus calamus* [33.11] after 72 h. Some possible reasons for these differences are insect strains, test conditions or test material. Larval mortality was increased with the increase of plant extracts concentrations. These effects are attributed to some well known toxic compounds such as piperine, caryophyllene, limonene, α-pinene, and β-pinene in *P. nigrum*, and oleic acid and linoleic acid in *Jatropha curcas*

According to many authors, any substance that reduces food consumption by an insect can be considered as an antifeedant. Isman [18] defined antifeedants as behavior-modifying substances that deter through a direct action on taste organs in insects. This definition excludes chemicals that suppress feeding by acting on the central nervous system, or a substance that has sublethal toxicity to the insect. Feeding inhibition in insect pests is the most important in the search for new and safer methods for pest control in stored grains. The high antifeedant effects of *P. nigrum* powder against *Callosobruchus maculatus* at 25 and 30 g/kg on black gram were reported by [17]. They attributed the antifeedant properties of *P. nigrum* to the piperine that killed the beetles earlier. Pepper seed extracts had also been shown to deterred Lily leaf beetles *Liliocerus lilii* Scopoli and *Acalymma vittatum* from damaging leaves of lily and cucumber plants respectively at concentrations in the 0.1-0.5 range [9]. The finding of the present study are also is in agreement with that of [40] who reported good protection of cowpea seeds from *Callosobruchus maculatus* damage in storage due to use of *J. curcas* seed oil as a repellent and antifeedant. They also reported doses of 1.0 ml/150 g grains and above gave superior mortality of the pest in cowpea. The mosquitocidal assay against fourth instar *Aedes aegypti* larvae showed that both linoleic and oleic acids had an LD50 of 100 μg/ml. In caterpillar bioassays, linoleic and oleic acids reduced the growth of *Helicoverpa zea* by 88 and 85%, *Lymantria dispar* by 93 and 91%, *Orgyia leucostigma* by 81 and 80% and *Malacosoma disstria* by 77 and 75%, respectively [48]. The petroleum ether extract of *P. nigrum* and *J. curcas* showed a positive dose dependent antifeedant activity. The reduced consumption of rice kernels treated with both plant extract by *C. cephalonica* larvae are likely to be the main cause of growth inhibition (Table 2). Both plant extract showed harmful effect on *C. cephalonica* larvae growth and development.

Significant reductions in egg hatchability revealed the harmful effects of petroleum ether extract of *P. nigrum* and *J. curcas* towards *C. cephalonica* eggs. This observation is in agreement with that of [25] who reported that food treated with *Jatropha gossypifolia* seed extract strongly inhibit the fecundity of *Tribolium castaneum* compared with *Tribolium confusum* at doses of 8000 and 16000 ppm.

These results are attributed to the physico-chemical action of the compounds including piperine, caryophyllene, limonene, oleic acid, linoleic acid, menthone, menthol, α-pinene and β-pinene. Inhibition in egg hatching of the pulse beetle, *Callosobruchus chinensis* with *P. nigrum* essential oils were reported by [49], who observed that egg hatching was inhibited significantly when fumigated with sublethal concentration of the essential oil. Inhibition in adult emergence with *P. nigrum* was reported by [17] who observed lesser number of *Callosobruchus maculatus* adults emerging in black gram seeds treated with *P. nigrum* powder at doses of 25 and 30 g/kg. In treatments with *P. nigrum* oils, regression analysis showed a dose-dependent significant correlation with adult emergence [F=160.15], with the number of adults emerging from the fumigated larvae decreasing in concentration dependent manner [49]. High efficacy of the hexane extract of *P. nigrum* on 2nd instar larvae of *Spodoptera litura* was obtained with adult emergence of 19.79% for treatments of up to 40 mg/ml compared to the control [83.12%] [50].

Inhibition of progeny emergence with *J. curcas* extracts were studied by [38]. It was shown that water extracts of dried ground seeds of *J. curcas* at a dose of 5% [w/w] significantly reduced *S. zeamais* progeny emergence in treated grains of maize and cowpea.

References

1. Lucas E, Riudavets J (2002) Biological and mechanical control of Sitophilus oryzae (Coleoptera: Curculionidae) in rice. J Stored Prod Res 38: 293-304.

2. Allotey J, Azalekor W (2000) Some aspects of the biology and control using botanicals of the rice moth, Corcyra cephalonica (Stainton), on some pulses. J of Stored Prod Res 36: 235-243.

3. Huang F, Subramanyam B (2004) Responses of Corcyra cephalonica (Stainton) to pirimiphos-methyl, spinosad, and combinations of pirimiphos-methyl and synergized pyrethrins. Pest management science 60: 191-198.

4. Coelho MB, Marangoni S, Macedo ML (2007) Insecticidal action of Annona coriacea lectin against the flour moth Anagasta kuehniella and the rice moth Corcyra cephalonica (Lepidoptera: Pyralidae). Comparative Biochemistry and Physiology Part C: Toxicology & Pharmacology146: 406-414.

5. Ayyar PNK (1934) A very destructive pest of stored products in South India, Corcyra cephalonica, Staint. (Lep.) B Entomol Res 25: 155-169.

6. Prakash A, Rao J (1996) Botanical pesticides in agriculture. Taylor & Francis, USA.

7. Johnnie R Hayes, Mari S Stavanja, Brain M Lawrence (2006) Biological and Toxicological Properties of Mint Oils and Their Major Isolates.CRC Press USA.

8. Salimon J, Abdullah R (2008) Physicochemical properties of Malaysian Jatropha curcas seed oil. Sains Malaysiana 37: 379-382.

9. Ian M Scott, Helen R Jensen, Bernard j.R Philogene, John T Arnason (2008) A review of Piper spp. (Piperaceae) phytochemistry, insecticidal activity and mode of action. Phytochemistry Reviews 7: 65-75.

10. Satyajit D Sarker, Latif Z, Alexandar I Gray (2006) Natural Products Isolation. Methods in Biotechnology. 20. Humana press New Jersey.

11. Khani M, Awang R.M, Omar D, Rahmani M, Rezazadeh S (2011) Tropical medicinal plant extracts against rice weevil, Sitophilus oryzae L. Journal of Medicinal Plants Research 5: 259-265.

12. Gupta S, Jafar S, Jaiwal V, Raman Singh P, Maithani M (2011) Review on titrimetric analaysis. International Journal of Comprehensive Pharmacy 2 : 1-6.

13. Robertson JL, Russell RM, Savin NE (1980) Polo: Poloplus A User's Guide to Probit or Logit Analysis. Pacific Southwest Forest and Range Experiment Station Berkely California USA.

14. Finney D (1971) Probit analysis (3rd ed.) London: Cambridge University Press.

15. Huang Y, Ho Shuit.-Hung, Lee Hsien-Chieh, Yap Yen.-Ling (2002) Insecticidal properties of eugenol, isoeugenol and methyleugenol and their effects on nutrition of Sitophilus zeamais Motsch. (Coleoptera: Curculionidae) and Tribolium castaneum (Herbst) (Coleoptera: Tenebrionidae). J Stored Prod Res 38: 403-412.

16. Gomah E, Nenaah (2011) Toxic and antifeedant activities of potato glycoalkaloids against Trogoderma granarium (Coleoptera: Dermestidae). J Stored Prod Res 47: 185-190.

17. Mahdi S Rahman M (2008) Insecticidal effect of some spices on Callosobruchus maculatus (Fabricius) in black gram seeds. University Journal of Zoology, Rajshahi University. 27: 47-50.

18. Isman MB, Koul O, Luczynski A, Kaminski J (1990) Insecticidal and antifeedant bioactivities of neem oils and their relationship to azadirachtin content. J agric food chem 38: 1406-1411.

19. Hung Ho S, Wang J, Sim KY, Ee GCL, Imiyabir Z, Yap KF, Shaari K, Hock Goh S (2003) Meliternatin: a feeding deterrent and larvicidal polyoxygenated flavone from Melicope subunifoliolata. Phytochemistry. 62: 1121-1124.

20. Abbott W (1925) A method of computing the effectiveness of an insecticide. J Am Mosq control Assoc 18: 302-303.

21. SAS Institute (2006) Base SAS 9.1. 3 procedures guide: SAS Publishing.

22. Senguttuvan T, Kareem A.A, Rajendran R (1995) Effects of plant products and edible oils against rice moth Corcyra cephalonica Stainton in stored groundnuts. Journal Stored Prod Res 31: 207-210.

23. Aggarwal K, Tripathi A, Ahmad A, Prajapati V, Verma N et al. (2001) Toxicity of menthol and its derivatives against four storage insects. Insect Sci. Appl. 21: 229-236.

24. Pathak PH, Krishna SS (1991) Postembryonic development and reproduction in Corcyra cephalonica (Stainton) (Lepidoptera: Pyralidae) on exposure to eucalyptus and neem oil volatiles. J Chem Ecol 17: 2553-2558.

25. LA Muslima Khanam,AR Khan,M khalequzzaman,SM Rahman (2008) Effect of Sapium indicum, Thevetia neriifolia and Jatropha gossypifolia seed extract on the fecundity and fertility of Tribolium castaneum and Tribolium confusum. Bangladesh J Sci Ind Res 43: 55-66.

26. Adebowale KO, Adedire CO (2006) Chemical composition and insecticidal properties of the underutilized Jatropha curcas seed oil. Afr J Biotechnol 5: 901-906.

27. Chieng T, Assim Z, Fasihuddin B (2008) Toxicity and Antitermite activities of the essential oils from Piper sarmentosum. Malaysian Journal of Analytical Sciences 12: 234-239.

28. Awoyinka O, Oyewole I, Amos B, Onasoga O (2006) Comparative pesticidal activity of dichloromethane extracts of Piper nigrum against Sitophilus zeamais and Callosobruchus maculatus. African Journal of Biotechnology 5: 2446-2449.

29. Scott I, Gagnon N, Lesage L, Philogene B, Arnason J (2005) Efficacy of botanical insecticides from Piper species (Piperaceae) extracts for control of ruropean chafer (Coleoptera: Scarabaeidae). J Econ Entomol 98: 845-855.

30. Sighamony S, Anees I, Chandrakala T, Osmani Z (1986) Efficacy of certain indigenous plant products as grain protectants against Sitophilus oryzae (L.) and Rhyzopertha dominica (F.) Journal of Stored Products Research 22: 21-23.

31. Scott IM, Jensen H, Nicol R, Lesage L, Bradbury R, et al. (2004) Efficacy of Piper (Piperaceae) extracts for control of common home and garden insect pests. J Econ Entomol 97: 1390-1403.

32. Scott IM, Helson BV, Strunz GM, Finlay H, Sanchez-Vindas PE, Poveda L, Lyons DB, Philogène BJR, Arnason JT (2007). Efficacy of Piper nigrum (Piperaceae) extract for control of insect defoliators of forest and ornamental trees. Can Entomol 139: 513-522.

33. Openshaw K (2000) A review of Jatropha curcas: an oil plant of unfulfilled promise. Biomass and Bioenergy 19: 1-15.

34. Igbinosa O, Igbinosa E, vincent N Chigor, Olohirere E Uzunuigbe, Sunday O Oyedemi et. al. (2009) Polyphenolic Contents and Antimicrobial activity and phytochemical screening of stem bark extracts from Jatropha curcas (Linn). African Journal of Pharmacy and Pharmacology 3: 58-62.

35. Gübitz, G.M, Mittelbach M, Trabi M. (1999) Exploitation of the tropical oil seed plant Jatropha curcas L. Bioresource Technology. 67: 73-82.

36. Shah S, Sharma A, Gupta MN (2005) Extraction of oil from Jatropha curcas L seed kernels by combination of ultrasonication and aqueous enzymatic oil extraction. Bioresour Technol 96: 121-123.

37. Jing LI,Fen-hong,yan-yan CHEN, Fang CHEN (2006) Insecticidal activity of Jatropha curcas seed extracts against several insect pest species. J Agrochemicals 1.

38. Adabie-Gomez D, Monford KG, Agyir-Yawson A, Owusu-Biney A, Osae M (2006) Evaluation of four local plant species for insecticidal activity against Sitophilus zeamais Motsch.(Coleoptera: Curculionidae) and Callosobruchus maculatus (F)(Coleoptera: Bruchidae). Ghana Journal of Agricultural Science 39:147-154.

39. Sirisomboon P, Kitchaiya P, Pholpho T, Mahuttanyavanitch W (2007) Physical and mechanical properties of Jatropha curcas L. fruits, nuts and kernels. Biosystems Engineering. 97: 201-207.

40. Boateng BA, Kusi F (2008) Toxicity of Jatropha seed oil to Callosobruchus maculatus (Coleoptera: Bruchidae) and its parasitoid, Dinarmus basalis (Hymenoptera: Pteromalidae). J Appl Sci Res 4: 945-951.

41. Kumar A, Sharma S (2008) An evaluation of multipurpose oil seed crop for industrial uses (Jatropha curcas L.): A review. Ind crops prod 28:1-10.

42. Dowlathabad MR, Sreeyapureddy A, Adhikari A, Bezawada K, Nayakanti D (2010) Pharmaceutical investigation and biopesticidal activity of Jatropha curcas L. seed oil on digestive enzymic profiles of Cnaphalocrocis medinalis (rice leaf folder) and Helicoverpa armigera (cotton boll worm). International Research Journal of Pharmacy 1:194-200.

43. Kshirsagar R.V (2010) Insecticidal activity of Jatropha curcas seed oil against Callosobruchus maculatus (FABRICIUS) infesting Phaseolus aconitifolius JACQ. The Bioscan 5: 415-418.

44. Zahir AA, Rahuman AA, Ba-gavan A, Elango G, Kamaraj C (2010) Adult emergence inhibition and adulticidal activities of medicinal plant extracts against Anopheles stephensi Liston. Asian Pac J Trop Med 3: 878-883.

45. Kovendan K, Murugan K, Vincent S, Kamalakannan S (2011) Larvicidal efficacy of Jatropha curcas and bacterial insecticide, Bacillus thuringiensis, against lymphatic filarial vector, Culex quinquefasciatus Say (Diptera: Culicidae).j Parasitol Res 102: 1251-1257.

46. Tomass Z, Hadis M, Taye A, Beyene Y.M (2011) Larvicidal effects of Jatropha curcas L against Anopheles arabiensis (Diptera: Culicidea). MEJS 3: 52-64.

47. Rahuman AA, Gopalakrishnan G, Venkatesan P, Geetha K (2008) Larvicidal activity of some Euphorbiaceae plant extracts against Aedes aegypti and Culex quinquefasciatus (Diptera: Culicidae). Parasitol Res 102: 867-873.

48. Ramsewak RS, Nair MG, Murugesan S, Mattson WJ, Zasada J (2001) Insecticidal fatty acids and triglycerides from Dirca palustris. J Agric Food Chem 49: 5852-5856.

49. Chaubey MK (2008) Fumigant toxicity of essential oils from some common spices against pulse beetle, Callosobruchus chinensis (Coleoptera: Bruchidae). J Oleo Sci 57: 171-179.

50. Fan LOHS, Muhamad R, Omar D, Rahmani M (2011) Insecticidal properties of Piper nigrum fruit extracts and essential oils against Spodoptera litura. International Journal of Agriculture and Biology 13: 517-522.

15

Effect of Variety and Practice of Cultivation on Yield of Spring Maize in Terai of Nepal

Ghimire S[1]*, Sherchan DP[2], Andersen P[3], Pokhrel C[4], Ghimire S[5] and Khanal D[6]

[1]Technical officer, Nepal Agricultural Research council, Nepal
[2]ARTC Manager, Cereal System Initiative for South Asia, Nepal
[3]Associate Professor, Departrment of Geography, University of Bergen, Norway
[4]Associate Professor, Central Departrment of Botany, Tribhuvan University, Nepal
[5]Assistant Manager, Rastriya Banijya Bank, Nepal
[6]Senior Program Officer, Karuna Foundation Nepal

Abstract

A field experiment was conducted in farmer's field of MainaPokhar and Deudakala Village Development Committee in Bardiya District of Nepal. The objective of study was to identify the appropriate combination of variety and cultivation practice of maize in spring season. Two maize varieties Rajkumar (hybrid) and Arun2 (Open Pollinated Variety-OPV) were sown at the field of 6 different farmer's field. The experimental plot design was Randomized Complete Block Design with 6 replication and 4 treatments considering each farmer as a replication. There were 4 treatment combinations consisting two varieties and two practice of cultivations namely P1V1 (improved practice + Rajkumar), P1V2 (Improved practice+Arun2), P2V1 (Farmers practice + Rajkumar) and P2V2 (Farmers practice + Arun2). Result showed a significant (P<0.01) differences among varieties. But no significant difference was found in yield by the interaction of variety and practice. The statistically analyzed results revealed that the effect of cultivation practice and their interaction effect on grain yield were found non-significant but the response of the variety were found highly significant difference on grain yield, where the Rajkumar variety produced the highest average grain yield of 5.13 t/ha. It indicated that Rajkumar variety performs better than Arun2 in both improved and farmers practice of cultivation. Maximum grain yield ranging from (3.17 to 7.25 t/ha) and (1.60 to 6.32 t/ha) was produced by Rajkumar in improved practice and farmers practice of cultivation respectively while minimum grain yield was found in Arun2 ranging from (0.95 to 4.43 t/ha) and (0.81 to 4.09 t/ha) in improved practice and farmers practice of cultivation respectively. P1V1 scored the highest score followed by P2V1, P1V2 and P2V2 in farmers' preference ranking. Rajkumar variety cultivated with improved practice was found giving the best yield along with highest net return and Benefit Cost ratio of Rs. 30047.7 and 1.41 respectively.

Keywords: Maize; Hybrid; Open pollinated variety

Introduction

Agriculture is the mainstay of Nepalese economy. It also supplies about 80% of the country's total industrial raw materials and contributes about 70% of the total export earning of the country [1]. In Nepal, there is about 9,06,253 ha. of area under maize cultivation and annual production is about 20,67,722 smt. with productivity of 2.28 mt/ha [2]. Among cereals, maize is an important food and feed crop which ranks third after wheat and rice in world. As food producers are experiencing greater competition for land, water and energy, and the need to curb the many negative effects of food production on environment is becoming increasingly clear and this challenge requires changes in the way food is produced, stored, processed, distributed and accessed that are as radical as those that occurred in 18th and 19th century [3].

Nearly half the area under maize is planted with traditional varieties home saved seeds, which are continuously at the risk of degenerating (due to open pollination). The seed replacement rate is only 1%. Manures and fertilizers are not applied in sufficient quantities [4]. Limited and irregular access of improved seeds and quality fertilizers specifically to the small holders in the remote villages is the main constraint for maize production [5]. Most of the farmers are not aware about information on crop management aspects particularly balances use of fertilizers and maintaining optimum plant population per hectare.

There is a big yield gap in maize for both mid hills and Tarai of Nepal. The experimental yield of OPV maize is 6.70 t/ha whereas attainable yield is 5.70 t/ha. Attainable yield is the maximum experimental yields

in farmers' fields. The national average of maize is 2.51 t/ha [6]. So the yield gap at present is 3.50 t/ha. Similarly, the experimental yield of hybrid maize is 8.15 t/ha and attainable yield is 7.27 t/ha, so the actual yield gap is 5.64 t/ha [7]. If we narrow down the yield gaps for both in OPVs and hybrids by 2 t/ha, then yield would be double and the demand for grains and feeds will easily be met and fulfilled with this increment. Thus, the focus should be directed towards the arrowing of gaps through increasing access of improved seeds to the farmers and improved crop management practices.

Enhancing productivity of maize through identification of the best combination of management practice with preferred variety to add a brick in food security was the major goal of the experiment. In terai region cultivation of maize is done mostly in spring season; therefore this experiment was conducted with the objective:

1. To study the effect of varieties, cultivation practice and their interaction effect on yield of maize.

*Corresponding author: Ghimire S, Technical officer, Nepal Agricultural Research council, Nepal, E-mail: shantwana@narc.gov.np

2. To find out the farmer's preference for different combination of variety and practice.

3. To study and analyze the economic yield difference between the local variety and hybrid variety of maize under different management practice.

Materials and Methods

Location

The experiment was carried at 6 farmer's field 3 from Mainapokhar VDC and 3 from Deudakala VDC of Bardiya district from February, 2013 to July 2013 which was the command area of CSISA (Cereal System Initiatives for South Asia). The farmer's field was identified by the CSISA team. Geographically the study area is located in between 28°07" to 2839" N in latitude and 81°03" to 81°41" E in longitude.

Cropping history and soil character of the field

The experimental plot had potato-maize-rice sequential cropping pattern of one year rotation.

Soil samples were taken randomly from four different spots of each replication at a depth of 0-15cm and found to be sandy loam and loam.

Weather condition during experimentation

The climate of the experimental site was characterized as tropical and dry. It was characterized by three distinct season i.e. rainy monsoon (June-October), cool winter (November-February), and hot spring (March-May). The maximum temperature during hot spring rises up to 39°C (May 2013) with minimum rainfall so the experimental plot faced problem of draught so that the germination was uneven because of drought and lack of irrigation facility. The draught problem also occurred during tasseling stage. The meteorological data of the cropping period was recorded from the meteorological station of Regional Agriculture Research Station, Khajura, Banke.

Experimental design and data collection

Two varieties of maize, Rajkumar and Arun2 were selected for the experiment. The seeds of these varieties were sown in the field of six different farmers, three each from two different VDCs of Bardiya. The experimental plot design was Randomized Complete Block Design with six replication and four treatments. Each farmer field was considered as a replication for the experiment whiles the four different combinations of varieties and practice of farming cultivation was considered as treatments. The four treatment combinations consisted of combination of two varieties and two practices of cultivations (Table 1). The area of individual plots for each treatment was 12 m ˙ 7 m. The total plot area for all four treatment was 336 m² (12˙7˙4 m²). The net experimental area including all four treatments and six replications was 2,016 m² (12˙7˙4˙6 m²).

Fertilizer @120:60:60 N:P$_2$0$_5$:K$_2$O kg/ha (This dosage of fertilizer is the recommendation from NARC (National Agricultural Research Center)).

For Basal dose: @ 60:60:60 N:P$_2$0$_5$:K$_2$O kg/ha i.e (79.43:130.34:100 Urea:DAP:MOP kg/ha).

1st Top dress: @ 30 N kg/ha i.e. (65.21 Urea kg/ha).

2nd Top dress: @ 30 N kg/ha i.e. (65.21 Urea kg/ha).

Weeding: 1 time

Irrigation: 3 times

Row to row and plant to plant distance of 70˙20 cm was maintained.

P$_2$=farmers practice

Fertilizer @53.2:30.4:0 N:P$_2$0$_5$:K$_2$O kg/ha

Basal dose: @ 26.6:30.4:0 N:P$_2$0$_5$:K$_2$O kg/ha i.e. (31.98:66.08:0 Urea:DAP:MOPKg/ha)

Top dress: @ 26.6 N kg/ha i.e. (57.82:0:0 Urea Kg/ha)

Weeding: 1 time

Irrigation: 3 times

Row to row and plant to plant distance of 70˙20cm maintained.

3.5. Factor B (Varieties)

V1: RajKumar (Hybrid)

V2: Arun2 (Open Pollinated Variety)

Cultural Practice

Field preparation

The field was ploughed 15 days before sowing to incorporate the weed and crop residue in to the soil.

Fertilizer application

In improved practice, the maize crop was fertilized with @120:60:60 N: P: K Kg/ha through Urea, DAP and MOP in 3 split dose. The recommended amount of nitrogen, phosphatic and potassic fertilizers @ 60:60 kg/ha were calculated and weighed separately for and were applied in all experimental plots.

In farmers practice, the maize crop was fertilized with @ 53.2:30.4:0 N:P:K kg/ha in two split dose through Urea, DAP and MOP. The recommended amount of nitrogen fertilizer was calculated and weighed separately and was applied in all experimental plots. Fertilizer application was done @120:60:60 N:P$_2$0$_5$:K$_2$O kg/ha in three split dose in improved practice.

For Basal dose: @ 60:60:60 N:P$_2$0$_5$:K$_2$O kg/ha i.e (79.43:130.34:100 Urea:DAP:MOP kg/ha)

1st Top dress: @ 30 N kg/ha i.e. (65.21 Urea kg/ha)

2nd Top dress: @ 30 N kg/ha i.e. (65.21 Urea kg/ha)

And application of fertilizer in farmers practice was done @53.2:30.4:0 N:P$_2$0$_5$:K$_2$O kg/ha in two split dose.

Basal dose: @ 26.6:30.4:0 N:P$_2$0$_5$:K$_2$O kg/ha i.e. (31.98:66.08:0 Urea:DAP:MOP Kg/ha)

Top dress: @ 26.6: N kg/ha i.e. (57.82 Urea Kg/ha).

Seed sowing

The required amount of seed of all varieties each individual plot was calculated. The seed rate of @ 30 kg/ha, 252gm for each individual

S.No	Cultivation Practice (P)	Variety (V)	Notation
1	Improved Practice 120:60:60NPK kg/ha	Rajkumar	P1V1
2	Improved Practice 120:60:60NPK kg/ha	Arun2	P1V2
3	Farmers Practice 53.2:30.4:0 NPK kg/ha	Rajkumar	P2V1
4	Farmers Practice 53.2:30.4:0 NPK kg/ha	Arun2	P2V2
Treatment Details: Factor A (Cultivation Practice), P$_1$ = improved practice			

Table 1: Treatment Combination.

plot (84 m²) was used. Bold, biophysically good, healthy seeds of both varieties were selected and two seeds per hill were dropped manually in the row line. Maize seeds were sown maintaining 20cm between plant to plant and row to row distance 70cm. The sowing was done in 2013, February 27 (Table 2).

Thinning

Thinning was done on the 20th Days after Seeding (DAS) for all treatment to maintain a single plant per hill spot by removing all the other extra and weaker maize plant.

Plant protection

Stem borer was found problematic in the field at earlier 8 to 10 leaves stage of maize plant. Furadon @ 1 kg/ Kattha (333.3 m²) was applied against the stem borer which was applied in all part of leaf at grand growth stage (40 DAS). Grasshopper was also found to be problematic in the field. Chlorpyriphos 50% + Cypermethrin 5% at interval of 1 week for 3 times.

Weeding and irrigation

Manual weeding was done at knee high stage. Irrigation was provided just for 3 times at critical growth stages because of poor irrigation facility.

Harvesting and threshing

The whole plot was harvested manually when the plant turned to yellowish, ear husk turned into the brown and appearance of black layer at the base of each kernel when scratched by cutting plants with sickles near the ground level. Threshing of grain was done manually after the sun drying of harvested crop and grains were cleaned by winnowing and separately dried by maintaining 15% moisture. Harvesting of Arun2 (June 5th, 2013) was done 7-13 days before Rajkumar variety (18th June, 2013) (Table 3).

Preference ranking

A survey was done for ranking the preference of the farmers regarding the cultivation practice and variety. The survey was done by conducting Farmer's Field Day where farmers were given the form for evaluating the experiment by 8 different attributes (Days to maturity, leaf color, plant quality, cob per plant, length of cob, probable green cob, insect tolerance and overall quality). 27 farmers were taken for conducting this survey, forms were compiled and Garret ranking was done. (Form can be found in annex).

Garret analysis was done by using the formula

$$PercentagePosition = 100\left(\frac{Rij - 0.5}{Nj}\right)$$

Where,

R_{ij} = Rank given for i th item by jth individual

N_j = Number of items ranked by jth individual

Varieties	Maturity days	Yield			
		Improved practice (Average) t/ha	Farmers practice (Average) t/ha	Maximum (Improved)	Minimum (Farmers)
Rajkumar	100	5.14 (1.67)	3.96 (1.75)	7.25 t/ha	3.17 t/ha
Arun2	85	2.93 (2.10)	2.11 (1.38)	4.43 t/ha	0.71 t/ha
Note: Value in ( ) indicates standard deviation.					

Table 2: Varieties their maturity days with maximum, minimum and mean yield with standard deviation.

S.N	Cultivation Practice	Variety	Treatment	Ranking
1	Improved Practice	Rajkumar	P1V1	I
2	Improved Practice	Arun2	P1V2	III
3	Farmers Practice	Rajkumar	P2V1	II
4	Farmers Practice	Arun2	P2V2	IV

Table 3: Preference Ranking Table.

Measurement of yields

Grain yield: After drying and shelling of the harvested produce of each sample plant of individual plot, the grain yield was recorded. Grain yield was calculated on hectare basis by using following formula:

$$\text{Grain yield (kg/ha)}=\frac{\frac{Fresh\ wt}{Area\ in\ m^2}\times(100-moisture\%)}{(100-15\%\ adjusted)}\times10000$$

Economic Analysis

1. Cost of Cultivation: Cost of cultivation was calculated on the basis of available local charges for different agro-inputs viz. price of seed, labor charge, fertilizers, bullocks and other necessary materials.

2. Gross return: Economic yield was converted into gross return on the basis of market price of maize grain. The grain yield was calculated as per local prices Rs. 20/kg grain.

3. Net return: It was calculated by deducting the cost of cultivation from the gross return.

Net return (Rs/Ha) = Gross return-Total cost

4. Benefit cost ratio: Benefit cost ratio was calculated as below:

$$\text{Benefit:cost ratio(B:C ratio)}=\frac{Gross\ income(Rs/Ha)}{Total\ cost(Rs/Ha)}$$

Statistical Analysis

Microsoft Office package (MS Excel and MS word) were used extensively to feed in the primary data, make basic tables, charts and type in the overall text. The data was first tabulated in Microsoft Excel and effect of variety on yield and effect of practice of cultivation on yield were statistically analyzed by using R (64 x 2.15.1 version) program (Table 4).

Results

The experimental results were analyzed and presented in this chapter with figures and tables where necessary (Figure 1-4).

Effect of cultivation practice and varieties on grain yield (t/ha)

Grain yield of a crop is the result of combined effect of growth, development and yield attributes. These parameters are governed by the heredity of the particular variety but at the same time these are also modified by the level of management and the environmental to which the crop is exposed.

Effect of practice of cultivation in yield: Grain yield was not significantly influenced by practice of cultivation. However, higher yield was found in Rajkumar variety with improved practice. The hybrid variety Rajkumar gave higher yield of 4.54 t/ha in average where as Arun2 produced the mean grain yield 2.52 t/ha. The effect of farmers practice in yield of maize was found to be non-significant (p>0.05).

Effect of variety in yield: There was significant (P<0.05) difference in the grain yield between varieties. Grain yield was significantly

Treatments	Gross return per hectare Rs	Cost per hectare Rs	Net return per hectare Rs	B:C ratio
T1(P1V1)	102600	72552.3	30047.7	1.41
T2(P1V2)	58200	66402.3	-8202.3	0.88
T3(P2V1)	78600	82038	-3438	0.96
T4(P2V2)	42000	75888	-33888	0.55
Mean	70350	74220.15	-3870.15	0.95

Table 4: Revenue, Cost, Net Return and B:C ratio.

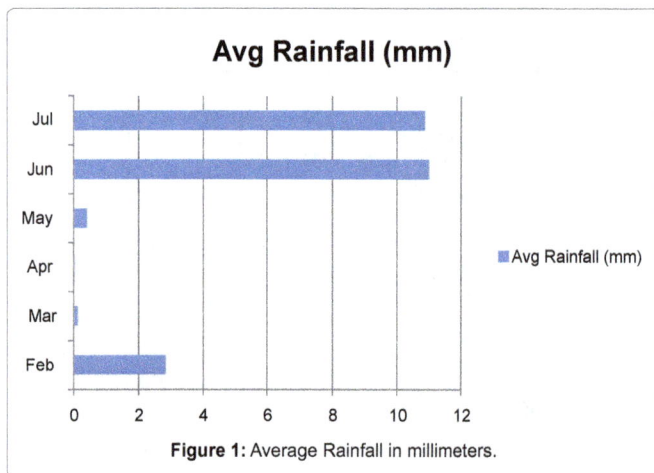

Figure 1: Average Rainfall in millimeters.

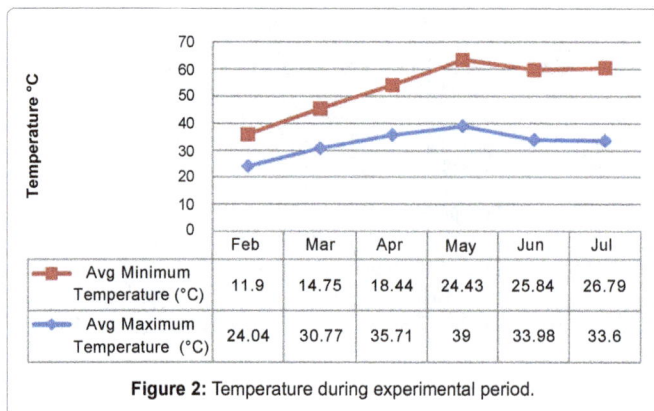

Figure 2: Temperature during experimental period.

	Feb	Mar	Apr	May	Jun	Jul
Avg Minimum Temperature (°C)	11.9	14.75	18.44	24.43	25.84	26.79
Avg Maximum Temperature (°C)	24.04	30.77	35.71	39	33.98	33.6

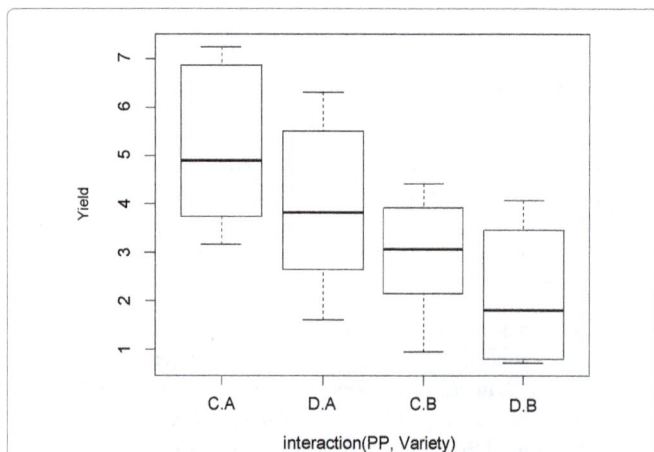

Figure 3: Boxplot showing difference in yield (t/ha) for different practice of cultivation. C.A=Improved practice of cultivation+Rajkumar; D.A= Farmers practice of cultivation+Rajkumar; C.B= Improved practice of cultivation+Arun2; D.B= Farmers practice of cultivation+Arun2).

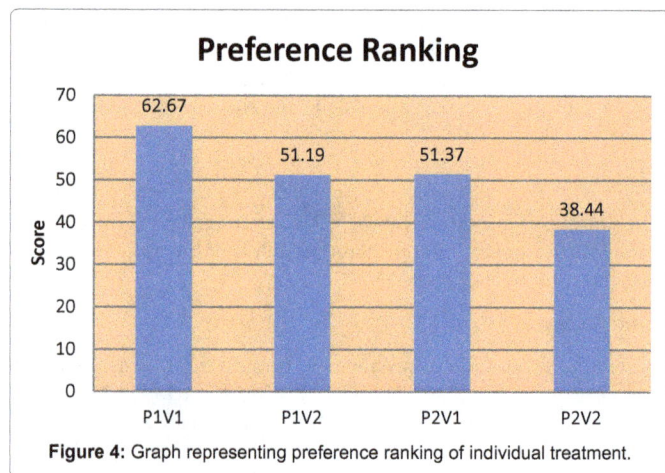

Figure 4: Graph representing preference ranking of individual treatment.

affected by crop varieties sown. Irrespective of fertilizer dose and irrigation maximum grain yield ranging from (3.17 to 7.25 t/ha) and (1.60 to 6.32 t/ha) was produced by Rajkumar in improved practice and farmers practice of cultivation respectively. While minimum grain yield was found in Arun2 ranging from (0.95 to 4.43 t/ha) and (0.81 to 4.09 t/ha) in improved practice and farmers practice of cultivation respectively.

While performing two-way ANOVA with the data having two predicting variable with two categorical values of each; we get the result that the interaction affect between the practices and varieties was not significantly performing to give the result in yield. And after running the linear model with interaction effect eliminated; we found that not practice but varieties of the species are significantly different to give different result in yield.

Preference ranking

Preference ranking was done conducting farmers field day. Analysis of the collected data was done through Garret ranking. Average of each treatment was calculated and percent and score were calculated accordingly. And the average of the score of each treatment was calculated to get the final result of ranking which is presented below.

Treatment P1V1 was at the first rank which means it is the most preferred and P2V1, P1V2, and P2V2 on the second, third and fourth respectively.

From the result of garret ranking we came to know that the combination of Rajkumar variety and improved practice got the first rank considering the 8 attributes (Days to maturity, leaf color, plant quality, cob per plant, length of cob, probable green cob, insect tolerance and overall quality). P1V1 is in the first rank with highest score which is the combination of Rajkumar variety and improved practice. Similarly, P2V1 got the 2nd rank which is the combination of Rajkumar variety and farmers practice followed by P1V2 and P2V2. Here the treatment which got the highest and second highest score differs only in variety which means if hybrid variety is adopted for farming we can get the good yield.

Economic Analysis

The economics of various treatments under study was worked out to evaluate the most beneficial combination of cultivation practice and maize cultivars.

Cost of cultivation

The cost of cultivation for Improved farming practice was higher than the cost of cultivation for Farmers Practice but the return was also higher in improved farming. The cost of cultivation is the total expenditure incurred for raising crops in cropping system. Cost of cultivation was calculated on the basis of local charges for different agro-inputs viz. labor, fertilizer, compost and other necessary materials.

Gross return

The total monetary economic produce and by products obtained from the crop is called gross return. It is calculated based on the local market price of the produce. The mean gross return of the experiment was Rs 70350 per hectare and ranging from Rs 42000 to Rs 102600.

Net return

Net return is the ultimate product obtained by subtracting cost of cultivation from gross return. It is a good indicator of sustainability of crop since this represents the actual income of the famer. It was calculated by subtracting the cost of cultivation from the gross return. The mean net return of the experiment was (-Rs 3870.15) and it ranged from Rs (-33888 to 30047.70) per hectare.

B:C ratio

Benefit cost ratios the ratio of gross return to cost of cultivation which can also be expressed as returns per rupee invested. Any value greater than 2 is considered safe as the farmer gets 2.00 for every rupee invested. On the other hand B/C ratios of 1 for the agricultural sector have been fixed for any enterprises to be economically viable. Therefore any crop enterprise must maintain a 1 B:C ratio to be economically sustainable (Bhandari, 1993). The mean B:C ratio in the experiment was 0.95 and ranged from 0.55 10 1.41. Rajkumar variety grown under improved cultivation practice was the most cost effective farming practice because only P1V1 had a ratio >1, based on the price relations used in the calculation.

Since B:C ratio is higher in T1 (1.41) with net return of Rs 30047.7, we can conclude that the combination of improved cultivation practice with Rajkumar variety is the best and most cost effective. Since B:C ratio and net return of T1 and T3 are higher as compared to T2 and T4 we can say that Rajkumar is the superior variety as it is performing better in farmers cultivation practice also. Talking about Arun2, its combination with improved practice also didn't give the satisfactory results, not only because Rajkumar cultivated with farmers method cultivation is performing better than Arun2 with improved method of cultivation, but also because the economic return is not satisfactory.

Discussion

Significantly higher yield by hybrid maize cultivated with improved practice is due to exploitation of hybrid vigor when growth environment is provided during the life cycle [8,9]. Reported that the genotypic constitution largely determine the response of a variety to chemical fertilizers. Terai and inner Terai of Nepal is highly potential for hybrid cultivation particularly in spring and winter seasons [10]. Development and use of hybrid seeds can enhance crop yields and performance in ways that are different from and not necessarily dependent on heterosis by itself [11]. OPV's may be a valuable option for maize producers under some circumstances, but the use of an OPV or recycled seed would be a step backward for grain yield. Generally, a hybrid will produce 18% more grain than most of the better OPV [12]. In this experiment, yield of Rajkumar was found to be 42.99% and 46.71% more than Arun2

in improved practice and farmers practice respectively. Response by farmers in farmer's field day for preference ranking was very much positive about Rajkumar variety as it performed best in poor irrigated condition also. Hybrid maize technology has made significantly yield advances and increased productivity in both developed and developing countries [13]. The replacement of open pollinated varieties by hybrids is an effective way to enhance productivity [10].

Yield is not determined nor only by the variety neither the practice of cultivation but with the perfect combination of both. Poor variety with best practice may be performing far better than best variety with poor practice. But the best variety with best practice will surely perform the best in grain yield which is P1V1 in our research which is the combination of Rajkumar variety and improved cultivation practice. Farmers' preference was towards Rajkumar in different attributes and overall quality considered. Farmers' preference to green cob was also higher in Rajkumar variety. Rajkumar variety performed best in both practice of cultivation with poor irrigation facility. The maize hybrids performed exceptionally better during drought stress [14]. It is well recognized fact that cultivation of hybrids maize cultivars is one of the best alternatives to increase the production and productivity of maize in Nepal [10]. The statistically analyzed results revealed that the effect of cultivation practice and their interaction effect on grain yield were found non-significant but the response of the variety were found highly significant difference on grain yield, where the hybrid variety , Rajkumar produced the highest average grain yield of 5.13 t/ha. Similar result was found by [15] while comparing Rajkumar with other varieties. Among the two varieties Arun2 and Rajkumar, variety Rajkumar has high benefit: cost ratio under both improved and farmers practice 1.41 and 0.96 respectively. However, Stover yield was not taken into consideration because farmers deny to harvest due to storage problem. The higher yields and revenues (including the market value of home-consumed maize) of hybrids outweighed the higher costs of hybrid maize cultivated with improved practice of cultivation method.

Conclusion

Hybrid maize is suitable for higher production and has higher potential than OPVs in the Terai condition. Choosing wrong variety may result loss of yield, resulting in food insecurity and loss of profits. Irrespective cultivation practices Rajkumar hybrid was found to be superior during spring season in Terai region. Economic analysis depicted that the highest net return was observed in Rajkumar Hybrid in improved practice of Rs. 30047.7 with B:C ratio 1.41 which scored highest in preference ranking and gave 42.99% more yield than Arun2. Finally, the most important thing is to develop crop production technology with emphasis on cost reducing, input efficient seed production technology and soil fertility improvement.

Acknowledgement

I would like to express my deepest gratitude Dr. Medha Devare and CSISA-Np family. I am ever indebted to Dr. Ole R. Vetaas and Dr. Ram Prasad Chaudhary. I want to thank Head of Department Prof. Pramod K. Jha and all the Department of Botany family for their cooperation. Special thanks farmers from research site who provided the field to conduct research and NARC family.

Reference

1. Gurung DB (2006) Genetic diversity, heterosis and combining ability with Nepalese maize varieties. Science City of Mufloz, Nueva Ecija, Phillipines: Central Luzon State University.

2. MOAC (2011) In Krishi Diary 2069. Hariharbhawan: Agriculture Information and Communication Centre.

3. Charles JGH, Beddington RJ, Crute RI, Lawrence H, David L et al. (2010) Food Security: The Challenge of Feeding 9 Billion people. Science, 327: 812-818.

4. Koirala GP (2001) Factors Affecting Maize Production, productivity and Trade in Nepal. Kathmandu Nepal: CIMMYT, NMRP.

5. NMRP (2011) Annual Report. Rampur Chitwan Nepal: National Maize Research Program.

6. Nepal GO (2013) Krishi Diary. Hariharbhawan Lalitpur: Agriculture Information and Communication Center.

7. Acharya SR (2011) Production potential and economic viability of winter maize cultivars under different nitrogen levels in irrigated condition. Postgraduate Dissertation. Rampur, Chitwan: Institute of Agriculture and Animal Sciences.

8. Pokhrel B (2006) Response to two maize cultivars to different level of nitrogen during winter season at rampur, Chitwan. Rampur, Chitwan, Nepal: Institution of Agriculture and Animal Sciences, Tribhuvan University.

9. Singh K, Prasad P (1990) Response of promising rainfed maize (Zea mays) varieties to nitrogen application in north-western Himalayan region. Indian Journal of Agriculture Sciences, 7: 475-477.

10. Kunwar CB, Shrestha J (2014) Evaluating Performance of Maize hybrids in Terai Region of Nepal. World Journal of Agricultural Research, 2: 22-25.

11. Duvick DN (1999) Heterosis: Feeding people and protecting natural resources. American Society of Agronomy, 19-29.

12. Anonymus (2011) Hybrid or open pollinated variety seed-weigh up the option. Bothaville, 9660: GRAIN SA.

13. Katuwal RB (2012) Influence of variety and planting date on cold tolerance and yield for winter maize in Nepal. Chitwan, Rampur: Institute of Agriculture and Animal Science.

14. Oyekale KO, Dainel IO, Kamara AY, Akintobi DA, Adedbite A E Evaluation of Maize Hybrids for Yield Characteristics under Drought Stress. Ibadan, Nigeria: International Institute of Tropical Agriculture.

15. Dawadi DR, Sah SK (2012) Growth and Yield of Hybrid Maize (Zea mays L.) In Relation to Planting Density and Nitrogen Levels during Winter Season in Nepal. Tropical Agricultural Research 23: 218-227.

16

Calibration and Validation of DSSAT V.4.6.1, CERES and CROPGRO-Models for Simulating No-Tillage in Central Delta, Egypt

Harb OM, Abd El-Hay GH, Hager MA and Abou El-Enin MM*

Agronomy Department, Faculty of Agriculture, Al-Azhar University, Cairo, Egypt

Abstract

Crop simulation programs allow analyzing and exploring various tillage-rotation combinations and management. This study was conducted to apply and evaluate the DSSAT program under Egyptian conditions. The study was carried out to investigate the effect of tillage system, fertilizer rates and cereal/legume rotation on the crop yield and soil quality. The CERES-maize and CROPGRO-broad bean models were used to simulate the studied crop yield.

Field observations showed that, the effect of tillage systems during the summer season of 2013 did not differ significantly due to studied maize traits. Regarding, winter season of 2013/2014), the results showed that, CA tillage system increased significantly all studied broad bean traits as compared with the other tillage systems. Referring to, the summer season of 2014, CA system scored the significant high values for the studied maize traits.

As for the effect of studied NPK fertilizer levels, results indicated that, 100% of the recommended doses of NPK favored the values of the studied maize and broad bean traits significantly during summer 2013 and winter 2013/2014, as compared by 50% of the recommended dose of NPK fertilizers, while, there are no-significant difference between the two fertilizer levels for maize traits in the third season (summer, 2014).

With regard to, the first order interaction effect between the tested factors, results of the three trial seasons revealed that, growing maize or broad bean under the condition of conservation agriculture (CA) and fed by 100% or 50% of the recommended dose of NPK fertilizers scored the greatest values for most of maize studied traits and broad bean, and the differences between them did not reach the significant level. On contrast, the lowest value were resulted under the condition of Conventional agriculture (CT) and fed by the 50% of the recommended dose of NPK fertilizers.

The models that were used in this study also reflected this trend. The CERES-maize and CROPGRO-broad bean models greatly discovered stimulation for grain yield/fed., harvest index as affected by interaction effect between tillage systems and fertilizer rats, which their RMSE ranged between excellent and good, RMSE = (8.44, 12.19) and (11.70,16.79) and (0.15, 12.02) for summer 2013, winter 2013/2014 and summer 2014 seasons respectively, through (maize→ broad bean→ maize) crop sequence.

Keywords: Conservation agriculture; DSSAT v.4.5; NPK fertilizer; Crop sequence

Introduction

Heavy and continuous conventional agriculture can cause loss of soil organic carbon, as well as increase soil erosion and deterioration of soil structure [1]. In the last few years, the search for practices that improve soil fertility and productivity and agricultural sustainability has increased. Interest in conservation agriculture technique (such as reduce and no-tillage) is growing be. Because these practices reduce soil erosion, therefore preserving soil structure and fertility [2]. Improve in the soil structure and increase its productivity by applying conservation agriculture technique has been reported in numerous studies [3]. In 1973/74 Conservation agriculture, synonymous of zero tillage, was used only on 2.8 million ha worldwide. In 1999 zero tillage was adopted on about 45 million ha worldwide, growing to 72 million ha in 2003, and to 154.81 million ha by 2014 [4]. Fastest adoption rates have been experienced in South America where some countries are using no-tillage on about 70% of the total cultivated area.

Many countries talk the direction for applying the conservation agriculture. Conservation Agriculture emerged as new agricultural technique successfully applied over the last years mainly in American countries. However, African agricultural systems have triggered a controversy on CA adoption and its suitability in smallholders' environments. Assessment of CA adoption requires a detailed revision

of several social and economic factors and conditions. Figure 1 shows CA adoption in selected countries of Africa.

The crop simulation program such as DSSAT V.4.6.1 could be used to evaluate different tillage practices and crop rotation [5]. This program includes tillage routines that modify soil structure and mix soil constituents. It combines several crop simulation models, soil carbon and nitrogen (N) models, daily soil water model, and field management options to simulate crop productivity and environmental effects. The DSSAT V.4.6.1includes CERES-based Soil organic matter model and CENTURY model [1,6,7].

Our objective of this study was to evaluate the capacity of DSSATv.4.6.1-CERES and CROPGRO models to predict yield and its components traits of some Egyptian maize and broad bean varieties

***Corresponding author:** Abou El-Enin MM, Faculty of agriculture, Universitas Al-Azhar, Cairo, Egypt, E-mail: Magro_modeller@yahoo.com

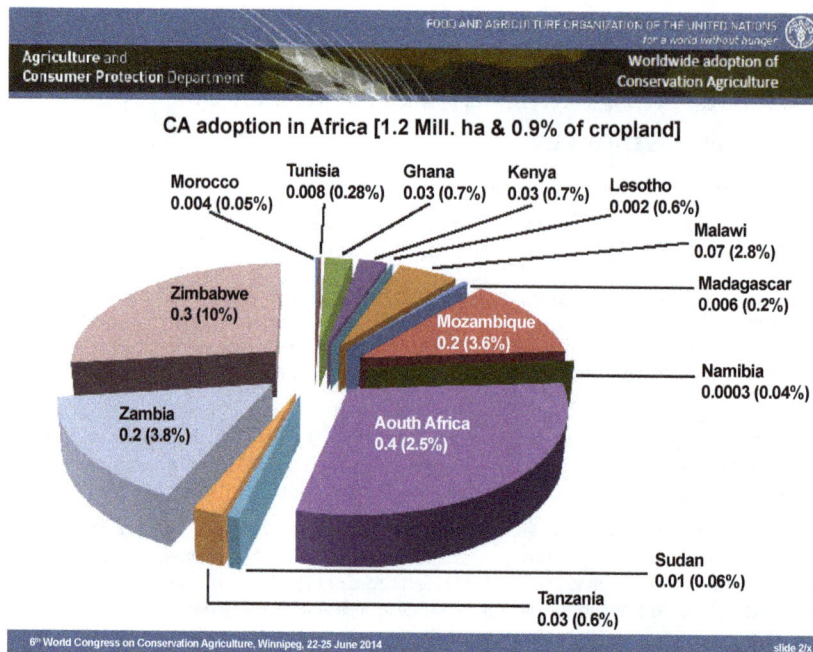

Figure 1: Shows CA adoption in selected countries of Africa.

grown in clay soil under different tillage systems as well as fertilizer rats through cropping system of (maize→ broad bean→ maize).

Materials and Methods

The materials and methods of this investigation are presented as follows:-

Field experiment

The field experiment started in summer 2013 and continued for 3 seasons in Gemmieza agricultural experimental research station, Egyptian Agricultural Research Center (ARC).

The studied experimental treatments:

Tillage systems treatments (TS):

• *Conventional agriculture (CT)*

In this system, the normal conventional agricultural practises of growing crop were done such as tillage.

• *Conservation agriculture (CA)*

Under the conditions of this system, the soil was left without any land preparation and the previous crop residuals was hammered and left on soil surface and the seed was growing by hand drilled around hills.

• *Semi-conservation agriculture (SCA)*

This method as the same conservation agriculture method without hand drilled around hills.

Fertilizer treatments:

• the recommended fertilizer (NPK)

• half of the recommended fertilizer (1/2 NPK)

The phosphorus and potassium fertilizer rate of each crop were applied as, single calcium super phosphate (15.5% P2o5) and potassium sulphate (48% K_2O) during soil preparation for (CT) tillage treatments while that fertilizers were added broadcasting through (SCA) and (CA) tillage systems.

Regarding to, nitrogen fertilizer rate for each crop as shown before in Table 1 was applied in the form of urea (46%N) before water irrigation as follow:

• *Maize crop:*

The total amount was devoted in to two equal portions as follow:

1. Before the first irrigation at plant ages of 20 days from sowing date

2. Before the second irrigation at plant age of 35 days from sowing date.

In reference to, broad bean success inoculation for its seed were done by *R. leguminosarum* bacteria respectively and the nitrogen fertilization take place after 10 days from sowing date at the rate of 15 kg N /fed.

Single hybrid-10 maize, it was sown at the recommended seeding rate (15 kg/fed), in hills, 2-3 grains were hand affair planted in each hill spaced at 20 cm apart, on the 5 and 7th April in 2013 and 2014 seasons respectively and harvested on 16 and 20th August 2013 and 2014 respectively.

As for, Broad bean seeds were planted by affair method by hand at the rate of 2-3 seeds/hill spaced at 20 cm apart. Sowing date on the 22th October 2013/2014 seasons, and harvested on 26th March 2014.

Experimental design

In the three studied seasons, each field experiment included six treatments, which were the combination of three systems of tillage practice, and two levels of NPK fertilizer, the treatments were arranged in a split- plot design with three replicates. The main plots were

Fertilizer crops	Crop variety	Nitrogen (kg N/fad)	P$_2$O$_{5\ 15\%}$ (kg/fad) Before planting	K$_2$SO$_4$ (kg/fad)	Seeding rate (kg/fad)
Broad bean	Egypt-1	15	150	50	60
Corn/maize	Single hybrid-10	120	200	50	15

Table 1: Shows the recommended nitrogen, phosphorous, potassium fertilizer rates, and seeding rates for the studied crops variety.

randomly devoted to the tillage treatments, regarding to, the sub-plots were randomly devoted to the fertilizer rates. 1 m alleys separated these plots from each other.

All plots were irrigated by surface irrigation system every 10 day for maize crop and 20 days intervals for broad bean crop according to region conditions.

Statistical analysis: All data were exposed to the proper statistical analysis according to Gomez [8]. The mean values were compared at 5% level of significance using least significant differences (L.S.D) test.

Studied attributes

Corn /maize crop

Five plant samples were taken randomly from each plot to measure the following traits:

• Cone length (cm).

• Weight of Cone (g).

• Biological yield (kg/fed): whole plants of each plot were harvested then weighted and transformed to biological yield per fed. According the plot area.

• Grain yield (kg/fed) It was determined by weighting the total grain yield of each plot, then converted to kg/fed.

• Harvest index (HI) was calculated according to the following formula:

HI = Seed yield (kg/fed.)/ Total biological yield (kg/fed.).

Broad bean, Lentil and Maize crops

Ten plants samples were taken randomly from each plot to measure the following traits:

• Number of pods/plant

• Biological yield (kg/fed): whole plants of each plot were harvested then weighted and transformed to biological yield per fed according the plot area

• Seed yield (kg/fed) It was determined by weighting the total seed yield of each plot, then converted to kg/fed.

• 100-Seeds weight (g) was obtained from the weight of 100 seed taken at random sample from each plot.

Crop simulation methods

Model description: The crop simulation model DSSAT (Decision Support System for Agro Technology) was chosen because it has been successfully used worldwide in a broad range of conditions and for multipurpose: as an aid to crop management. More than 18 different crops simulated with CSM, including maize, wheat, rice, barley, sorghum, millet, maize, peanut, dry bean, chickpea, cowpea, faba bean, velvet bean, potato, tomato, bell pepper, cabbage, Bahia and brachiaria and bare fallow. We used DSSAT version 4.6.1 which includes the

new tillage model based on the improved CROPGRO and CERES-Till [9]. A model used to predict the influence of crop residue cover and tillage on soil surface properties and plant development. CROPGRO and CERES-Till has been tested for broad bean and maize and has demonstrated the ability to simulate differences in soil properties and broad bean, maize yield under several tillage systems.

Input files for both of CERES-maize model and CROPGRO module requires an experimental details file, a weather data file, a soil data file and a genotype data file.

Experimental details file: Such as: field characteristics, soil analysis data, initial soil water, irrigation and water management, fertilizer management, tillage operations, environmental modifications, harvest management and simulation controls. Details of irrigation events for all the experiments.

Weather data file: The model requires daily weather data for the duration of the growing season. The minimum data required for above two models are solar radiation, minimum and maximum air temperature and rainfall [10].

Soil data file: The data related to soil profile, soil water, soil nitrogen and root growth characteristics, soil taxonomic classification, soil texture and other descriptive data of the experimental site were used to develop the soil file for the experimental station.

Genotype data file: Farmers can change cultivars in order to maximize yield. The DSSAT crop models also have the ability to take that source of variability into account. For each model, the cultivars are characterized by a specific set of genetic coefficients. These coefficients express the genetic potential of each genotype independently of all environmental constraints: soil; weather, etc. by simulating the yield of different cultivars in different conditions, it is possible to select the one (s) that best explore the available resources.

Calibration of models: Model calibration or parameterization is the adjustment of parameters so that simulated values compare well with observed ones. Genetic coefficients of CERES- maize and CROPGRO model are related to photoperiod sensitivity, duration of grain filling, conversion of mass to grain number, grain-filling rates, Maximum weight per seed (g), Time between first flowers and first pod, vernalization requirement, stem size and cold harden. The genetic coefficients used in two models characterize the growth and development of crop varieties differing in maturity as following Table 2 and 3.

Crop model validation: The comparison between actual data and predicted data were done through CERES-wheat, maize and CROPGRO-maize, faba bean, lentil models under DSSAT interface in three steps, i.e. retrieval data (converting data to CERES and CROPGRO model), validation data (comparing between predicted and observed data) and run the model.

Evaluation of applying CERES and CROPGRO model: The two models were evaluated through three methods:

•   The normalized root mean square error (RMSE) that is expressed in percent, calculated as explained by Loague [11]. with the

help of following Equation: $RMSE = \sqrt{\dfrac{\sum_{i=1}^{n}(P_i - O_i)^2}{n}} \times \dfrac{100}{M}$

Where n is the number of observations, Pi and Oi are predicted and observed values respectively, M is the observed mean value. The

Coefficients	Definition	Cultivar
		Single Cross - 10
P1	Thermal time from seedling emergence to the end of the juvenile phase (expressed in degree days above a base temperature of 8°C) during which the plant is not responsive to changes in photoperiod.	190
P2	Extent to which development (expressed as days) is delayed for each hour increase in photoperiod above the longest photoperiod at which development proceeds at a maximum rate (which is considered to be 12.5 hours).	1
P5	Thermal time from silking to physiological maturity (expressed in degree days above a base temperature of 8°C).	1000
G2	Maximum possible number of kernels per plant.	850
G3	Kernel filling rate during the linear grain filling stage and under optimum conditions (mg/day).	7
PHINT	Phylochron interval; the interval in thermal time (degree days) between successive leaf tip appearances.	49

Table 2: Genetic coefficients used in CSM-CERES-model characterize the growth and development of maize variety after Model calibration and validation.

Coefficients	Definition	Cultivar	
		Giza-111	Egypt-1
EM-FL	Time between plant emergence and flower appearance (R) photo thermal days	16.25	18.00
FL-SH	Time between first flower and first pod (R3) (photo thermal days)	10.00	10.90
FL-SD	Time between first flower and first seed (R5) (photo thermal days)	14.00	24.00
SD-PM	Time between first seed (R5) and physiological maturity (R7) photo thermal days	33.35	34.50
FL-LF	Time between first flower (R1) and end of leaf expansion photo thermal days	18.00	45.00
LFMAX	Maximum leaf photosynthesis rate at 30 C, 350 vpm CO2, and high light mg CO2/m2	1.05	1.00
SLAVR	Specific leaf area of cultivar under standard growth conditions cm2/g	350.00	285.0
SIZLF	Maximum size of full leaf (three leaflets) (cm2)	185.00	110.00
XFRT	Maximum fraction of daily growth that is partitioned to seed + shell	1.00	1.00
WTPSD	Maximum weight per seed (g)	0.176	1.10
SFDUR	Seed filling duration for pod cohort at standard growth conditions photo thermal days	42.50	21.00
SDPDV	Average seed per pod under standard growing conditions (#/pod)	2.07	2.40
PODUR	Time required for cultivar to reach final pod load under optimal conditions (photo thermal days)	10.00	18.00
THRSH	The maximum ratio of (seed/ (seed+ shell)) at maturity	78.00	77.00
SDPRO	Fraction protein in seeds (g (protein)/g (seed))	0.40	0.315

Table 3: Genetic coefficients used in CROPGRO-model characterize the growth and development of maize, broad bean and lentil varieties, which were obtained from Model calibration.

simulation is considered excellent with RMSE<10%, good if 10–20%, fair if 20–30%, and poor >30% for yield and yield components, the mean square error (MSE) was calculated into a systematic (MSEs).

• The Index of agreement (d) as described by Wilmott et al. [4] was estimated as shown in the following equation:

$$d=1-\left[\frac{\sum_{i=1}^{n}(P_i-O_i)^2}{\sum_{i=1}^{n}(|P_i|-|O_i|)^2}\right]$$

Where n is the number of observations, Pi the predicted observation, Oi is a measured observation, P'i = Pi −M and O'i = Oi −M (M is the mean of the observed variable). So if the d-statistic value is closer to one, then there is good agreement between the two variables that are being compared and vice versa.

The correlation coefficient between observed and predicted data was calculated to show the trend in observed and predicted data. Correlation coefficient: the measure of liner relationship between two variables x and y.

Characteristics studied by CERES and CROPGRO models: At the end of that study, comparison study between the observed and predicted data for the seed or grain yield (Kg/ha) and the harvest index of each studied crop according to the crop simulation program of DSSAT V.4.6.1 program (CERES-Cereal model and CROPGRO-Legumes model) because that traits is the best parameter to observe about the treatment crop effort done under the condition of thread heeds of conservation agriculture triangle (No tillage, permanent soil cover with crop residuals and crop rotations with different plant species).

Results

Maize after broad bean (summer 2013)

Results presented in Table 4 show the effect of tillage systems, NPK fertilizer levels as well as the interaction between them on studied traits of maize during summer 2013 season through (maize→ broad bean→ maize) crop sequence. It is worthy to mention that, insignificant differences had been achieved between Conventional agriculture (CT), semi-conservation agriculture (SCA) and conservation agriculture (CA) for the studied maize traits Figure 2.

Referring to, fertilizer levels, results in the same previous Table 5 indicated that, the recommended doses of NPK significantly favored maize cone length (cm), cone weight (cm), biological yield/fed and grain yield/fed as compared by 1/2 dose of recommended NPK fertilizers by 28.84%, 10.98%, 30.37% and 27.35% respectively. On the other side, the results indicated that, the 1/2 recommended doses of NPK significantly favored maize harvest index as compared by the recommended doses of it by 2.34%.

Concerning to the interaction between studied treatments, results recorded in the same previous Table cleared that, the application of conservation agriculture (CA) and fed by the recommended dose of NPK or half recommended dose of NPK fertilizers scored the greatest value for all maize traits as compared with the other treatments. On

Treatments		Cone length (cm)	Cone weight (g)	Biological yield (kg/fad)	Grain yield (kg/fad)	Harvest index
Tillage systems	Fertilizer level					
Conventional agriculture (CT)	100% NPK	25.00	391.70	12959.10	4975.00	0.384
	50% NPK	18.33	350.33	9956.00	3918.00	0.394
	Mean	**21.67**	**371.00**	**11457.60**	**4446.50**	**0.389**
Semi-conservation agriculture (SCA)	100% NPK	25.33	388.33	12724.30	4842.67	0.381
	50% NPK	20.33	345.00	9761.79	3783.33	0.388
	Mean	**22.83**	**366.67**	**11243.00**	**4313.00**	**0.384**
Conservation agriculture (CA)	100% NPK	25.67	392.00	13061.30	5042.67	0.386
	50% NPK	20.33	360.67	10000.40	3966.67	0.397
	Mean	**23.00**	**376.34**	**11530.90**	**4504.67**	**0.391**
	General Mean TS	**22.50**	**371.33**	**11410.50**	**4421.39**	**0.392**
	Mean of NPK					
	100% NPK	25.33	390.67	12914.90	4953.44	0.384
	50% NPK	19.66	352.00	9906.07	3889.33	0.393
LSD at 5%						
Tillage systems (TS) =		NS	NS	NS	NS	NS
Fertilizer (F) =		3.93	28.80	377.05	139.01	0.013
TS x F =		7.11	39.34	776.80	286.39	0.027

Table 4: Effect of tillage system and fertilizer levels as well as the interaction between them on yield and yield component of maize through cropping system of (maize→ brad bean→ maize) in season, 2013.

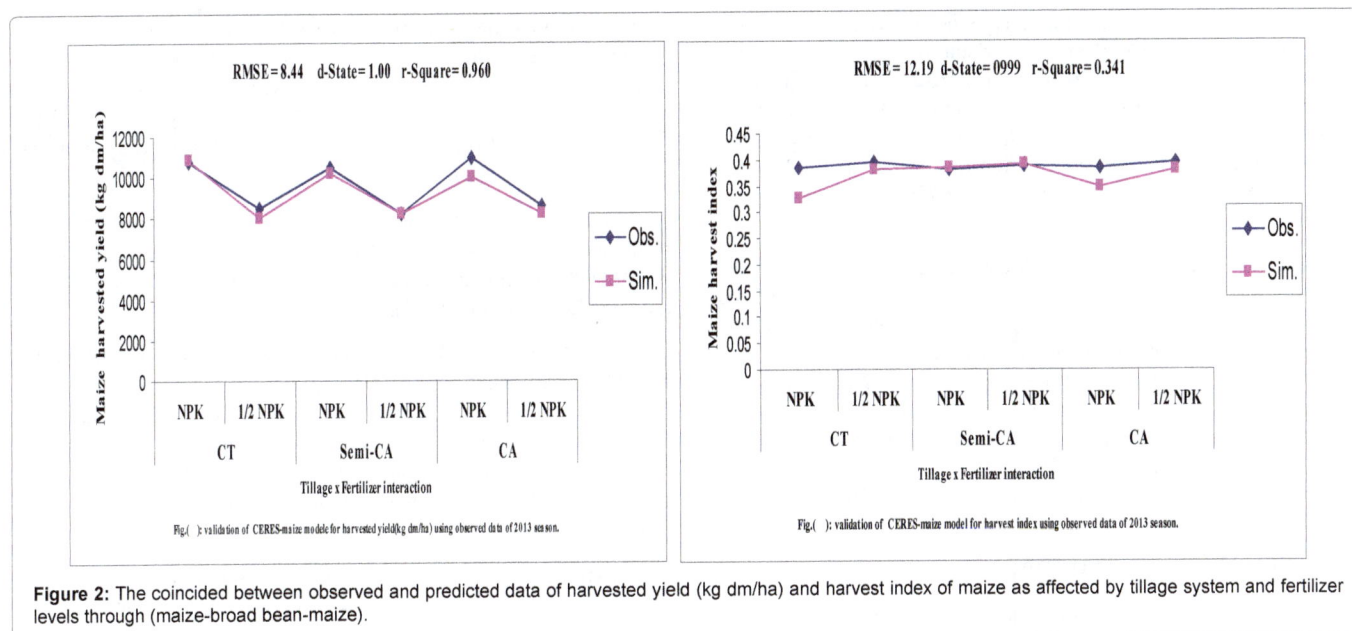

Figure 2: The coincided between observed and predicted data of harvested yield (kg dm/ha) and harvest index of maize as affected by tillage system and fertilizer levels through (maize-broad bean-maize).

Crop sequence		Maize after broad bean (Summer,2013)			
		Grain yield (kg dm/ha)		Harvest index	
Treatments		Observed	Simulated	Observed	Simulated
CT	NPK	10775	10820	0.384	0.328
	1/2 NPK	8486	8014	0.394	0.380
Semi-CA	NPK	10488	10212	0.381	0.385
	1/2 NPK	8194	8220	0.388	0.391
CA	NPK	10921	10000	0.386	0.350
	1/2 NPK	8591	8210	0.397	0.380
Validation CERES-Model					
RMSE=		8.44		12.19	
d-State=		1.000		0.999	
r -Square		0.960		0.341	
Coincided degree		Excellent		Good	

Table 5: The coincided between observed and predicted data of seed yield (kg dm/ha) and harvest index of maize as affected by tillage system and fertilizer levels through (maize-broad bean-maize) crop sequences.

Treatments		No. of pods/plant	Biological yield (kg/fad)	Seed yield (kg/fad)	100- seed Weight(g)
Tillage systems	Fertilizer level				
Conventional agriculture (CT)	100% NPK	18.00	4623	1182	73.28
	50% NPK	12.67	2693	1062	72.72
	Mean	**15.33**	**3658**	**1122**	**73.00**
Semi-conservation agriculture (SCA)	100% NPK	19.67	4783	1194	74.73
	50% NPK	18.00	4737	1172	74.61
	Mean	**18.83**	**4760**	**1183**	**75.00**
Conservation agriculture (CA)	100% NPK	30.00	7290	1517	91.6
	50% NPK	28.67	6137	1403	88.24
	Mean	**29.33**	**6714**	**1460**	**90.00**
	General Mean TS	**21.17**	**5044**	**1255**	**79.2**
	Mean of NPK				
	100% NPK	22.56	5566	1298	79.87
	50% NPK	19.78	4522	1212	78.52
LSD at 5%					
Tillage systems (TS) =		**0.46**	**110**	**49**	**2.66**
Fertilizer (F) =		**2.13**	**263**	**NS**	**NS**
TS x F =		**8.63**	**543**	**388**	**14.33**

Table 6: Effect of tillage system and fertilizer levels as well as the interaction between them on yield and yield component of broad bean through cropping system of (maize → broad bean→ maize) in season, 2013/2014.

contrast, the lowest values for maize cone length (cm), was resulted under the condition of Conventional agriculture (CT) and fed by the half-recommended dose of NPK fertilizers (18.33 cm). On the other hand, the lowest values for maize cone weight (cm), biological yield/fed and grain yield/fed were resulted under the condition of semi-conservation agriculture (SCA) and fed by the half recommended dose of NPK fertilizers (345 g), (9761.79 kg/fad), (2819.00 kg/fad) respectively.

Validation data by CERES-maize model (predicted data): The values of (RMSE), (D-state) and (r-Square) parameter, which used to make a judgment of the coinciding degree, between observed and predicted data of maize traits as affected by the interaction effect between tillage and fertilizer treatments, showed different levels of the coinciding degree grain yield (kg/ha) and harvest index in summer 2013 season, showed excellent and good compliance (RMSE =8.44 and 12.19), D-state were (1.00 and 0.999) and (r-Square = 0.960 and 0.341) between the observed and predicted data as affected by the previous interaction.

Broad bean after maize (2013/2014)

As shown in the Table 6, shows that broad bean no. of pods/plant, biological yield/fed, seed yield/fed) and 100-seed weight (g) as affected by tillage systems, fertilizer level and the interaction effect between them through (maize→ brad bean→ maize) crop sequence in winter 2013/2014 season. As a matter of fact, results revealed that, conservation agriculture (CA) significantly pronounced its superiority reflected on increase broad bean no. of pods/plant by 91.32%, biological yield/fed by 83.52%, seed yield/fed by 30.16% and 100-seed weight (g) by 23.17% as compare by Conventional agriculture (CT) system.

In relation to, fertilizer levels, results in the previous Table showed that, the recommended doses of NPK significantly favored broad bean no. of pods/plant and biological yield/fed as compared by 1/2 dose of recommended NPK fertilizers by 14.05% and 23.07% respectively. On the opposite of, there are no significant effect between the recommended doses of NPK and 1/2 recommended doses of it for seed yield/fed) and 100-seed weight (g).

In reference to, the interaction effect between studied treatments,

results indicated that, cultivating broad bean under the condition of conservation agriculture (CA) and fed by the recommended dose or half dose of NPK fertilizers scored the greatest value for no. of pods/plant (30, 28.67), biological yield/fed (7290, 6137 kg), seed yield/fed (1517, 1403 kg) and 100-seed weight (91.60, 88.24g) and the differences between them not reach to the significant level.

On the contrary, the lowest value for above mentioned traits was resulted under the condition of Conventional agriculture (CT) and fed by the half recommended dose of NPK fertilizers (12.67), (2693 kg/fad.), (1062 kg/fad) and (72.72 g) respectively.

Validation data by CROPGRO-faba bean model(winter 2013/2014 season): Results recorded in the Table 7 and Figure 3 show that, validation indexes which used to measure the simulation accuracy for faba bean characters ranged between good simulation accuracy for both of seed yield (kg/ha) and harvest index (RMSE=11.70 and 16.79), D-state =0.999 and 0.999. As for r-Square was 0.974 and 0.926 respectively.

This trend is in harmony with previous results reported by Hassanein MK [12] who studied faba bean yield and growth predictability using CROPGRO-legume model. It could be concluded that CROPGRO legume model could be used to predict yield and growth of faba bean under Egyptian conditions. In addition to, Oliveira et al. [13] who evaluates the CROPGRO-Dry bean model for simulating dry bean yield. The results show that the crop model can correctly reproduce the observed yield. This finding may indicate that the model is a useful tool to evaluate the crop response to variability and changing climate.

Maize after broad bean (summer 2014)

Results presented in Table 8 described, maize cone length (cm), cone weight (g), biological yield (kg/fad), grain yield (kg/fad) and harvest index as affected by tillage systems, fertilizer level and the interaction effect between them through (maize→ broad bean→ maize) crop sequence in 2014 season. Results indicated that, conservation agriculture (CA) significantly pronounced its superiority reflected on increase maize cone length (cm) by (13.48%, 1.29%), cone weight (g) by (33.62%, 9.5%), biological yield (kg/fad.) by (69.18%, 14.14%) and

Crop sequence		Broad bean after maize (winter, 2013/2014)			
		Seed yield (kg dm/ha)		Harvest index	
Treatments		Observed	Simulated	Observed	Simulated
CT	NPK	2503	2288	0.256	0.252
	1/2 NPK	2249	2000	0.394	0.32
Semi-CA	NPK	2530	2500	0.25	0.26
	1/2 NPK	2482	2400	0.247	0.241
CA	NPK	3213	3050	0.208	0.202
	1/2 NPK	2973	2950	0.229	0.200
Validation CROPGRO-Model					
RMSE=		11.70		16.79	
d-State=		0.999		0.999	
r -Square		0.974		0.926	
Coincided degree		Good		Good	

The simulation is considered excellent with RMSE<10%, good if 10–20%, fair if 20–30%, poor >30%

Table 7: The coincided between observed and predicted data of seed yield (kg dm/ha) and harvest index of broad bean as affected by tillage system and fertilizer levels through (maize-broad bean-maize) crop sequences.

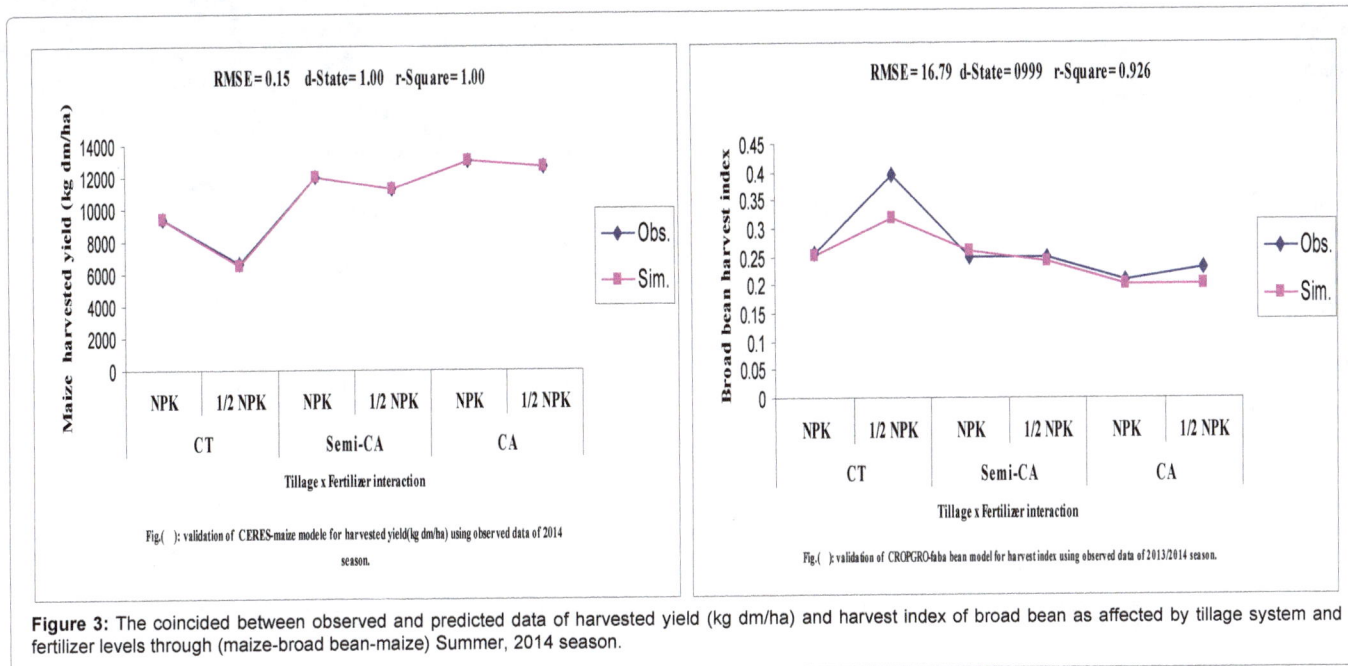

Figure 3: The coincided between observed and predicted data of harvested yield (kg dm/ha) and harvest index of broad bean as affected by tillage system and fertilizer levels through (maize-broad bean-maize) Summer, 2014 season.

grain yield (kg/fad) by (60.75%, 9.73%) as compared with either of Conventional agriculture (CT) or semi-CA respectively.

As for, fertilizer levels, results showed that, there are no significance effect between the recommended doses and the half dose of NPK for maize cone length (cm), cone weight (g). On the other side, the results also, indicated that, the recommended doses of NPK significantly favored maize biological yield (kg/fad), grain yield (kg/fad) and harvest index as compared by 1/2 dose of recommended NPK fertilizers by 5.94%, 12.72% and 6.35% respectively.

In respect of, the effect of first order interaction between tillage system and fertilizer levels, results revealed that, cultivating maize under the condition of conservation agriculture (CA) and fed by the recommended dose or half dose of NPK fertilizers exposed its superiority over than the same level of treatments reflected on gave the greatest value for cone length (27-26.33 cm), cone weight (386.67-381.67 g), biological yield/fed (18661.98-18600.20 kg), grain yield/fed (5981.67-5833.33 kg).

As for harvest index results revealed that the application of Conventional agriculture (CT) + recommended dose of NPK gave, the greatest value (0.358) for that trait and the differences between them reached to the significant level.

On the opposite side, the lowest values for maize pervious traits were resulted under the condition of Conventional agriculture (CT) and fed by the half recommended dose of NPK fertilizers (22.67 cm), (258 g), (9993.33 kg/fad) and (3037.33 kg/fad)and (0.304) respectively.

Validation data by CERES-maize model (summer 2014 season): Results recorded in Table 9 and Figure 4 show that simulation accuracy for maize characters as affected by (tillage x fertilizer) interaction. The results cleared that the calibration indexes (RMSE, D- state and r-Square) showed excellent and good simulation accuracy for both of seed yield (kg/ha) and harvest index (RMSE =0.15 and 12.02), (D-state= 1.00 and 0.999) and (r-Square = 1.00 and 0.790), respectively.

These results in agreement with El-Marsafawy [14] who found that

Treatments		Cone length (cm)	Cone weight (g)	Biological yield (kg/fad)	Grain yield (kg/fad)	Harvest index
Tillage systems	Fertilizer level					
Conventional agriculture (CT)	100% NPK	24.33	316.67	12031.70	4312.33	0.358
	50% NPK	22.67	258.33	9993.33	3037.33	0.304
	Mean	23.50	287.50	11012.50	3674.83	0.331
Semi-conservation agriculture (SCA)	100% NPK	27.67	385.00	16599.90	5566.67	0.335
	50% NPK	25.00	316.67	16046.70	5200.00	0.324
	Mean	26.33	350.83	16323.30	5383.33	0.33
Conservation agriculture (CA)	100% NPK	27.00	386.67	18662.00	5981.67	0.321
	50% NPK	26.33	381.67	18600.20	5833.33	0.314
	Mean	26.67	384.17	18631.10	5907.50	0.318
General Mean TS		25.50	340.83	15322.30	4988.56	0.326
Mean of NPK						
	100% NPK	26.33	362.78	15764.50	5286.89	0.335
	50% NPK	24.67	318.89	14880.10	4690.22	0.315
LSD at 5%						
Tillage systems (TS) =		1. 03	26.71	39.85	71.14	0.004
Fertilizer (F) =		NS	NS	292.11	187.02	0.012
TS x F =		4.06	97.49	601.81	385.3	0.024

Table 8: Effect of tillage system and fertilizer levels as well as the interaction between them on yield and yield component of maize through cropping system of (maize→ broad bean→ maize) in season, 2014.

Crop sequence		Maize after broad bean (Summer, 2014)			
		Grain yield (kg dm/ha)		Harvest index	
Treatments		Observed	Simulated	Observed	Simulated
CT	NPK	9340	9332	0.358	0.357
	1/2 NPK	6578	6560	0.304	0.210
Semi-CA	NPK	12056	12050	0.335	0.335
	1/2 NPK	11262	11258	0.324	0.323
CA	NPK	12955	12950	0.321	0.321
	1/2 NPK	12634	12634	0.314	0.314
Validation CERES-Model					
RMSE=		0.15		12.02	
d-State=		1.000		0.999	
r -Square		1.000		0.790	
Coincided degree		Excellent		Good	

The simulation is considered excellent with RMSE<10%, good if 10–20%, fair if 20–30%, poor >30%

Table 9: The coincided between observed and predicted data of seed yield (kg dm/ha) and harvest index of maize as affected by tillage system and fertilizer levels through (maize-broad bean-maize) crop sequences.

CERES maize-model could be used in the Delta region to simulate maize productivity, water needs through different cropping patterns in Egypt.

Also, Abdrabbo, et al. [15] who found that, the Calibration and validation of CERES-Maize crop simulation model using experimental datasets of years 2011 and 2012 were done successfully giving excellent values for RMSE and d-Stat evaluation indexes.

Discussion

The results revealed that, by applying the conservation agriculture instructions (1. minimum soil disturbance, 2- permanent soil cover with crop residuals or cover crops and 3. Crop rotation with different plant species, which include legumes) starting from summer season 2013 with maize crop through winter season 2013/2014 with broad bean and summer season of 2014 with maize in the same cites, the results recorded gradually improvement started from non-significant differences between the three tested tillage systems on maize studied traits that agree with Peigne [2] who found that, zero tillage with residue retention is characterized by slower initial maize growth, compensated for by an increased growth in the later stages, positively influencing final maize grain yield. They added that, zero tillage with retention of crop residue resulted in time efficient use of resources as opposed to Conventional agriculture. Also, Paz [16] who revealed that, maize yield when cropped under no-till system present higher productivity combined with crop rotation than under continuous cropping; lower productivity tends to occur under Conventional agriculture and the difference in productivity under no-till using crop rotation and continuous cropping is 1,000 kg/ha for maize. Moreover, Zheng [17] founded that, CA practices were significantly higher in maize yield (7.5%) as compared with Conventional agriculture (CT).

In addition, that may be due to improved soil aggregate stability, soil health and quality, reduce erosion and improve water use under CA as reported by Grigoras [18]. Through winter 2013/2014 season with broad bean, started CA or SCA (semi-CA) pronounced their superiority reflecting an increase of almost broad bean traits such as, No. of pods/plant, biological yield/fed, seed yield/fed, and 100-seed

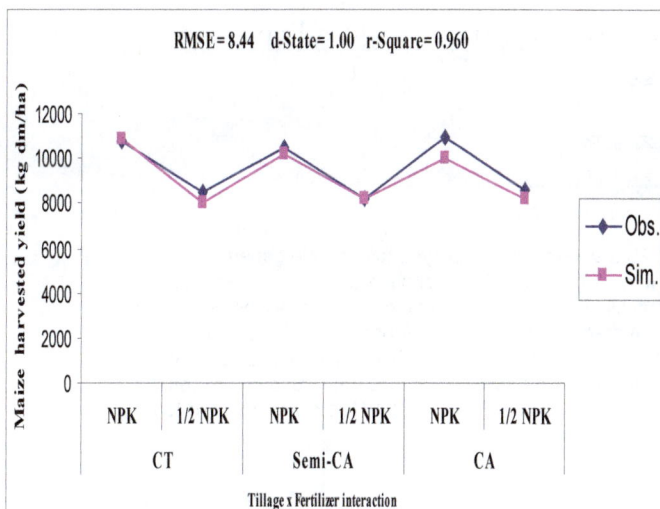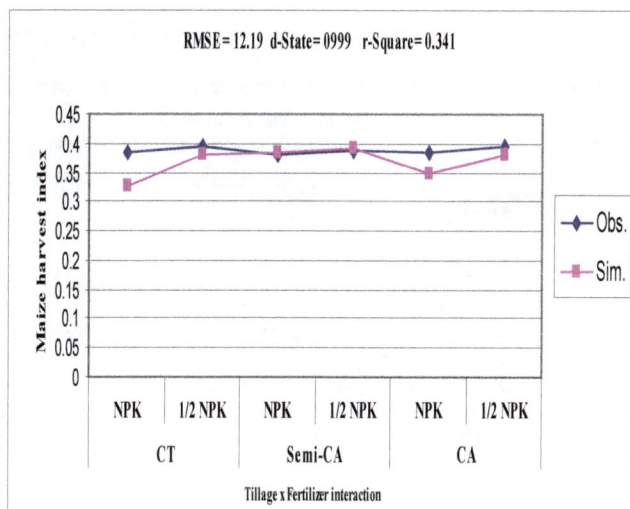

Figure 4: The coincided between observed and predicted data of harvested yield (kg dm/ha) and harvest index of maize as affected by tillage system and fertilizer levels through (maize-broad bean-maize).

weight (g) these results may be attributed to the accumulate effect of nutrients in the soil as appositive effect of CA or SCA compared by (CT) system [19].

After harvesting broad bean and by applying the three tillage systems and cultivate maize, also CA or SCA tillage system led to more positive effect on the studied maize traits, these results probably attributed to the role of the residual organic nitrogen as constructive element come from planting broad bean before.

As for, the results of first order interaction effect between tillage system and fertilizer NPK rate through the crop sequences maize → broad bean→ maize for each crop. It is very interesting to mention that, CA or SCA led to save half dose of NPK fertilizer rate for each crop and that gained by the greatest values of studied traits for maize, broad bean and maize through 2013, 2013/2014 and 2014.

References

1. Melero S, López-Garrido R, Murillo JM, Moreno F (2009) Conservation tillage: short- and long-term effects on soil carbon fractions and enzymatic activities under Mediterranean conditions. Soil Till Res 104: 292-298.

2. Peigne J, Ball BC, Roger-Estrade J, David C (2007) Is conservation tillage suitable for organic farming? A review. Soil Use and Management 23: 129-144.

3. Verhulst N, Govaerts B, Verachtertb E, Castellanos AN, Mezzalamaa M, et al. (2010) Conservation agriculture, improving soil quality for sustainable production systems? In: Lal R, Stewart BA (eds.) Advances in Soil Science: Food Security and Soil Quality. CRC Press, Boca Raton, pp. 137-208.

4. Willmott CJ, Ackleson SG, Davis RE, Feddema JJ, Klink KM, et al. (1985) Statistics for the evaluation and comparison of models. J Geophys Res 90: 8995-9005.

5. Hoogenboom G, Jones JW, Porter CH, Wilkens PW, Boote KJ, et al. (2010) Decision support system for agrotechnology transfer, Version 4.5. Overview. University of Hawaii, Honolulu, HI, USA 1.

6. Godwin DC, Singh U (1998) Nitrogen balance and crop response to nitrogen in upland and lowland cropping systems. In: Tsuji GY, Hoogenboom G, Thornton (eds.) Understanding options for agricultural production Kluwer Acad Publ, Boston. pp. 55-77.

7. Gijsman AJ, Hoogenboom G, Parton WJ, Kerridge PC (2002) Modifying DSSAT crop models for low input agricultural systems using a soil organic matter-residue module from CENTURM. Agron J 94: 462-474.

8. Gomez KA, Gomez AA (1984) Statistical Procedures for Agricultural Research. (2ndedn) Wily, New York 704.

9. Dadoun FA (1993) Modeling tillage effects on soil physical properties and maize (Zea mays, L.) development and growth. Unpublished PhD thesis, Michigan State University, MI.

10. Panda RK, Behera SK, Kashyap PS (2003) Effective management of irrigation water for wheat under stressed conditions. Agricultural Water Management 63: 37-56.

11. Loague K, Green RE (1991) Statistical and graphical methods for evaluating solute transport models: overview and application. J Contam Hydrol 7: 51-73.

12. Hassanein MK, Medany MA, Haggag ME, Bayome SS (2003) Prediction of yield and growth of faba bean using cropgro legume model under Egyptian conditions. ISHS Acta Horticulture 729: 215-219.

13. Oliveira Ch, Costa JMN, Junior TJP, Ferreira WPM, Justino FB, et al. (2012) The performance of the CROPGRO model for bean (Phaseolus vulgaris L.) yield simulation. Acta Sci Agron Maringa 34: 239-246.

14. El-Marsafawy SM, Eid HM, Moustafa AT, El-Rayes S (2000) Simulation of maize yield under different sowing dates using crop system model. 5th Conference Meteorology and Sustainable Development 226-234.

15. Abdrabbo MAA, Hashem FA, Elsayed ML, Farag AA, Abul-Soud MA, et al. (2013) Evaluation of CSM-Ceres-Maize Model for Simulating Maize Production in Northern Delta of Egypt. Life Sci J 10: 3179-3192.

16. Paz CE (1999) Program of Agricultural Sustainable (pas) in Santa Cruz of the Mountain Range. Published in Memories of National Meeting of Wheat and Smaller Cereals 18-189.

17. Zheng C, Jiang Y, Chen C, Sun Y, Feng J, et al. (2014) The impacts of conservation agriculture on crop yield in China depend on specific practices, crops and cropping regions. Crop J 2: 289-296.

18. Grigoras MA, Popescu A, Pamfil D, Has I, Cota LC (2011) Effect of Conservation Agriculture on Maize Yield in the Transilvanian Plain, Romania. Int J of Biological Veterinary and Agric and Food Eng 5.

19. Parton WJ, McKeown B, Kirchner V, Ojima DS (1992) CENTURY user's manual. Colorado State Univ, NREL Publ, Fort Collins, CO, USA.

17

Genetic Diversity Analysis in Chickpea Employing ISSR Markers

Gautam AK[1], Gupta N[1], Bhadkariya R[1], Srivastava N[2] and Bhagyawant SS[1]*

[1]*School of Studies in Biotechnology, Jiwaji University, Gwalior, 474011, India*
[2]*Department of Bioscience and Biotechnology, Banasthali University, Banasthali, 304022, India*

Abstract

In the present study, inter simple sequence repeat (ISSR) markers were employed to estimate genetic diversity in 13 accessions of chickpea including cultivated and wild. Among all these anchored ISSR primers tested, pentanucleotide repeat primer UBC-879 produced better amplification patterns. A total of 150 bands were amplified in a molecular weight range of 100-2000 bps revealing an average of 21.4 bands per primers and 1.64 bands per primer per genotype. The repeats $(GA)_8C$, $(AG)_8YT$, $(GA)_8YC$, $(AG)_8C$, $(GTT)_6$ and $(GT)_8YC$ give least amplification. UPGMA dendrogram constructed between these accessions depicted three major clusters. Based on genetic origin and diversity index viz. ICC-14051, ICC-13441, ICC-15518, ICC-12537, and ICC-17121 recommended to be selected as a parent in future breeding programmes for chickpea.

Keywords: Chickpea; ISSR; Legumes; PCR

Introduction

Genetic diversity is the essence of crop improvement. Traditionally, diversity is examined by measuring variation in morphological parameters, biochemical parameters or using molecular markers. Environmental factors influence the expression of morphological and biochemical traits. The DNA based molecular markers are currently used for Marker Assisted Selection (MAS). ISSR or Simple Sequence Repeats (SSRs) are short tandem repetitive DNA sequences with a repeat length of few (1-6) base pairs [1]. These sequences are abundant, dispersed throughout the genome and are highly polymorphic in comparison with other molecular markers [2].

The applications of ISSR markers in plants are well known [3]. These are PCR based markers that permits detection of polymorphism in microsatellite and inter-microsatellite loci, without prior knowledge of DNA sequence [4]. In this technology, the SSR primers have either 5' or 3' extension of one or more bases including anchor sequences thereby making the profile easier for analysis [5].

Chickpea (*Cicer arietinum* L.) belongs to the family of leguminosae, is the third most important pulse crop of the world and India is the largest producer country [6]. Technically, it belongs to the tribe *Cicereae* Alef of the family *Leguminosae* with genus *Cicer* and species *arietinum*. Chickpea seeds are nutritious due to the presence of protein, carbohydrates, vitamins and minerals [7]. It contains well-balanced amino acids and low levels of anti-nutritional factors in comparison to other grain legumes [8]. Some of the biotic and abiotic factors reduce chickpea production worldwide and search for elite genotypes are being continuously carried out employing PCR-based markers [3]. The Indian Institute of Pulses Research, Kanpur, India maintains more than 3000 chickpea accessions as a primary gene pool stock (www.iipr.res.in). Assessment of genetic diversity from the available genetic stocks is a key factor aimed at improvement of crop performance [9]. In our earlier reports, ISSR-PCR analysis showed correlation with pedigree data amongst the 20 accessions of chickpea which included the species of *C. arietinum* and *C. reticulatium* [10]. Present investigation was undertaken to analyse genetic diversity between the 13 chickpea accession of *C. arietinum* and *C. reticulatum* available in the primary gene pool accessions.

Materials and Methods

Chemcials

Agarose, CTAB, Tris-HCl, EDTA, SDS, sodium chloride were purchased from Hi Media. dNTPs and Taq DNA polymerase was obtained from MBI, Fermentas, Richlands B.C. ISSR primers were obtained from Operon Technologies Ltd., (Alameda, California). All other chemicals and reagents used were of analytical grade.

Seed materials

A representative set of 13 accessions of *Cicer* that included cultivated and wild. Agronomic details of these accessions are given in Table 1. The collections encompassing eight geographic origins of world are shown in the Figure 1. All accessions were obtained from International Crops Research Institute for the Semi-Arid Tropics (ICRISAT) Patancheru, Hyderabad, India, under MTA understanding.

DNA isolation

DNA was isolated using the CTAB extraction method of Talebi et al. [11] with minor modifications. Hundred milligram seed material was ground in liquid nitrogen followed by homogenization with 1 ml freshly prepared extraction buffer. To this, 20% SDS was added and incubated at 60°C for 30 min. Then after, 92 µl of 5 M NaCl was added and subsequently, 75 µl of CTAB solution was added and re-incubated at 65°C for 15 min. To this cocktail, 300 µl of chloroform: Isoamyl alcohol mix (24:1) was added. This was followed by centrifugation at 12,000 g for 15 min at 4°C in a Sigma centrifuge 3-16 K. Chloroform: Isoamyl alcohol mix was readded to the supernatant in 1:1 volume and re-centrifuged at 12,000 g for 15 min at 4ºC. Subsequently, precipitation

*Corresponding author: Bhagyawant SS, School of Studies in Biotechnology, Jiwaji University, Gwalior, 474011, India
E-mail: sameerbhagyawant@gmail.com

S. No.	Chickpea accessions	Type	Origin	Agronomic details	Geolocation (Latitude, Longitude)
1	ICC-14051	*Cicer arietinum* (L.)	Ethiopia	Drought tolerant	8.97°N, 42.22°E
2	ICC-7554	*Cicer arietinum* (L.)	Iran	Resistant to fungal disease	32.42°N, 53.68°E
3	ICC -2307	*Cicer arietinum* (L.)	Myanmar	*Helicoverpa* resistant	20.78°N, 97.02°E
4	ICC-13441	*Cicer arietinum* (L.)	Iran	*Helicoverpa* resistant	32.42°N, 53.68°E
5	ICC-15518	*Cicer arietinum* (L.)	Morocco	Drought tolerant	34.34°N, 5.32°W
6	ICC-1422	*Cicer arietinum* (L.)	India	Water use efficiency	17.38°N, 78.48°E
7	ICC-11627	*Cicer arietinum* (L.)	India	*Ascochyta blight* resistant	31.53°N, 75.91°E
8	ICC-6263	*Cicer arietinum* (L.)	Russia	Salinity stress tolerant	61.52°N, 105.31°E
9	ICC-12537	*Cicer arietinum* (L.)	Ethiopia	*Helicoverpa* resistant	8.85°N, 38.77°E
10	ICC-11944	*Cicer arietinum* (L.)	Nepal	*Ascochytablight* resistant	28.12°N, 82.30°E
11	ICC-306	*Cicer arietinum* (L.)	Russia	Resistant to pod borer	61.52°N, 105.31°E
12	ICC-4958	*Cicer arietinum* (L.)	India	Drought tolerant	17.38°N, 78.48°E
13	ICC-17121	*Cicer reticulatum*	Turkey	Higher branches	37.53°N, 40.88°E

Table 1: Agronomic details of *Cicer* accessions used in the study.

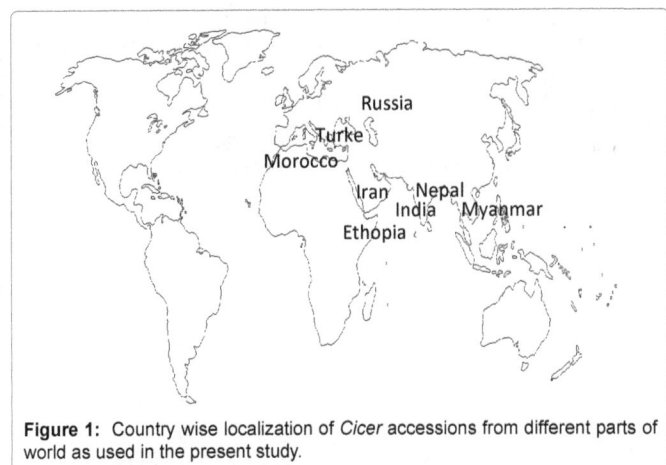

Figure 1: Country wise localization of *Cicer* accessions from different parts of world as used in the present study.

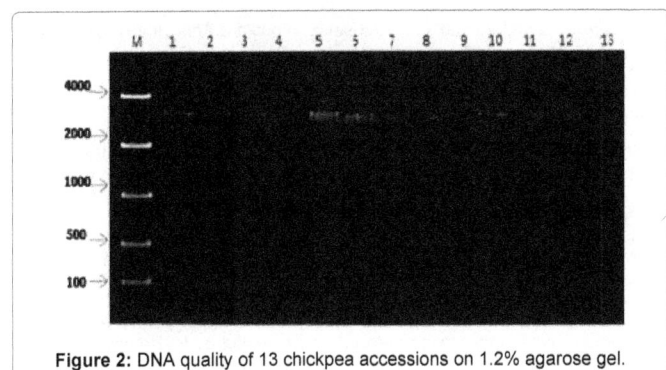

Figure 2: DNA quality of 13 chickpea accessions on 1.2% agarose gel.

was done by adding chilled isopropanol 40% v/v as final concentration. The precipitated DNA was then centrifuged as a pellet and cleared with 70% ethanol. The ethanol washed DNA was air dried and dissolved in 100 µl of Tris-EDTA buffer (19 mM Tris-HCl pH 8.0; 1 mM EDTA pH 8.0). The quality of extracted DNA was checked on 1.2% agarose gel (Figure 2) and further stored at -20°C until use.

ISSR-PCR amplifications

The PCR procedure for DNA amplification was employed according to the method of Welsh and Mcclelland [4]. The reaction cocktail contained 25 µl reaction volumes containing 10 mM Tris-HCl pH 9.0; 50 mM KCl; 0.1% Triton-x-100; 1.5 mM $MgCl_2$; 0.1 mM dNTP; 2 mM primer; 0.5 unit of Taq DNA polymerase and 25 ng template DNA. Amplifications were carried out in a Bio-Rad 3.03 version thermo-

cycler. Programme was set for 35 cycles with an initial melting at 94°C for 4 min, followed by denaturation at 94°C for 1 min. The annealing was performed at 56°C for 1 min, followed by polymerization at 72°C for 2 min. Final extension was carried out at 72°C for 7 min. Details of ISSR primers are represented in the Table 2.

Agarose gel electrophoresis

The PCR products were separated on 1.5% agarose gels in 1X TAE by electrophoresis at 100 V for 3 h and bands were detected by ethidium bromide staining. 4 kb standard DNA molecular weight was used as a marker (Figure 2). Clearly resolved bands were scored visually for their presence or absence. Jaccard's similarity coefficient [12] was estimated from these binary data using Past [13] software.

Results and Discussion

The chickpea accessions possessing important agronomic characters like higher yield potential with early maturity and wilt resistance were analyzed for estimating genetic diversity using 10 ISSR primers. The primers tested were two pentanucleotide viz. $(CTTCA)_3$ and $(GGAGA)_3$, three trinucleotide $(ATG)_6$, $(GTT)_6$ $(TAT)_6$ and five dinucleotide anchor repeat primers. Among all these anchored ISSR primers tested, pentanucleotide repeat primer UBC-879 produced better amplification patterns. The 3 primers did not amplify DNA of any chickpea genotypes tested. Such non-amplifying primers were also been reported earlier [14]. A total of 150 bands were amplified across the 13 genotypes with these primers in a molecular weight range of 100-2000 bps revealing an average of 21.4 bands per primers and 1.64 bands per primer per genotype. The total number of bands amplified by 3' anchored primers varied from 20-50. The primer $(CTTCA)_3$ has amplified maximum number of 80 bands across the these accessions analyzed. Whereas the primer sequence $(AG)_8C$ has amplified least number of bands (40). The repeats $(GA)_8C$, $(AG)_8YT$, $(GA)_8YC$, $(AG)_8C$, $(GTT)_6$ and $(GT)_8YC$ give least amplification in the present study. This may be due to the region that chickpea genomic sequences may not be in the range of amplification by Taq DNA polymerase.

The cluster matrix method UPGMA was used and a dendrogram was constructed from a similarity matrix based on Jaccard's similarity coefficient values. The cophenetic correlation between ultrametric similarities of the tree and similarity matrix was high, indicating that the cluster analysis strongly represents the similarity data. The cluster analysis revealed the genetic relationships between the chickpea genotypes with similarity index values ranging from 0.11 to 0.19 with an average value of 0.15. Three broad clusters of chickpea genotypes were obtained from dendrogram as shown in Figure 3. Cluster one

S. No.	Primer	Anchor sequence	Tm (°C)	MW (Dalton)
1	UBC-809	5'-AGAGAGAGAGAGAGAGAGC-3'	47°C	5359
2	UBC-811	5'-GAGAGAGAGAGAGAGAGAC-3'	52°C	5359
3	UBC-834	5'-AGAGAGAGAGAGAGAGAGYT-3'	45°C	5374
4	UBC-841	5'-GAGAGAGAGAGAGAGAGAYC-3'	47°C	5359
5	UBC-850	5'-GTGTGTGTGTGTGTGTGTYC-3'	47°C	5287
6	UBC-864	5'-ATGATGATGATGATGATG-3'	48°C	5613
7	UBC-869	5'-GTTGTTGTTGTTGTTGTT-3'	41°C	5559
8	UBC-871	5'-TATTATTATTATTATTAT-3'	28°C	5469
9	UBC-879	5'-CTTCACTTCACTTCA-3'	36 °C	4439
10	UBC-880	5'-GGAGAGGAGAGGAGA-3'	45°C	4772

Table 2: ISSR primers used in the present study.

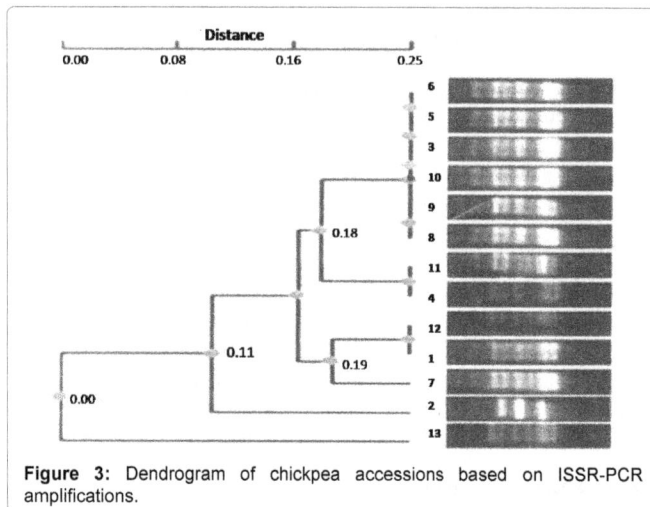

Figure 3: Dendrogram of chickpea accessions based on ISSR-PCR amplifications.

included chickpea accessions of ICC-12307, ICC-15518, ICC-1422, ICC-6263, ICC-12537, ICC-11944 and subclad of this cluster include accession ICC-6306 and ICC-13441. Cluster two includes accession ICC-14051 and ICC-4958 and subclade of this cluster include ICC-11627. Cluster three includes ICC-17121 as out group while genotype ICC-7554 has given polymorphy. The well-defined clustering was not observed between desi and kabuli genotype.

Applications of ISSR markers in genetic diversity analysis are becoming more popular for marker-assisted selection. Earlier studies of Ratnaparkhe et al. [15] revealed the inheritance of inter-simple sequence repeat polymorphisms and linkage analysis with *Fusarium* resistance gene in chickpea. The results shown that the ISSR loci they studied expressed virtually in complete agreement with the exception of mendalian segregation.

Literature perusal suggests that PCR can successfully forecast the growth habitats and geographic origin. Studies of Iruela et al. [16] in the genus *Cicer* and cultivated chickpea using combination of RAPD and ISSR markers of 26 accessions including kabuli and desi types were employed. Rao et al. [17], developed RAPD and ISSR fingerprinting in cultivated chickpea and its wild progenitor *Cicer reticulatum* L. They concluded ISSR analysis as a reliable attributes for estimation of genetic diversity than of RAPD. Fatemeh et al. [18], employed ISSR markers to fingerprint genetic diversity and conservation of landrace chickpea from north-west of Iran. ISSR markers are effectively scored for genetic diversity analysis of various crop plants. Amirul et al. [19], estimated genetic diversity among collected purslane accessions. Based on their assessment they recommended few of the accession that can be used as parents in future breeding program. In the present investigation, the

polymorphic band as produced by primer UBC-879 may disseminate genetic diversity. The bands specific for accession no ICC-14051, ICC-13441, ICC-15518, ICC-12537, and ICC-17121 could be further developed into a SCAR marker. The present analysis also reflects the known phylogenetic relationships in chickpea. Understanding genetic diversity in the primary gene pool collection is crucial for any crop improvement program [20]. Chickpea being a self-pollinated crop has a narrow genetic base. Looking at huge chickpea germplasm available and to reduce costs of field experimentation, genotype screening using PCR based markers is a prerequisite for breeding programme. Based on genetic origin and diversity index viz. ICC-14051, ICC-13441, ICC-15518, ICC-12537 and ICC-17121 recommended to be selected as a parent in future breeding programmes for chickpea. More efforts are needed to screen chickpea accessions which can estimate genetic diversity within the *Cicer* species to identify elite genetic stocks.

Acknowledgments

Authors are grateful to Dr. H. D. Upadhyay Head, Genebank, International Crops Research Institute for the Semi-Arid Tropics (ICRISAT) Patancheru, Hyderabad, India, for providing the seed materials.

References

1. Litt M, Luty JA (1989) A hypervariable microsatellite revealed *in vitro* amplification of dinucleotide repeat within cardiac muscle actin gene. Am J Hum Genet 44: 397-401.

2. Wang Z, Weber JL, Zhon G, Tanksley SD (1994) Survey of plant short tendem DNA repeats. Theor Appl Genet 88: 1-6.

3. Singh R, Sharma P, Varshney RK, Sharma SK, Singh NK (2008) Chickpea Improvement: Role of Wild Species and Genetic Markers. J Biotechnol Genet Eng Rev 25: 267-314.

4. Welsh J, Mcclelland M (1990) Fingerprinting genomes using PCR with arbitrary primers. Nucl Acids Res 18: 7213-7218.

5. Gupta M, Chy YS, Romero-Severson J, Owen JL (1994) Amplification of DNA markers from evolutionarily diverse genomes using single primers of simple-sequence repeats. Theor Appl Genet 89: 998-1006.

6. FAO, Agriculture Data (2010) United Nations Food and Agriculture Organization.

7. Wang N, Hatcher DW, Tyler RT, Toews R, Gawalko EJ (2010) Effect of cooking on the composition of beans (*Phaseolus vulgaris* L.) and chickpeas (*Cicer arietinum* L.). Food Res Int 43: 589-594.

8. Santiago CRA, Moreira-Araújo RSR, Pinto de Silva MEM, Arêas JAG (2001) The potential of extruded chickpea, corn and bovine lung for malnutrition programs. Innov Food Sci Emerg Technol 2: 203-209.

9. Renganayaki K, Read JC, Fritz AK (2001) Genetic diversity among Texas bluegrass (*Poa arachnifera torr.*) revealed by AFLP and RAPD markers. Theor Appl Genet 102: 1037-1045.

10. Yadav P, Koul KK, Shrivastava N, Mendaki MJ, Bhagyawant SS (2015) DNA polymorphisms in chickpea accessions as revealed by PCR-based markers. Cell Mol Biol 61: 84-90.

11. Talebi R, Naji AM, Fayaz F (2008) Geographical patterns of genetic diversity in cultivated chickpea (*Cicer arietinum* L.) characterized by amplified fragment length polymorphism. Plant Soil Environ 54: 447-452.

12. Jaccard P (1908) Nouvelles Researches Sur Ladistribution Florale. Bulletin dela Societe Vaudoise des Sciences Naturelles 44: 223-270.

13. Pavlicek A, Hrda S, Flegr JFT (1999) Freeware Program for Construction of Phylogenetic Trees on the Basis of Distance Data and Bootstrap/Jackknife Analysis of the Tree Robustness. Application in the RAPD Analysis of Genus Frenkelia. Folia Biologica (Praha) 45: 97-99.

14. Bhagyawant SS, Srivastava N (2008) Genetic fingerprinting of chickpea (*Cicer arietinum* L.) germplasm using ISSR markers and their relationships. African J of Biotechnology 7: 4428-4431.

15. Ratnaparkhe MB, Santra DK, Tullu A, Muehlbauer FJ (1998) Inheritance of inter simple sequence repeat polymorphism and linkage with *fusarium* wilt resistance gene in chickpea. Theor Appl Genet 96: 348-353.

16. Iruela M, Rubio J, Cubero JI, Gil J, Millán T (2002) Phylogenetic Analysis in the Genus *Cicer* and Cultivated Chickpea Using RAPD and ISSR Markers. Theor Appl Genet 104: 643-651.

17. Rao L, Usha RP, Deshmukh P, Kumar P, Panguluri S (2007) RAPD and ISSR Finger-Printing in Cultivated Chickpea (*Cicer arietinum* L.) and Its Wild Progenitor *Cicer reticulatum* Ladizinsky. Genetic Resources and Crop Evolution 54: 1235-1244.

18. Pakseresht F, Talebi R, Karami E (2013) Comparative Assessment of ISSR, DAMD and SCoT Markers for Evaluation of Genetic Diversity and Conservation of Landrace Chickpea (*Cicer arietinum* L.) Genotypes Collected from North-West of Iran. Physiol Mol Biol Plants 19: 563-574.

19. Md Amirul A, Abdul SJ, Rafii MY, Azizah AH, Arolu IW, et al. (2015) Genetic diversity analysis among collected purslane (*Portulaca oleracea* L.) accessions using ISSR markers. Comptes Rendus Biol 338: 1-11.

20. Choudhary P, Khanna SM, Jain PK, Bharadwaj C, Pramesh JK et al. (2013) Molecular Characterization of Primary Gene Pool of Chickpea Based on ISSR Markers. Biochem Genet 51: 306-322.

18

Registration of a Newly Released Sweet Potato Variety "Hawassa-09" for Production in Ethiopia

Gurmu F* and Mekonen S

South Agricultural Research Institute, Hawassa Research Center, Hawassa, Ethiopia

Abstract

Hawassa-09 (TIS-8250-1) is a white fleshed sweet potato variety that was selected from 12 genotypes and one local and three previously released varieties used as checks. The variety was developed by Hawassa Agricultural Research Center in southern part of Ethiopia. Hawassa-09 was released in 2017 for adaptation to low and mid altitude areas of southern and similar agro-ecologies of Ethiopia. Hawassa-09, along with the rest genotypes, has been evaluated in national variety trials across three locations, Hawassa, Halaba and Dilla for two consecutive years, 2014 and 2015. This variety gave a mean storage root yield of 49.2 t ha^{-1} with 56% and 283% yield advantage over the standard and local check, respectively. Then after selecting Hawassa-09 as a best variety, a variety verification trial was conducted for one more season in order to see the performance of the variety along with the checks across locations both on-station and on-farmers' fields on a 10 × 10 m plots. Finally, Hawassa-09 was officially released and registered as a new variety due to is outstanding performance. It is a stable, best adapted variety with medium sized roots and good resistance to sweet potato virus disease, the major sweet potato disease in Ethiopia.

Keywords: Hawassa-09; Registration; Storage root yield; Variety; verification

Introduction

Sweet potato [*Ipomoea batatas* (L.) Lam] is among world's most important food crops, especially in developing countries where it is produced as a staple food crop having adapted to a wide range of environmental conditions [1,2]. In Ethiopia, sweet potato is grown by smallholder farmers as one of the food security crops [3-5]. Sweet potato covered over 59,000 ha with production of over 2.7 million tons in the 2014/15 cropping season [6]. Sweet potato is an integral part of the cropping system in the eastern, southern and south western parts of the country [4,6-11]. Sweet potato is considered as life vest in its major production areas since it can give some yields when other cereal and legume crops fail due to drought and other weather related constraints. The adaptability of the crop to marginal environments and its production with minimal inputs makes it an important food security crop in the country [5].

There are about 18 white fleshed and six orange fleshed sweet potato varieties released in Ethiopia [4,12]. However, most of these varieties are obsolete and are not under production. Currently, only three varieties, namely Awassa-83, Kulfo and Tula are being produced by farmers. These varieties give relatively better root yield in areas where sweet potato virus diseases pressure is low. However, in hot spot areas, they are affected by sweet potato virus disease and the root yield is highly affected [13,14]. Resistance to sweet potato virus disease is one of the major breeding objectives since sweet potato virus disease is the major sweet potato disease in East Africa [15,18].

Hawassa-09 (TIS-8250-1) was selected from a number of genotypes introduced from the International Potato Center (CIP) and the Asian Vegetable Research and Development Center (AVRDC). The variety was mainly selected for resistance to sweet potato virus disease and high root yield. After performing a multi-stage performance trial over locations and years, Hawassa-09 was officially released and registered in 2017 as a new variety due to is outstanding performance.

Materials and Methods

Plant materials

Twelve genotypes along with one local (farmer variety) and three standards (released varieties) were used for the national variety trials. A variety verification trial consisting of two candidate and two check varieties was conducted for one year. The list of the 16 genotypes (12 genotypes and four checks) used in the national variety trials is given in Table 1.

Test environments

Over 114 genotypes were evaluated in 2012 and 2013 and 12 genotypes were selected based on resistance to sweet potato virus disease and root yield and promoted to the national variety trial. Then the 12 genotypes and the four checks were evaluated at three locations, namely Hawassa, Halaba and Dilla for two consecutive years, 2014 and 2015. Finally, a variety verification trial consisting of two candidate varieties, TIS-8250-1 and CN-1754-5 and two check varieties, farmers' variety and released variety (Beletech) was conducted for one year (2016). The variety verification trial was conducted at the above three locations both on-station and on six farmers' fields (two on-farm trials at each location). The trials were evaluated by the national variety verification technical committee and Hawassa-09 (TIS-8250-1) was approved by the committee for release and registration as a new variety for low and mid altitude areas of Ethiopia.

*Corresponding author: Gurmu F, South Agricultural Research Institute, Hawassa Research Center, P.O. Box 1226, Hawassa, Ethiopia
E-mail: fekadugurmu@yahoo.com

Results and Discussion

Characteristics of the variety

Hawassa-09 is a white fleshed sweet potato variety with cream skin colour. It has high levels of resistance to sweet potato virus disease and produces high root yields. The agronomic and morphological characteristics of the variety are displayed in Table 2.

Yield performance

Hawassa-09 gave a mean storage root yield of 49.2 t ha^{-1} with 56% and 283% yield advantage over the standard and local check, respectively. The storage root performance of the 16 sweet potato genotypes tested across locations and years is given in Table 3.

Reaction to sweet potato virus disease

Sweet potato virus disease is the major sweet potato disease in Ethiopia and Hawassa-09 was found to be free from viruses with visual observation and disease scoring. Serological test was also conducted and the result revealed negative reaction of the variety to the common sweet potato viruses in Ethiopia.

Quality attributes

Hawassa-09 is a white fleshed sweet potato variety with attractive storage root color and medium marketable root size. This variety can reach maturity in less than four months' time. It is resistant to the major sweet potato viruses in Ethiopia and is a high yielding variety with good overall acceptance after boiling.

S. No.	Genotypes	Status
1	TIS-841-6	Accession
2	TIS-82/0602-12	Accession
3	TIS-82/0602-2	Accession
4	TIS-9068-6	Accession
5	TIS-82-0602-6	Accession
6	TIS-8250-1	Accession
7	TIS-70357-2	Accession
8	Mae	Released variety
9	CN-2063-6	Accession
10	CN-2066-2	Accession
11	CN-1754-5	Accession
12	CN-1752-6	Accession
13	Becule-type-1	Accession
14	Bercume	Released variety
15	Local	Farmer variety
16	Guntute	Released variety

Table 1: List of sweet potato varieties used in the national variety trials.

Growth characteristics	
Growth habit	Spreading
Petiole length	16-20 cm
Petiole pigmentation	Green with purple petiole near to the leaf
Abaxial leaf vein pigmentation	Lower surface of the veins totally purple
Leaf colour at maturity	Green upper surface, purple vein lower surface
Shape of central leaf lobe	Semi-elliptic
General outline of the leaf lobe	Lobed
Leaf lobe number	03-May
Flowering habit	Not flowered in most cases in the test environments or 5% white flower
Major agronomic attributes	
Adaptability	Low to mid altitude areas: 1500- 2000 meters above sea level for optimum yield
Soil texture	Loam and sandy loam
Planting date	Mainly June and July. But can be planted any time if irrigated.
Spacing	60 cm b/n rows and 30 cm b/n plants
Seed rate	55,555 cuttings per ha
Resistance to diseases	Resistant to major sweet potato viruses in Ethiopia
Resistance to insect pests	Tolerant to sweet potato weevil
Days to maturity	100-120
Root yield	49 t ha^{-1}
Storage root characteristics	
Root shape	Round elliptic
Root formation	Dispersal
Predominant root skin color	Cream
Predominant root flesh color	White
Root length	16-20 cm
Root diameter	7-10 cm
Individual root weight	0.8-1.5 kg
Sensory attributes	
Texture of boiled roots	Moderate dry
Colour of boiled roots	Cream
Taste	Intermediate sweet
Overall acceptance	Accepted by most of the panelist farmers

Table 2: Agronomic and morphological characteristics of Hawassa-09 (TIS-8250-1).

Genotypes	2014				2015				Overall mean	YAOSC (%)	YAOLC (%)
	Hawassa	Dilla	Hallaba	Mean	Hawassa	Dilla	Hallaba	Mean			
TIS-8441-6	16.7	8	9.8	11.5	5.1	6	3.3	4.8	8.1	-74.2	-36.6
TIS-82/0602-12	10.4	17.2	10.2	12.6	5	7.9	1.9	4.9	8.7	-72.3	-31.9
TIS-82/0602-2	9.5	7.8	9.6	9	3.7	6	2.2	4	6.5	-79.5	-49.6
TIS-9068-6	15.4	19.1	14.8	16.4	7.5	7.1	2.5	5.7	11.1	-65	-13.9
TIS-82/0602-6	19.6	16.5	8.7	14.9	13.4	4.7	8.9	9	12	-62	-6.7
TIS-8250-1	57.4	55.9	49.8	54.4	47	47.9	37	44	49.2	56	283.3
TIS70357-2	50.2	53.5	51.9	51.9	41.4	36.6	38.2	38.7	45.3	43.6	253
Mae	37	32.8	38.9	36.2	41.5	19.1	19.9	26.8	31.5	0	145.8
CN-2063-6	13.7	10	12.6	12.1	5.2	13.1	16.2	11.5	11.8	62.6	-8
CN-2066-2	12.1	16.3	11.9	13.4	8.2	9.3	4.1	7.2	10.3	67.3	-19.7
CN-1754-5	55.6	54.6	53.2	54.5	46.8	32.4	38.3	39.2	46.8	48.5	264.8
CN-1752-6	58.7	55.4	49.8	54.6	41.2	38.9	35.4	38.5	46.6	47.7	262.9
Becule-type-1	7.8	6.4	8.9	7.7	6.2	3.7	3.2	4.4	6	-80.8	-52.9
Berkume	18	7.2	11.1	12.1	6.3	9	5.4	6.9	9.5	-69.9	-26
Local	18.3	14.8	13	15.4	5.9	10.4	14.6	10.3	12.8	-59.3	0
Guntute	16.3	24.5	10.2	17	7.4	9.5	4.1	7	12	-74.2	-6.5
Mean	26.1	25	22.8	24.6	18.2	16.3	14.7	16.4	20.5	-	-
LSD (5%)	13.4	10.9	7.2	6.1	9	7.9	9.5	5	3.9	-	-
CV (%)	31.8	25.3	19	26.4	33.1	26.2	38.6	32.3	28.9	-	-
R^2	89.9	92.3	96.4	92.6	91.3	95.5	90.7	92.9	93.2	-	-

*YAOSC: Yield advantage over standard check (%), YAOLC: Yield advantage over local check (%)

Table 3: Combined mean storage root yield (t ha⁻¹) performance of sweet potato genotypes tested over three locations and two years.

Variety maintenance and dissemination

The breeder seed of the variety is maintained by Hawassa Agricultural Research Center. Seed multiplication of the variety is underway and dissemination of the variety to the beneficiary farmers will be facilitated in collaboration with the bureau of agriculture.

Acknowledgement

The authors thank the Southern Agricultural Research Institute and the Ethiopian Institute Agricultural Research for the financial support and for provision of facilities during execution of the field works.

References

1. Lebot V (2010) Sweet potato: root and tuber crops. Springer Science+Business Media, LLC, New York, USA.

2. Wang Z, Li J, Luo Z, Huang L, Chen X, et al. (2011) Characterization and development of EST-derived SSR markers in cultivated sweetpotato (*Ipomoea batatas*). BMC Plant Biol 11: 131-139.

3. Tadesse T (2006) Evaluation of root yield and carotene content of orange-fleshed sweetpotato clones across locations in southern region of Ethiopia. MSc Thesis, Hawassa University, Hawassa.

4. Tofu A, Anshebo T, Tsegaye E, Tadesse T (2007) Summary of progress on orange-fleshed sweetpotato research and development in Ethiopia. In: Proceedings of the 13th International Society for Tropical Root Crops (ISTRC) Symposium, Arusha. ISTRC, Arusha, Tanzania.

5. Gurmu F, Hussein S, Laing M (2015) Diagnostic assessment of sweetpotato production in Ethiopia: constraints, post-harvest handling and farmers' preferences. Res Crops 16: 104-115.

6. CSA (2015) Ethiopia Agricultural Sample Survey 2014/2015: Report on Land Utilization (Private Peasant Holdings, Meher Season). Central Statistical Agency (CSA), Federal Democratic Republic of Ethiopia, Addis Ababa, Ethiopia.

7. CSA (2010) Crop Production Forecast Sample Survey, 2010/11 (2003 E.C.). Report on Area and Crop Production Forecast for Major Crops. Central Statistical Agency of Ethiopia, Addis Ababa, Ethiopia.

8. CSA (2011) Agricultural Sample Survey 2010/2011 (2003 E.C.). Report on Area and Production of Major Crops. Central Statistical Agency of Ethiopia, Addis Ababa, Ethiopia.

9. CSA (2012) Agricultural Sample Survey 2011/2012. Report on Area and Production of Major Crops. Central Statistical Agency of Ethiopia, Addis Ababa, Ethiopia.

10. CSA (2013) Agricultural Sample Survey 2012/2013. Report on Area and Production of Major Crops. Central Statistical Agency of Ethiopia, Adis Ababa, Ethiopia.

11. CSA (2014) Ethiopia Agricultural Sample Survey 2013/2014: Report on Land Utilization (Private Peasant Holdings, Meher Season). Central Statistical Agency, Addis Ababa, Ethiopia.

12. ARC (2015) Sweet potato production and field management in Ethiopia, Production Manual. Awassa Agricultural Research Center, Hawassa, Ethiopia.

13. Mekonen S, Handoro F, Gurmu F, Urage E (2014) Sweetpotato diseases research in Ethiopia. Int J Agric Inn Res 2: 2319-1473.

14. Mekonen S, Bekele B, Tadesse T, Gurmu F (2016) Evaluation of exotic and locally adapted sweetpotato cultivars to major viruses in Ethiopia. Greener J Agric Sci 6: 69-78.

15. Mwanga ROM, Kriegner A, Cervantes-Flores JC, Zhang DP, Moyer JW, et al. (2002) Resistance to sweetpotato chlorotic stunt virus and sweetpotato feathery mottle virus is mediated by two separate recessive genes in sweetpotato. J Am Soc Horti Sci 127: 798-806.

16. Settumba BM, Patrick RR, Jari PTV (2003) Incidence of viruses and viruslike diseases of sweetpotato in Uganda. Plant Disease 87: 329-335.

17. Mwololo JK, Mburu MWK, Njeru RW, Ateka EM, Kiarie N, et al. (2008) Resistance of sweetpotato genotypes to sweetpotato virus disease in Coastal Kenya. In: Kasem, Z.A., editors. The 8th African Crop Science Society (ACSS) Conference Proceeding, El-Minia, Egypt. 27-31 October, 2007 Afr Crop Sci Soc p 2083-2086.

18. Ndunguru J, Kapinga R, Sseruwagi P, Sayi B, Mwanga R, et al. (2009) Assessing the sweetpotato virus disease and its associated vectors in northwestern Tanzania and central Uganda. Afr J Agric Res 4: 334-343.

19

In Vitro Regeneration of Dalle Khursani, an Important Chilli Cultivar of Sikkim, using Various Explants

Karma Landup Bhutia, NG Tombisana Meetei and VK Khanna*

School of Crop Improvement, College of Post Graduate Studies, CAU, Umiam, Meghalaya, India

Abstract

An efficient micro propagation protocol was developed for Dalle Khursani, an important chilli cultivar of Sikkim. Aseptic cotyledon, shoot tip and hypocotyl explants of Dalle Khursani (*Capsicum annuum*) were cultured on Murashige and Skoog medium containing 16 different combinations of plant growth regulators for *in vitro* regeneration. Regeneration was observed only in 8 combinations of growth regulators among which medium containing 4 mg/l Thidiazuron (TDZ) showed the best result with an average of 2.95 shoots per explant and explant response of 73.95%. This was followed by MS medium containing 4 mg/l TDZ + 0.5 mg/l Gibberrellic Acid 3 (GA$_3$) + 0.5 mg/l Indole Acetic Acid (IAA) with an average of 1.94 shoots per explant and 66.66% explant response. Among the three explants used, cotyledons showed the best response in terms of number of shoots per explant with an average of 1.76. Regenerated shoots elongated and rooted well on MS medium containing 2 mg/l GA$_3$ + 1 mg/l IAA with an average shoot length of 3.10 cm and 6.35 ± 0.98 roots per shoot with explants response of 85% and 75%, respectively. The regenerated plants were acclimatized on a mixture of normal soil and artificial soil with 78% survival.

Keywords: Dalle Khursani; *In vitro* regeneration; Explants; cotyledon; Shoot tip; Hypocotyls; Thidiazuron; Artificial soil

Introduction

Chilli (*Capsicum sp.*) is a self-pollinated dicot plant and belongs to the family Solanaceae. Chilli has its centre of origin in American tropics. Capsicum is derived from the Greek word 'kapsimo', meaning 'to bite'. There are thought to be 25-30 species of Capsicum, of which 5 species; *C. annuum* L, *C. frutescens* mill, *C. chinense, C. baccatum* L. and *C. pubescens* have been domesticated and cultivated [1]. Capsicum is grown in the world on an area of 1.5 million hectare with the production of 10.60 million tons. In India, it is grown in an area of 0.775 million hectare with an average yield of 1.6 metric tonnes/hectare Indiastat.com, 2015.

Dalle Khursani belongs to *Capsicum annuum*. It is mostly grown in Sikkim and its surrounding regions like Darjeeling for its pungent fruits. It is one of the hottest chilli pepper with a Scoville rating of 100,000 to 350,000 SHU (For comparison Naga King chilli has Scoville rating of 330,000-1,000,000 SHU, Tabasco red pepper sauces has rating of 2500-5000 SHU and pure capsaicin has Scoville rating of 16,000,000 SHU).

In vitro regeneration is the process where a cell or group of cells differentiates to form organs in media containing required elements in aseptic environment. *In vitro* regeneration technique is generally used for micro propagation of elite crops and for the improvement of crops via genetic transformation as well to produce distant hybrids by using special techniques such as embryo culture, protoplast culture etc. The conventional method of chilli plant propagation using seeds is restricted by the short span of viability and low germination rate of seeds. Chilli plants are also highly susceptible to fungal and viral pathogens [2]. Capsicum species also has inherited problems like genotype dependence and recalcitrant nature (inability of plant cells, tissues and organs to respond to *in vitro* culture), etc. associated with *in vitro* regeneration system. Nevertheless, several reports are available for *in vitro* regeneration of different species / cultivars of chilli. Therefore, with the aim of developing an efficient *in vitro* regeneration protocol

for Dalle Khursani, this research program was taken up as it will further help in micro propagation and improvement of the crop.

Materials and Methods

The seeds of the chilli cultivar Dalle Khursani were collected from the farmers' field at Temi, South Sikkim, and India. The seeds were first treated with 0.2% bavistin and washed with distilled water. The seeds are then taken inside the Laminar Flow chamber and immersed in 70%ethanol for 10 to 15 sec, followed by sterilization with 2% sodium hypochlorite (NaClO) for 15 to 20 minutes. They were then washed thoroughly with sterile distilled water for 5 times to remove the traces of NaClO. Sterilized seeds were then inoculated in culture bottles containing MS basal media without any growth regulators for germination. After 15 to 20 days of germination, the explants namely, the cotyledon, shoot tip and hypocotyl were excised from the seedlings.

Murashige and Skoog (MS) medium supplemented with different plant growth regulators (PGRs) was used in this experiment. The media used for shoot induction is given in (Table 1).

After 4 weeks of culture, the regenerated shoots were sub-cultured for elongation as well as for rooting on MS media supplemented with various concentrations and combinations of growth regulators as given in (Table 2). The cultures were maintained in the culture room at 25°C ± 0.5°C temperature, photoperiod regime of 16 hrs light and 8 hrs dark and with a relative humidity (RH) of 60%. The observations that were

*Corresponding author: VK Khanna, School of Crop Improvement, College of Post Graduate Studies, CAU, Umiam, Meghalaya, India
E-mail: khannavk@rediffmail.com

MEDIA	Plant growth regulators (mg/l)					
	TDZ	BAP	KIN	IAA	NAA	GA_3
MS_1	2.0	-	-	-	0.5	-
MS_2	-	2.0	-	-	0.5	-
MS_3	-	-	2.0	-	0.5	-
MS_4	4.0	-	-	-	-	-
MS_5	-	4.0	-	-	-	-
MS_6	-	-	40	-	-	-
MS_7	2.0	2.0	-	-	-	-
MS_8	2.0	-	2.0	-	-	-
MS_9	-	2.0	2.0	-	-	-
MS_{10}	4.0	4.0	-	-	-	-
MS_{11}	4.0	-	4.0	-	-	-
MS_{12}	-	4.0	4.0	-	-	-
MS_{13}	6.0	-	2.0	-	-	-
MS_{14}	-	8.0	-	-	-	2.0
MS_{15}	-	4.0	-	0.5	-	0.5
MS_{16}	4.0	-	-	0.5	-	0.5
MS_{17}	Control without any growth hormones					

(TDZ = Thidiazuron; BAP = Benzyl Amino Purine; IAA = Indole Acetic Acid; KIN = Kinetin; NAA = Naphthalene Acetic Acid, GA_3 = Giberrellic Acid 3)

Table 1: Different combinations of growth regulators for direct regeneration.

MS MEDIA	Plant growth regulators (mg/l)			
	KIN	GA_3	NAA	IAA
ER_1	0.0	0.0	0.0	0.0
ER_2	2.0	0.5	0.0	0.0
ER_3	0.5	2.0	0.0	0.0
ER_4	0.0	2.0	0.5	0.0
ER_5	0.0	0.0	2.0	0.0
ER_6	0.0	2.0	0.0	0.5
ER_7	0.0	0.5	0.0	2.0

Table 2: Different concentrations and combinations of growth regulators for shoot elongation and rooting.

recorded from the in vitro culture included the explants' response i.e., the percentage of explants responding to in vitro culture, the number of shoots per explants, shoot length, root length and the number of roots. After 8 to 9 weeks of culture, the seedlings were transferred into artificial soil for acclimatization.

Artificial soil was prepared by mixing Perlite, Vermiculite and Peat (1:1:1 ratio) and was further mixed with autoclaved soil at 1:1 ratio. Fully developed elongated shoots with roots were taken out from the culture tubes or bottles and the base portion was washed thoroughly with running water to remove the medium attached on it. Regenerated seedlings were transplanted on disposable cups containing artificial soil and sterilized soil and then transferred to greenhouse for hardening.

For each treatment, 10 replications were used and the experiment was repeated three times. Completely Randomized Design (CRD) was used as the experimental design and for statistical analysis statistical software SPSS version 17.0 was used.

Results and Discussion

Among the 18 different media combinations used initially for in vitro culture, adventitious shoot induction was observed in only eight combinations of the growth regulators listed in (Table 3). Only these eight media combinations were used for further study.

Effect of explants

The present study indicated the effect of explants on the in vitro

regeneration of Dalle Khursani to be highly significant (Table 4). Among three types of explants used viz. cotyledon, shoot tip and hypocotyl, the maximum number of shoots per explant was observed in cotyledons (3.76 ± 0.58 in MS_4). On the other hand, shoot tip explants showed better results in terms of percentage of explants response (81% in MS_4) (Figure 1).

In earlier available reports, the best response in terms of number of shoots per explant was obtained from cotyledon explants as in the case of sweet pepper [3] and Capsicum annuum L. cv. Kaddi B [4]. Other reports are also available which show successful regeneration from cotyledon explants [4-8]. Shoot tips have also successfully used as explants [9-11]. In some reports, the hypocotyl explants showed better result both in terms of number of shoots per explant as well as percentage of explant which responded [7,12,13]. These results suggested that the response of the different explants to in vitro culture varies for different cultivars/species. In the case of Kharsani D, also, the cotyledon was found to be the explant of choice as it gave the maximum number of shoots per explant (3.76 ± 0.58) when 4 mg/L TDZ alone was used in the medium.

Effects of media

The effect of media for in vitro culture of Dalle Khursani was also found to highly significant (Table 5). Types and levels of plant growth regulators in growing media play an important role during in vitro regeneration of plants. In the present study, among different concentrations and combinations of growth regulators used, MS medium supplemented with 4 mg/l TDZ (MS_4) showed the best results both in terms of number of shoots per explant as well as explant

Sl. No.		Mean no. of shoots per explant (Mean ± SE_M)			Media
	Media	Cotyledon	Shoot tip	Hypocotyl	Mean
1	MS_1	0.95 ± 0.33	0.78 ± 0.13	0.28 ± 0.12	0.667
2	MS_4	3.76 ± 0.58	2.70 ± 0.29	2.40 ± 0.33	2.946
3	MS_7	1.90 ± 0.34	1.20 ± 0.19	0.70 ± 0.20	1.278
4	MS_8	1.27 ± 0.32	1.20 ± 0.17	0.70 ± 0.19	1.056
5	MS_{10}	1.77 ± 0.32	0.83 ± 0.14	1.40 ± 0.29	1.333
6	MS_{13}	1.00 ± 0.40	0.87 ± 0.15	0.43 ± 0.18	0.767
7	MS_{14}	1.40 ± 0.29	1.30 ± 0.16	0.83 ± 0.22	1.189
8	MS_{16}	2.00 ± 0.42	1.97 ± 0.21	1.83 ± 0.28`	1.944
Explant Mean		**1.759**	**1.358**	**1.076**	

C.D	0.05% or 5%	0.01% or 1%
Explant	1.05	1.67
Media	0.7	1

Table 3: Effect of different media treatments on regeneration of shoots from cotyledon, shoot-tip and hypocotyl explants after 4 weeks of culture.

Sl. No.	Media	Explant Response (%)			Mean
		Cotyledon	Shoot tip	Hypocotyl	
1	MS_1	22.22	61.11	16.66	33.35
2	MS_4	65.6	81.25	75	73.95
3	MS_7	56.66	70	30	52.22
4	MS_8	36.66	73.33	33.33	47.77
5	MS_{10}	56.66	66.66	50	57.77
6	MS_{13}	20	60	23.33	34.44
7	MS_{14}	46.66	76.66	36.66	53.33
8	MS_{16}	50	86.66	63.33	66.66
	Mean	44.3	71.95	41.03	

Table 4: Response percentage of cotyledon, shoot-tip and hypocotyls explants to different growth hormones for direct regeneration after 4 weeks of culture.

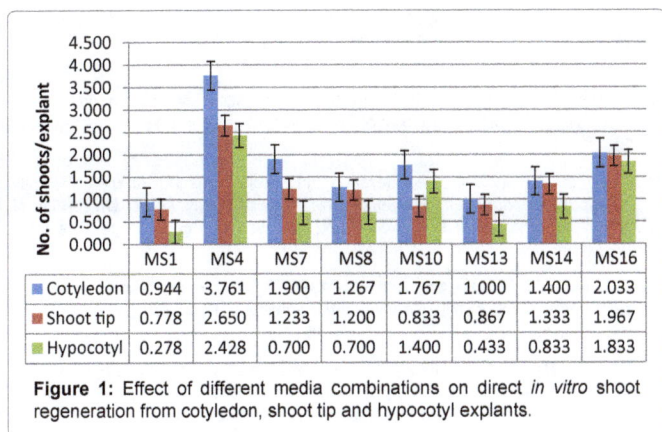

Figure 1: Effect of different media combinations on direct *in vitro* shoot regeneration from cotyledon, shoot tip and hypocotyl explants.

	MS1	MS4	MS7	MS8	MS10	MS13	MS14	MS16
Cotyledon	0.944	3.761	1.900	1.267	1.767	1.000	1.400	2.033
Shoot tip	0.778	2.650	1.233	1.200	0.833	0.867	1.333	1.967
Hypocotyl	0.278	2.428	0.700	0.700	1.400	0.433	0.833	1.833

Source of Variation	Degrees of freedom	Sum of square	Mean sum of square	$F_{calculated}$	$F_{tabular}$ 5%	$F_{tabular}$ 1%
Replication	2	0.558	0.279			
Treatment	23	43.299	1.882			
Explant (E)	2	5.66	2.83	17.453	3.2	5.1
Media (M)	7	34.279	4.897	30.2	2.22	3.05
(E x M)	14	3.359	0.239	1.479	1.91	2.5
Error	46	7.458	0.162			
Total	71	51.316				

Table 5: Analysis of variance (ANOVA) among different explants, combination of growth regulator and interaction between explants and media.

response (Tables 3 and 4). This concentration of TDZ was many times higher than the concentration of TDZ reported by Channappagaudar [4] where the highest frequency of shoot regeneration was obtained on MS medium containing 0.5 mg/l TDZ. Results of present study show that callus formation at petiolar end of cotyledon (Figure 2A) basal end of shoot tip and cut ends of hypocotyl explants on media containing lower concentration of BAP interferes with the shoot formation. However, higher concentration of BAP (8 mg/L) showed direct regeneration of shoots from explants. These results are similar to the previous reports where higher concentration of BAP showed enhanced shoot induction [6,14-20]. TDZ has a high efficiency in stimulating cytokinin dependent shoot regeneration from a wide variety of plants [17]. Previous reports suggest that MS media supplemented with 0.5 to 2 mg/L TDZ are best for shoot induction [17,21-25]. In contrast to the previous reports, results of the present study show that, supplementing MS media with a slightly higher concentration of TDZ (4 to 6 mg/L) enhanced shoot induction.

Effect of explant x media (interaction)

Irrespective of the growth media, shoot tip explants showed high explants response ranging from 61% to 81% response with an average of 1.36 shoots per explant and irrespective of explants, MS media containing 4 mg/L TDZ showed the highest average number of shoots (2.95) per explant in all the three types of explants used. Analysis of variance showed no significant differences among the interaction between explants x media at 1% and 5% level of significance (Table 5).

Shoot elongation

In the present study, among different combinations of growth regulators tried, medium containing a combination of GA_3 (2 mg/L) and IAA (0.5 mg/L) was found to be the best for shoot elongation with an average shoot length of 3.1 ± 0.33 and 85% response. Similar combination was used by Kumar et al. [26], but the concentration of

growth regulators was higher in the present case. Other findings also suggested the addition of GA_3 alone or in combination with either cytokinin or auxin in media to be suitable for enhancing elongation of regenerated shoots [4,6,7,13,26-28]. Another study reported that sub culturing of regenerated shoots on MS media without any growth regulators enhanced the shoot elongation [18,21,22,29]. In our study also, shoot elongation was observed in media devoid of growth regulators with an adequate shoot length of 3.00 ± 0.36 cm and 80% response.

Rooting

In tissue culture generally shoots are allowed to regenerate first than roots by manipulating growth regulators in the growing media. In the present study, results showed that MS media containing 2 mg/L GA_3 and 0.5 mg/L IAA was best for root induction with an average of 6.35 ± 0.98 roots per explant and 70% response (Table 6). Similar results were reported earlier where presence of IAA in media was suitable for root induction [27,30,31]. In previous reports MS media amended with NAA was best for root induction [18,31]. In the present study, MS medium containing 2 mg/L NAA showed the second best results in terms of number of roots per shoot (5.55 ± 0.94) with a 55% response. MS medium without any growth regulators also gave good results for root induction with an average of 4.60 ± 0.87 roots per shoot and 65% response. This result was similar to those reported by Song et al. [21] and Arous et al. [32].

Hardening

In vitro regenerated, elongated and rooted seedlings of Dalle Khursani (*Capsicum annuum*) were taken out from the culture tubes after 8-9 weeks of culture and the traces of media from the basal region or roots were removed by washing thoroughly with running tap water. Plantlets were than transplanted in cups containing a mixture of artificial soil and normal soil at 1:1 ratio. The cultures were kept inside the culture room under controlled conditions for at least 2 - 3 weeks after which they were transferred to the Greenhouse (Figure 3). In this experiment, 78% of *in vitro* regenerated seedlings survived hardening.

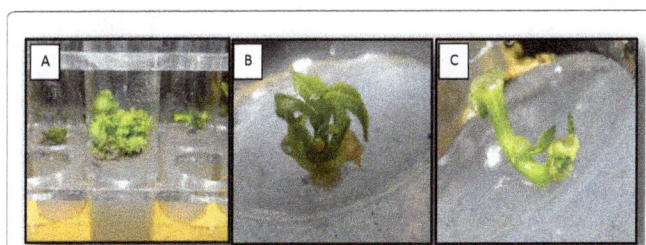

Figure 2: *In vitro* direct regeneration. (A) Rosette like structure formed at petiolar end of cotyledon explants. (B) Shoots regenerated from shoot tip explants. (C) Hypocotyl explants showing direct regeneration.

Media	Shoot length (cm) (Mean ± SEM)	Response (%)	Root No. (Mean ± SEM)	Root length (cm) (Mean ± SEM)	Response (%)
ER_1	3.00 ± 0.36	80	4.60 ± 0.87	1.30 ± 0.25	65
ER_2	1.10 ± 0.25	55	2.70 ± 0.62	1.15 ± 0.26	55
ER_3	2.00 ± 0.30	75	1.45 ± 0.53	0.55 ± 0.18	35
ER_4	2.00 ± 0.32	65	4.05 ± 0.92	1.38 ± 0.30	55
ER_5	0.80 ± 0.22	45	5.55 ± 0.94	1.80 ± 0.34	65
ER_6	3.10 ± 0.33	85	6.35 ± 0.98	1.68 ± 0.29	70
ER_7	1.55 ± 0.24	75	4.20 ± 0.74	2.15 ± 0.41	65

Table 6: Effect of different combinations and concentration of growth hormones on shoot elongation, root induction and root elongation.

(A)   (B)

(C)

Figure 3: Shoot elongation (A), rooting (B) and hardening (C).

References

1. Kothari SL, Joshi A, Kachhawaha S, Ochoa-Alejo N (2010) Chilli Pepper A review on tissue culture and transgenics. Biotechnology advances 28: 35-48.

2. Morrison RA, Koning RE, Evans DA (1986) Anther culture of an interspecific hybrid of Capsicum. Journal of Plant Physiology 126: 1-9.

3. Otroshi M, Ghehsareh NM (2013) The effect of plant growth regulators and different type of explants on organogenesis of pepper (*Capsicum annuum* L.) in in-vitro. Journal of Applied Science and Engineering Technology 3: 271-278.

4. Channappagoudar SB (2007) Studies on in vitro regeneration and genetic transformation in chilli (*Capsicum annuum* L.). Ph.D thesis submitted to University of Agriculture Science (Dept of Gen and Pl Breeding, Agri College, Dharwad, UAS) India.

5. Kehie M, Kumaria S, Tandon P (2013) *In vitro* plantlet regeneration from cotyledon segments of *Capsicum chinense* Jacq. cv. Naga King Chili and determination of capsaicin content in fruits of in vitro propagated plants by high performance liquid chromatography. Scientia Horticulturae 164: 1-8.

6. Verma S, Dhiman K, Srivastava DK (2013) Efficient *in vitro* regeneration from cotyledon explants in bell pepper (*Capsicum annuum* L. cv. california wonder). International Journal of Advanced Biotechnology and Research 4: 391-396.

7. Pishbin N, Mousavi A, Kalatejari S, Shariatpanahi M, Jahromi BB (2014) The effect of plant growth regulators and different types of explants on *in vitro* regeneration of sweet pepper (*Capsicum annuum* L.). International Journal of Bioscience 5: 139-146.

8. Husain S, Jain A, Kothari S L (1999) Phenylacetic acid improves bud elongation and In vitro plant regeneration efficiency in *Capsicum annuum* L. Plant Cell Reports 19: 64-68.

9. Gururaj HB, Giridhar P, Sharma A, Prasad BCN, Ravishankar GA (2004) *In vitro* clonal propagation of bird eye chilli (*Capsicum frutescens* Mill.). Indian Journal of Experimental Biology 42: 1136-1140.

10. Sanatombi K, Sharma GJ (2008) *In vitro* plant regeneration in six cultivars of *Capsicum* sp. using different explants. Biologia Plantarum 52: 141-145.

11. Sanatombi K, Sharma GJ (2012) *In vitro* regeneration of *Capsicum chinense* Jacq. Current Trends in Biotechnology and Pharmacy 6: 66-72.

12. Bustos MGV, Aguado-Santacruz GA, Carrillo-Castañeda C, Aguilar-Rincón VH, Espitia-Rangel E, et al. (2009) *In vitro* propagation and agronomic performance of regenerated chili pepper (*Capsicum* sp.) plants from commercially important genotypes. In Vitro Cellular and Developmental Biology-Plant 45: 650-658.

13. Kumar V, Gururaj HB, Prasad BC, Giridhar P, Ravishankar GA (2005) Direct shoot organogenesis on shoot apex from seedling explants of *Capsicum annuum* L. Scientia Horticulturae 106: 237-246.

14. Rizwan M, Sharma R, Soni P, Gupta NK, Singh G (2013) Regeneration protocol for chili (*Capsicum annuum* L.) variety Mathania. *Journal of Cell and Tissue* Research 13: 3513-3517.

15. Sanatombi K, Sharma GJ (2006) *In vitro* regeneration and mass multiplication of *Capsicum annuum* L. Journal of Food, Agriculture and Environment 4: 205-206.

16. Ashajyothi SS (2004) Regeneration and transformation studies in chilli (*Capsicum annuum* L.). M. Sc. (Agri.) Thesis, submitted to University of Agricultural Sciences (Dept. of Gen. and Pl. Breeding, Agri. College, Dharwad, UAS) India.

17. Manoharan M, Vidya CSS, Sita GL (1998) Agrobacterium-mediated genetic transformation in hot chilli (*Capsicum annuum* L. var. Pusajwala). Plant Science 131: 77-83.

18. Hyde CL, Phillips GC (1996) Silver nitrate promotes shoot development and plant regeneration of Chilli pepper (*Capsicum annuum* L.) *via.* organogenesis. *In Vitro* Cellular and Developmental Biology-Plant 32:72-80.

19. Christopher T, Rajam MV (1994) *In vitro* clonal propagation of *Capsicum* spp. Plant Cell, Tissue and Organ Culture 38: 25-29.

20. Gunay AL, Rao PS (1978) *In vitro* plant regeneration from hypocotyl and cotyledon explants of red pepper (*Capsicum* Spp). Plant Science Letters 11: 365-72.

21. Song JY, Sivanesan I, An CG, Jeong BR (2010) Adventitious shoot regeneration from leaf explants of miniature paprika (*Capsicum annuum*) 'Hivita Red' and 'Hivita Yellow'. African Journal of Biotechnology 9: 2768-2773.

22. Ahmad N, Siddique I and Anis M (2006) Improved plant regeneration in *Capsicum annuum* L. from nodal segments. Biologia Plantarum 50: 701-704.

23. Santana N, Canto A, Barahona F, Miranda M (2005) Regeneration of Habarno peppers (*Capsicum chinense* Jacq.) *via.* organogenesis. HortScience 40: 1829-1831.

24. Dabauza M, Pena L (2001) High efficiency organogenesis in sweet pepper (*Capsicum annuum* L.) tissues from different seedling explants. Plant Growth Regulation 33: 221-229.

25. Szasz A, Nervo G, Fasi M (1995) Screening for *In vitro* shoot forming capacity of seedling explants in bell pepper (*Capsicum annuum* L.) genotypes and efficient plant regeneration using thidiazuron. Plant Cell Reports 14: 666-669.

26. Kumar OA, Rupavathi T, Tata SS (2012) Adventitious shoot bud induction in chili pepper (*Capsicum annuum* L. cv. x-235). International Journal of Science and Nature 3: 192-196.

27. Joshi A, Kothari SL (2006) High copper levels in the medium improves shoot bud differentiation and elongation from the cultured cotyledons of *Capsicum annuum* L. Plant Cell, Tissue Organ culture 12: 71-76.

28. Golegaonkar PG, Kantharajah GS (2006) High-frequency adventitious shoot bud induction and shoot elongation of chilli pepper (*Capsicum annuum* L.). *In Vitro* Cellular and Developmental Biology-Plant 42:341-344.

29. Siddique I, Anis M (2006) Thidiazuron induced high frequency shoot bud formation and plant regeneration from cotyledonary node explant of *Capsicum annuum* L. Indian Journal of Biotechnology 5: 303-308.

30. Kehie M, Kumaria S, Tandon P (2012) *In vitro* plantlet regeneration from nodal segments and shoot tips of *Capsicum chinense* Jacq. Cv. Naga King Chilli. Biotech 2: 31-35.

31. Dafadar A, Das A, Bandopadhyay B, Jha TB (2012) In vitro propagation and molecular evaluation of *Capsicum annuum* L. cultivar with high chromosome number (2n = 48). Scientia Horticulturae 140: 119-124.

32. Arous S, Boussaid M, Marrakchi M (2001) Plant regeneration from zygotic embryo hypocotyls of Tunisian chili (*Capsicum annuum* L.) Journal of Applied Horticulture 3:17-22.

20

Performance of Bt Cotton Hybrids to Plant Population and Soil Types Under Rainfed Condition

Rajeshwar Malavath*, Ravinder Naik, Pradeep T and Sreedhar Chuhan

Acharya N.G Ranga Agricultural University, District Agricultural Advisory and Transfer of Technology Center, KVK,ARS, Adilabad-504 001, RARS, Jagitial , India

Abstract

Afield experiment was conducted both in black cotton and red chalka soils during kharif 2008-09 and 2009-10 seasons in Adilabad District of Andhra Pradesh at six different locations through farmer's participatory mode to find out the response of BG-II cotton hybrids under two different spacing's in rainfed conditions. These experiments were carried out by the District Agricultural Advisory and Transfer of Technology Center, Adilabad in collaboration with ATMA project functioning at Adilabad. Three cotton hybrids viz., Mallika BG-II (Boll Guard), Rasi BG-II and Paras Brahma BG–II which are most popular among the farmers were sown under two different spacing's in different soils. The data revealed that, hybrids did not differ significantly in plant height, number of sympodial branches/plant, number of bolls/plant, boll weight and seed cottonyield in both the years of testing and also in both the soils. But, spacing's had significantly influenced number of bolls/plant, boll weight and seed cottonyield. However, interaction effect was significant only for plant height. Closer spacing of 60 x 60 cm in red chalka soils (2033 and 2253 kg ha^{-1}) and 90 x 60 cm in BC soils (2300 and 2450 kg ha^{-1}) gave significantly higher seed cottonyield than wider spacing of 90 x 90 cm (1500 and 1863 kg ha^{-1}) and 120 x 90 cm (1767 and 1983 kg ha^{-1}) during both the years of investigation respectively. Thus it is concluded that Bt hybrids need to be planted with higher plant density to realize good yields.

Keywords: Bt Cotton Hybrids; Plant population; Rainfed condition red sandy soils; Black cotton soils

Introduction

The black cotton soils of Adilabad district of Telangana are moderately deep to very deep, clayey and moderate to moderately well drained with slow permeability and low hydraulic conductivity. Soils have swell shrink characteristic and crack during summer. The moisture retention capacity is high. The early rainfall enters into the soil through cracks and once the cracks are closed the water stagnation occurs due to slow permeability. The Red sandy loam soils have light surface texture and gravelliness with kaolinite clay mineralogy resulting in poor water holding capacity. Surface crusting is common problem in this soil. The low water-holding capacity does not permit post-rainy season cropping without irrigation. Surface soils are denuded and subject to serious erosion problems by runoff process [1].

Cotton is an important fiber crop, which is cultivated in more than 80 countries of the world and play a key role in economic and social affairs of the farming community. In India, nearly 65 per cent of the cotton crop is grown under rainfed conditions on a variety of soils ranging from well drained deep alluvial soils in the north to black clayey soils of varying depth in central region, black, mixed black and red soils in south zone. Cotton is semi-tolerant to salinity and sensitive to water logging and thus prefers well- drained soils [2]. Cotton cultivation in Adilabad district of A.P has gone up after the introduction of Bt Cotton hybrids and is presently grown in an area of about 2.794 lakh hectares (2010-2011) mainly due to the Bt cotton, genetically modified to make insecticidal protein(s) from the soil bacterium *Bacillus thuringiensis* was first commercialized in 2002 against bollworms. Bt cotton spread rapidly, resulting in greatly increased productivity and reduced insecticide use [3].

Adilabad District is located in Northern Telangana Zone of Andhra Pradesh, situated between 77° 46' and 80° 01' of the Eastern Longitude and 18° 40' and 19° 56' of Northern Latitudes. The soils of the district are predominantly black which constitutes about 80 percent. Cotton is one of the most important commercial crops grown in the district for the last 50 years under rainfed conditions. Cotton yields under rainfed ecosystem are low owing to erratic rainfall and hence the crop suffers from moisture stress during post monsoon season which coincides with flowering and boll development stages. Majority of the farmers in the district do not follow the recommended spacing which is most important agronomic practice under rainfed conditions to get good crop yield.

It is in this context that a systematic study was undertaken to assess the performance of three Bt cotton hybrids in both black cotton (BC) and red chalka soils with two different spacings. Six locations appropriate to farming situation were chosen in the district.

Materials and Methods

The experiment was conducted on farmer's fields as farmers participatory approach during Kharif 2008-09 and 2009-10 seasons at six different locations such as Mannur, Borigam, Bansupalli, Utnoor, Kerameri and Rajura. An average rainfall of 1093.6 mm in 2008 and 1137.0 mm during 2009 season was received in 55 and 57 rainy days respectively.

The experiment was laid out in split plot design with Bt. Hybrids (Mallika BG-II, Rasi BG-II and Paras Brahma BG -II) as main plots and spacing as sub plots with 7 replications. The main plots were same both in BC soils and Red chalka soils. However, spacing varied in subplots.

*Corresponding author: Rajeshwar Malavath, Acharya N.G Ranga Agricultural University, District Agricultural Advisory and Transfer of Technology Center, KVK, ARS, Adilabad-504 001, RARS, Jagitial
E-mail: rajeshoct31naik@gmail.com

In Red chalka soils (sandy clay loam), treatmental spacing adopted was 60 × 60 cm which is the recommended spacing and was tested against a spacing of 90 × 90 cm, adopted by farmers. In BC soils, treatmental spacing was taken as 90 × 60 cm which is the recommended spacing and tested against the farmers practice of 120 × 90 cm.

Three popular hybrids viz., Mallika BG-II, Rasi BG-II and Paras Brahma BG–II were included and the crop was sown in second fort night of June during both the years of study. Standard crop management practices were adopted to raise a good crop. Observations were recorded on yield attributes and the crop was harvested periodically in three pickings and the yield was recorded.

Results and Discussion

There was no significant difference observed among the hybrids tested for plant height at harvest, number of sympodial branches per plant, number of bolls per plant and seed cotton yield (kg ha⁻¹). With regard to boll weight, pooled mean over two years indicated that Mallika BG II recorded significantly higher boll weight (5.37 g) compared to Raasi BG II in black cotton soils (Table 1).

Where as in red chalka soils mean over two years (Table 2) indicated that Mallika BG II gave significantly higher plant height (113.7 cm) compared to Raasi BG II (109.7 cm) but it was on par with Brahma BG II (113.5 cm). No significant difference was observed among the varieties tested for number of sympodial branches per plant. The number of bolls per plant and boll weight was significantly higher in Mallika BG II (34.0 and 4.61 respectively) and it was on par with Brahma BG II (32.7 and 4.57 respectively). Similarly there was no significant difference among the varieties for seed cotton yield except in 2009-10 where Mallika BG II exhibited significantly highest seed cotton yield of 2140 kg ha⁻¹ but it was on par with Brahma BG II (2070 kg ha⁻¹).

Effect of Spacing

Spacing did not have any significant influence on either the plant height at harvest or the number of sympodial branches per plant during individual years and also mean over two years in black cotton soils (Table 1). Wider spacing of 120 × 90 cm produced significantly higher number of bolls per plant (37.5) and boll weight (5.39 g) compared to closer spacing of 90 × 60 cm (35.5 and 5.22 g respectively). It might be due to better aeration and adequate interception of light as well as lesser competition for nutrients due to low plant population per unit area. On the contrary significantly higher seed cotton yield was recorded in closer spacing (2375 kg ha⁻¹) compared to wider spacing (1874 kg ha⁻¹).

For red chalka soils closer spacing of 60 × 60 cm resulted in significantly higher plant height of 114.8 cm compared to wider spacing of 90 × 90 cm (109.8 cm) (Table 2). Similar results were observed by Rajendran et al. [4]. The effect of spacing on number of sympodial branches per plant was non significant during individual years and also mean over two years. As regards number of bolls per plant, boll weight and seed cotton yield the effect of spacing is similar in red chalka soils as that of black cotton soils where hybrids in closer spacing had lesser number bolls per plant, lesser boll weight and more seed cottonyield compared to wider spacing. Similar results were obtained by Bhalerao et al. [5] under rainfed condition. The seed cottonyield of black cotton soils obtained was higher when compared to red chalka soils and this could be mainly due to the high water holding capacity in Black cotton soils.

Interaction Effect

Interaction among the hybrids tested and spacing was non significant for all the parameters in both the soils except for plant height at harvest in both the soils.

Overall, the cotton yields recorded during 2008-09 was comparatively low as compared to 2009-10 even though the quantity of rainfall was high during 2008-09 (894.5 mm with 42 rainy days) might be due to high intensity and uneven distribution. Whereas during 2009-10, though rainfall was comparatively low (683.0 mm with 56 rainy days), its uniform distribution for longer period with low intensity enhanced the yield levels (Table 3).

Performance of non Bt cotton hybrids under recommended spacing

Treatments	Plant height at harvest (cm)			No. of sympodial branches/plant			No. of bolls/plant			Boll weight (g)			Seed cottonyield (kg/ha)		
	2008-09	2009-10	Mean	2008-09	2009-10	Mean	2008-09	2009-10	Mean	2008-09	2009-10	Mean	2008-09	2009-10	Mean
Main plot-Bt. Hybrid															
1. Mallika BG II	119.0	127.5	123.2	20.5	22.5	21.5	34.5	38.5	36.5	5.25	5.50	5.37	2090	2250	2170
2. Raasi BG II	118.0	120.0	119.0	19.5	21.0	20.2	35.5	37.0	36.2	5.04	5.35	5.20	1960	2190	2075
3. Brahma BG II	121.0	122.5	121.7	20.5	21.5	21.0	35.5	38.0	36.7	5.25	5.42	5.34	2050	2207	2129
Mean	119.3	123.3	121.3	20.2	21.7	20.9	35.2	37.8	36.5	5.18	5.42	5.30	2033	2216	2125
S Em ±	3.6	2.9	2.5	0.88	0.73	0.59	0.79	0.97	0.69	0.09	0.01	0.04	65.0	57.0	43.0
S Ed ±	5.1	4.1	3.5	1.24	1.04	0.84	1.11	1.38	0.98	0.13	0.02	0.06	91.0	81.0	60.0
CD (at 5%)	NS	NS	NS	NS	NS	NS	NS	NS	NS	NS	0.05	0.14	NS	NS	NS
Sub plots-Spacing															
1. 90 × 60 cm	121.7	123.3	122.5	19.6	21.0	20.3	34.3	36.6	35.5	5.10	5.33	5.22	2300	2450	2375
2. 120 × 90 cm	117.0	123.3	120.2	20.6	22.3	21.5	36.0	39.0	37.5	5.26	5.52	5.39	1767	1983	1874
Mean	119.4	123.3	121.4	20.1	21.7	20.9	35.2	37.8	36.5	5.2	5.4	5.3	2034	2216	2125
S Em ±	4.7	4.4	2.7	0.98	1.11	0.72	0.58	0.75	0.43	0.08	0.07	0.04	64.6	44.6	29.1
S Ed ±	6.7	6.3	3.8	1.39	1.57	1.02	0.82	1.06	0.61	0.11	0.10	0.07	91.3	63.0	41.2
CD (at 5%)	NS	NS	NS	NS	NS	NS	1.74	2.23	1.29	0.02	0.20	0.14	192.0	132.5	86.6
Interactions															
S Em ±	3.8	3.6	2.2	0.8	0.9	0.6	0.5	0.6	0.3	0.1	0.1	0.0	51.8	35.7	23.3
S Ed ±	5.3	5.0	3.1	1.1	1.3	0.8	0.7	0.9	0.5	0.1	0.1	0.1	73.2	50.5	33.0
CD (at 5%)	11.2	10.7	6.5	NS	NS	NS	1.4	NS	NS	NS	NS	NS	NS	NS	NS
CV (%)	12.9	11.8	12.3	15.8	16.6	16.2	15.4	16.4	15.9	14.7	14.1	13.4	10.3	16.5	13.4

Table 1: Effect of spacing on number of bolls per plant, boll weight and seed yield of cotton in Black cotton soils *Kharif* 2008-09 and 2009-10.

Treatments	Plant height at harvest (cm)			No. of sympodial branches/plant			No. of bolls/plant			Boll weight (g)			Seed cottonyield (kg/ha)		
	2008-09	2009-10	Mean	2008-09	2009-10	Mean	2008-09	2009-10	Mean	2008-09	2009-10	Mean	2008-09	2009-10	Mean
Main plot-Bt. Hybrid															
1. Mallika BG II	107.5	120.0	113.7	19.5	20.5	20.0	31.5	36.5	34.0	4.40	4.81	4.61	1755	2140	1948
2. Raasi BG II	104.5	115.0	109.7	19.5	19.0	19.2	29.5	35.0	32.2	4.41	4.55	4.48	1725	1965	1845
3. Brahma BG II	110.0	117.0	113.5	18.5	18.5	18.5	31.0	34.5	32.7	4.45	4.70	4.57	1820	2070	1945
Mean	107.3	117.3	112.3	19.2	19.3	19.2	30.7	35.3	33.0	4.42	4.69	4.55	1767	2058	1913
S Em ±	0.5	2.8	1.5	1.22	0.9	0.63	0.77	0.67	0.54	0.05	0.06	0.03	54.0	53.0	48.0
S Ed ±	0.7	4.0	2.1	1.72	1.27	0.9	1.09	0.95	0.77	0.07	0.09	0.05	76.0	75.0	68.0
CD (at 5%)	1.5	NS	4.5	NS	NS	NS	NS	2.06	1.67	NS	0.18	0.11	NS	163.3	NS
Sub plots-Spacing															
1. 90 × 60 cm	110.0	119.7	114.8	18.3	19.0	18.7	29.0	34.0	31.5	4.23	4.61	4.42	2033	2253	2143
2. 120 × 90 cm	104.7	115	109.8	20.0	19.6	19.8	32.3	36.6	34.5	4.61	4.77	4.69	1500	1863	1682
Mean	107.4	117.4	112.3	19.2	19.3	19.2	30.7	35.3	33.0	4.4	4.7	4.6	1767	2058	1913
S Em ±	1.5	2.8	1.34	1.11	0.93	0.65	0.49	0.78	0.35	0.04	0.05	0.04	40.7	58.9	35.8
S Ed ±	2.2	4	1.89	1.57	1.32	0.91	0.69	1.1	0.49	0.06	0.07	0.06	57.5	83.3	50.6
CD (at 5%)	4.7	NS	3.99	NS	NS	NS	1.46	2.32	1.04	0.13	0.15	1.30	121.0	175.1	106.4
Interactions															
S Em ±	1.2	2.3	1.1	0.9	0.7	0.5	0.4	0.6	0.3	0.03	0.04	0.03	32.6	47.2	28.7
S Ed ±	1.8	3.2	1.5	1.3	1.1	0.7	0.6	0.9	0.4	0.04	0.06	0.05	46.1	66.8	40.6
CD (at 5%)	3.8	6.8	3.2	NS	NS	NS	NS	NS	NS	0.10	NS	NS	NS	NS	NS
CV (%)	14.8	17.9	16.3	18.8	15.7	17.2	15.2	17.2	16.2	13.1	13.5	13.3	17.5	19.3	18.4

Table 2: Effect of spacing on number of bolls per plant, boll weight and seed yield of cotton in Red chalka soils *Kharif* 2008-09 and 2009-10

Month	2008-09		2009-10	
	Total Rainfall Received (mm)	No. of Rainy days	Total Rainfall Received (mm)	No. of Rainy days
June	126.3	6	101.2	8
July	220	14	166.5	18
August	378.8	16	185.3	15
September	141.87	4	122.5	5
October	12.2	1	39.6	5
November	14	1	26.4	3
December	0.0	0	-	0
January	0.0	0	16.1	1
February	0.0	0	7.5	1
March	2.0	0	0.8	0
April	0.0	0	0.0	0
May	0.5	0	1.0	0
Total	894.5	42	683.0	56

Table 3: Rainfall of the district during the year 2008-09 and 2009-10.

of 90 × 90 cm or 120 × 90 cm even under rainfed situation during eighties was more than satisfactory due to better control of sucking pests and boll worms with predicted behavior of weather conditions particularly the rainfall. However, due to fluctuations in weather parameters, spread of cotton crop to newer areas and indiscriminate use of insecticides leading to resurgence of certain pests drastically affected the yield potential of hybrids in due course of time. It was at this juncture i.e., a decade back Bt cotton hybrids were introduced and as a result the menace of boll worms was over come and relatively good retention of bolls was witnessed. Retention of maximum number of bolls also sometimes made the plants to change their growth habit (determinate/indeterminate) according to the prevailing seasonal conditions. Under closer spacing due to reduced canopy and more number of plants per unit area plants exhibited determinate growth particularly when monsoon ceased early and the farmers got good yields.

Therefore due to these frequent changes in macro and micro weather conditions the performance of Bt hybrids becomes unpredictable and the farmers suffered huge losses or benefits. It is in this context that the results of the present experimentation have got immense practical utility. Thus it is suggested to adopt closer spacing in both types of soil as it provides better opportunity for Bt hybrids to express their potential under rainfed conditions in Adilabad district.

Acknowledgement

The authors are thankful to authorities of Acharya N.G Ranga Agricultural University, Hyderabad and ATMA Project Adilabad, Andhra Pradesh, India for their technical as well as financial support during the course of investigation.

References

1. Ground Water Information (2007) Central Ground Water Board Ministry of Water Resources. Government of India, Adilabald District, Andhra Pradesh southern region, Hyderabad: 1-38.

2. Revolution in Indian Cotton (2009) Directorate of Cotton Development Department of Agriculture & Cooperation. Ministry of Agriculture, Govt. of India, Mumbai, National Center of Integrated Pest Management ICAR, Pusa Campus, New Delhi: 1-59.

3. Asia-Pacific Consortium on Agricultural Biotechnology (APCoAB) (2009) Bt cotton in India-A status report. Asia-Pacific Consortium on Agricultural Biotechnology, New Delhi, India: 1-37.

4. RajendranK, Mohamed Amanullah M, Vaiyapuri K (2010) Effect of Spacing and Nutrient Levels on Bt Cotton. Madras Agric J97:379-380.

5. Bhalerao PD, Patil BR, Ghatol PU, Gawande PP (2010) effect of spacing and fertilizer levels on seed cotton yield under rainfed condition. Indian J Agric Res 44: 74–76.

21

Phytochemical Screening and Anthelmintic Evaluations of the Stem Bark of *Afzelia Africana* 'SM' (Keay, 1989) against *Nippostrongylus Barziliensis* in Wistar Rats

Simon MK* and Jegede CO

Department of Veterinary Parasitology and Entomology, Faculty of Veterinary Medicine, University of Abuja, Nigeria

Abstract

The anthelmintic activity of partitioned portions of the crude methanolic extract of *Afzelia africana* was evaluated *in-vivo* in rat model, experimentally infected with *Nippostrongylus braziliensis*. Crude methanolic extract of the plant was obtained and further partitioned between three solvents (petroleum ether, chloroform and N-butanol). Four portions (i.e., petroleum ether, chloroform, N-butanol and the aqueous methanol portions) were obtained after the partitioning. The crude methanolic extract and all the portions (with the exception of petroleum ether) were tested for anthelmintic activity against *N. braziliensis* in rats. The anthelmintic activity was assessed by comparing the number of worms recovered from rats treated with the portions to those from non-treated infected controls. Deparasitization rate of 70% or greater was considered as significant. The chloroform and N-butanol portions produced significant deparasitization (p<0.05) when data were subjected to ANOVA. The chloroform and N-butanol portions caused deparitization at the rate of 79.20% and 72.72% respectively when a maximum tolerated dose (1000 mg/kg^{-1}) was administered. The crude methanolic and aqueous methanol extracts induced non-significant (p>0.05) deparasitization rate of 62.50% and 53.24% respectively. Phytochemical screening conducted on the crude methanolic extract and the four portions of the plant revealed constituents that has anthelmintic activity such as alkaloids; steroids, saponins, tannins, flavonoids and cardiac glycoside.

Keywords: In-vivo; Phytochemistry; *Anthelmintic; Afzelia africana; Nippostrongylus braziliensis*

Introduction

Parasitic nematodes are among the most common and economically important agents of infectious disease of grazing livestock, especially in small ruminants; in the tropics, subtropics and other parts of the world [1,2]. Livestock production in tropical climates suffers heavy economic losses due to gastrointestinal parasites [3,4]. The greatest losses associated with nematode parasite infections are sub-clinical, and economic assessment shows that the financial costs of internal parasitism are enormous [5,6]. The loss is characterized by lower output of animal products (meat, milk, hides and skins), low manure, poor traction power, poor carcass quality/organ condemnation, death and medication costs, which all impact negatively on the livelihood of small holder farmers [7,8].

Livestock producers have generally derived substantial benefit from the use of conventional anthelmintic drugs in controlling livestock parasitosis. In Africa, however, declining funding for veterinary services and the rising costs (occasioned by depreciating value of local currencies) of these services has made it difficult for resource-poor farmers to have access to such services. This has led to the increasing demand for effective and low cost anthelmintic drug by African smallholder livestock producers and pastoralist in order to reduce having to pay for the high cost of the services [9-12]. There is equally the need to produce animal products free from industrial chemical input [13,14] and avoid possible tendencies of environmental pollution [15]; also the need to discover new therapeutic substances of natural origin with low toxicity to man and animals [16], as well as overcoming the rapid escalation in anthelmintic resistance worldwide [17]. *Afzelia africana SM* is a tree species commonly found in savanna fringing forest and drier parts of forest regions. It is commonly referred to as; mahogany, kawo, apa and akpalata in English, Hausa, Yoruba and Igbo respectively [18]. The seed is widely used for medicinal purposes, for industrial production of soap, margarine, and candle making and as

diets such as condiments and seasoning in soup [19]. Atawodi [20] identified the use of herbal preparations including *Afzelia africana* in the treatment of helminthiasis. This study therefore asseed the *in vivo* anthelmintic effects of the crude methanolic extract and partitioned portions of the crude methanolic extract of the bark of *Afzelia africana* against adult *Nippostrongylus braziliensis* in experimentally infected rats and also identified the active fractions from the crude methanolic extracts of the plant via a separation process, with the view of providing scientific basis for their use in ethno-veterinary practices.

Materials and Methods

Plant collection, identification and preparations

The stem barks of *Afzelia africana* was collected in April, 2005 from New Bussa in Niger State, Nigeria. The plant was identified by a botanist from the department of biological sciences Ahmadu Bello University, Zaria and a specimen voucher number 2276 was deposited. The samples (5 kg) of the plant bark were sun dried, pulverized into powdered form using mortar and pestle and sieved as described by Onyeyili et al. [21]. The methanolic extract was obtained using 500 g the powdered plant material in a soxhlet apparatus (Quick fit corning Ltd, Stafford England) after which the solution was evaporated to dryness in vacuum using a rotary evaporator coupled to a thermo-regulator (RII-35). Twenty grammes (20 g) of the dried crude methanolic extract were partitioned

---

*Corresponding author: Simon MK, Department of Veterinary Parasitology and Entomology, University of Abuja, Nigeria, E-mail: kawesimon2002@yahoo.com

with petroleum ether, chloroform and N-butanol (150 ml each) using separating funnel as described by Suleiman et al. [22]. The portions were then referred to as petroleum ether, chloroform, N-butanol and aqueous methanol portion respectively. After evaporating the solvents, the portions were tested for anthelmintic activity on albino rat experimentally infected with L3 stage of *Nippostrongylus braziliensis*.

Phytochemical screening

Before the experimental treatment, the crude methanolic extract, petroleum ether, chloroform, N-butanol and the aqueous methanol portions of the extracts were subjected to phytochemical screening using standard techniques of Ciulei [23]; Sofowora [24] and Brain and Turner [25]. The three solvents used have different polarity, thus, it is expected that they extract the various active principles in the plant base on their polarity.

Experimental animal

Sixty three (63), six to seven weeks old albino Wistar rats of both sexes and weighing between 100 to 160 g were used. The rats were acclimatized for two weeks in the laboratory and fed commercially prepared feed; water was given ad lib. Rats for the anthelmintic study were dewormed using albendazole at 200 mgkg^{-1} two weeks before infection [22].

Toxicity test

Due to dearth of information on the precise dosage of the plant preparation, a maximum tolerated dose experiment described by Lorke [26] was determined using fifteen (15) rats. The established maximum tolerated dose was then used as the basis for the administration of the crude methanolic extract and the various portions of the plant extract in the anthelmintic activity studies.

Experimental infection/design

Forty two worm-free rats were infected subcutaneously in the cervical region with 200 viable L3 stage of *N. braziliensis* in 0.2 ml of distil water using an 18-gauge needle attached to an insulin syringe [22]. Five days post infection, fresh faecal sample from each infected rat was collected by squeezing it out from the rectum. Faecal samples were examined quantitatively for *N. braziliensis* egg using the simple floatation method [27]. Rats not shedding ova of *N. braziliensis* were excluded from the experiment. The infected rats were randomly allocated to three (3) experimental groups (A-C). Group A (positive control group) having six rats, were treated with albendazole at 200 mgkg^{-1} body weight [22], Group B (experimental group) having twenty four rats; were divided into four sub-groups of six rats each and treated with the crude methanolic, chloroform, N-butanol and aqueous methanol portions of the extract based on the maximum tolerated dose [28]; whereas group C (negative control) having twelve rats; were subdivided into two groups of six rats each and given distil water and propylene glycol as placebo based on the maximum convenient volume (MCV) of 5 mlkg^{-1} [22].

Treatment with crude methanolic extract and the various portions

Oral treatment with the crude methanolic extract and the various portions of the plant extract was administered on day seven (7) post infection. Before the treatment all rats were weighed to determine the appropriate dose and the maximum convenient volume (MCV) for individual rats. Observation was made daily for two days for abnormal behavioral signs.

Worm counts

Two days post treatment; the treated rats were fasted for 24 hours, salvaged for adult worm count using the WAAVP guides [29]. The first 15 cm of the small intestine was removed, cut longitudinally and placed between two clean 20 cm glass slides. The section was examined at 40X magnification of a dissecting microscope. Visible worms were counted and recorded. The fraction that caused the highest reduction in worm count and does not produce any behavioral changes in the rats was considered to be the most active portion.

Percentage deparasitization

Percentage efficacy (deparasitization) of the crude methanolic extract and the various portions of was calculated according to the method used by Cavier [30].

Statistical analysis

Means of data obtained were analyzed statistically using the software package for GraphPad prism (version 4.0-----2003). Statistical significance for the anthelmintic effect of crude methanol extract, chloroform, butanol portion and aqueous portion was assessed by ANOVA; subsequently Borferroni's multiple comparison tests. P value <0.05 was considered significant.

Results

Phytochemical screening

The crude methanolic extract of the plant had alkaloids, steroids, saponins, carbohydrate, flavonoids, tannins and cardiac glycosides as constituents; the petroleum ether portion of the plant revealed the presence of only alkaloids and steroids; while the chloroform portion, when screened, showed the presence of alkaloids, steroids, flavonoids, tannins and cardiac glycosides. The screening of the N-butanol portion revealed the presence of carbohydrate, in addition to the constituent seen in the chloroform portion. The screening of the aqueous methanol portion revealed the presence of steroids, flavonoids, tannins, carbohydrate and cardiac glycosides (Table 1).

Toxicity test

At a dose range of 10 to 1000 mgkg^{-1}, the crude methanol did not cause any visible toxic effect on the rats. On the other hand, from 1600 to 5000 mgkg^{-1}, the rats demonstrated varying degrees of signs of toxicity which manifested as visible body weakness, inability to move and reduced appetite from 24 hrs to 7 days post administration. Therefore the doses of 1600 to 5000 mgkg^{-1} were considered unsafe for the rats; and the dose of 1000 mgkg^{-1} body weight was chosen as the experimental treatment dose for both the crude methanol extracts and the various portions (Table 2).

Constituents	CME	Portions Pet ether	Chloroform	N-butanol	Aq. Methanol
Alkaloids	+	+	+	+	-
C. glycoside	+	-	+	+	+
Carbohydrate	+	-	-	+	-
Flavonoids	+	-	+	+	+
Saponins	+	-	-	+	+
Steroids	+	+	+	+	+
Tannins	+	-	+	+	+

Table 1: Phytochemical screening for the CME and various partitioned portions of *A. africana*.

	10	100	Dose (mg/kg⁻¹) 1,000	1,600	2,900	5,000
Initial No. of rat	3	3	3	3	3	3
Mortality	0	0	0	0	0	0
Observation	A	A	A	B	C	D
Inference	-	-	-	+	++	+++

Key
A=rats active 6-24 hrs and beyond   B=rats showed weakness for more than 24 hrs
C=rats showed weakness for more than 48 hrs   D=rats showed weakness for more than 7 days
-=no sign of toxicity +=slightly toxic ++=less toxic +++=toxic

Table 2: Maximum tolarated dose/toxicity of crude methanol extract of *A. africana* on rats.

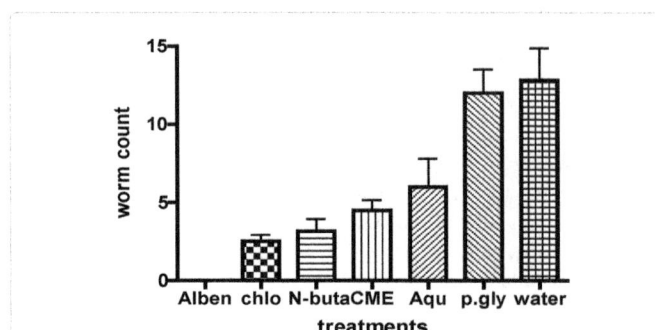

Figure 1: Mean ± SD of worm count after treatment with the various fractions of *A. africana* extracts, albendazole and the placebos.
alben-albendazole, **CME**-crude methanol extract, **N-but**-N-butanol, **Chlor**-chloroform, **Aqu**-aqueous methanol, **P.glycol**-propylene glycol.

The anthelmintic effect of the crude methanolic extract and the various portions

Rats that had oral infection of 200 L3 followed by treatment with crude methanolic extract at 1000 mg/kg⁻¹ had a mean worm count of 4.5; while those infected and treated with chloroform, N-butanol and aqueous methanol portion at 1000 mg/kg⁻¹ and albendazole at 200 mg/kg⁻¹ had respective mean worm count of 2.5, 3.5, 6.0 and 0; compared to the mean worm count 12.83 and 12.0 from the negative controls (water and p. glycol) treated rats (Figure 1). The crude methanol extract and the respective portion produced percentage deparasitization of 62.50% (crude methanol extract), 79.20% (chloroform), 72.72% (N-butanol) and 71.40% (aqueous methanol). Albendazole-treated rats (positive control), gave a 100% deparasitization, while water and p. glycol (negative controls) gave 0% deparasitization (Table 3). The deparasitization produced by the crude methanol extract and the chloroform, N-butanol and aqueous methanol fractions were significant ($p<0.05$) when compared to that produced by the placebo-treated negative control rats, while the deparasitization produced by the albendazole extract was highly-significant ($p<0.001$) (Table 3).

Discussion

Recent harmonizations on anthelmintic efficacy guidelines in ruminants have indicated that for a drug to be considered efficacious, a 90% reduction in total worm count (TWC) should be achieved [28]. However, it was considered 'a priori' that the efficacy of the plant extracts would be biologically significant if a reduction in total worm count (TWC) above 70% occurred [31]. Rats treated with chloroform and N-butanol portions of the plant showed anthelmintic activity. The chloroform had the highest reduction in TWC of 79.20%; this was followed by N-butanol portion with reduction in TWC by 72.20%. The

crude methanolic extract and aqueous portion did not produce the required biological significant reduction in TWC (62.50% and 53.24% respectively). However, the crude methanolic extract was statistically significant in comparison with the untreated control groups. The results of this study equally demonstrated that the parasite *N. braziliensis* was highly sensitive to albendazole with complete deparasitzation (100%) at a dose rate of 200 mg/kg⁻¹ body weight; this is similar to the findings reported by Suleiman et al. [22]. The N-butanol and aqueous portions were soluble in water and other polar solvents like alcohol, suggesting that the constituents of these portions are mainly polar compounds. However, the crude methanol extract and chloroform fractions were only soluble in propylene glycol (a non polar solvent). The outcome of the phytochemical screening revealed that the plant had constituents including tannins, alkaloids, flavovoids, cardiac glycosides and sterols which may have helminthic activities [14,21,32-37]. Tannins have been shown to have anthelmintic activities [32]. However, the anthelmintic effect of plants containing tannins depends on the type and content of tannins in the plant [32,38]. For instance, Kahiya et al. [39] in *in-vitro* studies reported that condensed tannins from the leave extract of *Acarcia nitotica* inhibited the development of *H. contortus* larvae in goat. In another study, tannins polyphenols from bryophytes were shown to have anthelmintic activity against *N. braziliensis* [34]. Tannins could binds to the free proteins available in the GIT of the host reducing nutrients available to the parasites resulting into starvation and death [40,41]. Also tannins are capable of binding with the glycoproteins on the cuticle of the parasites leading to their death of the parasite [42]. It is therefore probable to assume that the tannins contained in this plant could have had similar anthelmintic effect with the ones earlier described. Flavonoids are also believed to stimulate intestinal motility similar to that produced by acetylcholine [43], thereby causing rapid worm expulsion from the GIT. Lahlou [36] reported that flavonoid is one of the phytochemicals that have anthelmintic effect. Having identified flavonoids in almost all the portions used in this study, it is possible that it has had a significant anthelmintic effect on the *N. braziliensis* resulting in the observed deparasitization.The present study has shown that alkaloids are present in all the portions. It is possible that the presence of alkaloids has contributed to the significant deparasitization observed. Previous findings by Al-Qarawi et al. [44] reported that alkaloids extracted from both the latex and leaves of *Calotropis procera*, was effective in inhibiting the exsheatment of L_3 of *H. contortus* to L_4 in sheep, while Lateef et al. [37] also reported that alkaloids and their glycosides extracted from the root of *Adhatoda vestica* was effective against mixed gastrointestinal infections in sheep. Cardiac glycoside was identified and prominent in all the portions used in this study and therefore may have contributed to the observed anthelmintic effect of the plant. Cardiac glycoside has been shown to induce tonic contraction that resulted in the expulsion of the worms from rat GIT [37,45-47]. The dose determination studies (maximum tolerated dose) [28] was carried out on the premise that the plant extracts under investigation had nmgkg⁻¹) o alternative data to support any intended dosage. In this work, the injurious dose (1600 to 5000 and the MTD (1000 mgkg⁻¹) were determined. However, the plant extracts could not produce death even when the highest dose (5000 mgkg⁻¹) was given; thus suggesting the safety of the extract. The investigation of chemical compounds from natural products is fundamentally important for the development of new anthelmintic drugs, especially in view of the vast worldwide flora. Thus a quality controlled extraction of *A. africana* and the isolation of their bioactive compounds could be a promising alternative to conventional anthelmintic for the treatment of gastrointestinal helminthes of ruminant in the future. One problem associated with the use of these plants in traditional medicine is lack of

Rats	CME (1000 mgkg⁻¹)	Worm count after treatment with chloroform (1000 mgkg⁻¹)	N-buta (1000 mgkg⁻¹)	Aq. M (1000 mgkg⁻¹)	Alb. (1000 mgkg⁻¹)	Placebo1 (1000 mgkg⁻¹)	Placebo2 (1000 mgkg⁻¹)
1	4	2	4	14	0	4	6
2	2	3	3	4	0	17	16
3	7	4	2	8	0	18	15
4	5	1	5	3	0	12	10
5	5	3	1	5	0	14	11
6	4	2	6	2	0	12	14
Mean ± SD	4.5 ± 1.64b*	2.5 ± 1.05a*	3.5 ± 1.87a*	6.0 ± 4.43b	0.0 ± 0.0a*	12.83± 5.0	12.0 ± 3.74
% DPZ	62.50	79.20	72.72	53.24	100	0	0

Key
CME=crude methanol extract, Aq.M=aqueous methanol, Alb=albendazole,
Mean with *within the column are significantly different at p<0.001, while those with the letter a and b show no significant difference between their means at p>0.05 as determined by Borferroni's multiple comparison test.
%DPZ=percentage deparasitization

Table 3: Worm count and percentage deparasitization 7 days after treatment with crude methanol extract and fractions of A. Africana.

consistency of the dose. However, this was overcome by evaluating a maximum tolerated dose in order to reveal an appropriately non-toxic dose that was used in this study.

Conclusion

Result from this study demonstrated that the chloroform and N-butanol (partition portions of CME) of the plant are effective against experimental N. braziliensis infection in rats at a non-toxic dose of 1000 mgkg⁻¹. The chemicals believed to constitute the active principles in A. africana have significant anthelmintic efficacy whereas albendazole was found to be highly efficacious. However, the efficacies of these portions have no statistical significant difference to those of albendazole at a dose rate of 200 mgkg⁻¹. This demonstrates that the rat model was a reliable system for assessing in-vivo anthelmintic efficacy [48]. The result obtained in this study justifies further investigation into anthelmintic effects of the two portions of the plant extract in other animal species. More detailed studies are needed to isolate, characterized and evaluate the active components and the mechanism of action.

Acknowledgements

The authors thank the University of Abuja, Nigeria for granting the main author a staff development leave. The authors also like to thank all the technical staff of the Department of Veterinary Parasitology and Entomology for their assistance.

References

1. Prichard R (1994) Anthelmintic resistance. Vet Parasitol 54: 259-268.

2. Perry BD, Randolph TI, McDermoh JJ, Sones KR, Thornton PK (2002) Investing in animal health research to alleviate Poverty. International livestock research institute (ILRI) Kenya, 148.

3. Copeman DB (1980) The importance of helminth parasites in animal production system in the tropics [ruminants; South East Asia]. Asian-Australasian Animal Science Congress, Selangor, Malaysia.

4. Al-Quaisy HH, Al-Zubaidy AJ, Altaf KI, Makkwai TA (1987) The pathogenicity of haemonchosis in sheep and goats in Iraq: 1. clinical, parasitological and haematological findings. Vet Parasitol 24: 221-228.

5. Preston JM, Allonby EW (1979) The influence of breed on the susceptibility of sheep of Haemonchus contortus infection in Kenya. Res Vet Sci 26: 134-139.

6. McLeod RS (1995) Costs of major parasites to the Australian livestock industries. Int J Parasitol 25: 1363-1367.

7. Perry BD, Randolph TF (1999) Improving the assessment of the economic impact of parasitic diseases and of their control in production animals. Vet Parasitol 84: 145-168.

8. Chiezey NP, Gefu JO, Jagun AG, Abdu PA, Alawa CBI, et al. (2000) Evaluation of some Nigerian plants for anthelmintic activity in young cattle. Proceedings of International Workshop on Ethnoveterinary Practices, Kaduna, Nigeria.

9. Uza DV, Umunna NN, Oyedipe EO (1996) The productivity of muturu cattle (Bos brachycerus) under the traditional management system. Herd health. Bulletin of Animal Health and Productin in Africa 444: 151-152.

10. Sangster NC (1999) Anthelmintic resistance: past, present and future. Int J Parasitol 29: 115-124.

11. Abdu PA, Jagun AG, Gefu JO, Mohamme AK, Alawa CBI et al. (2000) A survey of ethnoventrinary practices of agropastoralist in Nigeria. In: Gefu, JO, Abdu PA, Alawa CBI (Eds) Ethnovet practices in Nigeria. Proceeding of the International workshop on ethnoveterinary practices Kaduna, Nigeria.

12. Ademola IO, Fagbemi BO, Idowu SO (2004) Evaluation of the anthelmintic activity of Khaya senegalensis extract against gastrointestinal nematodes of sheep: in vitro and in vivo studies. Vet Parasitol 112: 151-164.

13. Githiori JB, Höglund J, Waller PJ, Leyden Baker R (2003) Evaluation of anthelmintic properties of extracts from some plants used as livestock dewormers by pastoralist and smallholder farmers in Kenya against Heligmosomoides polygyrus infections in mice. Vet Parasitol 118: 215-226.

14. Athanasiadou S, Tzamaloukas O, Kyriazakis I, Jackson F, Coop RL (2005) Testing for direct anthelmintic effects of bioactive forages against Trichostrongylus colubriformis in grazing sheep. Vet Parasitol 127: 233-243.

15. Vierra LS, Calvalcante ACR, Pereira MF, Dantas LBA, Ximenes LJF (1999) Evaluation of anthelmintic efficacy of plants available in Ceara State west east Brazil. For the control of goat gastro intestinal nematodes. Review of Veterinary Medicine 150: 447-452.

16. Guarrera MP (1999) Traditional antihelmintic, antiparasitic and repellent uses of plants in Central Italy. J Ethnopharmacol 68: 183-192.

17. Jackson F, Coop RL (2000) The development of anthelmintic resistance in sheep nematodes. Parasitology 120 (suppl) s95-s107.

18. Keay RWY (1989) Tress of Nigeria. (3rdEdn) Clavendon press Oxford 146-216.

19. Ajah PO, Madubuike FN (1997) The proximate composition of some tropical legume seeds grown in two states in Nigeria. Food Chem 59: 361-365.

20. Atawodi SE, Usman M, Bulus ST, Atawodi JC, Wakawa L, et al. (2000) Herba treatment of some peotozoan and parasitic diseases of poultry in the middle of Nigeria. Procceedings of International Workshop on Ethnoveterinary Practice 79-84.

21. Onyeyili PA, Amin JD, Gambo HI, Nwosu CO, Jibike GI (2001) Toxicity and anthelmintic efficacy of ethanolic stem bark extract of Nauchlea latifolia. Nigeria Veterinay Journal 22: 74-79.

22. Suleiman MM, Mamman M, Aliu YO, Ajanusi JO (2005) Anthelmintic activity of the crude methanolic exract of Xylopia aethiopica against Nipposstrongylus braziliensis in rats Veterinarski arhiv 75: 487-495.

23. Ciulei T (1981) Methodology for the analysis of vegetable drugs. Chemical industries branch. Division of Industrial Operation, UNIDO 16-30.

24. Sofowora A (1982) Medicinal plants and traditional medicine is Africa. John Wiley and Sons. NewYork. 76-77.

25. Brain KR, Turner TD (1975) The practical evaluation of phytopharmaceuticals and therapeutics. Wright-Scientechnica, Bristol 10-30.

26. Lorke D (1983) A new approach to practical acute toxicity testing. Arch Toxicol 54: 275-287.

27. Soulsby EY (1982) Helminths, Arthropods and protozoa of domestic animals 7th (Edn.) FLBS Barrierve Tindal London.

28. Vencruysse J, Holdsworths P, Letonja T, Barth D, Conder, G, et al. (2001) International harmonization of anthelmintic efficacy guidelines. Vet Parasitol 99: 171-193.

29. Wood IB, Amaral NK, Bairden K, Duncan JL, Kassai T, et al. (1982) World Association for the Advancement of Veterinary Parasitology (W.A.A.V.P.) second edition of guidelines for evaluating the efficacy of anthelmintics in ruminants (bovine, ovine, caprine). Vet Parasitol 10: 265-284.

30. Cavier R (1973) Chemotherapy of internal nematode. In: Hawking F Chemotherapy of Helmithiasis (Edn.) International Encyclopedia of Pharmacology and Therapeutics. (1stEdn) Pergamon Press Ltd Headington Hill Hall, Oxford 1: 437-500.

31. Githiori JB, Hoglund J, Waller PJ, Baker RL (2003) The anthelmintic efficacy of the plant, Albizia anthelmintica, against the nematode parasites Haemonchus contortus, of sheep and Heligmosomoides polygyrus of mice. Vet Parasitol 116: 23-34.

32. Niezen JH, Waghorn GC, Charleston WAG (1998) Establishment and fecundity of *Ostertagia circumcinta* and *Trichostrongylus colubriformis* in lambs fed lotus (*Lotus pedenculanis*) or perenial rye grass (*Lolium perenne*). Vet Parasitol 78: 13-21.

33. Athanasiadou S, Kyriazakisi I, Jackson F, Coop RL (2000) Consequences of long-term feeding with condensed tannins on sheep parasitised with *Trichostrongylus colubriformis*. Int J Parasitol 30: 1025-1033.

34. Gamenara D, Pandolfi E, Saldaña J, Domínguez L, Martínez MM, et al. (2001) Nematocidal activity of natural polyphenols from bryophytes and their derivatives. Arzneimittelforschung 51: 506-510.

35. Prashanth D, Asha MK, Amit A, Padmaja R (2001) Anthelmintic activity of Butea monosperma. Fitoterapia 72: 421-422.

36. Lahlou M (2002) Potential of Origanum Compactum as a cercaricide in Morocco. Ann Trop Med Parasitol 96: 89-90.

37. Lateef M, Iqbal Z, Khan MN, Akhtar MS, Jabbar A (2003) Anthelmintic activity of *Adhatoda vesica* roots. Int J Agric Biol 5: 86-90.

38. Niezen JH, Waghorn TS, Charleston WAG, Waghorn GC (1995) Growth and gastrointestinal nematode parasitism in lambs grazing either lucerne (*Medicago sativa*) or sulla (*Hedysarum coronarium*) which contains condensed tannins. J Agric Sci 125: 281-289.

39. Kahiya C, Mukaratirwa S, Thamsborg SM (2003) Effects of Acarcia nilotica and Acacia karoo diets on Haemonchus contortus infection in goats. Vet Parasitol 115: 265-274.

40. Athanasiadou S, Kyriazakisi I, Jackson F, Coop RL (2001) Direct anthelmintic effects of condensed tannins towards different gastrointestinal nematodes of sheep. In vitro and in vivo studies. Vet Parasitol 99: 205-219.

41. Schultz JC (1989) Tannins-insect interaction. In: Hemingway RW, Karchesy J.J (Ed.), Chemistry and significance of condensed tannins. Plenum Press New York 417-433.

42. Thompson DP, Geary TG (1995) The structure and fuction of helminth surfaces. In: Marr JJ (Ed.), Biochemistry and Molecular Biology of Parasites (1st edn), Academic Press. New York 203-232.

43. Akendenque B (1992) Medicinal plants used by the Fang traditional healers in Equatorial Guinea. J Ethnopharmacol 37: 167-143.

44. Al-Qarawi AA, Mahmoud OM, Sobaih, Haroun EM, Adam SE (2001) A preliminary study on the anthelmintic activity of Calotropis procera latex against Haemonchus contortus infection in Najdi sheep. Vet Res Commun 25: 61-70.

45. Kim YK, Valdivia HH, Maryon EDB, Anderson P, Coranado R (1992) High molecular weight proteins in nematode C. elegans bind (H) ryanodine and form a large conductance channel. Biophy J 63: 1379-1384.

46. Hong SJ (1996) Inhibition of mouse neuromuscular transmission and contractile function by okadaic acid and canthasridin. Br J Pharmacol 130: 1211-1218.

47. Maryon EB, Saari B, Anderson P (1998) Muscle-specific action of ryanodine receptor channel in Caenorhabditis elegans. J Cell Sci 19: 2885-2895.

48. Wahid FN, Behnke JM, Conway DJ (1989) Factors affecting the efficacy of ivermectin against Heligmosomoides polygyrus (Nematospiroides dubius) in mice. Vet Parasitol 32: 325-340.

22

GIS Based Physical Land Suitability Evaluation for Crop Production in Eastern Ethiopia: A Case Study in Jello Watershed

Rediet Girma[1]*, Awdenegest Moges[1] and Shoeb Quraishi[2]

[1]*School of Biosystems and Environmental Engineering, Hawassa University, Ethiopia.*
[2]*School of Natural Resource and Environmental Engineering, Haramaya University, Ethiopia.*

Abstract

This study was aimed at identifying the current physical land suitability for maize, wheat and sorghum in Jello watershed under Chiro woreda in accordance to the FAO (1976) framework. The suitability mapping carried out with the help of GIS was compare with the LU being practiced. Relevant land quality (LQ) and land characteristics (LCs) data on climate, topography and soil following medium intensity survey technique were collected and the analysis was held after converting the data into a usable format for the LE process. Consequently through the querying analysis, the suitability rating process was run for individual LCs and based on the maximum limitation method, the overall suitability was assigned for specific land mapping units (LMUs) and displayed as suitability map with the integration of GIS. Results showed that out of the 1650ha, wheat production was moderately suitable (S2) on 6%; marginally suitable (S3) on 33% and not appropriate (N) on 61% of the land. 52% and 48% of the area was marginally suitable (S3) and unsuitable (N) for maize cultivation respectively. 33% of the area was marginally suitable (S3) and the rest (67%) was not suitable (N) for sorghum. Overall, presently none of the thirty three LMU fell under highly suitable (S1) class and based on the individual LCs, fertility status (exceedingly available P not assigned as S1) was found to be the most severe limiting factor. The comparison made between the existing land use being practiced and the findings from this study showed, 800ha (48%) and 1100ha (67%) area of land was mismatched (currently not suitable) for maize and sorghum cultivation respectively. Based on the analysis, wheat cultivation is relatively better (moderately suited) than the land use being practiced (maize and sorghum) on the bases of the present situation for 100ha (LMU23 and 30).

Keywords: Crop production; Ethiopia; GIS; Land evaluation; Limitation method

Introduction

Land needs careful and appropriate use that is vital to achieve optimum productivity and to ensure environmental sustainability for future generation. This requires an effective and operative management of land information on which such decisions should be based because land is one of the non-renewable natural resource. Decision on appropriate use includes the past and present human activities [1] and the status of physical and chemical properties of the land. Land evaluation (LE) is concerned with the assessment and valuation of land when used for specified purposes. It involves the execution and interpretation of basic surveys of data on climate, soils, vegetation and other aspects of land in terms of the requirements of alternative forms of land use. To be of value in planning, the range of land uses considered has to be limited to those which are relevant within the physical, economic and social context of the area considered [2].

According to [3] the utmost pertinent solutions for the utilization of land resources in sustainable way is land-use plan by proposing alternative measures and combine the different land characteristics to solve land misuse problems. Obviously to collect, store, incorporate and analysis the different land attributes that differ spatially, Geographic Information System (GIS) could be applied [4].

Ethiopia's social and economic development is highly dependent on agriculture. Leading industry and future overall country development is also expected to be driven by the progress in the agricultural sector [5]. Even if Ethiopia is endowed with rich biodiversity, throughout the country the speedy expansion of cultivation, settlements and other human activities in combination with unsustainable natural resource management even in unsuitable land has increasingly grown.

These expansions clearly exert pressure on the resource of land especially shifting of marginal and forestland in to cultivation purpose is a common practice. This is a great threat for resources as well on the resultant socio-economy, and environmental components since agriculture normally involves clearance of any natural vegetation present [2]. Due to improper land use, over exploitation and mismanagement of natural resources coupled with socio-economic factors, the problem of land degradation is on the rise and has become an issue of concern [6].

To combat land degradation, harmonizing the often-conflicting objectives of intensified human needs and socio-economic development, while maintaining and enhancing the ecology life support functions of land resources is a must. Land suitability evaluation is very important to provide information on the constraints and opportunities for the use of the land and therefore guides decisions on optimal utilizations of the resources [7]. This enables to guarantee the long-term productive potential of these resources all together by a compound effort which progressively brings the resource degradation under control [8].

At Jello watershed et al. [9] reported that an increase in cultivated

*Corresponding author: Rediet Girma, School of Biosystems and Environmental Engineering, Hawassa University, Ethiopia
E-mail: red8.girma@gmail.com

and settlement lands by 55% and 107% respectively with a decline in forest lands by 80% occurred over the 30-year period since 1966. Hence, human activities are expanded onto marginal areas because the local people are entirely dependent on natural resources for their livelihood along with other physical, socio-economic and political factors. These intense changes in land use/land cover may result a significant resource imbalance due to the incompatibility of the land with land use, over exploitation of the resource and mismanagement in terms of its capability and suitability.

Though crop production is dominant in the area, the land is not evaluated/assessed and used according to its natural capability and suitability for wheat, maize and sorghum. Such types of land use practice which may seem to be highly profitable in the short run will likely to take the lion sharing to cause soil erosion and resource degradation. Such trends in agricultural production and natural resource status of the land parcel require crucial efforts for the reason that proper use of land depends on its suitability for a specific purpose that integrates different measures in sustainable way [2]. Therefore, the principal goal of this study was to perform the actual qualitative land suitability evaluation and carry out suitability mapping for the existing land use types (wheat, maize and sorghum) with an understanding of the limiting factors by integrating different information using GIS tools. Comparison between the present land use being practiced and the findings from this study was also accomplished.

Materials and Methods

Study area

The study area is situated in Najabas kebele of Chiro Woreda of West Hararghe Zone in the Oromiya region (Figure 1) around 326 km east of Addis Ababa. Its altitude extends between 1780-2660 m.a.s.l and the average annual rainfall is 751.3 mm [10]. During the rainy months, farmers plant sorghum, maize, wheat, inter-cropping with chat and in some parts of the area vegetables like onion, tomato and cabbage and banana as fruit tree. The area of interest covers a total of 1650 ha and agriculture is the major livelihood of the people.

Data collection methods

Secondary data; climatic data records (Figure 2), topographic map and the LU practices were obtained from the Department of Land Resource and Environmental Protection of the woreda.

A medium intensity soil survey (1:50,000) was used, soil sampling density of one observation for 50 ha [11]. Consequently, a total of thirty three land mapping units (LMU) were prepared over the entire 1650 ha area (Figure 3a) and one representative profile pit for each LMU was also opened (Figure 3b) and geographically referenced by using GPS. A soil sampling technique in a zigzag pattern was implemented as recommended by [12], to make it more representative; twenty sub-samples of the same amount were collected from each LMU at two different fixed rooting depths (0-30 cm and 30-60 cm) separately and later, the sub-samples from similar depth were mixed carefully to made a composite sample. As a result, a total of sixty six composite soil samples over the entire area were prepared, and analyzed. Rooting depth was measured using a measuring stick; surface stoniness was estimated by selecting plots randomly to make it representative. Measurement was replicated five times and the average value was recorded in terms of areal percentage for each LMU [13]. Soil drainage class was assessed using soil profile color in combination with depth of mottling occurrence [14,15]. Flooding or inundation condition was characterized by flooding duration based on the information obtained

from local people [15]. The average slope gradient was measured using clinometer aimed in the direction of the steepest slope [16].

Data analysis methods

Soil analysis: Bouyoucos hydrometer method was used for textural analysis and according to USDA system textural triangle was used for grouping of soil textural classes [17]. Soil pH was determined using a pH-meter with soil to water suspension ratio of 1:2.5. The OC content was determined using the standard Walkley and Black's oxidation method [12] and organic matter (OM) was computed by multiplying the organic carbon (OC) value with a constant 1.724. Electrical conductivity meter was used to measure the EC of saturated extract of the soil [12]. TN was determined by Kjeldahl standard method [18].

According to [12], the concentration of Na and K was determined by flame photometer apparatus whereas Ca and Mg were determined by atomic absorption spectrophotometric techniques. CEC was determined on the basis of displacement after washing procedure using ammonium acetate [17]. Available phosphorus was determined using Olsen's method [12,19]. $CaCO_3$ was determined using titrimetric method with acid [20] and base saturation (BS%) was calculated by dividing the sum of extractable bases (Ca^{2+}, Mg^{2+}, K^+, and Na^+) by

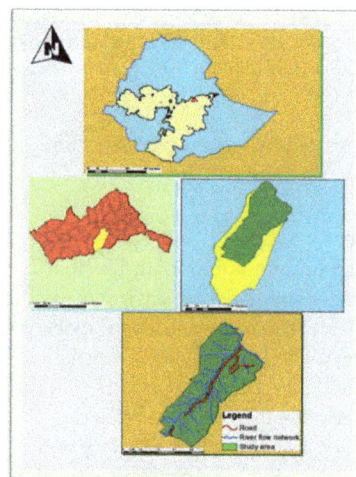

Figure 1: Location map of the study area.

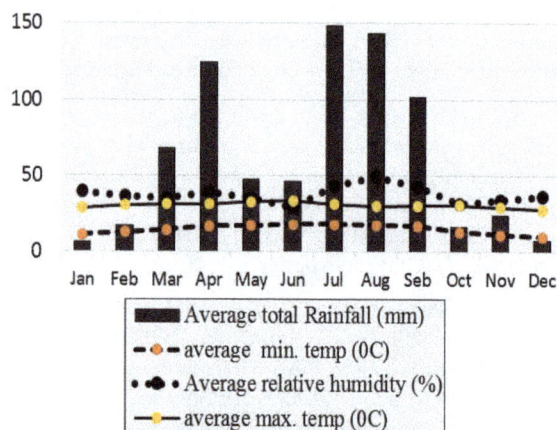

Figure 2: Climatic data.

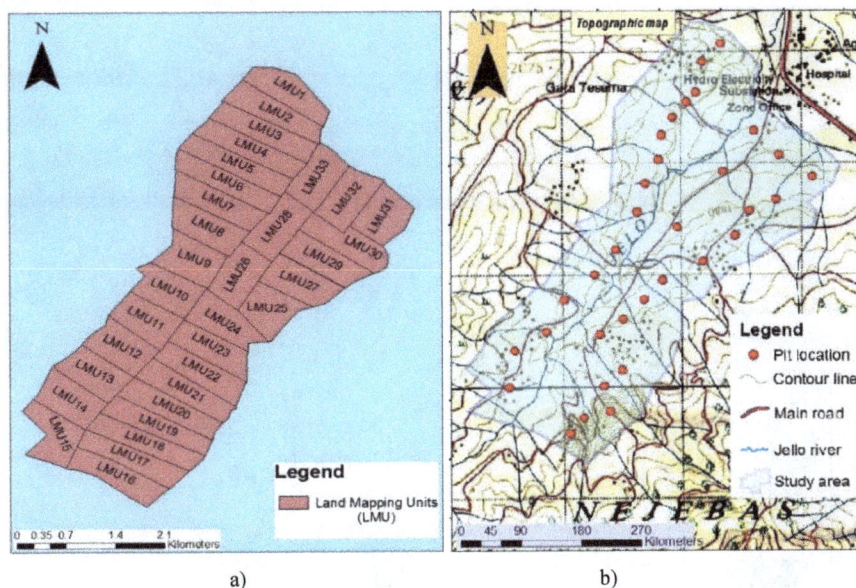

Figure 3: LMUs (a) and pit locations (b) of the study area.

cation exchange capacity (CEC) and multiplying it by 100 [21,22]. Exchangeable sodium percentage (ESP) was calculated by dividing the exchangeable Na to measured CEC values and multiplying it by 100 [12,15,21].

Land evaluation process: After assessing the suitability of the land for general cultivation use, the land evaluation (LE) process was proceed based on the maximum limitation method in terms of FAO's framework comparing the LCs or LQ values of each LMU with the requirements of the proposed LUT (maize, sorghum, and wheat) to identify the actual qualitative land suitability depending on physical environment data generated from topographic features, current soil characteristics, wetness condition and growing period climate data. The LE process comprised of computing the LCs values, suitability classification and land suitability mapping.

Computing the LCs values: The collected LCs data were processed and converted in to applicable LCs values (data base) using simple statistical approaches [23]. Climatic parameters during the crop growing cycle (for annual crops) was considered and an average value was calculated. In addition, the soil characteristic values changing with depth were also recalculated as depth weighted average over the 60 cm soil depth using three sections of equal thickness (20 cm) with a proportional weighting factors of 1.50, 1.00 and 0.50 from depth correction indices table (Van Ranst and Ann Verdoodt, 2005).

Depth correction

0-30cm: 0-20cm $\Rightarrow$ 1.5×(20-0) =30

20-30cm $\Rightarrow$ 1×(30-20) = 10

Sum= 40

30-60cm: 30-40cm $\Rightarrow$ 1×(40-30) = 10

40-60cm $\Rightarrow$ 0.5×(60-40) =10

Sum= 20

Therefore, the recalculated depth weighted average soil characteristic values over the total 60 cm soil depth was calculated by

dividing the summation of the product of depth correction and soil characteristic value from 0-30 cm and 30-60 cm soil depths by the total depth (60 cm).

Suitability classification: Once the database was created and prepared, for each LCs values layers were made using GIS (Figure 4a). For the accomplishment of GIS assisted land suitability evaluation, querying analysis (attribute queries) was used based on the attributes of every LMU to generate individual land suitability classification (LSC) for each LCs values/layers separately in reference to the suggested crop-specific requirement. After merging the individual LSC layers using the GIS Merge window, the overall LSC was assigned for each LMU by its most limiting characteristics (Figure 4b).

Suitability mapping: As an output, the land suitability map of the study area was displayed and shown on individual, transparent maps using different colors to indicate the suitability classes which had all its corresponding land qualities and land characteristics in its attribute table with the help of ArcGIS (Figure 4). Comparison was accomplished between the findings from this study and what is being practiced today to select relatively the better land use option.

Results and Discussion

Diagnostic LCs with set of values is illustrated in the table below for every LMU (Table 1). The soil characteristics values explained were the recalculated depth weighted average values. The deep rooting depth of representative soil profiles was measured to be more than 100cm for all LMU, which was treated as the ideal depth for annual crops as described by [12,13].

Land suitability evaluation for general cultivation

LMU32 (50 ha) is the only one assigned as highly suitable (S1); LMU1, 6, 7, 8, 18, 23 and 28 (350 ha) were grouped under moderately suitable (S2) class; nineteen (950ha) LMUs were classified as marginally suitable (S3) class and LMU2, 9, 14, 16, 17 and 19 (300 ha) were not suitable (N) due to fertility status and slope condition for cultivation purpose in general.

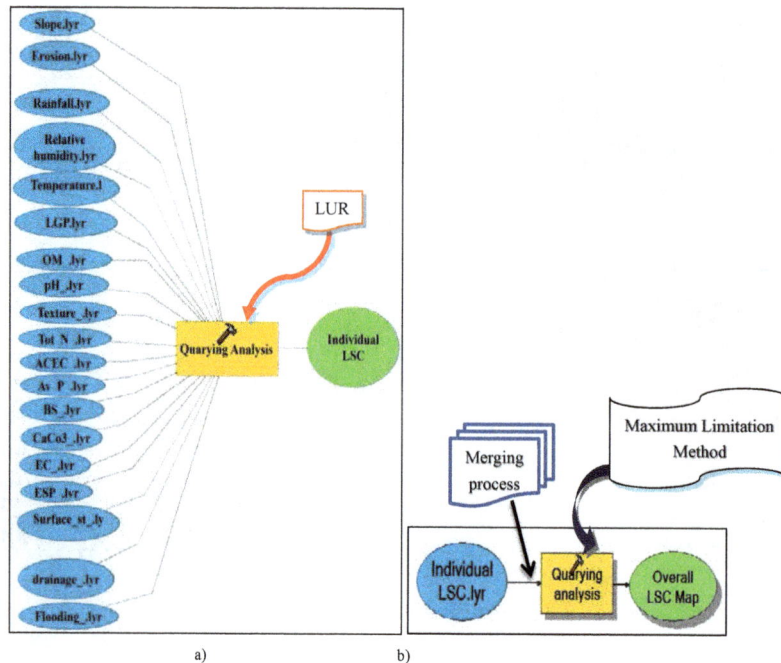

Figure 4: (a) GIS model used for individual LSC and (b) overall LSC.

LMU	Soil texture class	ESP %	EC dSm^{-1}	CaCO$_3$ %	CEC Cmol/kg	OM %	Av. P ppm	TN %	pH H$_2$O	BS %	Slope %	Drainage class	Flooding risk	Surface stoniness, %
1	SCL	5.9	0.1	5.4	57	2.49	1.75	0.13	7.5	75.2	5-8	Well	Nil	3
2	Sandy loam	7.72	0.1	5.3	47.6	0.45	0.49	0.04	7.1	82.4	8-15	Well	Nil	3.5
3	SCL	7.27	0.25	7.4	56.9	1.15	0.88	0.11	7.23	71.3	3-8	Well	Nil	12
4	SCL	7	0.17	4.8	49	0.65	1.27	0.05	6.93	79.9	3-8	Well	Nil	4.2
5	SCL	6.9	0.09	3.4	49.6	1.51	1.09	0.08	7.07	87.4	3-8	Well	Nil	10
6	Loam	10.29	0.18	5.13	43.5	1.99	3.3	0.14	7.73	87.6	3-8	Moderate	Nil	2.1
7	SCL	7.74	0.14	2.97	43.9	2.65	1.64	0.07	7.47	70.2	8-15	Well	Nil	2
8	SCL	6.16	0.26	6.6	56.5	3.32	2.56	0.12	7.33	83.1	3-8	Well	Nil	0.2
9	Sandy loam	8.34	0.14	3.87	41.5	0.53	0.5	0.05	6.97	71.8	5-8	Well	Nil	1
10	Sandy loam	6.73	0.11	7.7	45	2.22	1.83	0.13	6.97	92.1	3-5	Well	Nil	3.4
11	SCL	6.17	0.21	4	56.9	2	2.27	0.2	7.5	45.2	5-8	Well	Nil	4
12	SCL	6.48	0.15	3.87	52.2	1.96	1.56	0.12	7.27	64.8	8-15	Well	Nil	5.7
13	Loam	5.35	0.28	6.6	54	3.59	7.19	0.19	7.17	64.9	3-8	Moderate	Slight	8.2
14	SCL	9.65	0.14	4.47	46	1.14	3.52	0.08	6.93	77.7	>15	Well	Slight	9
15	SCL	12.38	0.21	8.4	41.8	3.38	4.22	0.15	7.1	81.2	3-5	Well	Nil	4.6
16	SCL	6.97	0.15	4.9	49.7	3.15	7.17	0.14	7.07	72.1	>15	Well	Nil	2
17	Clay loam	6.24	0.21	8	61.3	1.27	1.99	0.07	6.97	82.8	>15	Moderate	Nil	2.5
18	SCL	6.48	0.29	6.3	53.6	2.85	8.39	0.13	6.77	86.2	8-12	Well	Nil	3
19	Loam	4.61	0.14	9.03	61.8	3.59	5.81	0.13	7.3	78.9	>15	Moderate	Nil	2.9
20	Loam	7.09	0.16	3.6	47.8	2.31	7.07	0.12	7.57	78.1	8-15	Moderate	Nil	4.8
21	SCL	6.3	0.12	9.53	57.6	2.54	4.33	0.53	7.13	84.3	8-15	Well	Nil	5.2
22	SCL	8.13	0.07	11.27	43.1	2.87	4.8	0.14	7.73	89.9	3-5	Well	Nil	6
23	SCL	7.74	0.2	4.8	52.9	2.53	7.12	0.77	8.13	75.9	3-5	Well	Nil	2.8
24	SCL	6.05	0.1	4	43.6	2.93	7.33	0.59	6.93	88.9	3-8	Well	Nil	7.4
25	Clay loam	7.64	0.24	7.13	50.4	3.57	5.9	1.32	7.73	90.8	8-10	Moderate	Nil	4
26	Loam	6.21	0.14	6.5	53.5	3	5.9	1.13	8.13	69.1	5-8	Moderate	Nil	4.5
27	Clay loam	8.61	0.16	4.6	42.2	2.71	4.73	0.67	7.93	84.8	8-10	Moderate	Nil	12
28	SCL	8.07	0.12	6.1	52.1	2.87	5.83	0.54	7.37	83.8	3-8	Well	Nil	2.5
29	SCL	12.28	0.1	5.9	39.7	3.47	7.47	0.83	7.73	90.6	8-15	Well	Nil	4
30	SCL	6.69	0.12	4.8	52.6	3.34	6.9	0.67	7.8	89.7	3-5	Well	Nil	3.6
31	SCL	6.04	0.1	9.4	50.4	2.34	2	0.84	7.07	70	5-8	Well	Nil	6.2
32	SCL	5.71	0.12	6.9	49.4	2.98	3.2	1.31	7.04	88.6	0-3	Well	Nil	2.9
33	Clay loam	7.8	0.11	6.2	52.8	3.25	4.07	0.86	6.7	74.9	0-3	Moderate	Nil	3.4

Remark: "SCL"= Sandy clay loam

Table 1: The recalculated depth weighted average soil characteristics values.

Individual LSC for each LCs

On the bases of individual LCs values, a separate class was rated (except for the rooting depth /›100cm/ and temperature assigned as S1 concerning the three LUT; rainfall labeled as S1 for sorghum, S2 for wheat and maize; relative humidity was allocated as S2 for maize and sorghum).

Suitability ratings for wheat: Individual LCs were examined for wheat, accordingly the suitability percentage are shown in Figure 5. Suitability ratings for maize: In a similar way, individual LCs values were also matched with the requirement of maize. The suitability percentage is shown in Figure 6. Suitability ratings for sorghum: Individual LCs suitability percentage is also shown in Figure 7.

Overall land suitability classification

In general, land suitability classification of the mapping units centered on the most limiting land characteristics was classified and labeled into different suitability classes using the GIS quary builder technique (Figure 8). The overall suitability map for each LUTs was presented as an output after merging the individual LSC layers acquired. Overall land suitability classification for wheat: The overall suitability class of the study area for wheat cultivation was generally grouped into three ratings (Figure 9a) as moderately suitable (S2); marginally suitable (S3); and not suitable (N). Soil fertility, topographic feature, surface stoniness in conjunction with rainfall (LMU23 and 30) and pH (for LMU23) were considered as the limiting factors in general.

Overall land suitability classification for maize: The interpretation from the overall suitability map (Figure 9b) generated with the help of GIS tools shows that, no LMU was assigned as moderately suitable for maize production. The restrictive factors inducing the two suitability class (S3 and N) assigned were owed by the soil fertility status, pH and topographic factors.

Overall land suitability classification for sorghum: Similarly, the overall suitability class for sorghum cultivation falls under marginally suitable (S3) and unsuitable (N) (Figure 9c). The suitability class was brought by the dominant limiting factors as soil fertility, topographic condition, RH, together with textural class (for LMU2, 9 and 10), surface stoniness aimed at LMU13 and pH for LMU23 and 26. Concerning the three land utilization types, non-suitability class (N) was also observed as a mutual rating for 800ha (48% or LMU1, 2, 3, 4, 5, 7, 8, 9, 10, 11, 12, 14, 16, 17, 19 and 31). LMU25 and 27 were assembled to be marginally suitable for maize and wheat. For that of maize and sorghum, there was also 100 ha (LMU23 and 30) grouped under marginal suitability (S3). As far as the suitability map of wheat and sorghum was referred;

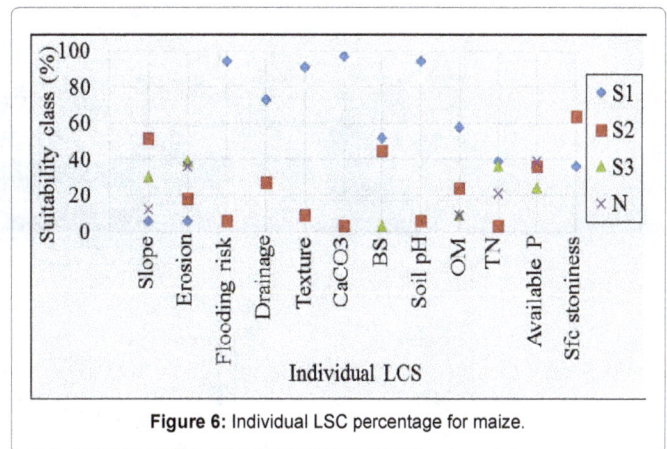

Figure 6: Individual LSC percentage for maize.

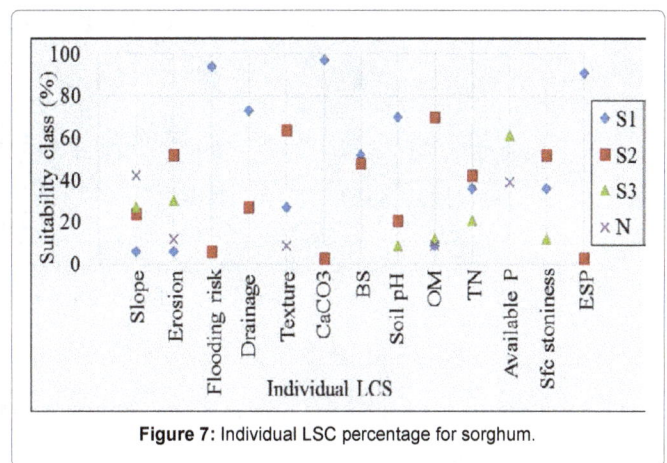

Figure 7: Individual LSC percentage for sorghum.

LMU18, 20, 21 and 29 (200 ha) was unsuitable (N) for both LUTs. The outcome of this study also tells that none of the thirty three LMU falls under highly suitable (S1) class.

Currently, farmers on the entire watershed cultivate maize and sorghum (more dominantly) intercropping with chat. According to the comparison made between the existing land use being practiced and the findings from this study, 800 ha (48%) and 1100 ha (67%) area of land was mismatched (currently not suitable) for maize and sorghum cultivation respectively.

Based on the analysis, wheat cultivation is relatively better (moderately suited) than the land use being practiced (maize and sorghum) on the bases of the present situation for 100 ha (LMU23 and 30). Comparatively maize cultivation is the other option (it is marginally suited) for LMU18, 20, 21 and 29 (200ha) and wheat or maize is better on LMU25 and 27 (100 ha) rather than sorghum cultivation at present. On the other hand, none of the three land utilization types are suitable for 800ha or 48% of the total area.

The continuing of existing land use (LU) practices beyond the natural ability of the land had been considered as a catalytic agent for the exponential depletion of the present scarce soil resource, the low soil fertility status for instance. The nonstop expansion of agricultural practice was the driving force for the decline of forest area and in some parts of the surveyed area long and steep slope is used for cultivation purposes.

Concurrently unless and other wise measures are taken, these

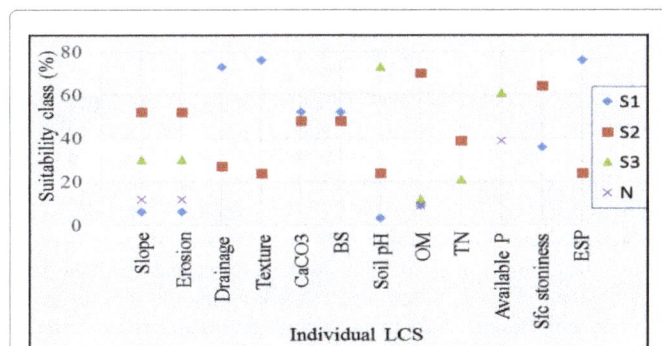

Figure 5: Individual LSC percentage for wheat.

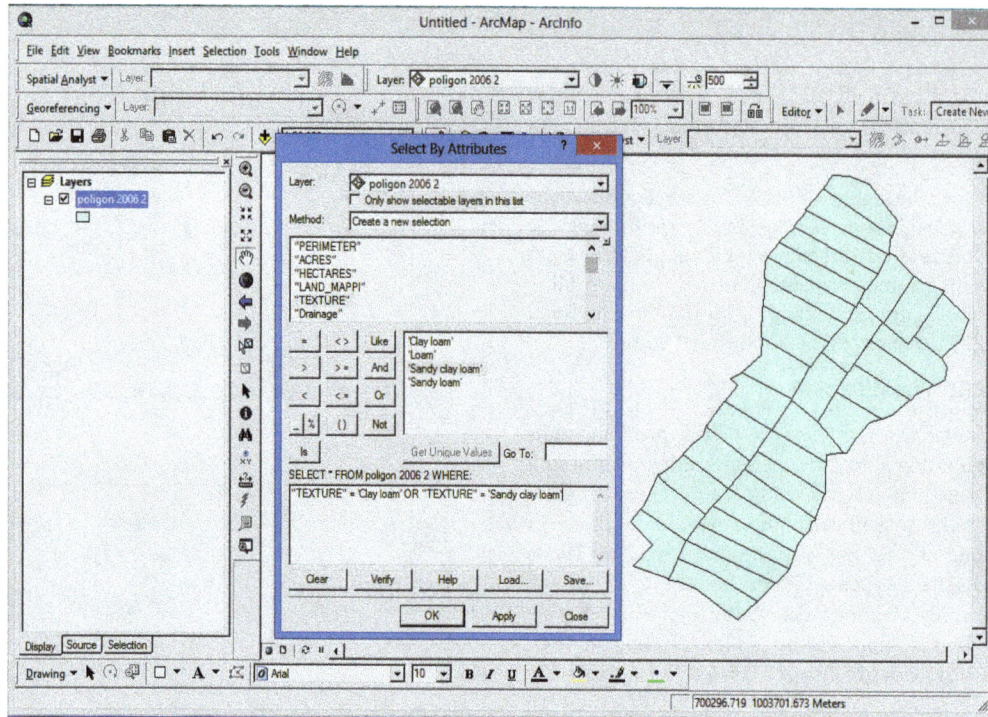

Figure 8: Quarying analysis window used.

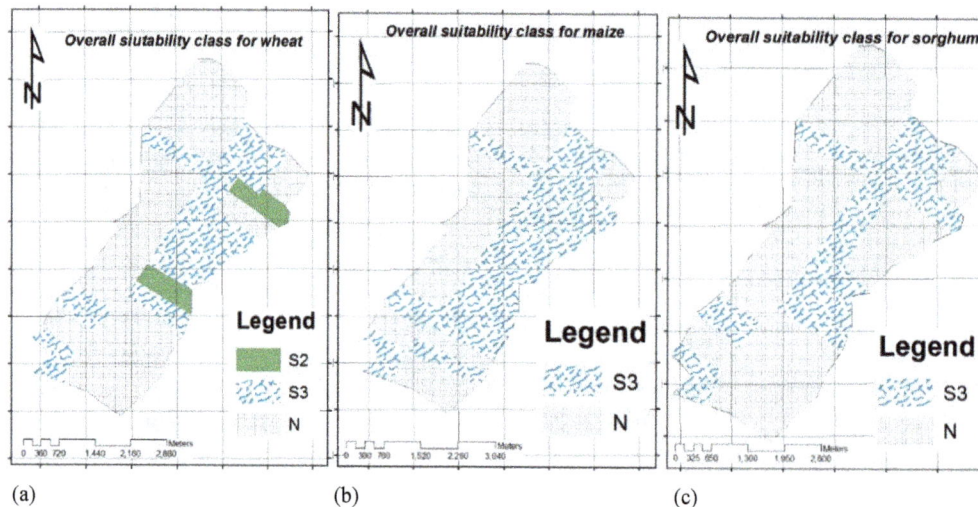

Figure 9: (a) Overall land suitability map for wheat, (b) maize and (c) sorghum.

augment soil erosion, resource degradation and adverse changes in river regimes (for instance, frequent flooding and absence of dry season stream flow) leading to irreversible degradation without hyperbole.

Conclusion

Currently, the surveyed area was assembled into three suitability classes as moderately suitable (S2) for 6%, marginally suitable (S3) for 33% and 61% unsuitable (N) for wheat. On the other hand two suitability classes were observed for maize (52% and 48%) and sorghum (33% and 67%) cultivation as marginally suitable (S3) and unsuitable

(N) respectively. From the overall suitability ratings attained, presently none of the thirty three land mapping units fell under highly suitable (S1) class to any of the three land utilization types. Based on the individual LCs, exceedingly available P (named as the most severe limiting factor) of the soil was the only one responsible for not to be characterized as high suitability (S1). Therefore, major limiting land characteristic for crop production in the district had been the low fertility status of the soil attributed to the low amount of available P. This also indicates that such lands can be degraded and easily loose the productive potential if the existing land use practices are ongoing and no well-timed appropriate measures are undertaken.

Acknowledgements

We are grateful for the support of the Federal Ministry of Agriculture. Land users and experts from bureau of Land Resource Management of Chiro woreda provided me information and participated in data collection are greatly acknowledged.

References

1. Keshavarzi A, Sarmadian F, Ahmadi A (2011) Spatially-based model of land suitability analysis using Block Kriging. Australian Journal of Crop Science 5:1533-1541.

2. FAO (1976) A framework for land evaluation. FAO soil bulletin 32, Rome.

3. Laosuwan T, Sangpradid S, Chunpang P (2013) Suitable areas for economic crops based on GIS and physical land evaluation model. International Journal of Soft Computing and Engineering (IJSCE) 3.

4. Mahmoud A, Shendi M.M, Pradhan B, Attia F (2009) Utilization of remote sensing data and GIS tools for land use sustainability analysis: case study in El-Hammam area, Egypt. Central European Journal of Geosciences 1: 347-367.

5. Teshome Yibarek, Kibebew Kibret, Heluf Gebrekidan, Sheleme Beyene (2013) Physical land suitability evaluation for rain fed production of cotton, maize, upland rice and sorghum in Abobo Area, western Ethiopia. American Journal of Research Communication 1: 296-318.

6. Panhalkar S (2011) Land capability classification for integrated watershed development by applying remote sensing and GIS techniques. Journal of Agricultural and Biological Science. Asian Research Publishing Network (ARPN) 6: 4.

7. Rabia AH (2012) A GIS based land suitability assessment for agricultural planning in Kilte Awulaelo district, Ethiopia. The 4th International Congress of ECSSS, EUROSOIL, Bari, Italy.

8. Mitiku H, Herweg K, Stillhardt B (2006) Sustainable Land Management-A New Approach to Soil and Water Conservation in Ethiopia.

9. Mohammed A (2006) Land use/cover dynamics over a period of three decades in the Jello micro-catchment, Chercher highlands, Ethiopia. Ethiopian Journal of Natural Resources.

10. BoA (2001) Result oriented community based participatory watershed development plan in Jello. Oromiya region, West Hararghe zone, Chiro woreda.

11. Deckers J, Spaargaren O, Dondeyne S (2002) Soil survey as a basis for land evaluation. Land Use, Land Cover and Soil Sciences. II: Encyclopedia of Life Support System.

12. FAO (2007) Methods of analysis for soils of arid and semi-arid regions. Rome, Italy.

13. USDA-NRCS (2008) Soil Survey Manual (Complete Print Friendly Version) NRCS Soils.

14. Briggs RD (1994) Site classification field guide. Cooperative Forestry Research Unit Technical Note 6. Miscellaneous publication 724.

15. Ritung S, Wahyunto Agus F, Hidayat H (2007) Land Suitability Evaluation with a case map of Aceh Barat District. Indonesian Soil Research Institute and World Agroforestry Centre, Bogor, Indonesia.

16. FAO (2006) Guidelines for soil description (4th edition), Rome.

17. Soil Survey Staff (2009) Soil Survey Field and Laboratory Methods Manual. Soil Survey Investigations Report No. 51, Version 1.0. R. Burt (ed.) US. Department of Agriculture, Natural Resources Conservation Service.

18. Sahlemedhin Sertsu, Taye Bekele (2000) Procedures for soil and plant analysis. National Soil Research center. Ethiopian Agricultural Research Organization. Technical paper No 24.

19. Taye Belachew, Yifru Abera (2010) Assessment of soil fertility status with depth in wheat growing highlands of Southeast Ethiopia. World Journal of Agricultural Sciences 6: 525-531.

20. USSLS (1954) Diagnosis and improvement of saline and alkali soil. Handbook 60, US Government Printing Office, Washington DC.

21. Moore G (2001) Soil guide: A handbook for understanding and managing agricultural soils. Agriculture Western Australia Bulletin No 4343.

22. USDA-NRCS (2004) Soil Survey Laboratory Methods Manual. Soil Survey Investigations Report No. 42 Version 4.0.

23. Van Ranst E, Ann verdoodt (2005) Laboratory of Soil Science. Land evaluation Part II: Qualitative Methods in Land Evaluation. International center for physical land resources. Belgium.

23

Response of Bell Pepper Crop Fertigated with Nitrogen and Potassium Doses in Protected Environment

Marcelo Zolin Lorenzoni*, Roberto Rezende, Álvaro Henrique Cândido De Souza, Cássio De Castro Seron, Tiago Luan Hachmann and Paulo Sérgio Lourenço De Freitas

*State University of Maringá, Maringá, Paraná, Brazil.

Abstract

Bell pepper is among the ten most economically important vegetables in the country. The proper management of water and fertilizers coupled with protected environment allows to obtain a quality agricultural production. This study aimed to evaluate the response of nitrogen and potassium doses applied through fertigation on the growth and yield of bell pepper crop, Magali R hybrid. A completely randomized design was used, with 16 treatments in a factorial scheme 4×4, with four replications. The treatments resulted from the combination of four doses of nitrogen (0; 73.4; 146.8 and 293.6 kg ha^{-1}) and potassium (0, 53.3; 106.7 and 213.4 kg ha^{-1}). The experimental plot consisted of a 25 L pot with a bell pepper plant. Seven harvests were made throughout the experiment, and fresh fruit number and mass were evaluated. Leaf area (LA) and total matter accumulation were evaluated as growth components. Regardless of the applied potassium doses, the variables LA and total dry matter showed higher results for N ranging from 155 to 194 kg ha^{-1}. The maximum fresh fruit matter (FFM) (1882 g plant^{-1}) occurred at the dose of 155 kg N ha^{-1} and 106.7 kg K ha^{-1} and the maximum number of fruits (NF) (16.3 fruits plant^{-1}) was obtained at the dose of 147 kg N ha^{-1} and 106.7 kg K ha^{-1}.

Keywords: *Capsicum annuum* L.; Mineral nutrition; Cultivation in pots; productivity; growth

Introduction

The bell pepper crop (*Capsicum annuum* L.) belongs to the Solanaceae family. In Brazil it is among the ten vegetables of greater economic and social importance. Its fruits have a high content of vitamin C and are consumed raw or ripe, and are used in the manufacture of condiments, pickles and sauces [20].

The use of bell pepper hybrid seeds and cultivation in protected environment are technologies that have been used to improve fruit quality and productivity. Among the vegetables conducted in protected cultivation, the main ones are peppers, tomatoes, cucumbers and leafy vegetables. The practice of cultivation in protected environment is a way to avoid the environmental adversities and may favor the production compared to the crop in the field, in addition to allowing increased cycle [2].

The bell pepper hybrid "Magali R" was released in 1995 and is able to combine productivity with oomycete resistance (*Phytophthora capsici*), better use of nutrients, easy adaptation to cultivation and marketing [3].

For good plant growth and to meet its nutritional and water needs in a protected environment, the use of irrigation systems is critical to enable higher frequency of irrigation and the use of fertigation [5]. Fertigation is a fertilizer application method along with irrigation water. It is an efficient technique of applying fertilizers to plants, allowing the use in smaller quantities at a time, as well as the ease incorporation of chemical in the ground, labor saving and convenience [15].

Nitrogen is the most important nutrient for bell pepper cultivation [6], and alongside with potassium, are the nutrients most required by the crop [7]. For this reason it is required greater attention with the supplementation of these nutrients. Almuktar et al. [8] noted the negative impacts on growth development of bell peppers, possibly due to the high concentrations of nutrients and minerals.

The information about the behavior and nutritional requirements of the bell pepper crop are based on work done in the field. Therefore it is fundamental to encourage research in order to generate information to help bell pepper producers make correct decisions regarding the conduction of the culture in protected cultivation.

Given the above, this study aimed to evaluate the growth and yield of bell pepper crop with different doses of nitrogen and potassium via fertigation in protected environment.

Materials and Methods

The experiment was conducted from February 2015 to August 2015 in a protected environment with the dimensions of 20 m long, 7 m wide and 3 m high, located in the Centro Técnico de Irrigação (CTI) of the State University of Maringá (UEM) in Maringá-PR, at the coordinates 23°25'57 "S, 51°57'08"W and 542 m altitude. The climate, according to Köppen classification, is CFA Mesothermal Humid, abundant rainfall in the summer and dry winters. The average annual temperature is 21.8°C.

The experimental design was completely randomized in a factorial 4×4 (four doses of N with 4 doses of K), with four replications, totaling 64 experimental units. Each unit is represented by a pot containing a plant and arranged in five longitudinal lines inside the protected environment, spaced 1.2 m between rows and 0.5 m between plants.

*Corresponding author: Marcelo Zolin Lorenzoni, State University of Maringá, Maringá, Paraná, Brazil, E-mail: marcelorenzoni@hotmail.com

The pots (Nutriplan®) of 25 L volume were filled with 25 kg of soil classified as Dystrophic Red Latosol (Oxisol), sandy texture, collected from the layer 0.0 to 0.20 m. The chemical characteristics were analyzed by the Laboratório Rural de Análise de Solos de Maringá and is shown in Table 1. The drainage system consisted of 4 kg of n° 1 crushed stone, enough to fill the bottom of the pot. On top of the gravel layer, a nonwoven fabric disk was placed to avoid losing the finer fraction of the soil.

Liming was done 60 days before the transplantation, in order to raise soil base saturation up to 80%. For planting fertilization, the fertilizers were mixed with the soil to homogenise the fertility condition, using 9.6 g K2O per pot, 28.8 g of P2O5 per pot and 500 g of organic matter per pot were applied 20 days before transplanting, following the recommendation of Trani PE [9.]

The treatments resulted from the combination of four nitrogen (N1=0, N2=73.4; N3=146.8 and N4=293.6 kg ha^{-1}) and potassium doses (K1=0, K2=53.3; K3=106.7 and K4=213.4 kg ha^{-1}), totaling 16 treatments, considering the maximum doses recommended by Trani PE [9], since there is no recommendation in the literature for the region in which this work was done. The amounts of nutrients were divided according to the absorption rate for bell pepper crop [10] and applied weekly via fertigation. The fertilizers used were urea Vitaplant® (45% N; 26% O; 21% C e 8% H) and potash Nutriplant® (60% K; 28% Cl; 12% O), because they are widely used.

Seedlings of bell pepper Magali R hybrid (Sakata Seed Sudameris) were produced in polyethylene trays Nutriplan® with 64 cells filled with commercial substrate Mecplant® proper for vegetables and transplanted when they had four to six true leaves, at 34 days after sowing.

Drip irrigation system was used with a 4 L h^{-1} flow emitter in each experimental unit. The replenishment of the water was controlled by daily weighing of pots using a digital scale Multivisi® with capacity of 40 kg. When soil moisture was approaching the critical moisture (0.09 g g^{-1}) the soil was irrigated until the moisture in the field capacity (0.2 g g^{-1}).

Seven fruit harvests were performed during the experiment, the first held at 66 days after transplanting (DAT) and the last at 136 DAT. The chosen harvest stage was when the fruits had maximum visual development, before acquiring the characteristic color of the variety (red).

Leaf area (LA), measured in cm^2 per plant, was quantified by the digital image method (Maller et al.) [11]. Total dry matter, without fruits, in grams per plant, was quantified when the samples, dried in an air circulation stove at 65°C, acquired constant mass.

Production variables were fresh mass and the number of commercial fruits. The fruits were classified as commercial when presented diameter and length greater than 4 and 6 cm, respectively, and unmarketable fruits those which had serious defects or different dimensions from the ones mentioned. The measures of length and diameter of fruit were taken with a digital caliper Mtx®. The fresh mass measurements were obtained using a digital scale Marte UX6200H with capacity of 6200 g.

The data were submitted to analysis of variance, at 1 and 5% probability, applying the F test. In case of significant interaction between levels, unfolding and study of regression took place, considering the linear and quadratic models. Statistical analyzes were performed using the statistical software Sisvar [12].

Results And Discussion

There was a significant interaction between nitrogen and potassium levels (p <0.01) for the variables LA, total dry matter (TDM), fresh fruit mass (FFM) and NF of the bell pepper plants (Table 2).

According to the unfolding of the N and K factors, for the variable LA, significant differences were found for the nitrogen doses in all potassium levels. Regarding potassium doses, there were significant differences in levels N2, N3 and N4. The quadratic model showed the best fit to the data for the application of nitrogen (Figure 1A). The increasing linear model presented better fit for the application of potassium in N2 and N3 levels, while the decreasing linear model showed the best fit for the N4 level (Figure 1B). Nitrogen doses above 220 kg ha^{-1} can cause disequilibrium with other macro or micronutrient with negative effects [13], which may be related to the decrease in LA value for all cases in which the highest dose of nitrogen was applied.

Larger LA values were obtained between nitrogen doses ranging from 155 to 173 kg ha^{-1}, regardless of level of potassium applied via fertigation. The maximum value of LA (7999.3 cm^2 per plant) was estimated at K4 level (213.4 kg ha^{-1}), requiring 155.3 kg ha^{-1} of nitrogen. With regard to the potassium application, the maximum value of LA (7840 cm^2 plant^{-1}) was estimated at the dose of 213.4 kg K ha^{-1} and applying 146.8 kg N ha^{-1} (Figure 1B).

Aragão et al. (2011) [1], studying the effect of different irrigation depths and nitrogen levels in bell pepper, Magali R cultivar, observed that in most applied depths LA increases with the increase in the dose of N [14], working with different managements of fertigation in bell pepper, observed that the LA presented a quadratic response with increasing N and K levels.

For the TDM variable, there were significant differences in the application of nitrogen at each level of the K factor. Subjected to regression analysis the TDM data showed better adjustment to the quadratic model (Figure 2A). With application of potassium there were significant differences for the N2, N3 and N4 levels. There were no significant differences for N1. This may be related to the application of K associated with the low amount of N, reducing dry mass production. N2 and N3 levels were better adjusted to the quadratic model, while the N4 level showed decreasing linear trend with increasing K doses (Figure 2B).

Larger TDM values were obtained within the range 162 to 194 kg N ha^{-1} (Figure 2A). Aragão et al. [1] verified an increase in TDM production with the increase in nitrogen doses.

For the potassium doses, the maximum value of TDM (66.7 g plant^{-1}) was found with the dose of 131 kg K ha^{-1} for N3 level (Figure 2B). Oliveira et al. [14] had a decrease in the production of TDM from 200% N and K with bell pepper crop undergoing different fertigation managements. This effect can be attributed to increased salinity of the soil above the tolerated by the crop resulting from the accumulation of ions in the soil due to the application of high amounts of nitrogen and potassium.

The variable FFM presented quadratic trend for the application of nitrogen and potassium, with the exception of the N1 and N4 levels which did not differ with the application of different doses of potassium (Figure 3). According to Malavolta [13] the excess nitrogen can cause reduction in fructification, directly affecting productivity.

The highest FFM value (1882 g plant^{-1}) was achieved with 106.7 kg K ha^{-1} and 155 kg N ha^{-1}. On average, the amount of nitrogen required to obtain maximum FFM, regardless of the K level applied, was set between the doses 155 to 168 kg ha^{-1}. Almuktar and Scholz [16] reported that with increasing nitrogen on bell pepper led increases total yield.

Regarding potassium fertigation, higher FFM values were found at the doses of 118 kg ha^{-1} for the N2 level and 115 kg ha^{-1} for the N3

pH	O.M.	P	Na⁺	K⁺	Ca⁺²	Mg⁺²	Al⁺³	H⁺	Sand	Silt	Clay
	g dm⁻³	---- mg dm⁻³ ----		------------------ cmol$_c$ dm⁻³ ------------------					---------- g kg⁻¹ ----------		
4.8	4.66	8.63	2.1	0.07	1.56	0.38	0.7	2.48	780	30	190

Table 1: Chemical characteristics of the soil used in the experiment.

Variation sources	Variables			
	LA	TDM	FFM	NF
	F values			
Nitrogen (N)	673.389 **	501.138 **	388.512 **	167.722 **
Potassium (K)	25.144 **	2.336 ns	61.239 **	18.354 **
N x K	17.841 **	11.375 **	13.934 **	5.949 **
N x K1	91.616 **	101.042 **	50.982 **	27.646 **
N x K2	127.045 **	115.343 **	84.400 **	30.861 **
N x K3	197.085 **	180.408 **	227.441 **	90.203 **
N x K4	311.166 **	138.470 **	67.491 **	36.861 **
K x N1	0.351 ns	0.305 ns	0.758 ns	1.291 ns
K x N2	29.736 **	10.047 **	35.736 **	22.481 **
K x N3	44.316 **	9.319 **	65.832 **	10.608 **
K x N4	4.262 **	16.791 **	0.715 ns	1.823 ns
CV (%)	10.97	9.19	6.88	8.41
General average	3521.83	41.20	1057.22	10.78

** Significant at 1% probability; * Significant at 5% probability; ns not significant

Table 2: Summary of the analysis of variance for leaf area (LA), total dry matter (TDM), fresh fruit mass (FFM) and number of fruits (NF) in bell pepper, fertigated with doses of nitrogen and potassium, cultivated in a protected environment.

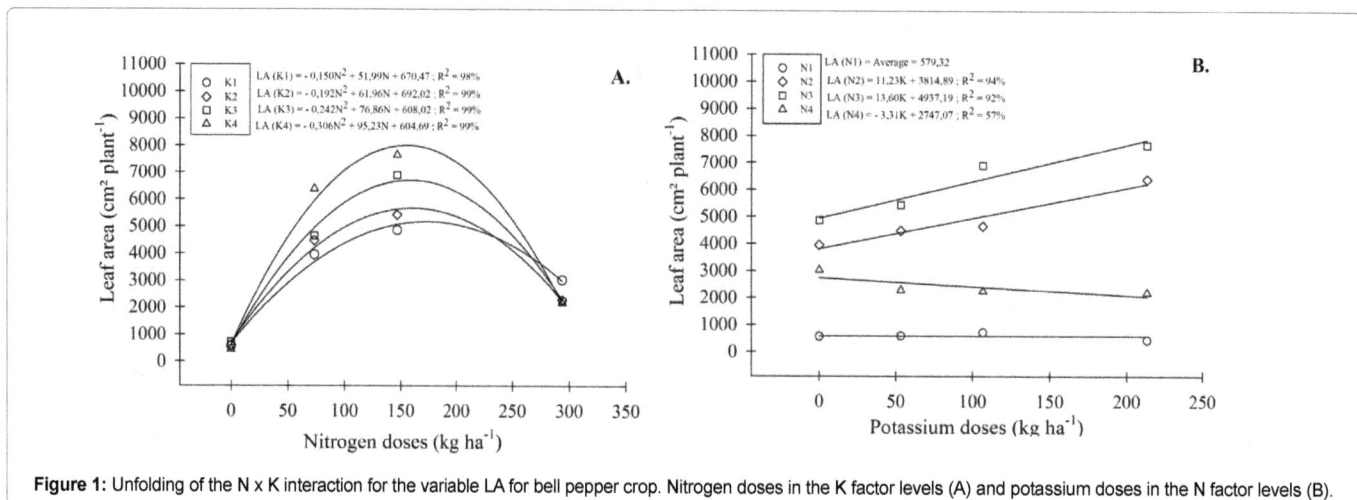

Figure 1: Unfolding of the N x K interaction for the variable LA for bell pepper crop. Nitrogen doses in the K factor levels (A) and potassium doses in the N factor levels (B).

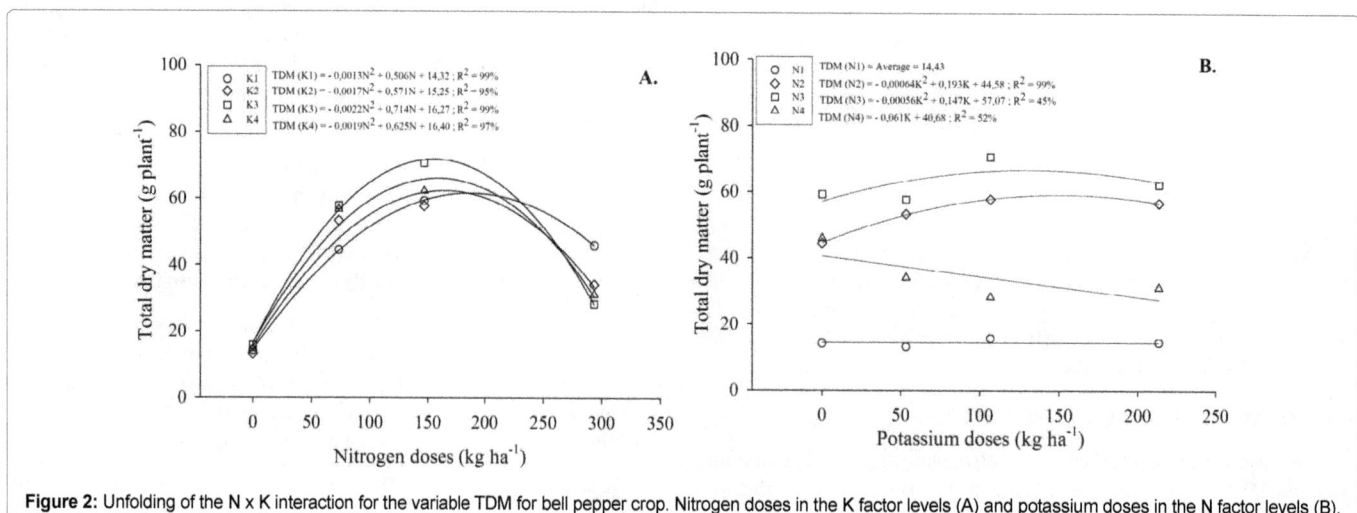

Figure 2: Unfolding of the N x K interaction for the variable TDM for bell pepper crop. Nitrogen doses in the K factor levels (A) and potassium doses in the N factor levels (B).

level (Figure 3B). According to Oliveira et al. [14], the reduction of the production of fruits per plant, when doses above the one which provided maximum performance were administered, can be attributed to the toxic effect of the fertilizers in the soil, resulting in reduced absorption of water and nutrients by plants.

It is possible to find in the literature studies that reported positive effect of fertilization with nitrogen and potassium on the yield of bell pepper crop, with linear or quadratic response [5,17-19].

For the variable NF, the unfolding of the N and K factors showed significant differences for nitrogen application at K levels, while for potassium application there were significant differences only in N2 and N3 levels, similar to what occurred to the FFM variable.

The fruits number data presented quadratic adjustment for nitrogen doses in all K levels and for potassium doses in N2 and N3 levels (Figure 4). This result was also verified by Melo et al. [5] who obtained quadratic adjustment for NF in relation to the application of potassium doses in the production of bell pepper fruits. Campos et al. [19] and Oliveira et al. [14] also found quadratic response for the NF variable due to the application of nitrogen and potassium.

The dose of nitrogen that promoted the greatest commercial NF (16.3 fruits per plant) was 147 kg ha[-1] for K3 level, whereas the dose of potassium which provided the highest commercial NF (16.1 fruits per

plant) was 128 kg ha[-1] for N2 level. Campos et al. [19] obtained higher value than the ones found in this study, for bell pepper cultivation, cultivar All Big, with 44 fruits per plant at a dose of 252 kg ha[-1] of nitrogen.

According to Malavolta [13] excess of nitrogen (doses above 220 kg ha[-1]) can cause reduction in fructification (Figure 4A), while the lack of potassium (doses below 60 kg ha[-1]) reduces the NF (Figure 4B).

The NF in this study was higher than that found by Araújo et al. [18-20]. These authors found higher commercial NF (12.8 fruits per plant) with the application of the maximum dose of N (400 kg ha[-1]).

Melo et al. [5] evaluating the effect of potassium doses in bell pepper crop obtained the maximum of three fruits per plant at the dose of 10 g per pot. Albuquerque et al. [4], studying the effects of different irrigation levels and potassium doses on the growth and yield of bell pepper, obtained an average of 6.5 fruits plant-1 at doses of 80 and 120 kg ha[-1].

The literature shows great variability in the NF in bell pepper crop that may be related to the lack of standardization of the number of harvests, cultivar or spacing used, nutritional management, conduction system and climate, which affects the number of harvested fruits.

The variables LA, TDM, FFM and NF were influenced by the application of nitrogen and potassium doses showing the interaction

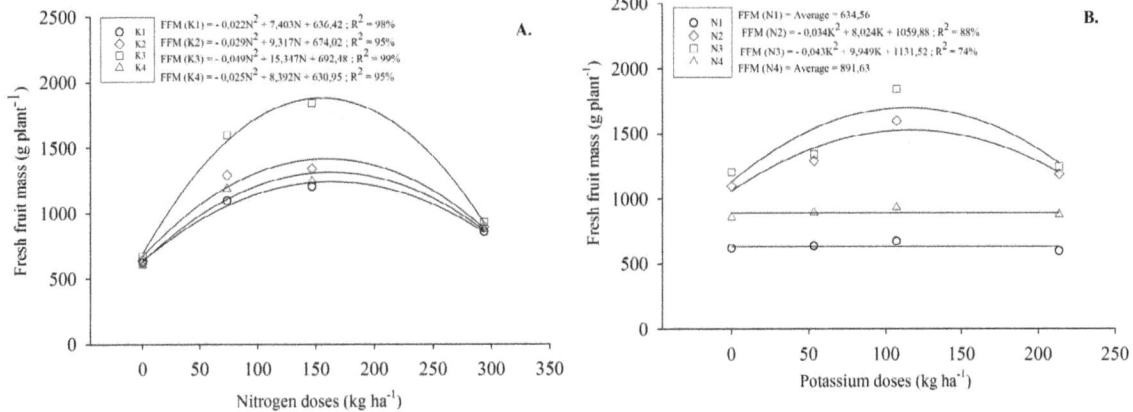

Figure 3: Unfolding of the N x K interaction for the variable FFM for bell pepper crop. Nitrogen doses in the K factor levels (A) and potassium doses in the N factor levels (B).

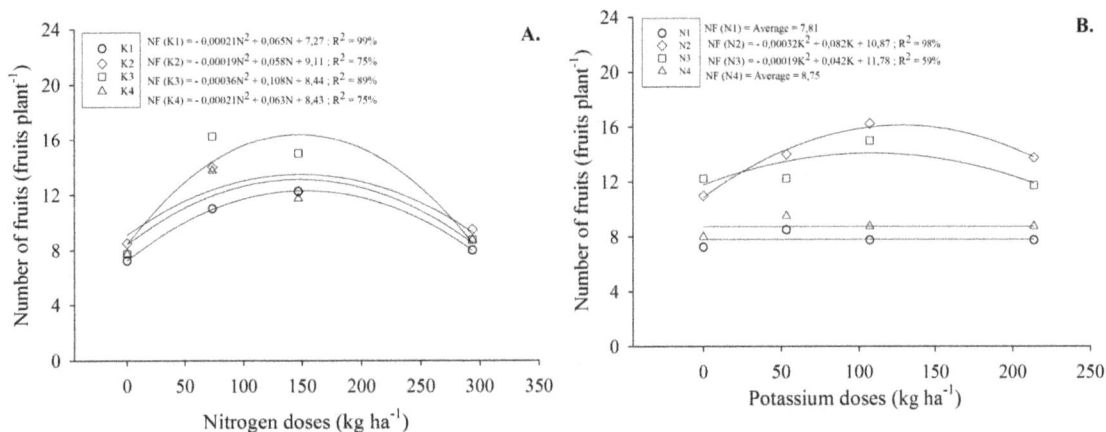

Figure 4: Unfolding of the N x K interaction for the variable NF for bell pepper crop. Nitrogen doses in the K factor levels (A) and potassium doses in the N factor levels (B).

of these fertilizers in the responses of the variables. Regardless of the potassium dose, the largest LA and TDM values were obtained by nitrogen fertilization with doses ranging from 155 to 194 kg ha^{-1}. The combination of the doses of 155 and 106.7 kg ha^{-1} of nitrogen and potassium, respectively, promoted higher production of FFM in bell pepper (1882 grams per plant) and the dose of 147 kg ha^{-1} of nitrogen and 106.7 kg ha^{-1} of potassium promoted the maximum NF per plant (16.3).

Acknowledgement

The Coordenação de Aperfeiçoamento de Pessoal de Nível Superior (CAPES) and State University of Maringá (UEM).

References

1. Aragão VF, Fernandes PD, Gomes RR, Santos Neto AM, Carvalho CM, et al. (2011) Effect of different irrigation and nitrogen levels in the vegetative phase pepper in a protected environment. RBAI 5: 361-375.

2. Lorentz LH, Lúcio AD (2009) Plot size and shape for chili pepper in plastic greenhouse. Rural science 39: 2380-2387.

3. Blat SF, Braz LT, Arruda AS (2007) Evaluation of sweet pepper hybrids. Brazilian horticulture 25: 350-354.

4. Albuquerque FS, Silva EFF, Albuquerque Filho JAC, Nunes MFN (2011) Growth and yield of pepper fertigated under different irrigation depths and potassium doses. Agriambi 15: 686-694.

5. Melo AS, Brito MEB, Dantas JDM, Silva Júnior CD, Fernandes PD, et al. (2009) Yellow pepper yield and quality under potassium levels in a greenhouse. RBCA 4: 17-21.

6. Aragão VF, Fernandes PD, Gomes Filho RR, Carvalho CM, Feitosa HO, et al. (2012) Production and efficiency in water use of chili submitted to different irrigation levels and nitrogen levels. RBAI 6: 207-216.

7. Epstein E, Bloom AJ (2006) Plants Mineral Nutrition : Principles and Perspectives . Plant, Londrina, Paraná.

8. Almuktar SAAAN, Scholz M, Al-Isawi RHK, Sani A (2015) Recycling of domestic wastewater treated by vertical-flow wetlands for irrigating chillies and sweet peppers. Agric Water Manage 149: 1-22.

9. Trani PE (2014) Liming and fertilization for vegetables under protected cultivation . Agronomic Institute of Campinas, Campinas, Sao Paulo.

10. Fontes PCR, Dias EN, Graça RN (2005) Nutrient uptake curves and a method to estimate nitrogen and potassium rates in sweet pepper fertigation. Brazilian Horticulture 23: 275-280.

11. Maller A, Rezende, Freitas PSL , Hara AT, Oliveira JM (2013) Comparison between leaf area predictive models using zucchini var. Novita plus. Encyclopedia Biosphere 9: 71-81.

12. Ferreira, DF (2008) Sisvar: a program for analysis and statistics education. Rev Scientific Symposium 6: 36-41.

13. Malavolta E (2006) Manual of plants mineral nutrition. Agronomic Publishing Ceres, São Paulo, São Paulo.

14. Oliveira, A. F. ; Duarte, S. N .; Medeiros, J. F .; Dias, N. S .; Silva, A. C. P, et al . (2013) Management of fertigation and doses of N and K in the cultivation of pepper in greenhouse. Agriambi 17: 1152-1159.

15. Frizzone JA Freitas PSL , Rezende, Faria MA (2012) micro irrigation : drip and micro sprinkler . Eduem , Maringa , Parana.

16. Almuktar SAAAN, Scholz M (2016) Experimental assessment of recycled diesel spill-contaminated domestic wastewater treated by reed beds for irrigation of sweet peppers. International journal of environmental research and public health 13: 1-20.

17. Marcussi FFN , Godoy LG , Villas Boas RL (2004) Nitrogen and potassium fertigation in sweet pepper culture based on N and K accumulation by plants. Irriga 9: 41-51.

18. Araújo JS, Andrade AP , Ramalho CI , Azevedo CAV (2009) Characteristics of bell pepper fruits cultivated in greenhouse under doses of nitrogen via fertigation. Agriambi 13: 152-157.

19. Campos VB , Oliveira AP , Cavalcante LF , Prazeres (2008) Yield of pepper submitted at nitrogen applied through irrigation water in protected environment. Journal of Biology and Earth Science 8: 72-79.

20. Carvalho, JA , Rezende FC, Aquino RF, Freitas WA, Oliveira EC (2011) Productive and economic analysis of red-pepper under different irrigation depths cultivated in greenhouse. Agriambi 569-574.

24

Perspectives on the Potential of Silvopastoral Systems

C Devendra*

Consulting Tropical Animal Production Systems Specialist, 130A Jalan Awan Jawa, Kuala Lumpur, Malaysia

Abstract

The importance of animal agriculture on productivity enhancement, nutritional and food security for economic rural growth in Asia is discussed in the context of the biophysical environment, available natural resources, preponderance of small farm systems, and opportunities for increasing the potential contribution. Arable land is a critical limiting factor, and an alternative to consider are the rainfed areas. The rainfed humid/sub-humid areas found mainly in South East Asia (99 million ha), and arid/semi-arid tropical systems found in South Asia (116 million ha) are priority agro-ecological zones (AEZs). They have been widely referred to as the less favored areas (LFAs), and low or high potential. The LFAs are characterised by very variable biophysical elements, notably poor soil quality, rainfall, length of growing season and dry periods, extreme poverty, and very poor people who continuously face hunger and vulnerability. There also exist large populations of ruminant animals, notably goats and sheep. About 43-88% of the total human population depends on agriculture for their livelihoods, of which 12-93% live in rainfed areas and 26-84% on arable land. In India for example, the ecosystem occupies 68% of the total cultivated area and supports 40% of the human and 65% of the livestock populations. Revitalised development of the LFAs is justified by the demand for agricultural land to meet human needs e.g. housing, recreation and industrialization; use of arable land to expand crop production to ceiling levels; very high animal densities. Animals play a multifunctional role and more importantly they can serve as the entry point for the development of LFAs. Efficient production systems are important and silvopastoral systems are underestimated and also underutilized throughout the developing countries, and especially where tree plantations are abundant such as with oil palm in Indonesia, Malaysia and Colombia. Concreted development attention together with research and development is necessary to promote its many economic advantages e.g. total factor productivity per unit land or labour, and value addition. Additionally, the system also promotes stratification, which provides an important opportunity to intensify NRM The strategies for promoting productivity growth will require concerted R and D on improved use of LFAs, application of systems perspectives for technology delivery, increased investments, a policy framework and improved farmer-researcher-extension linkages. These challenges and their resolution in rainfed areas can forcefully impact on increased productivity, improved livelihoods and human welfare, and environmental sustainability in the future.

Keywords: Animal agriculture; Diversification; Food security; Feed resources; Technology application; Systems perspectives; Farming systems research; Sustainability; Integration; Strategy; Impacts; Investments

Introduction

An assessment of the pathways of prevailing food systems, and investigations to ways of increasing current supplies of food production are compelling challenges of our times. The challenges are directly associated with a number of serious concerns which include: waning agricultural growth and reduced investments in research and development (R and D); rapid demographic pressures and urban migration; social pressures and consequences; increasing prices of household needs and services; social unrest and civil disorder; deprivation, survival and vulnerability; globalization and reduced capacity of small farm systems to maximize production, climate change, and stress on natural resources (land, crops, animals and water).

Asian farming systems, with their wide diversity of crops and animals, traditional methods, multiple crop-animal-environment interactions and numerous problems of farmers present increasingly complex issues associated with natural resource management (NRM) and the environment. The numerous interactions, and emerging complex constraints, underline the fact that these issues can only be fully resolved-not by any one discipline, but by community-based participation and interdisciplinary R and D, which will enable improved understanding. Education and training is a powerful and important driver of community-based participation which can sustain food security, poverty reduction and social equity.

In this context the empowerment of women in activities that support self-reliance can also be significant in increased contribution from agriculture.

Food systems are currently the defining and serious future concerns, rising incomes which are exerting tremendous pressure on the natural resources, and the capacity of science and technology to address these issues. The terms *Food Security* and *Food Insecurity* have now come to be used more widely, together with their links to poverty. Most governments see the need for a strong policy framework to support maximum food production to the extent possible, and can respond to changing consumer preferences.

A previous article published in the same journal Agro technology, a very high quality publication by OMICS, the focus of the discussion. That article emphasized the huge opportunities for improved R and D, implications for land-crops-animal interactions and intensification,

*Corresponding author: Dr. C. Devendra, PhD, DSc (Nott. UK) was honoured with the Third International Animal Agriculture award at the Eleventh World Association for Animal Production Conference held in Beijing, China on 15th October 2013, OMICS Publishing Group and the Editors congratulates him on this outstanding recognition, E-mail: cdev@pc.jaring.my

yield-enhancing increased productivity, impacts and environmental integrity.

The intent in this article is to complement the earlier article [1] by focusing specifically on the relevance of tree crops, and the development of integrate systems with ruminants and land use. As climate change takes effect, the land available for crop production contracts and this will place much stress on food production systems. The decreased availability of arable land in the future can increase the vulnerability of the poor who live by the land, which is of great concern and likely to be a major issue for discussion. Of equal concern is the loss of about 5.7 million hectares of arable land annually through soil degradation, and a further 1.5 million hectares as a result of water logging, salinisation and alkanisation [2].

Alternative options for food production: Development of rainfed areas to expand land use

Expanding the land area to lands outside of the irrigated areas is an important option for increasing food production in the future. This is especially justified by the following reasons:-

• Increasing demand for agricultural land to meet human needs e.g. housing, recreation an industrialisation

• Expansion of crop production to ceiling levels

• Increasing and very high animal densities

• Increased resettlement schemes and use of arable land

• Growing environmental concerns due to very intensive crop production e.g. acidification and salinisation with rice cultivation

• Human health risks due to expanding and often very intensive peri- urban poultry and pig production

• Land fragmentation

• Use of arable land by expanding perennial tree crops areas, and

• Urbanization.

By definition, rainfed areas refer to all those lands outside of the irrigated, more favored or high or low potential areas. They have been variously referred to as fragile, marginal, and dry, waste, problem, and threatened, range, less favored, low potential lands, forests and woodlands, and include a reference to lowlands and uplands. Of these terms, less favored areas (LFAs), low or high potential are quite widely used and is also been used in this paper.

Potential for food production systems

Mention must be made and to draw attention to the lands immediately outside of the irrigated areas, Due to their proximity and the benefit of residual soil moisture from the irrigated areas, these same areas are designated high potential areas, primarily very valuable for crop cultivation, and now also extensively used for cultivating tree crops like coconuts, citrus, coffee, cocoa, oil palm and rubber. In Malaysia for example, over 90% of the land area is used by oil palm, rubber and cocoa.

Despite these perceptions and concerns, the tide is changing In Sabah and Sarawak, more than Peninsular Malaysia. A number of plantations have already embarked on integration with consistently demonstrable economic benefits. One very large oil palm estate not only benefits from increased yield of fresh fruit bunches (FFB), but also breeds and supplies excellent good quality Brahman cattle. The

increasing interest in integrated systems in Sabah and Sarawak augers well for the future to benefit from efficiency in NRM and environmental sustainability.

Among these, the oil palm is a particularly important 'golden crop' and Asia has about 84% of the total world land area under oil palm of about 10.6 million ha. The largest land areas of 8.4 million ha of oil palm are found in Malaysia and Indonesia, who together own over 79% of the world planted area, and produce about 87% of the total world output of palm oil, followed by much smaller areas being found in Thailand, Philippines, India and Papua New Guinea. The integration model with oil palm offers extension of the principles of integration involved with other tree crops like coconuts in the Philippines, Sri Lanka and South Asia, rubber in Indonesia, and citrus in Thailand and Vietnam.

The use and development of the areas under different types of tree crops in integrated tree crop-ruminant or silvopastoral systems is grossly underestimated despite their potential very importance for several reasons. Integrated systems link the natural resources; crops, animals, land and water to economic, social and ecological perspectives. The process is holistic, dynamic, interactive, and multidisciplinary. The relevance of these systems, potential economic value, and opportunities for urgent development attention are highlighted in this paper.

Extent of rainfed areas

Table 1 provides data by region globally, on the extent and distribution of the different categories of rainfed areas [3]. In South East Asia, the total rainfed area is 99 million ha and in South Asia 116 million ha.

Table 1 also gives data on the extent of rain fed areas in individual countries. In South East Asia. the rain fed area as a proportion of total land available ranges from 63% in Indonesia. 68.5% in Malaysia to 97% in Cambodia. In South Asia, the corresponding values are from 27% in Pakistan to 84% in Nepal. Only in Pakistan and Sri Lanka does the percentage of irrigated land exceed that of the rain fed area. In absolute terms however, the largest irrigated land area of 43.8 million ha is found in India.

Silvopastoral Systems

The process of integration involving trees and ruminants is very complex due to the many variable biophysical factors and the natural production resources. Several terms have been used, but concerning the integration with ruminants [4] within which the term silvopastoral systems is being increasingly recognised:-

• Agro-forestry: involves the use of various tree crop options, usually woody perennials very commonly in rainfed areas

Table 1: Distribution of land types by region [6].

Region	Land type (% of total land)				
	Favored	Marginal	Sparsely arid lands populated	Forest and wood lands	Rural population living in favored lands (%)
Asia	16.6	30.0	18.5	34.6	37.0
Latin America and Caribbean	9.6	20.3	8.1	61.9	34.0
Sub-Saharan Africa	8.5	23.1	24.6	43.7	27.0
Near East and N. America	7.8	22.6	65.8	3.9	24.0
Total (105 countries)	10.7	24.0	25.9	39.4	35.0

Table 1: Distribution of land types by region

• Silvopastoral systems: involves trees (e.g. coconuts, oil palm and rubber) and animals

• Agro-pastoral systems: integrates crops, animals and trees.

There is a school of thought that believes that agro-forestry with its emphasis on tree crop options is a very good strategy to cope with climate change. The belief is reinforced by the view that the International Panel for Climate Change [5] have indicated that agro forestry is involved with plant biodiversity, plant species, permanent soil cover, seeds and diversification, which together with other economic, environmental and socioeconomic benefits for poor farmers and their livelihoods. The benefits are consistent with the finding earlier by CGIAR/TAC [6] that in a review of agro -forestry practices in 21 projects, 75% had net present value. In Indonesia, a case in point is growing the leguminous tree turi (*Sesbania gandiflora*) very commonly in the rice bunds. The legume increases soil fertility, provides fodder for animals which needs to be harvested for animals, and also young shoots and leaves for human consumption.

Silvopastoral systems by comparison are directly involved with the parent tree crop. In the case of oil palm for example, the whole system is a self—contained unit, going from breeding animals, beef production from either gazing or in situ feeding on oil palm by-products, and sale of beef for urban markets. In addition, grazing enables control of surplus grass and weeds with attendant savings in the cost of weedicides and fertilizers. More importantly, several studies have shown value addition in the output of FFB and therefore palm oil [7]. The net result is clear economic benefits, increased additional income over and above the wages from the plantations, food security; sustainable agriculture and environmental integrity. Under the circumstances, it is hard to ignore silvopasroral systems with all its challenges.

Types of interactions and benefits

There are many benefits of crop–animal–soil interactions [8], and result from the synergistic interactions of the system components. The following interaction is common, almost all of them resulting in tangible benefits:

• Beneficial effects of shade and available feeds on livestock;

• Draught animal power for land preparation and crop dung and urine for soil fertility and crop growth;

• Crop residues and agro-industrial by-products (AIBP) from trees *in situ*;

• Effects of native vegetation on the cost of weed control, crop management and crop growth

•Type of animal production systems (extensive systems combined with arable cropping, and systems integrated with tree cropping) leading to increased income and environmental integrity.

Table 2 indicates the locations of the main tree or perennial crop in many parts of the tropics, and the opportunities for adaptive research, for example sheep and coconuts in the Philippines and the development of sustainable silvopastoral systems. Table 2 also gives an indication of preferred animal options that are appropriate for individual tree crops. Small ruminants appear to be favored in most cases. The fact

Table 2: Potentially important perennial crops and their locations for use in integrated systems in the tropics.
*Small ruminants-Goats and sheep. **Large ruminants-Buffaloes and cattle

Crop	Location	Preferred animal species
1. Cashew	S. India, Vietnam	Small ruminants*
2. Citrus	India, Philippines, Thailand, Vietnam	Small ruminants
3. Cocoa	Malaysia, Papua New Guinea. Cote de Ivoire, Indonesia, Nigeria	Small ruminants
4. Coconuts	S.China, S.India, Indonesia, Philippines, SriLanka, Thailand	Large and small ruminants
5. Fruit trees e.g; Mango, plantain	N.India, Philippines Thailand, Costa Rica	Small ruminants
6. Oil palm	China, Indonesia, Malaysia Papua New Guinea, Thailand Columbia, Nicaragua	Large small and small ruminants
7. Rubber	Schema,Indonesia, Philippines, Malaysia, Thailand, Brazil	Small ruminants
8.Teak	Lao PDR, Myanmar	Large ruminants**

Table 3: Estimated gross margin of profits for meat production using crossbred Katjang goats with varying levels of fertility in Malaysia.

[1]15% among kids.
[2]25% culling per annum.
[3]$110 US per goat weighing about 30 kg.
[4]Cost of cultivated grass is 3.6 US$ cents per kg. fresh weight.
[5]No costs attached; both components are considered free.

Fertility level (% kids weaned/does mated)	80	100	120	140	160
I. The goat flock					
Flock size (breeding does)	10	10	10	10	10
Increase due to kids born	8	10	12	14	16
Less mortality[1]	7	8	10	12	14
Net increase in numbers less culls[2]	17	18	20	22	24
Cost of goats[3] ($ US)	1870	1980	2200	2420	2440
Cost of cull goats[3] ($ US)	110	220	220	220	220
Less cost of foundation does[3]	1100	1100	1100	1100	1110
Total gross revenue per year ($ US)	880	1100	1320	1540	1760
II. Cost of production					
(i) On cultivated forages, conc. with labour[4]	68.2	74.1	82.5	90.5	99.4
(ii) On uncultivated forages with family labour[5]	-	-	-	-	

Table 4: Effects of climate change on land use and livelihood systems of the poor (26).

Land use systems	Livelihood systems of the poor Including the landless
• Reduce soil moisture	• Reduced income
• Expansion of semi-arid and arid AEZs	• Increased povertys
• Increased droughts	• Increased vulnerability
• Increased rangelands	• Inability to adapt to heat stress
• Woody encroachment	• Increased food and nutritional
• **Desertification**	• Insecurity
• **Increased overstocking of animals e.g. with resultant soil degradation**	• Increased susceptibility to diseases
	• Reduced self-reliance
• Increase salinization	• Increased urban migration
• Reduced biodiversity	
• Effects on the systems	
• Educed systems services	
• Drift out of agriculture	

remains that ruminants provide the entry point for the development of integrated tree crops-ruminant systems (Figures 1-4).

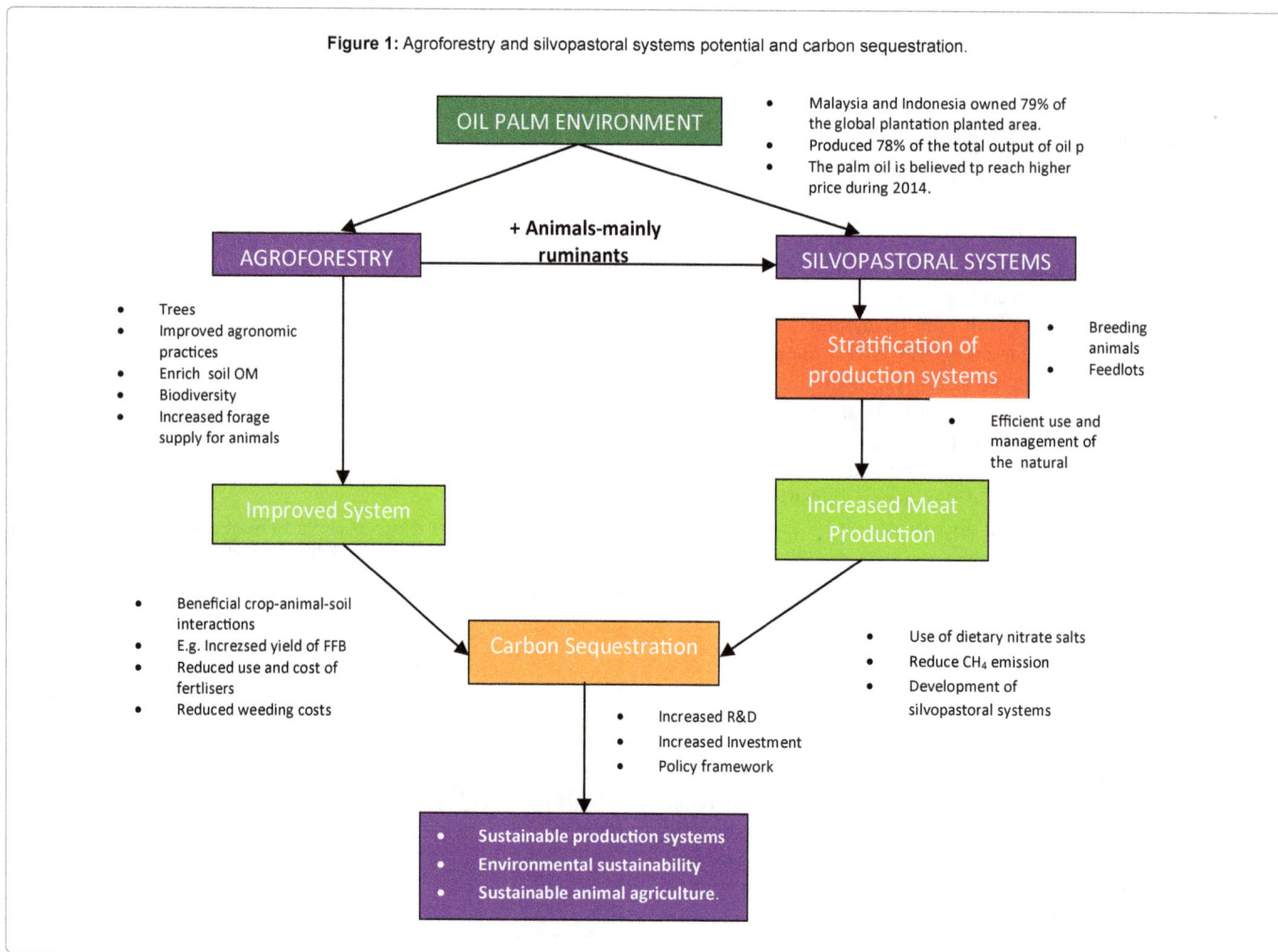

Figure 1: Agroforestry and silvopastoral systems potential and carbon sequestration.

The potential production options are many and are as follows:-

• Breeding ruminants (buffaloes, cattle, goats and sheep) for production systems

Table 2: Potentially important perennial crops and their locations for use in integrated systems in the tropics.

Table 3: Estimated gross margin of profits for meat production using crossbred

Katjang goats with varying levels of fertility in Malaysia.

Table 4: Effects of climate change on land use and livelihood systems of the poor 913).

• Zero-grazing systems (feedlots, goats and sheep)

• Rearing ruminants to use the available oil palm by-products

• Rearing ruminants for grazing and controlling weeds

• Rearing ruminants for draught and haulage operations

• As an entry point for development of integrated NRM and sustainable production systems

• Value addition and total productivity returns and

• As a hedge for possible reductions in the price of crude palm oil.

The oil palm environment offers a number of favorable production to enhance productivity to include:

➢ Forage dry matter availability: 2.99-2.16 mt/ha for three- and five-year-old palms, reducing to 435-628 kg/ha for 10-29 year-old palms [9]

➢ 60-70 forage species among young palms, which are reduced by about 66% among older palms [10]

➢ Forage categories: grasses, dicotyledons, legumes and ferns of 60, 2, 11 and 17% for 3-10 year old palms and 50% for grasses, 13% for dicotyledons. 2% for legumes and 35% for ferns in over 10 year trees

➢ 2 and 35% for 3-5-year-old, 6-10-year-old and over 10-year-old palms respectively [11] about 72–93% of the forages are palatable and of value to ruminants.

➢ Indigenous Kedah–Kelantan and Bali cattle are well suited to integration with oil palm

➢ Carrying capacity: 3 steers/ha in 3-4-year-old palms, with average daily gain of about 260-320 g/day.

The strategy of stratification within silvopastoral systems enables the development of other production systems including specialization and intensification. Breeding animals to meet the needs of zero grazing system is one example. Likewise there could be a separate production

Figure 2: Brahman and Hereford cattle integrated with oil palm in Sabah, Peninsular Malaysia.

Figure 3: Goats integrated with oil palm and seeking shelter under banana trees in Perak in Malaysia.

system for goats. The diversified operations also enable the development of sustainable production systems as well as carbon sequestration in the long term.

Economic Impacts

South Asia, South East Asia and the Pacific Islands

There exist 26 case studies which provide convincing information on economic impacts of the systems involving cattle. The review gave the following key results [12].

Increased animal production and income: This arises from increased productivity and meat off takes

Increased yield of FFB and income: By about 30% with measures of between 0.49-3.52 tones/ha/yr.

Savings in weeding costs: By about 47-60%, equivalent to 21–62 RM/ha/yr.

Internal rate of return: The IRR of cattle under integration was 19% based on actual field data.

Figure 4: Typical photograph of sheep integrated with coconuts in the Laguna area, Philippines.

Several theoretical calculations approximate to this value.

Integrated goats-oil palm production system in Malaysia

The concept of promoting the benefits of integrated systems presents an alternative model here which highlights the role of goats and the potential economic benefits of their integration specifically with the oil palm. There benefits are associated with are two main attributes:-

• The ability of the goat to digest fibrous feeds, and especially coarse roughages efficiently and use to advantage the mixed forage biomass of which 72-93% are suitable to feed ruminants [13] in oil palm plantations. In other words the non-competitive use of the forage biomass with a variety of feeds and the resultant advantages is a remarkable low input system of goat production which is associated with high profit margins. The available forage biomass under oil palm can be more fully utilized if mixed grazing is involved, such as goats and cattle

• The versatile ability of the goat to convert the forage biomass to precious meat, milk and skins together with the beneficial advantages to the oil palms, gives economic margins that are quite significant.

Goats are versatile animals with an ability to convert the forage biomass to precious meat, milk and skins that Goat meat fetches the highest prices among all meats in the open market. These attributes, together with the beneficial advantages to the oil palms, provide significant economic margins. Table 3 gives data and calculations on the scale and extent of the monetary gains, keeping in mind that goat meat fetches the highest prices among all meats in the open market. In this context, an important management issue influencing productivity and income generation from goats, concerns their fertility and the productive lifespan. The higher fertility implies more number of kids born. Calculations on the importance of good management influencing fertility and productivity on the scale, impact and income generation clearly showed that over for a 10-doe unit over seven years of productive life, with varying fertility levels of between 80-160%.(% kids weaned/does mated) gave a total gross revenue of between US $860-1760 per year.

West Africa

Coconuts, oil palm, cocoa and fruit trees are very common throughout West Africa, but R and D to integrate the animal component into some of these have been very weak, due probably to priorities and lack of resources.

The value of alley farming combining maize and supplemented with *Leucaena* or *Gliricidia* forages fed to livestock have been shown to be economically beneficial. West African djallonke sheep and West

African dwarf goats increased live weight. Continuous alley farming was more profitable than alley farnmig with fallow, or conventional no-tree farming [14].

Latin America and the Caribbean

Silvopastoral systems are part of traditional farming systems throughout Latin America and the Caribbean for the benefits, including sustainability, which have been previously highlighted. They are especially common in farms engaged in cattle production. In Columbia it has been found that trees were grown in 26-69% of the farms in which there was diversification with the plating of larger trees gave more shade and also produced timber [15].

In Costa Rica the profitability of milk production in small farms was higher when high value trees were planted than those without trees, especially when labor costs increased. Cashew and other high value timber trees have also been used in the development of silvopastoral systems [16].

Climate Change: Carbon Sequestration and Greenhouse Gases

A discussion on silvopastoral systems is incomplete without a reference to carbon sequestration and greenhouse gases. The first point to make is that it is an area with a paucity of R and D an needs urgent attention It is defined as the complex and secure storage of carbon that would otherwise be emitted or remain in the atmosphere [17]. The expanding land areas under oil palm provide good opportunities for carbon sequestration through more widespread use of grasses and tree legumes, and improved forage management practices, with resultant decreased carbon atmospheric emissions and global warming. It has been calculated [18] that in mixed farming systems, the carbon sequestered per hectare was 0.32 tc/ha/yr. The practical implication of this is that agronomic practices need to enhance these carbon sinks through enrichment of soil organic matter and the forage biomass under the oil palm.

Associated with above is the issue of greenhouse gas emissions (GHG), mainly CH_4, N_2O and CO_2 and their effects on climate change or global warming. Improved grass-legume pastures to feed grazing ruminants will have the beneficial effect of enhancing carbon sequestration and releasing more O_2 into the atmosphere. On the other hand, the presence of grazing ruminants will mean emissions of more CH_4 into the atmosphere, and their possible effects. In Brazil, Zebu cattle grazing tropical pastures produced a larger methane loss of 27 g/kg compared to either Holstein or Nellore cattle fed sorghum silage-concentrate diets that averaged 22 g/kg. Holstein or Nellore cattle on Brach aria or *Panicum* pastures consuming sorghum had methane losses that were close to the temperate forage-based diet of 20 g/kg [19].

Mitigation and adaptation are important aspects of ways to cope with climate change. Strategies to reduce GHG have largely focused on methanogen inhibitors and substrate levels, rather than at the feed quantity and quality end In practice, strategies will need to be developed that can have a balance between the two types of emissions which is consistent with minimal effects on climate change. Recent studies suggest that the fermentable nitrogen requirements of ruminants on diets based on low protein cellulosic materials can be met from nitrate salts {18} and this potentially reduces methane production to minimal levels [20,21] demonstrated that with adaptation, young goats given a diet of straw, tree foliage and molasses grew faster with nitrate as the fermentable N source by comparison [22] have shown a 60% reduction in methane production by sheep fed nitrate in a corn silage based diet.

They have shown persistent reduction of 16% methane in dairy cows supplemented with nitrate [23] quoted by Hulshof et al. [24] and a 32% reduction in methane production in beef cattle in Brazil when 2.2% nitrate replaced urea in a sugarcane/concentrate.

The Asian Development Bank [25] has produced a regional study on South East Asia on the economics of climate change highlighting perspectives on the regional interdependencies of climate change impacts and policies, and pooling of resources to address shared challenges with reference to Indonesia, Philippines, Thailand and Vietnam. The study has indicated that the agricultural dependent economies will contract by as much as 6.7% annually. The economic cost according to the report would be 2.2% of GDP by 2010 if only the impact on markets is considered, 5.7% if health costs and biodiversity losses are factored in, and 6.7% if losses from climate-related disasters are also included. The latter far exceeds the projected cost globally of climate change, estimated at 2.6% of GDP each year to the end of the century. Table 4 gives summary of the Effects of climate change on land use and livelihood systems of the poor.

Policy Framework

The task of stimulating waning agriculture, encouraging growth and promoting silvopastoral system to enhance productivity, food security and a clean environment provides major challenges for R and D. These need to be supported by strong policy requirements and are reflected in the following:-

• Affirmation of official policy to address waning agriculture [26,27], its revitalization efficiency in integrated NRM [28], and intensification of production systems (29)

• Priority for food insecurity and increased self-reliance

• Priority for concerted R and D on rain fed agriculture and small farm systems that are progressive and can intensify [30].

• Priority for pro-poor community-based activities that adapt to climate change.

• Promotion of ways and means to enhance C sequestration and reduce emissions of GHG in silvopastoral systems.

• Integrated tree crops-ruminant systems are underestimated. Policy interventions are required to stimulate their development.

• Empowerment of women is central to enhance their effective contribution to the use of productive resources for food and nutritional security, and the stability of farm households [30].

• Building systems R and D capacity to deal with complex problems of crop-animal-soil-water interactions and effects of climate change.

• Micro-credits should be made more accessible to small farms and the landless.

• Improvements are necessary for rural-urban market linkages, collection and processing centers to reduce transaction costs and in the value chain.

• Increase investments and promote public-private sector partnerships for greater engagement and agricultural productivity.

Investing in Animal Agriculture

Increased investments are justified and urgently needed to develop the LFUs, in view of the potential impact on increased productivity, poverty and food security, improved livelihoods and the environment.

It is especially important to note that in studies in India [31] and China [32] the returns to investments are very much higher in these areas in comparison to areas that have benefited from the Green Revolution. In the Indian context, improving agrarian prosperity and rural development focusing on the five pillars of public investment, credit, infrastructure (roads, transport and agro-processing), stable markets and knowledge transformation of farmers have been proposed [33]. Similarly, it has been reported that the estimated returns to agricultural R and D are high, and high enough to justify an even greater investment of public funds [34] as was reflected in the investments and policies on the use of high-yielding rice varieties that resulted in the success of the Green Revolution in India.

Adoption of Silvopastoral Systems

Overall and across regions, R and D on the development of this system has been very meagre. For a variety of reasons, the adoption rates of the system and methodologies continue to be very low. Adoptions for integrating ruminants with in general, and cattle and goats with oil palm are not surprising, and can make a very poor contribution to production household requirements.

Given the very low adoption of integrating ruminants with oil palm for example, it is relevant to enquire into the reasons for this situation. There are many reasons for this, and these are enumerated below:-

• Poor awareness of the potential of integration and the benefits and value addition of silvopastoral systems e.g. oil palm and ruminants.

• Associated with above, is the strong resistance by the crop-oriented plantation sector, and are least interested in introducing animals to the system.

• Crop scientists and plantation managers all have a plant production background without any interest in animals

• This perception is also fuelled by high prices for the key commodities e.g. palm oil.

• Additionally, they are also not interested in making more capital investments e.g. fencing and fodder production.

• Inadequate technology application and week understanding of systems R and D.

• Unattractive investment climate.

• Shortage of estate workers

• Onset of climate change and reservations about the effects on the tree crop, and

• Absence of policies to encourage integrated systems.

Ruminants as the Entry for Development

The investments will also be enhanced by increased emphasis on the integration of ruminants in these AEZs. Ruminants are an integral part of the LFUs such as the presence of goats, sheep and camels. The wider and more intensive use of ruminants in these areas including in silvopastoral systems is reflected in such advantages in them such as non-competitive use of mixed forage biomass, improved cycling of soil nutrients, improved soil fertility and conservation of soil and water, enables increased productivity, increased ecological and economic efficiency, and encourages wider replication and development of the system. It is suggested that ruminants are the entry point for development of these areas in a manner that is consistent with the efficiency of NRM, and preservation of the agricultural landscape.

Figure 1 captures these aspects in tandem with the targets that are clearly important.

Climate change will take effect irrespective of predictable difficulties and the consequences. Two major possibilities are of grave concern: firstly that of reduced arable land for food production for humans and reduced grazing areas for ruminants. Secondly and more importantly, the climatic effects can push back several more million people along with the million resource-poor people who already live there, to become more vulnerable and poorer (26,35)r. While R and D can provide mitigation and adaptive methodologies and solutions, for the vulnerable people who live there, time is not in their hands. The priority for the application of innovative improved technologies is urgent,to remove current constraints in the immediate tomorrow. Both these concerns need to be reinforced by strong policies. The enduring hope is secure and sustained food supplies, decreased poverty, significantly improved livelihoods, sustainable animal agriculture, and an intact agricultural landscape that will be cherished by future generations [35].

Conclusions

Climate change will take effect irrespective of predictable difficulties of the consequences. Two major consequences are of grave concern: firstly that of reduced arable land for food production for humans and reduced grazing areas for ruminants. Secondly and more importantly, the climatic effects can push back several more million people along with the million resource-poor people who already live there to become more vulnerable and poorer. While R and D can provide mitigation and adaptive methodologies and solutions, for these, time is not in their hands. The priority for the application of innovative improved technologies is urgent to remove current constraints in the immediate tomorrow. These need a strong policy framework to support maximum food production and respond to the needs and satisfaction of consumer preferences.

The complexities of the environment, externalities, and the threats of climate change, these present major challenges for R and D that cannot be ignored, Their resolution is our collective responsibility and a moral obligation. A definition of clear priorities and pathways for predictable solutions provide many reasons for the development of, increased contribution from animal agriculture, viewed from the perspective of production to consumption systems. The rain-fed areas including silvopastoral systems, currently very underestimated, provide compelling and urgent opportunities in the presence of several million landless and very poor vulnerable people. Also, the relatively large populations of ruminants therein can serve as the entry point for development. Improved efficiency of NRM, concerted application of yield-inducing technologies, and ways of mitigating and adapting technologies to cope with the effects of climate changes which can significantly enhance the capacity of food production. Social and effective development policies are also needed to spur sustainable increased agricultural productivity, improved human welfare, and self- reliance in the future. The hope for small farm systems, the very poor and the landless is that they will be more secure and can produce sustained food supplies, with the full knowledge that sustainable animal- agriculture can demonstrate its full potential, to realize the objectives of poverty, hunger and environmental degradation.

References

1. Devendra C (2012) Intensification of Integrated Natural Resources Use and Agricultural Systems in the Developing World. Agrotechnol 1: e101.

2. FAO (Food and Agriculture Organisation) (1996) FAO AGROSTAT 1995, 49, FAO, Rome, Italy.

3. Devendra C, Sevilla C, Pezo D (2001) Food-feed systems in Asia. Asian Australasian Journal of Animal Science 14: 733-745.

4. Carangal VR, Sevilla C (1993) Crop-animal systems research in Asia Proceedings V11th World Conference on Animal Production. Edmonton, Canada: 367-386.

5. ESCAP (2008) Economic and social survey of Asia and the Pacific 2008. Sustaining growth and sharing prosperity. Bangkok, Thailand.

6. CGIAR/TAC (2000) (Consultative Group on International Agricultural Research/ Technical Advisory Committee) CGIAR priorities for marginal lands, CGIAR, Washington, USA.

7. Devendra C, Thomas D (2002) Crop–animal interactions in mixed farming systems in Asia. Agricultural Systems 7: 27-40.

8. David J. Griggs, Maria Noguer (2001) Climate change 2001: The scientific basis. Contribution of Working. Group I to the Third Assessment Report of the Intergovernmental Panel on Climate Change Climate change. Weather 57: 1-3.

9. Current D, Scherr S (1996) Farmer costs and benefits from agroforestry and farm Forestry projects in Central America and the Caribbean. Agroforestry Systems 30: 87-10

10. Devendra C, Shanmugavelu S, Wong H K (2007) Integrated tree-crops ruminant systems: expanding the research and development frontiers in the oil palm. In: Proceedings Workshop on Integrated tree crops-ruminant systems (ITCRS/Assessment of status and opportunities in Malaysia. Academy of Science Malaysia 1-23.

11. Chen CP, Wong HK, Dahalan I (1993) Herbivors and the plantations, Proceedings. Third International Symposium on the Nutrition of Herbivores (Ed. Ho,Y.W) Penang, Malaysia 71-81.

12. Wong CC, Chin FY (1998) Meeting nutritional requirement of cattle from natural forages in oil palm plantation. Paper presented at National Seminar on Livestock and Crop Integration in Oil Palm, Kluang, Johor.

13. Devendra C (2007) Goats: biology, production and development in Asia. Academy of Sciences Malaysia, Kuala Lumpur, Malaysia.

14. Reynolds SG (1995) Pastures and Cattle under Coconuts. FAO Plant production and Protection paper No.91, 33.

15. Cajas-Giron YS, Sinclair (2001) Cultivation of multistorey silvopastoral systems on seasonality on dry pastures in the Caribbean region, Agroforestry systems 53: 215-225.

16. Lascano GF, Pezo DA (1994) Agrofoestry systems in humid forest margins of tropical America from a livestock perpective. In Copland, J/W, Djajanegara. A. And M.Sabrani (Eds) Agroforestry for human welfare, ACIAR Proceedings No 55, Canberra, Australia 17-24.

17. Watson RI, Noble IR, Ravindranath B, Verado DJ, Dokken DJ (2000) Intergovernmental Panel on Climate Change (IPCC), Special report on land use change and forestry, IPCC Secretariat, Geneva, Switzerland (Mimeograph).

18. Pretty JN, Noble AD, Bossio D, Dixon J, Hine RE, et al. (2006) Resource-conserving agriculture increases yields in developing countries. Environ Sci Technol 40: 1114-1119.

19. Lima MA, Primavasi O, Dermachi JJ, Manella M, Frighetto RT (2004) Inventory improvements for methane emissions from ruminants in Brazil, Rpt. to the Environment Protection Agency (EPA), USA.

20. Trinh PH, Ho QD, Preston TR, Leng RA (2009) Nitrate as a fermentable nitrogen supplement for goats fed forage based diets low in true protein. Livestock Research for Rural Development 21.

21. Leng RA (2008) The potential of feeding nitrate to reduce enteric methane production in ruminants. A Report to the Department of Climate Change. Common wealth Government of Australia. Canberra, ACT. Australia.

22. van Zijderveld SM, Gerrits WJ, Apajalahti JA, Newbold JR, Dijkstra J, et al. (2010) Nitrate and sulfate: Effective alternative hydrogen sinks for mitigation of ruminal methane production in sheep. J Dairy Sci 93: 5856-5866.

23. Zijderveld SM van, Dijkstra J, Gerrits WJJ, Newbold JR, Perdok HB (2010) Dietary nitrate persistently reduces enteric methane production in lactating dairy cows.In :greenhouse gases and animal agriculture conference October 3-8, 2010 Bamff, Canada.

24. Hulshof RB, Berndt A, Gerrits WJ, Dijkstra J, van Zijderveld SM, et al. (2012) Dietary nitrate supplementation reduces methane emission in beef cattle fed sugarcane-based diets. J Anim Sci 90: 2317-2323.

25. Asian Development Bank (2009) The economics of climate change in South East Asia: a regional review. Asian Development Bank, Manila, Philippines 91.

26. Devendra C (2010) Small farms in Asia. Revitalizing agricultural production, food security and rural prosperity. Academy of Sciences Malaysia, Kuala Lumpur, Malaysia.

27. Devendra C (2004) Integrated tree crops-ruminants systems. The potential importance of the oil palm. Outlook on Agriculture 33: 157-166

28. Devendra C (2007) Perspectives on animal production systems in Asia. Livestock Science 106: 1-18.

29. Devendra C (2009) Intensification of integrated oil palm–ruminant systems: enhancing increased productivity and sustainability in South East Asia. Outlook on Agriculture 38: 71-82.

30. Devendra C, Gender equity in sustainable animal-agriculture: enhancing empowerment and the contribution of women for improved livelihoods, stable households and rural growth. Second. 2nd Asian-Australasian Dairy Goat Conference, Bogor, Indonesia (Extended Abstr).

31. Fan S, Hazell P, Thorat H (2000) Targeting public investments by agro ecological zone to achieve growth and poverty alleviation goals in rural India. Food Policy 20: 411-428.

32. Fan S, Zhang L, Zhang X (2000) Growth and poverty in rural China: The role of public investments. Environment and Production Technology Division Discussion Paper No.66, International Food Policy Research Institute, Washington DC, USA.

33. Shankar R, Maraty P (2009) Concerns of India's farmers. Outlook on Agriculture, 38: 96-100.

34. Pardey PG, Beintema NM (2001) Slow magic-Agricultural R and D: a century after Mendel, International Food Policy Research Institute Food Policy Report, Washington DC, USA.

35. Devendra C, Chantalakhana C (2002) Animals, poor people and food insecurity: opportunities for improved livelihoods through natural resource management. Outlook on Agriculture 31: 161-176.

25

Single and Combined Effects of Organic Selenium and Zinc on Egg, Fertility, Hatchability, and Embryonic Mortality of Exotic Cochin Hens

V.G. Stanley[1]*, K. Hickerson[1], M.B. Daley[1], M. Hume[2] and A. Hinton[3]

[1]*Prairie View A & M University, Prairie View, Texas, USA*

[2]*Food and Feed Safety Research Unit, Agricultural Research Service, United States Department of Agriculture, College Station, Texas, USA*

[3]*Poultry Processing and Swine Physiology Unit, Agricultural Research Service, United States Department of Agriculture, 950 College Station Road, Russell Research Center, Athens, Georgia*

Abstract

A study was conducted to examine the effects of diets supplemented with organic selenium, (Se) and zinc, (Zn) on the performance of Cochin exotic breeder hens. Forty-two week old hens (n=120) and males (n=12) were assigned to four treatment groups of 10 females and 1 male each. Birds with no mineral supplementation (Group 1); feed supplemented with .33 ppm Se (Group 2) feed supplemented with 20 ppm Zn (Group 3), and feed supplemented with .33 ppm Se and 20 ppm Zn (Group 4). Eggs were collected for 21 days to determine egg production and egg weight. Fertility and embryonic mortality were determined by candling eggs on days 12 and 18. Hatchability was calculated on day 21 based on fertility and egg set. Results showed that egg production did not increase significantly, although birds provided feed containing Se or Se and Zn produced 4% and 6%, respectively, more eggs than the control hens Egg fertility was similar for most treatments, but fertility of hens provided the Zn-supplemented diet, was significantly lower than fertility of other treatment groups. Hatchability based on fertile eggs was 4.6% and 3.0% higher than the control and for eggs from hens provided feed supplemented with Se or Se+Zn, respectively. Early and late embryonic mortality was significantly (P<0.05) lower in eggs from hens provided diets containing Se+Zn, than in eggs from the control hens or hens fed diets containing Se or Zn only. In conclusion, supplementing diets of exotic hens with Se and Zn increased egg production and significantly reduced early and late embryonic death. Addition of these minerals to the diets might provide exotic bird producers a method to increase the performance of these birds.

Keywords: Exotic birds; Egg production; Fertility; Embryonic mortality; Hatchability

Introduction

Due to relatively low egg and meat production as compared to most commercial breeds of broilers and layers there is little data on the production performance of exotic birds. Therefore limited data are available concerning the level of egg production, fertility and hatchability for exotic birds. Factors affecting egg production, fertility, and hatchability for exotic birds are genetics, temperature, environment, disease, incubation, sanitation, overcrowding, handling, and stress [1,2]. Fertility is based essentially on the genetic makeup of the bird [3]. Various attempts have been explored to increase fertility and hatchability in exotic birds, such as cross-breeding for hybrid vigor; increasing essential nutrients, including trace minerals in diets; UV radiation for disease control and contamination of the eggs [4].

The relationship between egg productions, the age of the hens at maturity, and egg size had been reported to affect hatchability of exotic breeds of birds [5,6]. Because of their low commercial values due to poor performance, it was hypothesized that the egg production performance of exotic birds could be improved with dietary trace mineral supplements. Therefore, the objective of the current study was to examine the effects of organic selenium (Se) and zinc (Zn), alone and combined on the overall performance of exotic breeder hens.

Materials and Methods

Experimental design

Experiments was conducted with one hundred and twenty 42-week-old exotic breeder hens and twelve males. Birds were separated into four treatment groups of 30 hens and three males. Birds were reared in a commercial-lighted laying hen facility. The birds were reared on floors covered with wood shavings at the Poultry Center, Prairie View, Texas.

The lighting in the house was set at 15 h of light and 9 h of dark at the start of the experiment. Birds were fed treatment diets supplemented with Se, or Zn, and Se+Zn in a 2×2 factorial arrangement design with two levels of Se (0 and 0.33 ppm), two levels of Zn (0 and 20 ppm) (Table 1). All diets were iso-caloric (2,830 ME/kg), iso-nitrogenous (17% CP), and contained 3% Ca. The four treatments were arranged into a randomized factorial design with 3 replicates of 10 hens and 1 male each. Birds were allowed to consume feed and water add libitum. The duration of the experiment was 21 days. The experiment was reviewed and approved by the Animal Use Protocol of the University, with reference number 2012-0901-107.

Data collection

Eggs were collected daily from each group and stored at 12.7°C before incubation. On the 12th day of incubation, the eggs were candled to determine fertility and early embryonic mortality. Eggs that casted no embryonic shadow were considered infertile. After eighteen days, the fertile eggs were candled again for late embryonic mortality and then transferred from the setter to the Hatcher for complete incubation. Hatchability was calculated on fertile eggs and the total eggs set.

*Corresponding author: V.G. Stanley, Prairie View A&M University, Prairie View, Texas, USA, E-mail: vgstanley@pvamu.edu

Ingredient	Percentage
Yellow Corn	55.93
Soybean Meal (44% CP)	22.10
Alfalfa Meal (17% CP)	5.00
Meat and Bone Meal (50% CP)	3.00
Animal and Vegetable Fat	3.00
Limestone	8.22
Di-calcium Phosphate	1.15
Iodine Salt	0.25
Vitamin Trace Mineral Premix[1]	1.50
Calculated Values	
Crude Protein (%)	17.00
ME kcal/kg	2830
Crude fat (%)	4.0
Phosphorus (available) (%)	0.35
Calcium (%)	3.20
Methionine (%)	0.34
Methionine and Cystine (%)	0.62
Lysine (%)	0.76

[1]As-fed basis
Per kilogram of diet: Vitamin A (as Vitamin A outtake) 12,000 IU; cholicalciferol (as-fed basis) 3000 IU; Vitamin E (as x-tocopheryl acetate) 20 IU; Menadione Sodium bisulfile , 2.0 mg; Thiamine, 1.5 mg; riboflavin, 8.0 mg; niacin, 3000 mg; pantothinic acid, 150 mg; pyridoxine, 40 mg; vitamin B12, 15 μg; folic acid, 1.0 mg; biotin, 150 μg; Cobalt, 0.2 mg; Copper, 10 mg; iron, 80 mg; Iodine, 1.0 mg; Manganese, 120 mg; zinc, 120 mg; Selenium, 0.2 mg; butylated hydroxybotuene (BHT), 150 mg; zinc bacitracin, 20 mg.

Table 1: Composition of diet.

Diets	Egg[1] Production (%)	Egg weight (g)	Fertility[2] (%)	Non-Fertile (%)
Control	53.00[b]	44.01[a]	85.00[a]	15.00[b]
Selenium	57.00[b]	45.20[a]	81.00[a]	19.00[a]
Zinc	52.00[b]	44.51[a]	75.00[b]	25.00[a]
Selenium+Zinc	59.00[a]	46.10[a]	85.00[a]	15.00[b]
SEM	3.30	1.22	7.46	4.73

[a-c]Means within a column and with no common superscripts differ significantly (P<0.05).

[1]Percent egg production is calculated over a 21-days period as the total number of eggs laid per treatment divided by the total number of eggs laid over 21 days.

[2]Percent fertility is based on the number of eggs set, and candled on day 12 of incubation.

Table 2: Effects of organic selenium and zinc on total egg production, egg size and fertility of exotic hens.

Statistical analysis

Data were subjected to analysis of variance using general linear models procedures [7] with each replicate pen as an experimental unit. The model included replication, Se and Zn treatments and interaction of Se+Zn. Significant differences in percentage egg production fertility, hatchability, early and late mortality were determined using Duncan's Multiple Range Test of means. The level of significance was set at P<0.05.

Results and Discussion

The results of the different variables being analyzed are presented in table 2. Egg production of hens provided the diet supplemented with Se+Zn was significantly higher (P<0.05) than egg production by hens provided the other treatments. Egg weights were not significantly different among the 4 treatment groups. Fertility was similar for all treatments, except for the Zn-supplemented group, which was significantly lower (75%) than the other three treatment groups. Non-fertile eggs for groups given Zn only (25%) and Se only (19%) were significantly (P<0.05) higher compared to 15% for the control and

Se+Zn groups, respectively.

Eggs were candled on the 12th day for early embryonic death and on the d 18 for late embryonic death (Table 3). Early and late embryonic death was significantly lower (P<0.05) in eggs of hens provided a diet containing Se+Zn, whereas late embryonic death was significantly lower in eggs of hens provided a diet containing Se only. Supplementing the diet of hens with Se, Zn or Se+Zn produced no significant changes in the hatchability of eggs; however, results show that hatchability on eggs set from the Se+Zn treated birds was 3.3% higher than the control eggs. The reported results suggested that mineral nutrition greatly affects reproductive performance of the exotic birds. The hatchability rate of total eggs set, which ranged between 48.2% to 56.6%, and fertile eggs as high as 66.6%, suggests improvement regardless of how small will result in significant profit for the industry [5].

Rapid growth and development of embryos can result in the production of free radicals and hence growing broilers need higher levels of antioxidants [8]. To insure protection from free radicals, antioxidants are required, of which selenium is a core element of GSH-Px. The selenium used in this study was in an organic form which is reported to have a higher bioavailability than non-organic Se. Similar to growing broilers, young growing embryos develop under intensive conditions that require a robust and efficient immune system, which is dependent on selenium status. Placha et al. [8] also suggested supplementation with organic selenium is essential for life, because it prevents the blood phagocytic suppression and changes the glutathione peroxidase activity in duodenal tissue in broilers caused by sub-toxic levels of deoxynivalenol in the feed.

Hatchability is related to egg fertility and embryonic mortality throughout the hatching process. In the current study, hatchability was calculated on total eggs set and on fertility (Table 3). Also differences in hatchability among flocks increases with flock age [9]. The age of the breeder hens in this study was 42 wk, and this may have been one of the factors that produced low hatchability [10]. Stated that when hens reach the age of 30 wk, there is a slightly lower hatchability, which could be related to less fertility and greater incidence of non-viable series of eggs. Hatchability for eggs of older breeders decreases because of change in egg quality and failure to adjust the incubation condition [11,12]. It was further stated that as the hen ages, the albumen quality, which is the main source protein to the embryo development deteriorates. The birds in the current study came available at 42nd week. The improvement in hatchability for eggs from hens fed selenium and zinc supplementation could be due to the effect of selenium and zinc in maintaining the integrity of the albumen. Eggs from older breeders are known to hatch

Diets	Embryonic Death		Hatchability	
	Early[1]	Late[2]	Eggs Fertile	Eggs Set
	---------------------------------- % ----------------------------------			
Control	3.30[a]	1.50[a]	63.6[a]	53.3[a]
Selenium	3.70[a]	0.75[b]	68.2[a]	48.2[a]
Zinc	3.20[a]	1.50[a]	61.8[a]	51.8[a]
Selenium+Zinc	2.60[b]	0.50[b]	66.6[a]	56.6[a]
SEM	0.45	0.89	2.87	3.48

[a-b]Means within a column and with no common superscripts differ significant (P<0.05).

[1]Early embryonic death was calculated on 12th day of incubation based on the number of eggs candled.

[2]Late embryonic death was calculated on the number of eggs candled on 18th of incubation based on fertility.

Table 3: Effects of organic selenium and zinc single and combined on early and late embryonic mortality and hatchability.

earlier and suffer more from post-emergent holding in the hatchery. In this study we noticed great variation (not recorded) in the arrival of the chicks at 21 days in the control eggs.

The low hatchability reported in the current study could also be due to egg storage length. It was concluded that the real egg storage length is at least 1 to 4 days for hatching eggs [11]. They also stated that egg storage at the hatcheries decreased hatchability significantly. The egg collection and storage length in the current study ranged from 1 to 21 days, because of the small flock size and the need to get a representative sample of eggs. The combination of selenium and zinc supplementation appeared to be beneficial; as the hatchability from selenium and zinc treatment improved the level of hatchability compared to birds provided a diet that was not supplemented with these minerals. Egg storage depresses albumen quality affects embryonic viability as the flock ages, and results in a lower percentage of healthy day-old chicks [11,12]. Another factor that may have influenced the low hatchability observed in the current study could be the breed of the hens. The literature reported that there is a significant difference in hatchability among birds of exotic birds. Breeds respond differently to hatchability, and it is known to vary for different strains [10]. This observation suggests that different birds require different management in the laying programs as well as the hatcheries. The hens in the current study were exotic breeds, not commercial.

Conclusion

The limited attention that is being placed on the availability of antioxidants in feeds for poultry. Antioxidants block the effects of free-oxygen radicals, which destroy the cellular membranes. The egg yolk which is the main source of energy to the developing embryo, being lined with phospholipids, is sensitive to oxidation. The erosion of the yolk membrane shortens the life of the developing embryo. Most feed formulation incorporates vitamin E as the primary antioxidant, however, selenium, which is the core element in the synthesis of the body's natural defense enzyme (GSP-Px), can serve as an additional source of antioxidant to scavenger free radicals. Zinc, which is the core element of insulin, increases the bioavailability of glucose, which is converted into energy for the embryo. Therefore, in conclusion; the present study suggests that it is possible to improve fertility and hatchability of exotic breeds by supplementing the feed provided to these birds with Se and Zn.

Acknowledgements

The authors wish to thank the students and colleagues for their assistance during the experiment. We also thank Alltech Company, Nicholasville, KY, USA for providing the organic selenium and zinc, and to Dr. Ted Sefton, from Alltech, Canada for his assistance.

References

1. Christensen VL (2001) Factors associated with early embryonic mortality. World's Poult Sci J 57: 259-373.

2. Hocking PM, Bernard R, Robertson GW (2002) Effects of low dietary protein and different allocations of food during rearing and restricted feeding after peak rate of lay on egg production, fertility and hatchability in female broiler breeders. Br Poult Sci 43: 94-103.

3. Hocking PM, Bernard R (2000) Effects of the age of male and female broiler breeders on sexual behavior, fertility and hatchability of eggs. Br Poult Sci 41: 370-376.

4. Obidi JA, Onyeanusi BI, Ayo JO, Rekwot PI, Abdullahi SJ (2008) Effect of timing of artificial insemination on fertility and hatchability of shikabrown breeder hens. Intl J Poult Sci 7: 1224-1226.

5. Fayeye TR, Adeshiyan AB, Olugbami AA (2005) Egg traits, hatchability and early growth performance of the Fulani-ecotype chicken. Livestock Research for Rural Development 17.

6. Dzoma, DN, Motshegwa K (2009) A retrospective study of egg production, fertility and hatchability of farmed ostriches in Botswana. Intl J Poult Sci 8: 660-664.

7. SAS Institute Inc. (2004) SAS/STAT User's Guide: Version 9.1. SAS Institute Inc.

8. Placha I, Borutova R, Gresakova L, Petrovic V, Faix S, et al. (2009) Effects of excessive selenium supplementation to diet contaminated with deoxynivalenol on blood phagocytic activity and antioxidative status of broilers. J Anim Physiol Anim Nutr 93: 695-702.

9. Tona K, Onagbesan O, De Ketelaere B, Bruggeman (2007) A model for predicting hatchability as a function of flock age, reference hatchability, storage time and season. Arch Geflügelk 71: 30-34.

10. Yassin H, Velthuis AG, Boerjan M, van Riel J, Huirne RB (2008) Field Study on broiler eggs hatchability. Poult Sci 87: 2408-2417.

11. Lapao, C, Gama LT, Soares MC (1999) Effects of broiler breeder age and length of egg storage on albumen characteristics and hatchability. Poult Sci 78: 640-645.

12. Tona K, Bruggeman V, Onagbesan O, Bamelis F, Gbeassor M, et al. (2005) Day-old chick quality: Relationship to hatching egg quality, adequate incubation practice and production of broiler performance. Avian Poult Biol Rev 16: 109-119.

26

Maternal Environment Modulates Dormancy and Germination in *Vaccaria hispanica*

Zahra Hosseini Cici S[1,2]*

[1]*School of Crop Science, University of Guelph, Ontario, Canada*
[2]*School of Sustainable Agriculture, University of Payame-Noor, Tehran, Iran*

Abstract

Offspring performance is affected by mother plants via genes and maternal environment. Seed characteristics such and dormancy and vigour are affected by the environmental resources during plant development. Intra and inter-variation in seed dormancy and longevity are considered as a bet-hedging strategy to reduce the recruitment failure across years under environmental uncertainty. In this study, the effects of drought and herbivory, two common environmental stresses, were investigated on biomass and seed quality in *Vaccaria hispanica* (Mill.) Rauschert, an annual forb. Plants were subjected to different levels of water and simulated herbivory stress. Maternal water stress suppressed seed mass, but it stimulated dormancy in seeds. Progenies from the maternal stress environment were more persistent than those from the maternal control environment after being exposed to 45°C and 100% RH for 8 days. The findings highlighted the importance of the water maternal effect versus herbivory on seed dormancy and longevity in this species. The results may help us understanding the life cycle and population dynamics of *V. hispanica* in successive years.

Keywords: *Vaccaria hispanica*; Vigour; Dormancy; Maternal effect; Water stress; Simulated herbivory

Introduction

Fitness of offspring can be affected by maternal environment in different ways [1]. Unpredictable biotic and abiotic maternal environment may have a crucial role in determining germination fractions and seedling recruitment. Bet-hedging is an important strategy to reduce the risk of failure under environmental uncertainty [2]. Spreading germination over time by producing seeds with different dormancy and longevity can help reducing the risk of recruitment failure in any one year [3]. It is well known that, seeds are less vulnerable to environmental stresses than seedlings [2].

Persistent seed banks buffer populations in unfavourable environmental condition [3,4]. Seed longevity and dormancy can increase the seed bank of populations. Both the environment and genetics of plants are responsible for recruitment variation. The environment of the parental plant can influence offspring fitness by changing dormancy and germinability. This is very important for annual wild populations to have longer lived seed-lots to ensure survival from one generation to the next.

The species selected as the experimental organism was *Vaccaria hispanica* (Mill.) Rauschert, which occurs in Europe, Asia and has been introduced to North America. The species is an annual forb, which produces gray to brown rounded seeds. In Iran, this species is found in both agricultural and pasture areas. In pastures it often experiences herbivory (being grazed by livestock) and/or drought stress and mostly germinates in September and flowers in May (Cici, pers.obs.).

The main objective of this study was to test the hypothesis that increasing stress causes changes in seed characteristics and seedling recruitment of *V. hispanica* species. This will help to provide a better understanding of the recruitment biology and seed plasticity of this species in an environmentally fluctuating habitat.

Materials and Methods

Source of seeds

The study site was located on semi-arid slopes in the vicinity of Mallard in northwest of Alborz (35°43'53"N; 51°0'5"E), Iran. It has a long-term average rainfall of 260 mm and temperature of 21°C. The area has vegetation dominated by Asteraceae and Chenopodiaceae. In early spring, seeds were harvested at the time of dispersal from 50 randomly selected maternal plants located within a 15 m radius and transported in paper bags. They were pooled together, dried at 15°C and stored at 4°C for 6 months.

Plant Growth

To avoid undesirable carry-over environmental effects on species performance two sequential generations of *Vaccaria hispanica* were grown in a controlled environment with no biotic or abiotic stress.

Initially, seeds were placed in 90 mm plastic Petri dishes with filter paper drenched with GA3 at 1.0 g L^{-1} to stimulate germination and kept 2 days in a dark room at 15°C and 5 days in alternating 12 h light (30 µ mol m^{-2} s^{-1}) and 12 h darkness [3]. When cotyledons were fully evident, seedlings were transplanted to 50 plastic pots (2 L) filled with a commercial potting soil and moved in the greenhouse with an average temperature of 21 ± 3 and 15 ± 2°C during day and night time,

*Corresponding author: Zahra Hosseini Cici S, School of Sustainable Agriculture, University of Payame-Noor, Tehran, Iran, E-mail: z_h_cici@yahoo.com

respectively. The daytime light intensity was about 800 μmol m^{-2} s^{-1}. The soil was watered to the field capacity every other day with tap water. Once established, seedlings were thinned to one plant per pot to ensure the highest similar initial size. Seeds from these plants were collected and kept in paper bags. One year later these seeds were planted in the same environmentally controlled greenhouse and the same procedure was repeated.

Plant growth under different maternal environment

Seeds of the second generation were used for this main experiment. The establishment and growth conditions were the same as previously for all plants up to 3 weeks and then the following experimental procedures were applied.

A complete random block design was considered for the experiment. Plants were grouped into two watering treatments: Control and Drought, with plants being watered to field capacity every 2 and 10 days, respectively. At the same time, half of the plants from each watering treatment were subjected to simulated herbivory (80% defoliation). The herbivory treatment performed only once. Undamaged plants had no sign of herbivory at the end of the experiment. Scissors were used to remove 80% of leaf area and leaves. This level of defoliation has been observed in the pasture (Cici, pers. obs.). There were 40 plants (pots) per group. Forty five days after treatments were initiated, number of branches, number of leaves and index of architecture (number of branches/main stem length) were determined for each plant. Each plant was harvested when there was a total lack of chlorophyll present. At harvest, plants in each treatment were collected separately and divided into roots, shoots and seeds. Seeds of each plant were collected in paper bags and later counted. The roots were gently washed free of soil. Root and shoot biomass was oven-dried at 75°C for 48 h in paper bags and weighed.

Seed vigour experiment

An accelerated aging test was conducted to study vigour of freshly collected seeds from different maternal sources in the above experiment. The viability of seeds prior to the start of the experiment was checked by cutting seeds (100/seedlot) vertically and examining of the endosperm. Seeds were scored as unviable if dark or mushy endosperm was present. White and firm endosperms were scored as viable. The percentage of unviable seeds was below 2%, irrespective of maternal seed source.

Seeds were placed in an incubator (SMI4E; Sheldon Manufacturing, Cornelius, USA) operating at 45°C and 100% relative humidity [5]. Four incubation periods were tested: 0, 2, 3, 4 and 8 days. After incubation, seeds were washed and immediately placed into a germination test. Four replicates per incubation time and per maternal treatment were put in completely randomized blocks and exposed to 30 days of altering day/night temperature (20/10°C) under dark [6]. Fifty seeds per replicate were placed on two layers of Whatman No.1 filter paper moistened with 2.5 mL of distilled water in 5 cm-diameter plastic Petri dishes. Germinated seeds were counted and removed every 2 days during a period of 30 days. Seeds were considered germinated when their seedlings produced healthy cotyledons.

The viability of ungerminated seeds was tested by the TTC method [5]. Embryos were placed in a 1% solution of 2,3,5-triphenyl-2 H-tetrazolium chloride and those that turned pink were considered viable. The experiment was run twice.

Seedling recruitment

To investigate the maternal effect (of both drought and herbivory) on seedling recruitment in the next autumn, freshly collected seeds from each maternal source were sown in micro plots outside. A randomized block design with ten replicates was used in a seedbed filled with a potting soil. The size of each micro plot was 20 cm long × 10 cm wide. One hundred seeds were planted at a depth of 1 cm in each micro plot. Counts of seedling emergence were made at weekly intervals from mid-September 2012 to mid-May 2013. Between May and June 2013, all remaining seeds were exhumed using a wet sieving technique and their viability was tested by the TTC method as mentioned above [5].

Statistical analysis

All the data were analyzed for normality and homogeneity of variance prior to analysis. Variables were analysed by a two-way ANOVA (fixed factors: Herbivory and drought) using Minitab version 14 [7]. Three-way ANOVA was used to determine the effects of maternal environment on seed vigour. Tukey's HSD test was performed for multiple comparisons to determine significant ($p<0.05$) differences among levels in each treatment. For the seed vigour experiment, the two runs were not significantly different so the data were pooled over runs.

Results

Morphological traits

Neither drought nor herbivory damage affected the survival of plants (data are not shown). Final stem height, number of stems and leaves, mass accumulation and seed production were significantly affected by treatments (Table 1). Water supply had the major influence on maternal plants, with plants grown in drought conditions being significantly shorter, and producing less leaves and stems compared with plants subjected to herbivory (Figure 1). Total plant biomass decreased more significantly in response to drought in undamaged and damaged plants. It was found that the herbivory treatment neutralized plant responses to water shortage. Significant reduction in shoot: root ratio was observed in undamaged plants under water shortage. This response to water shortage was not obvious in plants subjected to damage (Figure 1).

Seed vigour (Seed longevity)

Vigour of seeds decreased with increasing time at 45°C and 100% relative humidity. It was significantly affected by maternal soil water but not by maternal herbivory (damage). About 65% and 100% of seeds from well watered maternal plants were dead after an incubation time of 4 and 8 days, respectively. Progeny from the drought environment were more persistent. Ninety and 10% of these seeds were alive after the incubation time of 4 and 8 days, respectively (Figure 2).

Seedling recruitment

Seeds produced by plants grown under different treatments were tested for germination and dormancy status in order to determine possible maternal effects. There were significant differences in final emergence percentages between seeds produced by well watered maternal plants and drought exposed maternal plants, respectively. After one year, about 20% and 60% of seeds from well watered and drought exposed maternal plants were still dormant. Maternal herbivory damage did not have any significant effect on the proportion of dormant seeds produced by treated plants (Figure 3).

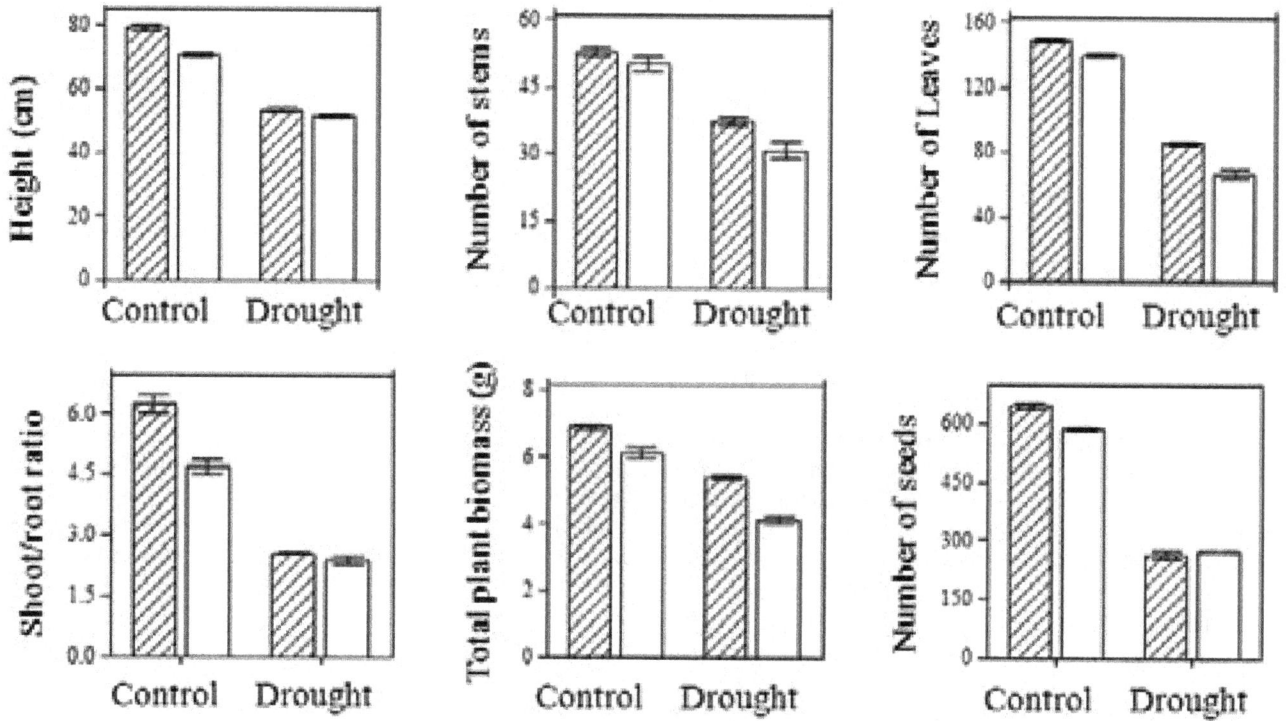

Figure 1: Effect of experimental drought and herbivory on different traits of *Vaccaria hispanica* plants. Hatched bars: undamaged plants; open bars: damaged plants.

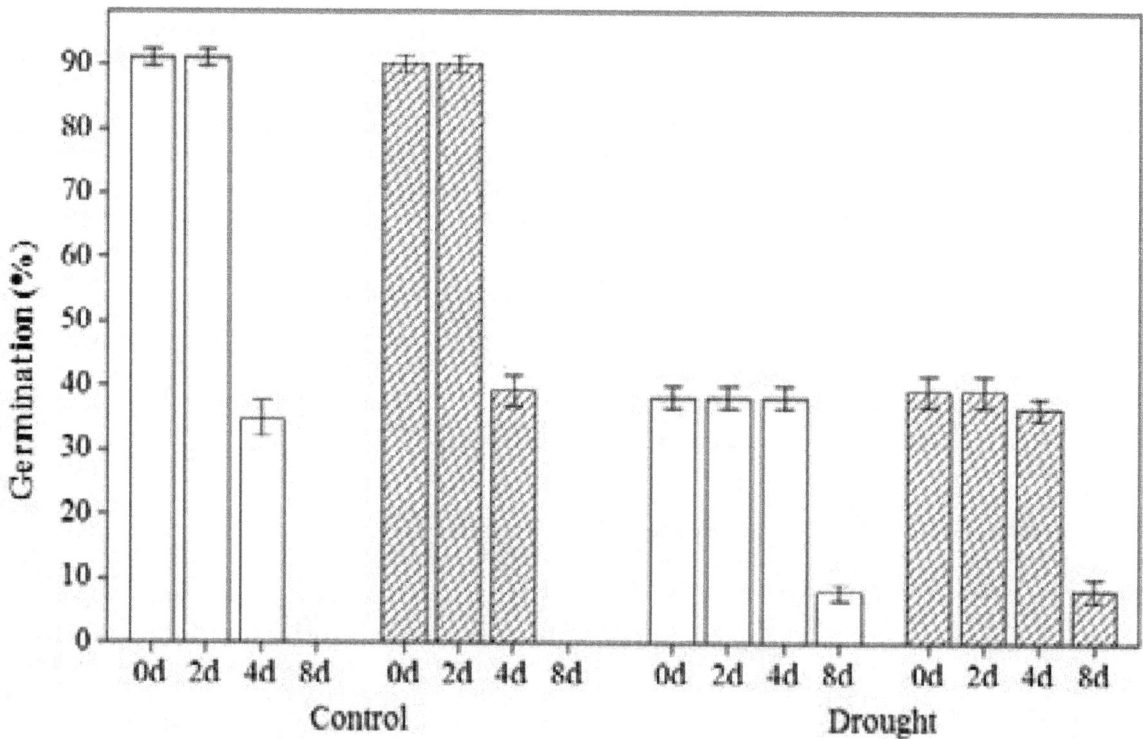

Figure 2: Effect of maternal plant environment (drought and herbivory) of *Vaccaria hispanica* on seed vigour of offspring using accelerated aging test (seeds were exposed to 45°C and 100% RH for 0, 2, 4, 8 days (d) in an incubator). Hatched bars: seeds from undamaged maternal plants; open bars: seeds from damaged maternal plants.

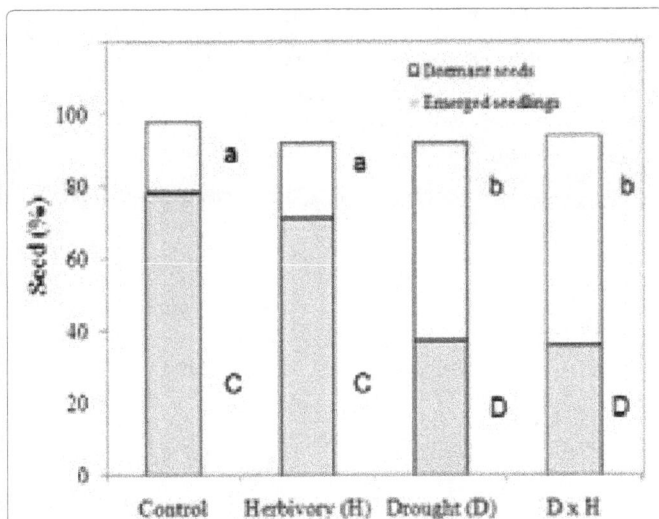

Figure 3: Effect of maternal plant environment (drought and herbivory) of *Vaccaria hispanica* on seedling emergence and seed dormancy in the following autumn. Bars with different letters are significantly different (Tukey's HSD, P=0.05).

	Drought (D)	Herbivory (H)	D × H
Final height (cm)	848[***]	19.5[*]	17.63[*]
Number of stems	300[***]	6.8[*]	4.6[*]
Index of architecture	3.35 [ns]	0.2 [ns]	11.7[*]
Number of leaves	1341[***]	130[**]	7[*]
Shoot/root ratio	422[***]	15[*]	22.12[*]
Total plant biomass	372[***]	58.11[*]	8.39[*]
Seed number	2344[***]	17.68[*]	24.23[*]

[*]P<0.05; [***]p<0.001; ns p>0.05. Main factors: Drought, Herbivory. F-values are shown along with statistical significance.

Table 1: Summary of two-way ANOVA showing the effects of drought and herbivory on different traits of *Vaccaria hispanica* plants.

Discussion

The population dynamics of plants is affected by climate change [8]. Diversity and composition of a plant community is affected by seeds as new offspring. Plants change their reproductive partitioning in terms of the mass, dormancy and longevity to acclimate to varying environments [9]. When offspring produced by each maternal plant shows different dormancy level, it may be an example of bet-hedging. Our study revealed that the dormant: non dormant seed ratio of *Vaccaria hispanica* is controlled by maternal water conditions but not by plant herbivory. While dormant seeds are more persistent and represent the low risk strategy, non-dormant seeds have high germinability and represent a high risk strategy [3,10]. A delay in germination of some of seeds would provide a reserve of seeds to germinate in future years. Such a bet-hedging strategy decreases the risk of reproductive failure under temporal environmental uncertainty [11].

Morphological traits

While maternal water limitation reduces the number of seeds, its effect on dormancy and emergence is more complex. Drought had the major influence on maternal plants. *V. hispanica* plants were highly plastic and displayed a phenotypic response to the environmental conditions, particularly drought, observed as changes in plant size, along with changes in seed mass [12].

Seed vigour

The changes induced by maternal growth environment were also passed on to the progeny [13]. While seeds from the water stressed maternal environment were all alive after being exposed to high temperature and humidity for 4 days (accelerated aging test), only 40% of the seeds from the control maternal plants remained alive. Seed vigour and longevity is a critical aspect in the population dynamics of a species. There was therefore evidence for a trade-off between fecundity and seed persistence in *V. hispanica* in a water shortage condition. Herbivory did not have any significant effect on seed vigour.

Seedling recruitment

Seeds from different maternal plants showed varying dormancy status. Variability in dormancy and germination of seeds from a plant reduces the risk from all siblings simultaneously suffering unfavorable conditions. Maternal environment drought had strong effects on seed dormancy in *V. hispanica*. Seeds produced by plants grown at low maternal water availability had lower germination percentage than those produced by plants that experienced no drought or herbivory alone. Previous work demonstrated that seeds from maternal plants grown in drought showed higher dormancy than seeds from a low water stress environment [14]. Maternal effects that have a significant influence on offspring fitness may confer a fitness advantage in environmental conditions similar to those experienced by the parents [15].

Conclusion

The current study shows that water status influences the population dynamics and seed persistence in *Vaccaria hispanica* and that variation in the seed bank (seed mass, seed vigour and dormancy) of *V. hispanica* are more dependent on water availability than herbivory. Seed dormancy and longevity as a bet-hedging strategy is likely to be an important contributor to stable species coexistence in the novel environmental conditions expected with climate change.

References

1. Roach DA, Wulff RD (1987) Maternal effects in plants. Annual Review of Ecology and Systematics 18: 209-235.

2. Gremer JR, Venable DL (2014) Bet hedging in desert winter annual plants: optimal germination strategies in a variable environment. Ecol lett pp: 1-7.

3. Cici SZH, Van ARC (2009) A review of the recruitment biology of winter annual weeds in Canada. Can J Plant Sci 89: 575-589.

4. Jain SK (1982) Variation and adaptive role of seed dormancy in some annual grassland species. Botanical Gazette 143: 101-106.

5. International Seed Testing Association (ISTA) (2014) International rules for seed testing. Seed Sci Technol 27.

6. Van AR, Cici SZH (2012) Timing of stinkweed and shepherd's pursr recruitment affects biological characteristic of progeny. Can J Plant Sci 92: 933-936.

7. Carver RH (2004) Doing Data Analysis with Minitab 14. Thomson, Victoria, Australia.

8. Willis SG, Hulme PE (2004) Environmental severity and variation in the reproductive traits of *Impatiens glandulifera*. Funct Ecol 18: 887-898.

9. Clauss MJ, Venable DL (2000) Seed germination in desert annuals: an empirical test of adaptive bet hedging. Am Nat 155: 168-186.

10. Wang L, Huang ZY, Bakin CC, Baskin JM, Dong M (2008) Germination of dimorphic seeds of the desert annual halophyte *Suaeda aralocaspica*, a C4 plant without Kranz anatomy. Ann Bot 102: 757-769.

11. Venable DL (1985) Ecology of achene dimorphism in *Heterotheca latifolia*. III. Consequences of varied water availability. J Ecol 73: 757-763.

12. Hughes L (2000) Biological consequences of global warming: is the signal already apparent? Trends Ecol Evol 15: 56-61.

13. Venable DL, Lawlor L (1980) Delayed germination and dispersal in desert annuals: Escape in space and time. Oecologia 46: 272-282.

14. Evans MEK, Ferriere R, Kane MJ, Venable DL (2007) Bet hedging via seed banking in desert evening primroses (*Oenothera, Onagraceae*): Demographic evidence from natural populations. Am Nat 169: 184-194.

15. Cici SZH, Van ARC (2011) Relative freezing tolerance of facultative winter annual weeds. Can J Plant Sci 91: 759-763.

27

Genetic Algorithm: A Veritable Tool for Solving Agricultural Extension Agents Travelling Problem

Adewumi IO*, Oluwatoyinbo FI, Omoyajowo AO, Ajisegiri GO and Akinsete AE

Department of Agricultural Engineering, Federal College of Agriculture, P.M.B. 5029, Moor Plantation, Ibadan, Oyo State Nigeria

Abstract

Genetic algorithm simulates the logic of Darwinian selection as observed in the biological evolutionary process (Cells' division, DNA, Mutation, etc.) to solve problems. They are based on one hand on a heuristic gradient ascension method (selection and crossover) and in another hand on a semi-random exploration method (Mutation). In this research work, application of genetic algorithms was explored for the optimization problem embodied in the transit problem of agricultural extension agents or workers in disseminating new innovation and technological advancement in agriculture. An order representation for the cost matrix for 10 cities and chromosomes was used. The result revealed that genetic algorithm can solve the routing problem of an agricultural extension agent in terms of time minimization in order to search for the shortest route, which will increase number of places that the extension agents can touch at reduced cost of transportation. This will help in achieving the nations' vision 2020 on food security.

Keywords: Genetic algorithm; Crossover; Mutation; Chromosome; Selection

Introduction

Sebusang [1] Odoemelam [2] that the crucial role of Information and Communication Technologies cannot be overemphasized in todays' world globalization, perceived as a major driver of the worlds' economies, is powered by an enhanced use of ICTs on which virtually every sector now relies. The current dispensation of competiveness within the global market space could not have been made possible without the interconnectedness of institutions and people. But many factors have continued to impinge on the increase drive towards the use of ICTs in developing economies particularly in Nigeria. Adewumi [3] has stated that the need to solve optimization problems is a dominant theme in the engineering world. A great number of analytic and numerical optimization techniques have been developed, and yet there are still large groups of functions that present significant difficulties for numerical techniques, and moreover, for analytical methods these are not continuous or differentiable everywhere; functions, which are non-convex, multi-modal, and functions which contain noise (Table 1). As a consequence, there is a continuing search for more robust optimization techniques that are able to overcome such problems. Oliver [4] has described Genetic Algorithm (GA) has one of the relatively new class of stochastic search algorithms. Stochastic algorithms are those that use probability to help their search. As the name implies, GA behaves much like biological genetics. It encodes information into strings just as living organisms encode characteristics into strands of DNA. Adewumi [3] stated that a string in a GA is analogous to a chromosome in biology. A population of strings competes and those strings that are the fittest procreate, the rest eventually die off childless. As with biological parents, two strings combine and contribute part of their characteristics to create their offspring, the new individual. This new string joins the population and fight to produce the next generation. If both parents contribute good building blocks (short section of the string) to the offspring, it will be fit and will procreate in its turn. If the building blocks are poor, then the offspring will die off without generating offspring (Table 2). A second-but-important process occurs in Gas: sometimes vary rarely. A mutation occurs and the offspring will incorporate a new building block that came from neither parent. The above cycle of death and birth repeats until an acceptable solution to the problem is found.

Goldberg [5] as reported by Prasanna [6] defines genetic algorithm (GA) as search algorithm based on mechanics of natural selection and natural genetics. Frick [7] gave a similar definition stating that GA is a software procedure modeled after genetics and evolution. Genetic algorithm is an adaptive heuristic search algorithm premised on the evolutionary ideas of natural selection and genetics. The basic concept of GA is designed to stimulate processes in natural system necessary for evolution, specifically those that follow the principle laid down by Charles Darwin of survival of the fittest. As such, it represents an intelligent exploitation of a random search within a defined search space to solve a problem. First pioneered by John Holland in the 1960s, genetic algorithm has been widely studied, experimented and applied in many fields of engineering world. Not only does it provide alternative methods to solving problem, it consistently outperforms other traditional methods in most of the problem links. Many of the real world problems involved finding optimal parameters, which might prove difficult for traditional methods, but ideal for GA. GA exploits the idea of the survival of the fittest and an interbreeding population to create a novel and innovative search strategy (Table 3). A population of strings, representing solutions to a specified problem is maintained by the GA.

N	1	2	3	4	6	10	20	50
No. Of Solutions	1	2	6	24	720	3 X 10⁶	2 X 10¹⁸	3 X 10⁶⁴

Source: Adewumi, 2009.

Table 1: Number of distinct solutions which exist for agricultural extension agents as a function of n, the number of cities.

---

*Corresponding author: Adewumi IO, Department of Agricultural Engineering, Federal College of Agriculture, P.M.B. 5029, Moor Plantation, Ibadan, Oyo State Nigeria, E-mail: adexio2010@gmail.com, adexio@yahoo.com

	1	2	3	4	5	...............J
1	C_{11}	C_{12}	C_{13}	C_{14}	C_{15}	...............C_{1j}
2	C_{21}	C_{22}	C_{23}	C_{24}	C_{25}	...............C_{2j}
3	C_{31}	C_{32}	C_{33}	C_{34}	C_{35}	...............C_{3j}
4	C_{41}	C_{42}	C_{43}	C_{44}	C_{45}	...............C_{4j}
5	C_{51}	C_{52}	C_{53}	C_{54}	C_{55}	...............C_{5j}
I	C_{i1}	C_{i2}	C_{i3}	C_{i4}	C_{i5}	...............C_{ij}

Generational Result

Table 2: The cost matrix.

Chromosomes						Raw Fitness	Relative Fitness
5	3	6	4	1	2	01.172	306.862
4	2	3	6	5	1	00.283	074.296
2	1	3	5	4	6	00.201	052.865
2	4	1	3	5	6	00.153	040.071
6	5	4	2	3	1	00.115	027.893
5	2	1	6	3	4	00.093	023.685
6	4	5	2	1	3	00.081	019.692
2	3	5	1	6	4	00.062	016.809
3	1	2	5	6	4	00.063	015.074
5	1	2	4	6	3	00.051	013.776
1	3	2	5	4	6	00.045	011.751
5	3	2	1	4	6	00.041	010.691
4	6	2	1	3	5	00.039	010.280
6	5	1	3	4	2	00.036	009.397
1	4	3	2	6	5	00.033	008.678
5	3	6	4	1	2	00.032	008.440
4	2	3	6	5	1	00.030	007.897
4	2	3	6	5	1	00.030	007.897
4	2	3	6	5	1	00.030	007.897
4	2	3	6	5	1	00.030	007.897
4	2	3	6	5	1	00.030	007.897
4	2	3	6	5	1	00.030	007.897
4	2	3	6	5	1	00.030	007.897
2	1	3	5	4	6	00.029	007.897

The Best Solution for This Generation
The best chromosome=5, 3, 6, 4, 1, 2.
Raw fitness=01.172.
Relative Fitness=306.862
The worst solution for this generation
The worst chromosome=2 1 3 5 4 5
Raw fitness=00.029
Relative fitness=007.455

Table 3: The population for generation number 7.

Representation of the problem domain

Genetic algorithm does not deal with the data contained in the problem (or its possible solutions directly, but uses a representation of these data. This representation is mostly an encoding of the original numerical (decimal) values into a binary string of '0' and '1'. In theory, the encoding can be done over any form of finite alphabets, but the binary representation is the most efficient as it is the most basic form of representation. Very often, the problem or its solution contains data not in numerical form. A behavior-oriented model for instance, will contain possible behavior rules the system is looking to optimize. A trading model for financial markets (such as Adaptive Portfolio Trading (APT) system) contains the behavior rules of 'buy' and 'sell' that the GA processes should learn to select once certain conditions are met. A GA-ready model will typically represent these rules through enumerative integer values (as simply as <enum> or <int> data types).

When the GA engine returns the selected integer value, the application object will execute the appropriate rule normally using some kind of a <switch> or nested <if> statement.

Biological metaphors for genetic algorithm

Genetics: Within most cells in the human body (and in most living organisms) are rod-like structures called Chromosomes. These chromosomes dictate various hereditary aspects of the individual. Within the chromosomes are individual genes: a gene encodes a specific feature of the individual, for example, a person's eye color is dictated by a specific gene (Table 4). The actual value of the gene is called an Allele; so, the eye color gene may produce brown eyes. A hierarchical picture is built up with alleles being encoded as genes with sequences of genes being chained together as chromosomes, which makes up the DNA of an individual [8]. When two individuals mate, both parents pass their chromosomes onto their offspring. In humans, who have 46-paired chromosomes in total, both parents pass on 23 chromosomes each to their child. Each chromosome passed to the child is an amalgamation of two chromosomes from a parent. The two chromosomes come together and swap genetic material, and only one of the new chromosome strands is passed to the child. So, the chromosome strands undergo a crossover of genetic material, which leads to a unique new individual. As if this were not enough, genetic material can undergo mutations, resulting from imperfect crossovers or other external stimuli. Although mutation is rare, it does lead to an even greater diversification in the population. It must be noted

Chromosomes						Raw Fitness	Relative Fitness
5	3	6	4	1	2	01.172	306.862
4	2	3	6	5	1	00.283	074.296
2	1	3	5	4	6	00.202	052.865
2	4	1	3	5	6	00.153	040.071
6	5	4	2	3	1	00.107	027.893
5	2	1	6	3	4	00.091	023.685
6	4	5	2	1	3	00.075	019.692
2	3	5	1	6	4	00.064	016.809
3	1	2	5	6	4	00.058	015.074
5	1	2	4	6	3	00.053	013.776
1	3	2	5	4	6	00.045	011.751
5	3	2	1	4	6	00.041	010.691
4	6	2	1	3	5	00.039	010.280
6	5	1	3	4	2	00.036	009.397
1	4	3	2	6	5	00.033	008.678
5	3	6	4	1	2	00.032	008.440
4	2	3	6	5	1	00.030	007.897
4	2	3	6	5	1	00.030	007.897
4	2	3	6	5	1	00.030	007.897
4	2	3	6	5	1	00.030	007.897
4	2	3	6	5	1	00.030	007.897
4	2	3	6	5	1	00.030	007.897
4	2	3	6	5	1	00.030	007.897
4	2	3	6	5	1	00.029	007.897

The best solution for this generation
The Best Chromosome=5 3 6 4 1 2
Raw fitness=01.172
Relative fitness=306.862
The worst solution for this generation
The worst chromosome=4 2 3 6 5 1
Raw fitness=00.030
Relative fitness=007.76

Table 4: The population for generation number 8.

however, that a significant number of mutations are harmful and can destroy good genetic code, so the rate of mutation must be low in order to prevent severe degradation of the genetic code.

Benefits of genetic algorithm

Martello [9] as cited by Goldberg [5] discovered that one of the GA's most important qualities is its ability to evaluate many possible solutions simultaneously. This ability, called Implicit Parallelism (Keith Grant) is the cornerstone of the GA's power. Implicit parallelism results from simultaneous evaluation of the numerous building blocks that comprises the string. Each string may contain millions of these building blocks, and the GA assesses them all simultaneously each time it calculates the string's fitness (Table 5). In effect, the algorithm selects for patterns inside the string that exhibit high worth, and passes these building blocks onto the next generation. This selection process enables genetic algorithm to perform well where traditional algorithm flounder such as in problems with huge search spaces [10]. The objective of this study is to use genetic algorithm (GA) to go through a number of cities that will be visited by agricultural extension agents or workers when disseminating new innovation and technology in agriculture to local farmers in order that.

The tour is completed within minimum time (time management).

Reduction in total cost value incurred on the tour.

Find the routes that minimize the total distance.

Profit maximization.

Chromosomes						Raw Fitness	Relative Fitness
5	3	6	4	1	2	01.172	306.862
4	2	3	6	5	1	00.283	074.296
2	1	3	5	4	6	00.202	052.865
2	4	1	3	5	6	00.153	040.071
6	5	4	2	3	1	00.107	027.893
5	2	1	6	3	4	00.091	023.685
6	4	5	2	1	3	00.075	019.692
2	3	5	1	6	4	00.064	016.809
3	1	2	5	6	4	00.058	015.074
5	1	2	4	6	3	00.053	013.776
1	3	2	5	4	6	00.045	011.751
5	3	2	1	4	6	00.041	010.691
4	6	2	1	3	5	00.039	010.280
6	5	1	3	4	2	00.036	009.397
1	4	3	2	6	5	00.033	008.678
5	3	6	4	1	2	00.032	008.440
4	2	3	6	5	1	00.030	007.897
4	2	3	6	5	1	00.030	007.897
4	2	3	6	5	1	00.030	007.897
4	2	3	6	5	1	00.030	007.897
4	2	3	6	5	1	00.030	007.897
4	2	3	6	5	1	00.030	007.897
4	2	3	6	5	1	00.030	007.897
4	2	3	6	5	1	00.030	000.000

The best solution for this generation
The best chromosome=5 3 6 4 1 2
Raw fitness=01.172
Relative fitness=306.349
The worst solution for this generation
The worst chromosome=4 2 3 6 5 1
Raw fitness=00.030
Relative fitness=000.000

Table 5: The population for generation number 9.

Due to time constraint, the number of cities is restricted to ten (10) to give 2 x 106 initial solutions from which the fitter chromosomes can be selected for further generation. The data type of chromosomes is also restricted to only integer numbers, binary and floating point numbers are not provided for in this study.

Outline of the basic genetic algorithm

[Start] Generate random population of n chromosomes (suitable solutions for the problem).

[Fitness] Evaluate the fitness $f(x)$ of each chromosome x in the population.

[New population] Create a new population by repeating following steps until the new population is complete.

[Selection] Select two parent chromosomes from a population according to their fitness (the better fitness, the bigger chance to be selected).

[Crossover] With a crossover probability cross over the parents to form a new offspring (children). If no crossover was performed, offspring is an exact copy of parents.

[Mutation] With a mutation probability mutate new offspring at each locus (position in chromosome).

[Accepting] Place new offspring in a new population.

[Replace] Use new generated population for a further run of algorithm.

[Test] If the end condition is satisfied, stop, and return the best solution in current population.

[Loop] Go to step 2.

Methodology

The ultimate goal of this research work is to determine a route through a set of n cities and back to starting point without repetition, which will maximize resources and save cost. N number of cities will be supplied from console for this project. The genetic algorithm search paradigm looks for the best permutation set, which will meet the objective of the study as stated earlier. In carrying out these work, n cities representing the number of cities was permuted, which will yield n! The fitter chromosomes among them are selected for future generation. The selected chromosomes are crossed over and mutated until an optimal solution is achieved.

Encoding the problem

The problem was coded using a sequence of integer values to represent the cities in a tour. For example: 1-2-3-4 represents a tour from city 1 through cities 2, 3 and 4 in that order.

The cost matrix

The cost matrix is represented by the distances between the two adjacent cities. For example C_{ij} represents the distance between city i and city j in cost table. The cost matrix is shown in the table below.

Analysis of genetic algorithm system for agricultural extension agent

The system of transit for agricultural extension agent can be analyzed using the following components:

Input: The input to the system is the number of cities to visit by an agricultural extension agent. This number of cities will be captured from the keyboard or from a remote location provided that there is accessibility to the system from that location. The input data is expected to be a valid integer number. Also, a user is expected to choose stopping criteria, which will be provided. Mutation probability and crossover probability are to be supplied from the console. They should be fraction number n, where $0 < n < 1$.

Process: The main transformation that is expected to take place within the system is the determination of a set of solutions to solve the problems. Once a valid input data is supplied from the console, the first process that will take place is to randomly generate a set of valid tours, which will be N! where N is the number of cities supplied by the user. After the generation of all possible random tours, fitness value of each tour will be determined. Each tour in genetic algorithm paradigm is what we refer to as Chromosomes. Part of the fitter chromosomes will be selected for future mating among them. Initial population will be displayed. Thereafter reproduction operators (selection, crossover, mutation merge) will be supplied on each generation thereby leading to the production of subsequent generations. This trend continues until stopping criteria is true.

Output: At the initial stage, the result of initial generation will be displayed before, and thereafter, the result of subsequent generations will be displayed from where the user can select whatever option he can afford.

Feedback: The feedback can either be positive or negative. The feedback from this system will be whether the system is optimizing or not. If it optimizes well, the user may have to increase the value of some of these parameters or reduce the parameters depending on the parameter in question.

Control: This depends on the result of the feedback. If the result is favorable, the generation number may have to be increased so that more optimal solutions can be achieved. Control is demonstrated with series of input data and result obtained in appendices A and B.

Environment: The environment of this system that worth mentioning here is minimum hardware requirement that was used to carry out this research work. This will flourish in an environment where search space is very wide and fuzzy.

Boundary and interface: This system can be interfaced with other areas of artificial intelligence like fuzzy logic, simulated annealing, evolutionary algorithm.

System development tool

The system development tool chosen for this project is dependent on the features provided. Visual Basic Version 6.0 is used for the implementation of this design. This is due to the flexibility, simplicity of features that enhances easy movement of data from console to the program and ability to easily print results into an output file and monitor simultaneously.

System description

This system deals with how best n cities can be toured at the lowest distance and cost possible. Visual Basic 6.0 was used to implement this and the entire program is explained below.

The main program: The main menu gives the user the opportunity to either optimize the traveling salesman problem using typical method, which will use just a method from each of the operators as follows: generational, integer representation, two-point crossover, uniform

mutation, and roulette selection and generation termination method. Customized optimization can also be done from this point by selecting from numbers of methods provided by each operator.

Generational algorithms: This module offers two options if customized is selected from the main menu. The generational option randomly generates the chromosomes automatically without using any city as a starting point or terminal point. On the other hand, steady state enables you to have options of selecting the city that will be origin or destination of a chromosome.

Data types: This customized option enables us to select what will be the format for representing cities in the chromosome. The data representations possible are binary representation, integer representation and floating-point representation. Once a data representation is chosen, it is this representation that will be used as object of manipulation for the subsequent period. The screenshot of this module is shown in Appendix B.

Crossover operators: This module allows the user to choose the type of crossover methods to be used in case we do not want the system to choose for us. The crossover methods available in this work are partially mapped crossover, order crossover and cycle crossover. The methods are actually expected to optimize the problem, but the solution provided by these methods are relatively the same. The screenshot of the module is shown in Appendix B. the results produced by these methods are shown in Appendix C.

Mutation operators: This module enables better result to be produced by carrying out some structural adjustment in the genes of the selected chromosome. The functionality of mutation depends on the data type selected. If the binary data type is only carried out by flipping over a bit from 0 to 1 or from 1 to 0, if the data type selected under data type module is integer representation, mutation is carried out by randomly selecting two mutation points. The genes in these two points are swapped to form other chromosomes. The mutation methods available are flip bit, boundary, uniform and Gaussian mutations.

Selection operation

Selection operation is the reproduction operator. This operator selects chromosomes that will form population for the subsequent generation(s). The methods under this module include roulette, tournament, top percent, best and random method. Whichever method selected here will copy chromosomes from both the old and the new populations to form population for the following generation.

Termination operator: This is the last module that determines when the genetic algorithm should stop. The methods available under this module include generation number, evolution time, fitness convergence, population convergence and gene convergence. The screenshot of this module is shown in Appendix B.

Parameter specification: This module takes all the necessary parameters that will be used by the operators. These parameters among other things include the following:

Population size: This parameter specifies the number of cities that will be involved in the problem. Due to time constraints, 10 cities were considered in this work. However, this can be improved upon for it to work on several numbers of cities. Whatever number of cities supplied will determine the number of solutions that will be provided. If population size is n, then, the population size will be n!

Termination parameter: This parameter determines the number of generations that genetic operation will pass through. This will depend on the type of termination method selected; for example, in generation number, if the generation number is 10, it implies that the genetic operation will go for 10 times.

Crossover probability: This probability specifies whether a crossover can be carried out on any selected chromosomes or not. The default is 0.6. The essence is to ensure that fitter chromosomes are able to survive to the following generation.

Mutation probability: This probability specifies the possibility of changing the internal structure of a chromosome. It is normally kept very low so that there will be no unnecessary changes in the structure of genes. The default is 0.01.

Program listing

The program listing of this research work is displayed in Appendix C.

System specification

The basic systems requirements to run this study are as given below:

Windows XP Professional, Window Vista, Linux Operating System.

Pentium 233 MMX Processor.

32MB SD RAM.

VGA Monitor.

The above requirement is the minimum requirement whereby we can use to run this research.

Testing

The approach adopted in testing the functionality of this program is mainly black box testing. Series of input genetic parameters are supplied from the console to get what the result will be. The operation is repeated for series of cities between 4 and 10 to see how the efficiency of the program as the search space continues to increase. Under this approach, we are only interested in the output and to what is actually happening within the program. Therefore, it may not be able to detect any redundant code that we may have within the program. In view of this, Glass Box Testing is also carried out to detect errors within the program. The methods adopted under glass box testing technique include the following: Loop testing, Branch testing, Path coverage and Statement coverage. The results generated as a result of this testing are shown in Appendix B of this work.

Result Analysis

Four different parameters are supplied and their results are displayed in Appendices A and B: the first set of parameters taken so small to show the limitation of genetic algorithms over a narrower search space. Notwithstanding, the population size, if the generation number is very small with lower crossover probability and mutation probability, the GA may go through series of generation with optimizing anything at all; for example agricultural extension parameters. As the generation number continues to increase with corresponding increase in the crossover probability and decrease in mutation probability, a search space will continue to increase thereby unleashing the capability of GA approach to solving a traveling salesman problem. Figure 1 in Appendix B shows the parameters supplied and their corresponding results are shown in Appendix B. However, care should be taken to vary the parameters of the transit for agricultural extension agent,

the reason being that while we are trying to exploit the benefit arising from increasing mutation probability, we may also loose the initial population entirely. Therefore, we should try as much as possible to keep the probability low. The default for mutation probability is 0.01 and crossover probability is 0.6. There is no default for population and stopping criteria; it is situational, that is, it is the problem at hand that determines their values.

Conclusion

Genetic algorithms balance exploitation with exploration. The crossover and mutation operators control exploration while the selection and fitness functions control exploitation. Mutation increases the ability to explore new areas of the search spaces but it also disrupts

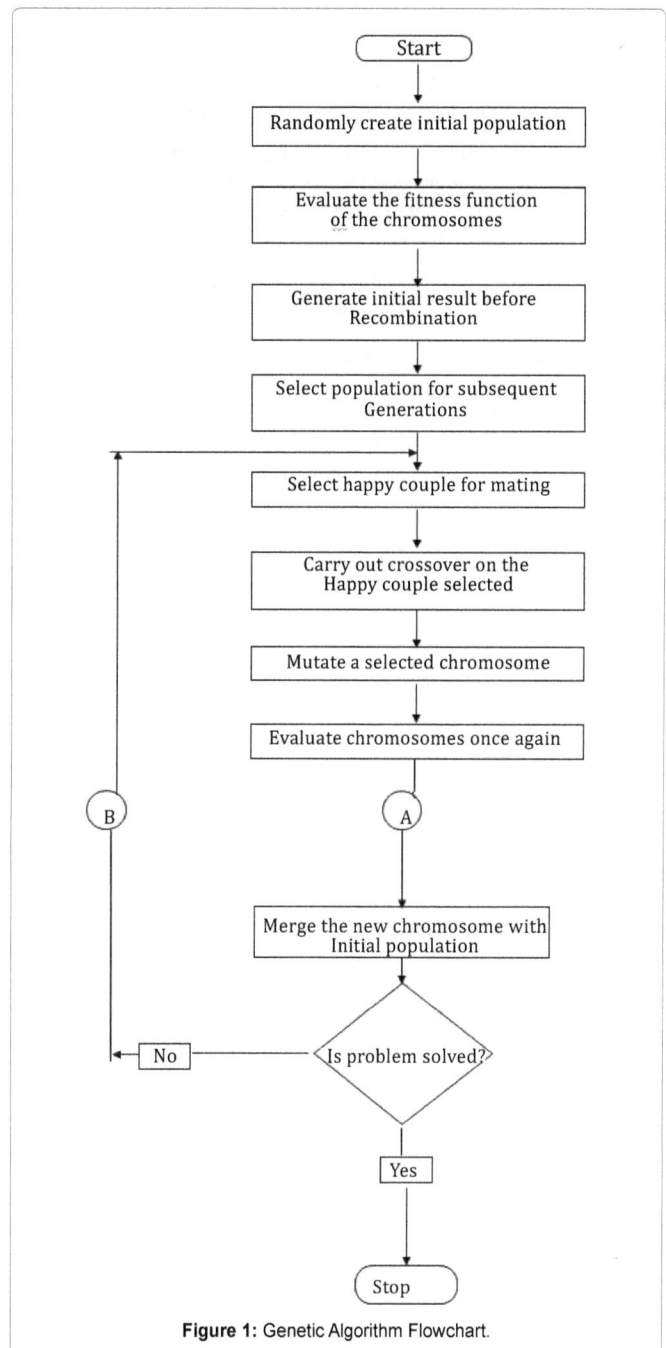

Figure 1: Genetic Algorithm Flowchart.

the exploitation of the previous generation. It has been clearly showed from the study so far that the objectives of the work has been achieved in terms of time management, shortest routes that minimize the total distance which will leads to reduction in total cost value incurred and profit maximization for the agricultural extension agents/firm during the course of disseminating new and improved technology to local farmers in order to increase the food security in the country. The ability to separate the search mechanism from the model representation makes genetic algorithms an ideal approach for the solution of very complex combinatorial optimization problems. Travelling problem of an agricultural extension agent is one of those complex problems with a very wide search space where genetic algorithm has proved very effective to find a lasting solution to the problem of determine the least possible cost of routing a number of cities.

Recommendation

For further work on genetic algorithm implementation in agriculture, the following must be taken into consideration:

Evaluation on a larger dataset, because in most time GA fails when it is applied to a very larger problems because it scales poorly in terms of complexity as the number of cities increases. So more research has to be carried out on the convergence of the chromosomes.

It also has to be evaluated for a longer time frame; this is because the solution quality degrades rapidly.

Proper fitness function has to be examined.

References

1. Sebusang SEM, Masupe (2003) ICT development in Botswana: connectivity for rural communities. Southern African Journal of Information and Communication (SAJIC) the Edge Institute/Research ICT Africa, Braamfortein ZA.

2. Odoemelam LE, Onuekwusi GC (2013) Effects of ICTS-Based Agricultural Extension Services in Enhancing the Well-Being of Farmers in Abia State, Nigeria. International Journal of Applied Research and Technology. 2: 15-22.

3. Adewumi IO (2009) Using Genetic Algorithm For Solving Travelling Salesman Problem. B.Sc. Project Submitted to Department of Industrial & Production Engineering, Faculty of Technology, University of Ibadan 2.

4. Oliver IM, Smith DJ, Holland JRC (1987) A study of permutation crossover operation on the Travelling Salesman Problem. Proceeding of the 2nd International Conference on Genetic Algorithms. 224-230.

5. Goldberg DE (1989) Genetic Algorithms in Search, Optimization and Machine Learning. Addison-Wesley.

6. Prasanna J, Jung YS, Dirk VG (1991) Parallel genetic algorithms applied to the traveling salesman problem. SIAM Journal of Optimization 1: 515-529.

7. Frick A (1998) TSPGA-An evolution program for the symmetric traveling Salesman Problem. In HJ. Zimmermann, editor, EUFIT98 - 6th European Congress on Intelligent Techniques and Soft Computing 513-517.

8. Whitley D (1994) A Genetic Algorithm Tutorial. International Journal of Statistics and Computing 4: 65-85.

9. Martello S (1983) An enumerative algorithm for finding Hamiltonian circuits in a directed graph. ACM Transactions on Mathematical Software 9: 131-138.

10. Torimiro DO, Kolawole OD (2008) In: Akinyemiju AO, Torimiro DO (eds) "New Partnership for Africa"s Development and Agricultural Technology transfer in a globalised world, Agricultural ExtensionL a comprehensive tratise. Lagos: Ikeja ABC Agricultural systems Ltd., pp. 402-414.

28

Stability of Soil Organic Matter and Soil Loss Dynamics under Short-term Soil Moisture Change Regimes

Parwada C[1,2]* and van Tol J[1,3]

[1]*Department of Agronomy, University of Fort Hare, Alice, South Africa*
[2]*Department of Crop Science, Bindura University of Science Education, Bindura, Zimbabwe*
[3]*Department of Soil-and Crop-and Climate Sciences, University of the Free State, Bloemfontein, South Africa*

Abstract

Soil properties are known to be influenced Soil Organic Matter (SOM) resident time. However, there is limited information on the interactive effects of SOM quality and soil moisture on SOC and Microbial Biomass Carbon (MBC) hence on soil losses. Therefore, this study investigated the effects of different organic matter and soil moisture in soils with low (<2%) initial SOC content on the SOC, MBC and soil loss with time of organic matter incorporation. Six soils were incubated for 34 weeks at 25°C after adding high quality (C/N=23) *Vachellia karroo* leaf litter and low quality (C/N=41) *Zea mays* stover. The effect of SOM quality and soil moisture on the SOC content, MBC and soil loss was significantly (P<0.05) the same within but varied across soils. Soils that were continuously wet lost more SOC than under alternating wet-dry moisture conditions. Microbial biomass carbon was controlled by the availability of organic matter and moist soil conditions. Low MBC values corresponded to high SOC and soil loss. Continuously wet soils with high sand particles promoted rapid loss of SOC compared to alternating wet-dry soils. Therefore, continuously wet sandy soils are likely to contribute more to the climate warming than alternating wet-dry soil moisture. In the wake of the climatic change, addition of OM in continuously wet soils need to be regulated but to reduce soil loss re-application of fresh OM has to be more frequent under continuously wet sandy soils than in alternating wet-dry moisture regimes.

Keywords: Carbon emission; Conservation; erosion; Mineralization; Texture

Introduction

The interaction between climate change and the global carbon cycle is an important aspect of the global environmental changes [1]. Soil is the largest pool of terrestrial organic carbon in biosphere, storing more Carbon (C) [2]. Therefore, the Soil Organic Carbon (SOC) stock has an irreplaceable function in mitigating climate change as a key component of the biosphere carbon cycle. Meaning that changes in SOC content significantly influence climate change and a slight change in the SOC stocks can have a considerable effects on atmospheric carbon dioxide concentration, contributing to climate warming [3]. Changes of the climate, particularly the temperature and rainfall have more pronounced effects on the resident period of the SOC by accelerating SOC decomposition offsetting a portion of the SOC losses. However, many researches relating the climate change to SOC are biased towards revealing the trends and future projection changes in the SOC and its effects on the environment, ignoring the current climatic scenarios. The climate change is manifested by changes in temperature, precipitation and length of the season [4]. Precipitations that are punctuated by prolonged mid-season dry spells are now a common phenomenon in many parts of the world and could result to detrimental effects on soil microbial action on SOC hence soil losses through erosion. The climatic factors affect the soil microbial activities on the Soil Organic Matter (SOM) thereby influencing the SOM resident period [5].

The climate change modifies soil temperature and moisture simultaneously and, although many researches have attempted to determine the effect of litter quality or moisture on soil microbial biomass, there is still limited attempts to explore the combined effects of both factors [6,7]. The influence of soil moisture on soil carbon stocks has received relatively little attention, although it has a key role in regulating the soil microbial activity [8]. There is also very limited understanding of how intermittent soil moisture affects the soil organic matter decomposition, as the relationships between the intermittent moisture content and quality of soil organic matter are not consistent across different soils.

In addition to the climatic factors, the quality of soil organic matter should also affect the rate of decomposition [9]. Several studies have indicated that the chemical and biochemical quality of litter affects mass loss during decomposition [10,11]. The addition of higher quality substrate (lower C/N ratio of <24 and lower lignin content) resulted to increased SOC mineralization compared to the addition of lower quality (C/N ratio >24) substrate [12-14]. Since the soil microbes need N (and other essential elements) as well as C, if there is little N in the residue, decomposition is slow. When immature legumes are ploughed into the soil that had lower dry matter but higher N concentration and low C/N, decomposition was faster [10]. On the other hand, the high cellulose, hemicellulose and lignin contents of legumes ploughed in at a matured age reduced the speed of decomposition. The C/N ratio in plant residues is highly variable and increases with maturity. An ideal substrate material was found to have C/N ratio=24 to satisfy the N requirement of microbes [12,13]. If the C/N ratio of residue >24, available soil N is consumed by microbes and this retards decomposition rate. A number of authors have reported linear increases in SOC related to the amount of organic matter applied, whilst others, have reported that the rate of SOC accumulation is dependent on the source of organic carbon [14-16]. This greatly suggests

*Corresponding author: Parwada C, Department of Agronomy, University of Fort Hare, Alice, South Africa, E-mail: cparwada@gmail.com

that litter quality is a major control factor of organic carbon content in various soils. Nevertheless, less is known about litter quality and SOC stabilization in different soils [17].

The SOC is one of the key factors influencing soil erodibility. This is due to the positive feedback of the SOC on soil quality as influenced by the Organic Matter (OM) [18]. The OM is the source of all SOC and the rate of OM decomposition in the soil has a direct effect on the amount of SOC present at any given time. Whilst any OM source can be used to enhance soil aggregation and stability, a major drawback is on ensuring selection of OM with clear and prolonged soil stabilizing effects. Different OM may have different effects on soil erodibility and organic carbon resident time depending on properties of the soil in question. Therefore, to maximize the benefits of OM in soil conservation, there is need to explore effects of various OM sources in stabilizing soils of different properties.

The patterns and controls of SOC storage are critical for our understanding of the biosphere, given the importance of SOC in the soil and the feedback to the atmosphere and the rate of climate change. The capacity to predict and ameliorate the consequences of climate change depends on a clear description of SOC content and controlling of SOC inputs and outputs [2]. One aspect of the organic carbon pool that remains poorly understood is its mineralization in different soils varying in moisture. What are the general patterns of SOC in different soils? Do the major determinants of SOC content differ with litter quality and soil moisture regime? How much SOC is stabilized in different soils especially under continuously wet and alternating wet-dry soil moisture conditions and what is the effect of litter quality on the SOC? This paper aim to provide preliminary answers to these and other questions, based on laboratory soil-litter incubation experiments.

We hypothesize that litter source and soil moisture are the major determinant of the abundance of soil microbes that influence the SOC content at any given time. Although soil temperature and the primary particle sizes distribution of the soil are influential in controlling the distribution of SOC, may normally be overruled by the effects of plant type under natural conditions [2].

The specific objectives of this study were to: (1) Determine the effect of litter quality and soil moisture on SOC content, microbial biomass carbon (MBC) soil loss changes with time in different areas of soil associations; and (2) Evaluate the relationship between the SOC content and Microbial Biomass Carbon (MBC) and soil loss dynamics in different areas of soil associations.

Materials and Methods

Description of the study area

The study was carried out in the soil physics laboratory at the University of Fort Hare (UFH), Eastern Cape Province (EC), and South Africa (SA). Soil used in the study were collected from the Ntabelanga area, EC, SA and is located about 380 km south east of the UFH and lies between 31° 7' 35.9" S and 28° 40' 30.6" E. The soils were low (<2%) organic carbon and unstable that are prone to erosion as evidenced by extensive areas of severe gully erosion on the inter-fluvial areas adjacent to stream channels. The Ntabelanga area is covered by land type Db344 that has only 10.2% of the soils with a low sensitivity to erosion. The majority, 71.3% of the area is covered by highly sensitive to erosion with the remaining 18.5% having moderate sensitivity to erosion [19,20]. The Ntabelanga area receives an annual rainfall total of about 749 mm, with most of it falling in December and January. The rain season is characterised by prolonged mid-season dry spells with

the lowest (15 mm) average rainfall received in June and the highest (108 mm) in January.

Site selection and soil sampling

This work followed on studies of Van Tol et al. [19,20]. In these studies the soils of land type Db344 were mapped in some detail. Based on these maps, areas of soil associations were identified and selected for sampling and incubation. Soils in an area of soil association are likely to behave the same to a certain treatment. Although this method included only nine sampling locations, these soils were representative of the majority of Db344. Twenty one soil samples were randomly collected using a zig zag pattern across a proposed Ntabelanga dam at the nine locations (Figure 1) representing the areas of soil associations (wet, melanic, apedal, semi-duplex, duplex and shallow). The soil samples were then compounded to six samples according to the existing areas of soil associations. The soils were collected basing on naturally existing horizons in each soil profile of the areas of soil associations. The soil profiles varied in depth; six were deeper than 30 cm and three shallow (i.e., <30 cm). Some of the sampling points were severely eroded and lacked the A horizon so that subsoils were exposed in many areas and others were rocky just below the A horizon meaning that the horizons were either in their natural or eroded conditions. The observed soil horizons were orthic A, melanic A, pedocutanic B, red apedal B, prismacutanic B, and G-horizon.

Laboratory soil incubation

The soil was air dried and was passed through a 2 mm sieve to homogenize the aggregates. The <2 mm soil aggregates were mixed to either *Vachellia karroo* leaves (low C/N) or *Zea mays* stover (high C/N). The *V. karroo* leaves were harvested at the beginning of winter season (May 2014) and *Z. mays* stover was from a harvested maize summer crop. The plant materials were shredded into very small segments and oven dried at 60°C then ground to pass through a 2 mm sieve. Subsamples of each ground litter were taken and measured for total C, N, lignin and polyphenols contents. The lignin (L) was determined by the Acid Detergent Fibre (ADF) method as outlined in Van Tol et al. [21]. The polyphenols were extracted in hot (80°C) 50% aqueous methanol and determined calorimetrically with tannic acid as standard [21,22].

A 600 g of each soil association was mixed with plant litter at a rate of 2.28 g OM (100 g soil)$^{-1}$ and 2.43 g OM (100 g soil)$^{-1}$ for *V. karroo* leaves and *Z. mays* stover respectively. The mixture constituted at least 2% Soil Organic Carbon (SOC) since initial SOC in the soil associations ranged from 0.7 to 1.35% (Table 1). Soil moisture was adjusted to 80% field capacity equivalent to 48 g H_2O 100 g^{-1} and put in 1000 mL jars then incubated at 25°C for 34 weeks alternating the moistures at three week interval from 80% field capacity (wet soil condition) to 40% field capacity equivalent to 24 g H_2O (dry soil condition) during the incubation period. In the experiment, soil amended with the *V. karroo* leaf or *Z. mays* stover but soil moisture was continuously maintained at the 80% field capacity (wet soil condition) for the entire incubation period were included and the treatments were triplicated. The changes in soil organic matter was obtained by repeatedly measuring the SOC and Microbial Biomass Carbon (MBC) from the jars at weeks 3, 14, 24 and 34 during incubation. Soil losses were also estimated at the same times.

Analyses of primary soil particle size distribution and SOC

The soils were analysed for primary particle size distribution by the

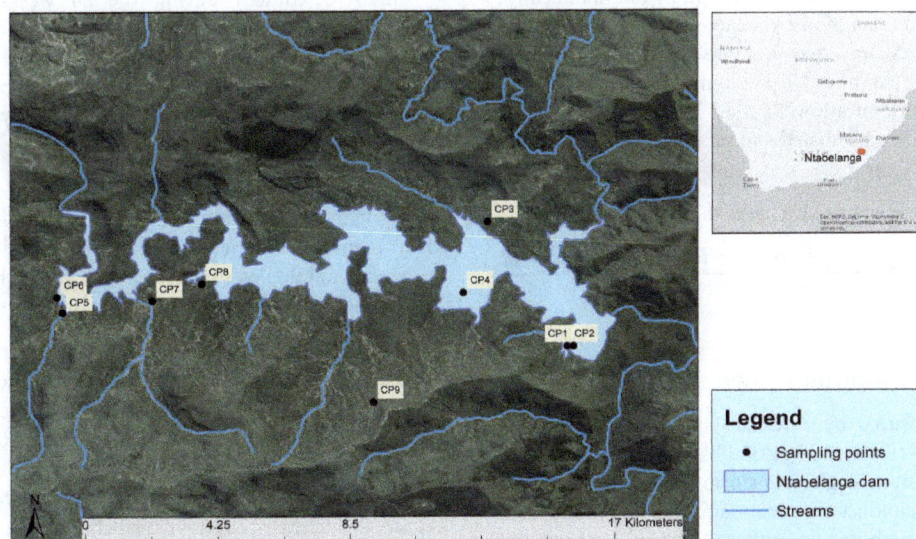

Figure 1: The randomly selected sampled points representing the areas of soil associations in the Ntabelanga area (Adapted from Parwada and Van Tol [40]).

Soil association	Horizon	South African soil form*	WRB reference**	Characteristics	Sand (%)	Clay (%)	Silt (%)	SOC (%)
Shallow	Orthic A (ot)	Glenrosa	Leptosols	Rock or rock horizon such as Lithocutanic B as second horizon	57.8	23.6	18.6	0.81
Wet	G-horizon (gh)	Katspruit	Gleysols	Water logged subsoil horizon	47.5	27.5	25	0.53
Melanic	Melanic A (ml)	Boheim	Phaozems	Pedocutanic B	18	62.5	19.5	0.39
Semi-duplex	Pedocutanic B (vp)	Valsrivier	Chromic Luvisols	Moderate degree of structure in the subsoil horizon	21	59	20	0.39
Apedal	Red apedal B (re)	Hutton	Ochric Ferralsols	Apedal subsoil horizon	49.5	31	19.5	1.35
Duplex	Prismacutanic B (pr)	Kroonstad	Stagnosols	Sandy topsoil on clayey prismacutanic B subsoil horizon	36	38	26	0.7
-	-	-	-	±SD	17.4	16.8	4.1	0.4

*Soil Classification Working Group, 1991; **IUSS Working Group WRB, 2015

Table 1: Descriptive statistics of mean soil particle size distribution, initial Soil Organic Carbon (SOC) content of the soil associations used in the incubation experiments.

hydrometer method as described by Okalebo et al. [23] and SOC was determined through the wet acid digestion modified Walkley Black method [24]. A gram of air-dried soil was transferred to a 500 cm³ Erlenmeyer flask and 10 cm³ $K_2Cr_2O_7$ solution was added by pipette to the soil sample. The flask was swirled to disperse the soil in the solution then 20 cm³ concentrated sulphuric acid was rapidly added, directing the stream into the solution. The flask was swirled again gently until soil and the reagents were mixed, and then swirled more vigorously for a total time of 1 min. The flask was allowed to cool on a sheet of asbestos for 30 min and then 150 cm³ de-ionised water, 10 cm³ concentrated ortho-phosphoric acid and 1 cm³ indicator were added. The excess dichromate with iron (II) ammonium sulphate solution was titrated, as the endpoint was approached, the solution colour changed to a dark violet brown. The iron (II) ammonium sulphate was added drop by drop until the colour changed sharply to green. Then the total organic carbon was calculated as follows:

$$Concetration\ of\ Fe(NH_4)_2(SO_4)_2\,mol\ dm^{-3} = \frac{10cm^3 K_2Cr_2O_7 \times 0.167 \times 6}{cm^3 Fe(NH_4)_2(SO_4)_2}$$

$$Organic\ C\% = \frac{\left[cm^3 Fe(NH_4)_2(SO_4)_2\,blank - cm^3 Fe(NH_4)_2(SO_4)_2\,sample\right] \times M \times 0.3 \times f}{soil\ mass(g)}$$

Where M=Concentration of the $Fe(NH_4)_2(SO_4)_2$ [mol dm⁻³] and f=1.3 (recovery factor).

Determination of Microbial Biomass Carbon (MBC)

The Chloroform Fumigation-Extract (CFE) procedure was used for determination of MBC during the soil incubation based on the methods of Anderson and Ingram [21]. Briefly, two 15 g of fresh soil of the incubated soil with known moisture content were weighed into a crucible and put into separate desiccators. In the other desiccator, a 100 mL beaker containing 25 mL of alcohol free chloroform with boiling chips was placed and a vacuum applied. The vacuum was applied to the fumigated soils samples until the chloroform was rapidly boiling, and then sealed and placed in a dark cupboard for 24 h at 25°C. The non-fumigated soil samples were also incubated the same as the fumigated soils in the dark but without a vacuum. The fumigated soil samples were evacuated using a vacuum pump for 5 min with each evacuation lasting at least 2 min. After evacuation, all the desiccators were opened and the organic carbon in the samples was extracted using 50 mh of 0.5 M potassium sulphate on a shaker at 180 rpm for 1 h and filtered using Whatman No. 42 filter paper. Organic C in the extracts was then determined using the dichromate oxidation method and calculated using an equation proposed by Anderson and Ingram; Joergensen and Emmerling [21,25].

$$c(\mu g / g - soil) = \frac{(HB - S) \times N \times E \times VD \times (VK + SW) \times 1000}{CB \times VS \times DM}$$

Where HB=Consumption of titration solution by hot blank (mL); S=Consumption of titration solution by sample (mL); N=Normality of the $K_2Cr_2O_7$; E=3; VD=Added volume of $K_2Cr_2O_7$ (mL); VS=Added volume of sample (mL); VK=Volume of K_2SO_4 extractant (mL); CB=Consumption of titration solution by cold blank (mL); SW=Amount of water in the incubated soil sample (mL); DM=Total mass dry soil (g).

The microbial biomass carbon (MBC) was then calculated as:

$$MBC = \frac{(organic\ C\ in\ fumigated\ soil - organic\ C\ in\ unfumigated\ soil)}{kEC}$$

Where kEC=0.38 [26].

Estimation of soil loss during incubation

Soil losses were measured by a rainfall simulator following a modified procedure by Nciizah and Wakindiki [27]. Briefly, rainfall was applied either as 8 min Single Rainstorm (SR). Three runs of rainfall simulations were conducted per soil sample. Splash cups filled with soil were saturated with distilled water. The samples were put under the rainfall simulator at an intensity of 360 mmh^{-1} ($\approx$60 mmh^{-1} natural rainstorm with time specific energy of 1440 J.m^{-2}.h^{-1}) [28]. The high intensity was to compensate for short falling distance of 0.4 m used when calibrating the rainfall simulator. After each rainstorm, the splashed sediments collected in the splash plate were washed into a jar, oven dried at 105°C for 24 h and weighed.

Data Analysis

Sampling was not destructive and thus the data was analysed using repeated measures Analysis of Variance (ANOVA) to compare treatment means. Where sphericity assumptions could not be met, the Greenhouse-Geisser correction of P was used. Pearson's test was used to determine whether significant correlations existed between soil organic carbon, soil loss and the observed selected soil properties were also done. All data were analysed using JMP version 11.0.0 statistical software [29].

Results and Discussion

The description of the areas of soil associations and analysed primary particle size distribution and soil organic carbon prior to the incubation is shown in Table 1. The shallow soils had most primary mineral particles in the sand (2-0.05 mm Ø) class and the melanic had most clay (<0.002 mm Ø) content. The total soil organic carbon content of the soil associations ranged from 0.39 to 1.35% with the least (0.39%) and highest (1.35%) observed in the melanic and apedal respectively.

The C/N ratio was used as an indicator for litter quality in this study. The V. karroo leaf had a lower (23.0) C/N ratio than the Z. mays stover (C/N=41.0). Besides the low C/N ratio, the V. karroo leaf litter had more of ADF lignin (L), Polyphenols (P) and higher (L+ P)/N ratio than the Z. mays stover (Table 2).

Parameter	Vachellia karroo leaf	Zea mays stover
Total C (%)	45.0 ± 1.45	49 ± 0.25
Total N (%)	2.0 ± 0.09	1.2 ± 0.38
C/N ratio	23.0 ± 1.78	41.0 ± 0.5
ADF lignin (L)	9.0 ± 0.02	3.0 ± 0.10
Polyphenols (P)	1.6 ± 0.01	1.3 ± 0.03
(L+P)/N ratio	5.3 ± 0.37	3.6 ± 0.5

Table 2: Selected chemical characteristics of Vachellia karroo leaf and Zea mays litter used in the study.

The SOC content during the 34 weeks of incubation was significantly (P<0.05) influenced by the soil association × litter and soil moisture, time × soil association and the time × litter and soil moisture interactions. Estimated soil loss (SL) and the microbial biomass carbon (MBC) were significantly (P<0.05) influenced by the time × soil type × litter and soil moisture interactions (Table 3).

The SOC was significantly (P<0.05) higher under the alternating wet-dry than in the continuously wet soil moisture conditions. The SOC content (%) was highest in the melanic and lowest in the shallow, wet, apedal and duplex soil associations under the alternating wet-dry and continuously wet soil moisture conditions respectively (Table 4).

The results of previous studies showed that SOC content rapidly decrease when microbial biomass is high [30]. The soil moisture regimes affected organic carbon mineralization where the organic matter decomposed and mineralized more rapidly in continuously wet than under alternating wet-dry moisture conditions. As the soil moisture level was maintained at 80% field capacity, the SOC was quickly mineralized because of increased oxygen availability that promoted microbial respiration which decreased with time of incubation [31]. Although, it is observed that the organic matter quality influences the SOC and microbial biomass carbon [32], in this study the litter quality effects was the same in a soil association but varied across the soil associations. These results are not similar to observations by Potthast et al. [14] who observed that addition of higher quality substrate (lower C/N ratio of<24 and lower lignin content) leads to a higher SOC mineralization than the lower quality (C/N>24) organic matter. This could be due to low quality V. karroo leaf compared to Z. mays stover classified as intermediate quality (Table 2) and associated with just a balance of immobilization and mineralization. This may have been caused by the aging of the V. karroo leaves, which were harvested at the beginning of winter period (May). However, the results agreed with [13], who observed that C/N ratio had no influence on different forest litter material and suggested other factors such as lignin/N ratio or secondary metabolites like polyphenols to influence decomposability of litter.

Soil organic carbon content in the soil associations significantly (P<0.05) decreased with time and was lowest at week 34 in all soil associations. The SOC was significantly (P<0.05) higher (1.56%) in the

Source of variation				
Between subjects		**SOC**	**SL**	**MBC**
Soil association (S)	$F_{4,62}$	14.41	558.4	112
	P	<0.0001	<0.0001	<0.0001
Litter (L)	$F_{3,62}$	168.76	650	8.9
	P	<0.0001	<0.0001	<0.0001
S × L	$F_{11,62}$	2.2656	46.8	5.8
	P	0.0253	<0.0001	0.0146
Within subject		**SOC**	**SL**	**MBC**
Time (T)	$F_{3,62}$	323.14	2360.6	240
	P	<0.0001	<0.0001	<0.0001
T × S	$F_{12,62}$	4.385	166.4	9.4
	P	<0.0001	<0.0001	<0.0001
T × L	$F_{9,62}$	4.7657	192.7	12.3
	P	<0.0001	<0.0001	<0.0001
T × S × L	$F_{36,62}$	1.5814	13.8	6.8
	P	0.0606^{ns}	<0.0001	<0.0001
ns: Not Significant at p=0.05				

Table 3: Repeated measures of Analysis of Variance (ANOVA) for Soil Organic Carbon (SOC), Soil Loss (SL) and Microbial Biomass Carbon (MBC) during the 34 weeks of incubation.

melanic at week 3 and lower (<0.6%) in the shallow, wet, semi-duplex, apedal and duplex at week 34 of incubation. At weeks 3 and 14, the SOC was significantly ($P<0.05$) the same in the shallow, wet and duplex soil associations (Figure 2).

The SOC content depends on the balance between C input and decomposition rates [33]. The general decrease in SOC content with

Soil association	Horizons	Litter × soil moisture	SOC (%)
Shallow	Orthic A	WD × Z. mays	1.06476[b]
		WD × V. karroo	0.99302[bc]
		W × Z. mays	0.71802[c]
		W × V. karroo	0.80671[c]
Wet	G-horizon	WD × Z. mays	1.03227[b]
		WD × V. karroo	1.02148[b]
		W × Z. mays	0.62905[c]
		W × V. karroo	0.63105[c]
Melanic	Melanic A	WD × Z. mays	1.44851[a]
		WD × V. karroo	1.48198[a]
		W × Z. mays	0.98719[bc]
		W × V. karroo	0.99271[bc]
Semi-duplex	Pedocutanic B	WD × Z. mays	1.30395[ab]
		WD × V. karroo	1.19019[ab]
		W × Z. mays	1.00040[bc]
		W × V. karroo	1.00490[bc]
Apedal	Red Apedal	WD × Z. mays	1.22038[ab]
		WD × V. karroo	1.30448[ab]
		W × Z. mays	0.80026[c]
		W × V. karroo	0.80036[c]
Duplex	Prismacutanic B	WD × Z. mays	1.01719[b]
		WD × V. karroo	1.00197[bc]
		W × Z. mays	0.81432[c]
		W × V. karroo	0.83170[c]

Table 4: Effects of litter source and soil moisture on the soil organic carbon content in the different areas of soil associations during the 34 week of incubation.

time of incubation could be due to the diminishing of the substrate (organic matter) due to microbial decomposition as there was no replacement. The higher SOC observed in the melanic at week 3 than in the other soil associations could be due to the higher pore volume due to more clay particles in the soils. Physical occlusion of SOM during microaggregates formation has been observed as the most important processes by which the SOM can circumvent decomposition. The exclusion of microbes and enzymes from pores is the key protection mechanism for occluded SOM in microaggregates. The pore volume at <0.1 μm Ø is about 21% of pore volume present in microaggregates [34]. The accessibility of microbes to the occluded SOM is restricted due to the ratio between dimensions of the microhabitat and the size of the organism. However, it is observed that higher diversity of microbial respiration in silt and clay (<0.002 mm Ø) fractions, which may be due to higher nutritional availability and exclusion from microbial decomposition [35].

The effect of litter source and soil moisture on the SOC content was significantly ($P<0.05$) the same at any incubation time. Same SOC content was observed under both continuously wet and alternating wet-dry soil moisture conditions at week 3 but the alternating wet-dry had more SOC content than continuously wet soil conditions in the subsequent weeks upto week 34 (Figure 3).

Estimated soil losses (t/ha) increased with time of incubation under both the continuously wet and alternating wet-dry soil conditions and were lowest and highest at week 3 and 34 of incubation respectively in all the soil associations. The soil losses were generally higher in the continuously wet (W × V. karroo and W × Z. mays) than in the alternating wet-dry (WD × V. karroo and WD × Z. mays) conditions in all the soil associations. The duplex and apedal had significantly ($P<0.05$) the highest estimated soil losses under all litter source and soil moisture treatments. The melanic had significantly ($P<0.05$) the least estimated soil losses compared to the other soil associations (Figure 4).

Microbial biomass during the incubation was measured by Microbial Biomass Carbon (MBC) (μg/g), and all the litter source and

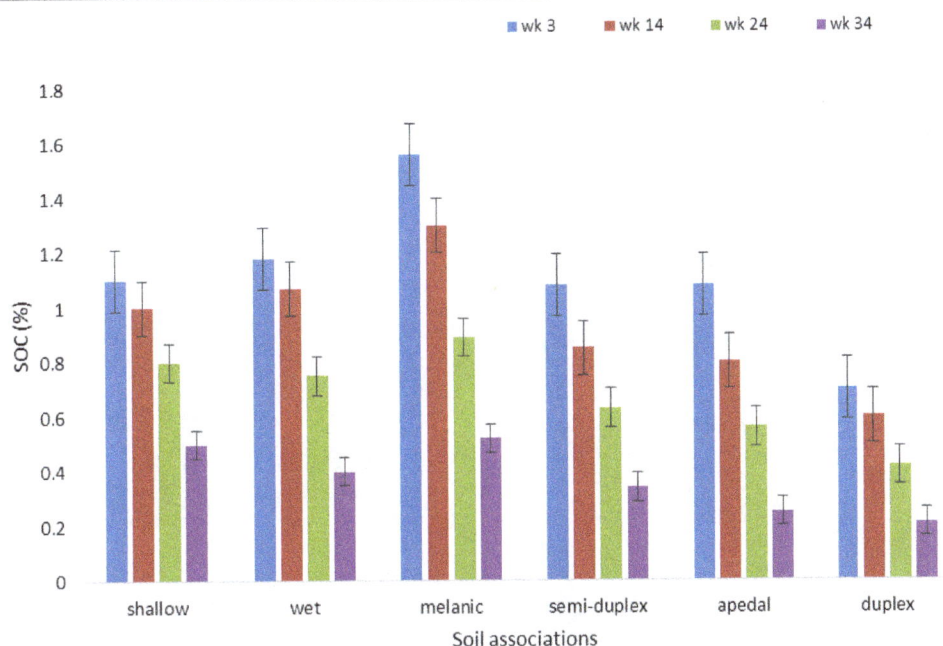

Figure 2: Trend of soil organic carbon content in the soil associations during the 34 weeks of incubation.

soil moisture treatment combinations significantly influenced changes in microbial biomass (P<0.05). The MBC (µg/g) varied according to the soil associations, decreased with time of incubation and ranged from highest at week 3 to lowest at week 34 of incubation in the shallow, wet and duplex, while was significantly (P<0.05) highest at week 24 and 34 in the melanic, semi-duplex and apedal soil associations. Same MBC was observed under both the continuously wet and alternating wet-dry soil moisture condition at weeks 3 and 14 but significantly (P<0.05) decreased under the continuously wet soils at weeks 24 and 34 of incubation in the shallow, wet and duplex (Figure 5).

The high contents of MBC generally indicate better soil quality. There is a general agreement that soil microbial activity has an optimal value when the soil is wet (but not saturated) and decreases at when the soil becomes dry. The reduction in microbial biomass carbon in alternating wet-dry soils observed as from week 24 of incubation could be mostly attributed to water and oxygen limitation of microbial activity: as matric suction increases soil water is held in pores inaccessible to microbes [36]. High microbial biomass carbon under the continuously wet soil moisture in the shallow, wet and duplex at week 3 and 14 may be ascribed to the presence of large soil pore spaces as the soil associations had higher sand content (Table 1) so the organic matter was readily available the soil microbes hence they proliferated. It is reported that MBC content has a positive relationship with soil moisture content [37]. The alternating wet-dry soil moisture could have resulted to reduction of the soil microbial action on the Organic Matter (OM) due to moisture and oxygen limitation during the dry phases therefore prolonged the resident time of the OM in the soils. In this study, the MBC was observed to be highly controlled by the availability of OM in the soil (Figure 5). In some studies dry soils generally has low SOC and that carbon mineralization is faster than accumulation in this case, indicating that soil moisture significantly influences SOC content and MBC (reference). This does agree to the findings in this study, as the wet-dry soil moisture did preserve the soil organic matter.

Estimated soil loss was significantly (P<0.05) inversely proportional to the SOC content, soil clay content and time of incubation in the soil associations. The estimated soil losses were positively correlated with the MBC (µg/g) however the MBC was significantly negatively correlated (P<0.05) with the SOC content and time of incubation and positively correlated with the quantities of clay particle in soil associations (Table 5).

The results of many studies showed a close correlation between MBC and SOC because most microorganisms are heterotrophic and their distribution and biological activity often depend on organic matter [30]. It is found that litter sources that easily decomposes could make the soil microbes to multiply rapidly and increase their activities, suggesting that the MBC content is effectively limited SOC [38]. These results agrees to a study by Wardle when he found that MBC was significantly positively correlated with SOC. Soil texture is an important soil physical parameter of soil structure, with higher clay particle content indicating well structured, resulting in more pore spaces and high porosity, which are related to increased soil moisture and MBC [39-41]. Increase clay content may indicate that SOM is more protected against soil microbial decomposition and this could explain the higher SOC content and MBC in the melanic, semi-duplex and apedal soil associations (Figures 2, 4 and 5).

	SL	SOC	MBC	Sand	Clay	Silt
SOC	-0.818***	-	-	-	-	-
MBC	0.673**	-0.738***	-	-	-	-
Sand	-0.31	-0.415	-0.302	-	-	-
Clay	-0.715***	0.623**	0.563**	0.008	-	-
Silt	0.421	0.392	0.043	0.21	-0.043	-
Time	-0.846***	-0.918***	-0.824***	0.031	-0.615**	0.31

SL: Stimated Soil Loss (t/ha); SOC: Soil Organic Carbon; MBC: Microbial Biomass Carbon; **Significant at P<0.05; ***Significant at P<0.001

Table 5: Pearson correlation matrix between soil loss (t/ha) and selected observed soil properties during the 34 weeks (time) of incubation.

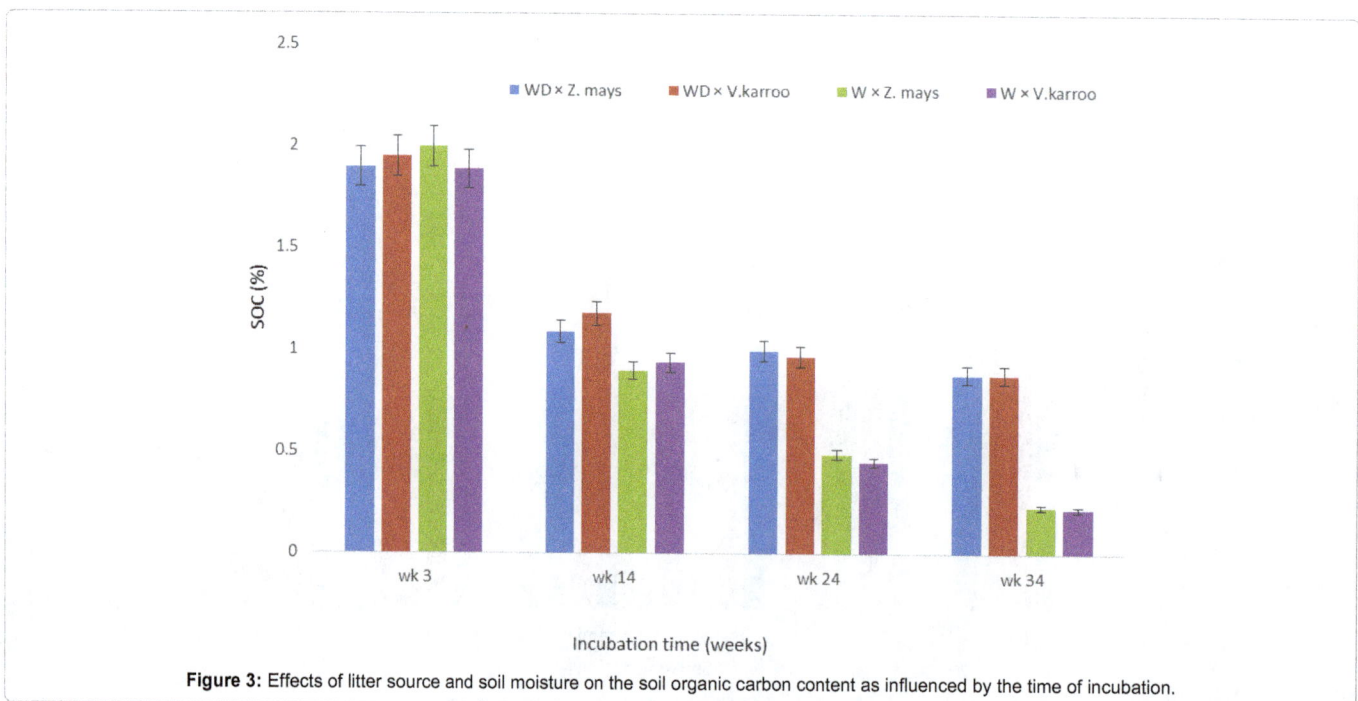

Figure 3: Effects of litter source and soil moisture on the soil organic carbon content as influenced by the time of incubation.

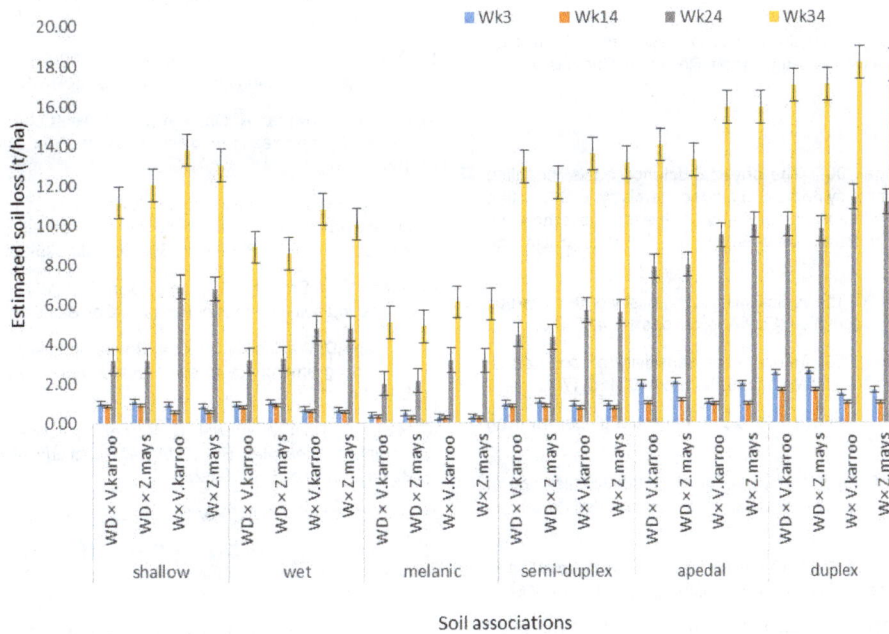

Figure 4: Estimated soil loss (t/ha) among the soil associations under different litter sources and soil moisture during the 34 weeks of incubation.

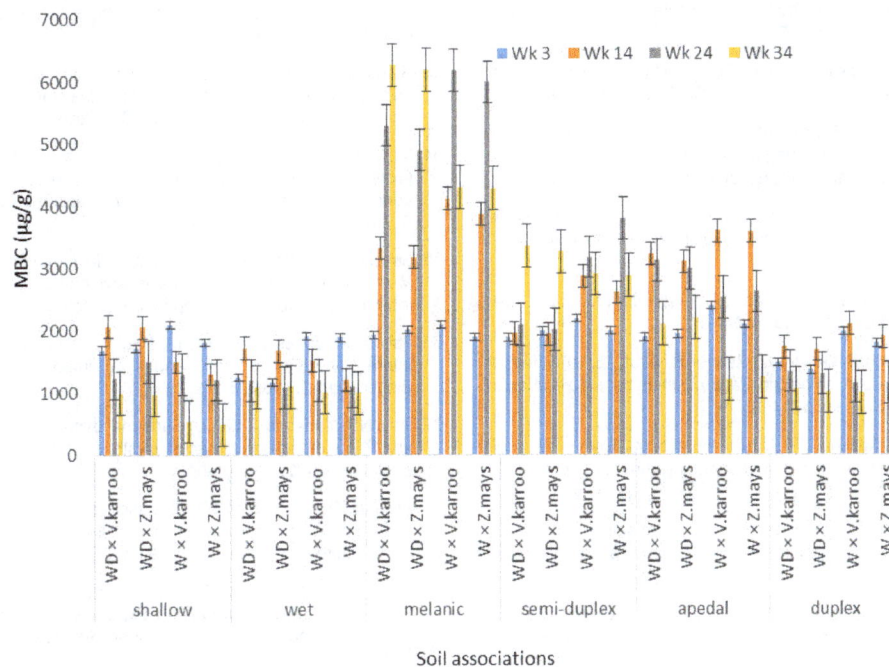

Figure 5: The microbial biomass carbon (μg/g) means during the 34 weeks of incubation.

Conclusion and Recommendations

The SOC content, MBC and soil loss in the soil associations were controlled by the soil texture and moisture content. Continuously wet sandy soils influenced more SOC loss, lower MBC and higher soil loss than soils higher in clay particles. The soil loss was proportional to the SOC and MBC at any time of incubation. Availability of OM under continuously wet sandy soil promoted rapid loss of SOC and soil from the soil associations. The alternating wet-dry moisture conditions showed to prolong the SOM resident time in all the soil associations though most SOM was conserved soils high in clay particles. Litter quality and soil moisture did not affect the SOC, MBC and soil loss within but affected across soil associations. Therefore addition of fast decomposing organic matter to the continuously wet sandy soils is likely to increase the climatic warming so has to be regulated. In conserving the soil from erosion, re-application of fresh organic matter has to be done after shorter times in the continuously wet sandy than under alternating wet-dry clay soil moisture regimes.

Acknowledgements

The authors gratefully acknowledge the Agricultural Research Council for funding received for the study as well as the Water Research Commission for financial support to the first author.

References

1. IPCC (2007) Climate change 2007. The physical science basis, coupling between changes in the climate systems and biogeochemistry. Contribution of working group 1 to the fourth assessment report of the intergovernmental panel on climate change. Cambridge university press: Cambridge, UK and NY, USA.

2. Jobbagy EG, Jackson RB (2000) The vertical distribution of soil organic carbon and its relation to climate and vegetation. Ecol Applications 10: 423-436.

3. Davidson EA, Janssens IA (2006) Temperature sensitivity of soil carbon decomposition and feedbacks to climate change. Nature 440: 165-173.

4. Smith P, Fang C, Dawson JJC, Moncrieff JB (2008) Impact of global warming on soil organic carbon. Adv Agron 97: 1-43.

5. Cox PM, Betts RA, Jones CD, Spall SA, Totterdell IJ (2000) Acceleration of global warming due to carbon-cycle feedbacks in a coupled climate model. Nature 408: 184-187.

6. Davidson EA, Janssens IA, Luo Y (2006) On the variability of respiration in terrestrial ecosystems: moving beyond Q_{10}. Global Change Biol 12: 154-164.

7. Curiel Yuste J, Baldocchi DD, Gershenson A, Goldstein A, Misson L, et al. (2007) Microbial soil respiration and its dependency on carbon inputs, soil temperature and moisture. Global Change Biol 13: 2018-2035.

8. Liu W, Zhang Z, Wan S (2009) Predominant role of water in regulating soil and microbial respiration and their responses to climate change in semiarid grassland. Global Change Biol 15: 184-195.

9. Guntinas ME, Gil-Sotres F, Leiros MC, Trasar-Cepeda C (2013) Sensitivity of soil respiration to moisture and temperature. J Soil Sci Plant Nutr 13: 445-461.

10. Chivenge P, Vanlauwe B, Gentile R, Six J (2011) Comparison of organic versus mineral resource effects on short-term aggregate carbon and nitrogen dynamics in a sandy soil versus a fine textured soil. Agric Ecosy Environ 140: 361-371.

11. Puttaso A, Vityakon P, Rasche F, Saenjan P, Trelo-ges V, et al. (2013) Does organic residue quality influence carbon stabilization in tropical sandy soil? Soil Sci Soc Am J 77: 1001-1011.

12. Fierer N, Jackson RB (2006) The diversity and biogeography of soil bacterial communities. Proc Natl Acad Sci USA 103: 626-631.

13. Jacobs A, Helfrich M, Hanisch S, Quendt U, Rauber R, et al. (2010) Effect of conventional and minimum tillage on physical and biochemical stabilization of soil organic matter. Bio Fert Soils 46: 671-680.

14. Potthast K, Hamer U, Makeschin F (2010) Impact of litter quality on mineralization processes in managed and abandoned pasture soils in Southern Ecuador. Soil Biol Biochem 42: 56-64.

15. Dick WA, Gregorich EG (2004) Developing and maintaining soil organic matter levels. In: Managing Soil Quality Challenges in Modern Agriculture. CABI Publishing, Wallingford.

16. Bhogal A, Nicholson FA, Chambers BJ (2009) Organic carbon additions: effects on soil bio-physical and physico-chemical properties. Eur J Soil Sci 60: 276-286.

17. Gentile R, Vanlauwe B, Six J (2011) Litter quality impacts short- but not long-term carbon dynamics in soil aggregate fractions. Ecol Applications 21: 695-703.

18. Cerda A (2000) Aggregate stability against water forces under different climates on agriculture land and scrubland in southern Bolivia. Soil Till Res 57: 159-166.

19. Van Tol JJ, Akpan W, Lange D, Bokuva C, Kanuka G, et al. (2014) Conceptualising long term monitoring to capture environmental, agricultural and socio-economic impacts of the Mzimvubu Water Project in the Tsitsa River. WRC project No: KV 328/14. Water Research Commission.

20. Van Tol JJ, Akpan W, Kanuka G, Ngesi S, Lange D (2014) Soil erosion and dam dividends: Science facts and rural 'fiction' around the Ntabelanga dam, Eastern Cape, South Africa. SA Geo J 98: 169-181.

21. Anderson JM, Ingram JJ (1993) Tropical soil biology and fertility (TSBF): A handbook of methods. Wallingford, UK, CAB International.

22. Hagerman AE (1988) Extraction of tannin from fresh and preserved leaves. J Chem Ecol 14: 453-461.

23. Okalebo JB, Gathua KW, Woomer PL (2000) Laboratory methods of soil and plant analysis: A Working Manual. TSBF-KARI-UNESCO, Nairobi, Kenya.

24. Chan KY, Bowman A, Oates A (2001) Oxidizible organic carbon fractions and soil quality changes in an oxic paleustalf under different pasture leys. Soil Sci 166: 61-67.

25. Joergensen RG, Emmerling C (2006) Methods for evaluating human impact on soil microorganisms based on their activity, biomass, and diversity in agricultural soils. J Plant Nutr Soil Sci 169: 295-309.

26. Vance ED, Brookes PC, Jenkinson DS (1987) An extraction method for measuring soil microbial biomass-C. Soil Biol Biochem 19: 703-707.

27. Nciizah AD, Wakindiki IIC (2014) Rainfall intensity effects on crusting and mode of seedling emergency in some quartz-dominated South African Soils. Water SA 40:4.

28. Martin C, Pohl M, Alewell C, Korner C, Rixen C (2010) Interrill erosion at disturbed alpine sites: Effects of plant functional diversity and vegetation cover. Basic Appl Ecol 11: 619-626.

29. SAS Institute Inc (2010) SAS campus drive. Cary, North Carolina, USA.

30. Yang Y, Guo J, Chen G (2009) Effects of forest conversion on soil labile organic carbon fractions and aggregate stability in subtropical China. Plant Soil 323: 153-162.

31. Mission L, Tang J, Xu M, McKay M, Goldstein A (2005) Influences of recovery from clear-cut, climate variability, and thinning on the carbon balance of a young ponderosa pine plantation. Agric Forest Met 130: 207-222.

32. Wang YF, Fu BJ, Lv YH, Song CJ, Luan Y (2010) Local-scale spatial variability of soil organic carbon and its stock in the hilly area of the Loess Plateau. China Quat Res 73: 70-76.

33. Huang J, Song C (2010) Effects of land use on soil water soluble organic C and microbial biomass C concentrations in the Sanjiang plain in Northeast China. Acta Agric Scand, Section B-Plant Soil Sci 60: 182-188.

34. McCarthy JF, Ilavsky J, Jastrow JD, Mayer LM, Perfect E, et al. (2008) Protection of organic carbon in soil microaggregates occur via restructuring of aggregate porosity and filling pores with accumulating organic matter. Geochi. Cosmoch. Acta 72: 4725-4744.

35. Sessitch A, Weilharter A, Gerzabek MH, Kirchmann H, Kandele E (2001) Microbial population structures in soil particle size fractions of a long-term fertilizer field experiment. App Environ Microbiol 67: 4215-4224.

36. Falloon P, Jones CD, Ades M, Paul K (2011) Direct soil moisture controls of future global soil carbon changes: An important source of uncertainty. Glob Biogeochem Cycles 25.

37. Devi NB, Yadava PS (2006) Seasonal dynamics in soil microbial biomass C, N and P in a mixed-oak forest ecosystem of Manipur, North-east India. Appl Soil Ecol 31: 220-227.

38. Landgraf D, Klose J (2002) Mobile and readily available C and N fractions and their relationship to microbial biomass and selected enzyme activities in a sandy soil under different management systems. J Plant Nutr Soil Sci 165: 9-16.

39. IUSS Working Group WRB (2015) World Reference Base for Soil Resources 2014, update 2015, International soil classification system for naming soils and creating legends for soil maps. World Soil Resources Reports No. 106. FAO, Rome.

40. Parwada C, Van Tol JJ (2016) Soil properties influencing erodibility of soils in the Ntabelanga area, Eastern Cape Province, South Africa. Acta Agr Scan Section B-Soil Plant Sci 67: 67-76.

41. Soil Classification Working Group (1991) Soil classification a taxonomic system for South Africa. Memoirs on the Agricultural Natural Resources of South Africa No. 15. Department of Agricultural Development, Pretoria.

29

Optimum Pattern of Compost used for Reducing Energy Consumption in Mushroom Production

Elham Hassanpour[1], Jamal-Ali Olfati[1]* and Mohammad Naqashzadegan[2]

[1]Faculty of Agriculture, Horticultural Department, University of Guilan, Rasht, Iran
[2]Faculty of Engineering, Mechanic Department, University of Guilan, Rasht, Iran

Abstract

In mushroom production it is necessary to design a growing pattern to balance yield against cost and to reduce energy use. Effects of use of multiple layer of compost in cultivation on energy consumption in mushroom cropping rooms were examined. Treatments included 1 layer (control), or 2, 3 or 4 layers of compost applied at the mycelium running stage. Numbers of compost layers did not affect fresh weight of mushrooms, number of mushrooms, yield and biological efficiency but hastened pinheads formation. Since there were no differences between layers of compost and control on yield and biological efficiency it appears that 2 layers will be sufficient to improve time to pinhead formation without negative effect on yield.

Keywords: *Agaricus bisporus*; Biological efficiency; Energy consumption; Yield

Introduction

Agaricus bisporus (Sing) is one of the most important cultivated, edible, mushrooms [1-4]. It contains high amounts of protein, minerals, vitamins D, K, B and sometimes A and C [4,5]. Button mushroom is a natural source of antioxidant agent against the free radicals superoxide radicals (O_2), hydroxyl radical (OH), hydrogen peroxide (H_2O_2) and lipid peroxide radicals and has potential as an anticancer factor [6].

Mushroom production can help contribute to grow food demand [7]. In Iran there are 1033 units for producing edible mushrooms, of which 704 are used for cultivating button mushroom [8] Iran Statistic Center. Button mushroom has vegetative and reproductive stages. Vegetative growth is on compost, or other substrates [4,9]. In the vegetative stage, temperature must be controlled, with the optimum temperature for mycelium growth being between 25°C and 28°C which require consumption of energy [10]. Agriculture and mushroom production (although it is only a small part of agriculture) is a large consumer of energy [11]. There are times when energy required for agriculture competes for that used by other activities [12]. Energy consumption in mushroom production has increased [13]. Energy consumption in Iran is higher than international standards [14].

It is necessary to design the optimum production pattern for mushrooms to conserve energy [15,16]. Moya et al. believed that energy consumption for heating relates to space among plants; densely spaced plants lead to lower energy consumption. There is no research in relation to reduction of energy consumption in mushroom production. In button mushroom, high yield depends on appropriate conditions of growth and the amount of energy used in cropping rooms needs to be considered. One layer of compost is normally used in button mushroom production. It may be that use of more than one layer of compost will affect growing conditions and consequently yield. This research was conducted to evaluate effects of compost layer number on cultivation on button mushroom development and biological efficiency and effects on energy consumption in spawn running rooms.

Materials and Methods

The experiment was conducted with a completely randomized design with 4 treatments and 3 replications in the faculty of Agriculture, University of Guilan, Rasht, Iran (37°16′ N, 51°3′ E). Treatments included 1 (control), or 2, 3 or 4 layers of compost.

Compost blocks (20 kg) were supplied by the Asian Mushroom Company, Karaj, Iran. Block length, width and height were 60, 40 and 20 cm, respectively. Fresh samples of the compost were oven dried at 75°C and dry weight determined. Temperature of compost blocks was measured at 9 AM each day during spawn running for 15 days. After completing spawn running, compost blocks were separated and casing to a 4 cm depth, applied [7] using commercial casing soil (Asian Mushroom Company). After casing, the substrate surface was covered with newspapers and sprayed with water to avoid drying. When the mycelium reached the surface of the casing layer the newspaper was removed and temperature in the room reduced from 25°C to 16°C to provide shocking to induce fruit body formation [17]. Time to pinhead formation was recorded. Mushroom yield was obtained from 2 flushes and mushrooms were harvested daily. During harvesting, numbers and fresh weight of mushrooms were determined. After the final harvest biological efficiency (BE) was determined. Yield and BE were determined according to formulae of Nogueira [18] de Andrade et al. Data were subjected to ANOVA in SAS (ver. 9.1, SAS Institute, Inc., Cary, N.C.). Means were separated with using Turkey's test.

Results

Changes of temperature during spawn running in compost treatments were similar (Figure 1). Temperature increased from day 1 to day 7 for all layers of compost. Temperature decreased, finally reaching 21°C and 23°C for 1 layer and 4 layers of compost, respectively; other

---

*Corresponding author: Olfati JA, Faculty of Agriculture, Horticultural Department, University of Guilan, Rasht, Iran
E-mail: jamalaliolfati@gmail.com

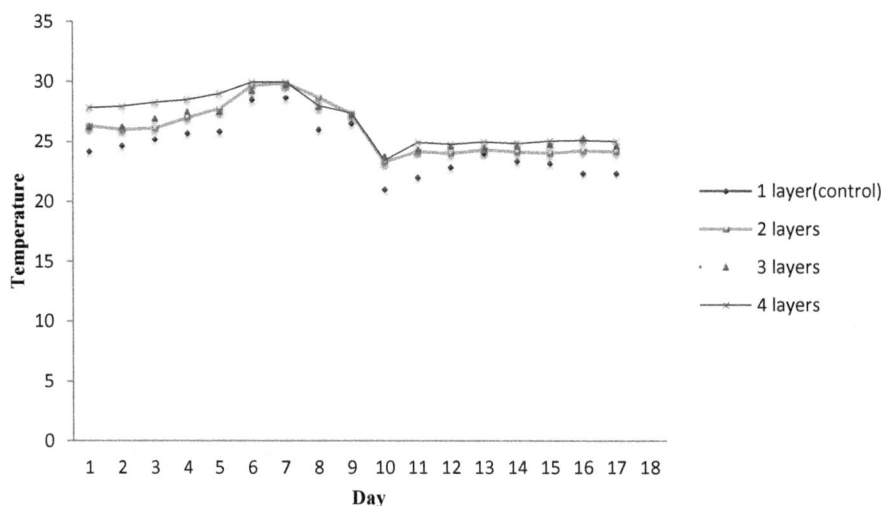

Figure 1: Temperature change during spawn running in treatments.

		Mean squares					
		Fresh weight			Number of mushroom		
Source of variation	df	Flush 1	Flush 2	total	Flush 1	Flush 2	total
Compost layer	3	164908.53 ns	1.88 ns	72692.22 ns	0.41 ns	2.81 ns	1591.16 ns
Error	8	66357.31	2.55	127106.7	0.54	1.49	831.96
CV (%)		18.6	20.94	18.54	18.92	9.55	17.44

ns = non-significant at P < 0.05.

Table 1: Effect of multi-layer cultivation on fresh weight and number of mushrooms in first, second and total flushes.

		Mean squares					
Source of variation	df	B.E. (%) Flush1	B.E. (%) Flush2	B.E. (%) Total	Y (%) Flush1	Y (%) Flush2	Days to pinhead formation
Compost layer	3	45.45 ns	0.19 ns	19.45 ns	4.10 ns	0.07 ns	13.92**
Error	8	17.59	0.32	33.63	1.66	0.07	0.38
CV (%)		18.51	10.16	18.51	18.61	19.19	4.69

ns, **, non-significant and significant at P < 0.01.

Table 2: Effect of multi-layer cultivation on biological efficiency (BE %), yield (Y %) and days to pinhead formation in first, second and total flushes.

treatments were intermediate. Increased temperature in first week may be due to reduced ventilation.

Numbers (average 163.72), fresh weight (average 1828.38 g/tray), and BE of mushrooms (29.79%) and yield (average 6.23) were not affected by treatment. Days to pinhead formation was affected by treatment (Table 1). One layer of compost (control) needed the most time for pinhead formation, 16.33 (average) days. The least time for pinhead formation was for the 3 layer treatment, 11.5 days. The multilayer treatments were similar to each other and less than for the control (Table 2).

Discussion

Growing mycelium produces heat and temperature increases during the first days of growth and then decreases [19]. Increasing layers of compost can increase temperature may be due to reduce ventilation and CO_2 exchange in running room. There was likely insufficient air between compost blocks so heat and CO_2 was retained [20]. Several layers of compost increases temperature and CO_2 stimulating mycelial growth during the mycelium running stage [10]. In other hand a positive correlation between the thermophiles population and amount of air supplied was recorded in previous research [21]. The reason for this needs additional study.

Optimum temperature for mycelium growth is 24°C-25°C and the maximum temperature is 28°C [7]. The temperature in the 3-4 layer treatment was at, or below this temperature. Appropriate ventilation results in moderate environment and CO_2 content stability in fruiting stage [10,19,20]. In compact growth substrate there is low CO_2, and respiration and growth decreases. Increased temperature increases evaporation and transpiration which results in decreased mycelial growth [20]. Growth and development of pathogens could increase because of anaerobic conditions [10].

Imbalanced mycelial running in substrate and weak, or inadequate, mycelial growth is the consequences of less than optimum compost temperature [7]. Temperature increasing to 30°C causes mycelial death and increased activity of microorganisms and insects *Megaselia nigra* M. and *Lycoriella auripila* W. feeding on mycelia [10]. However some authors Vedder and Hussain et al., [22,23] believe that for better mycelial growth average compost temperature should be 30°C and mycelium only die after a prolonged exposure to above 34°C.

Several layers of compost resulted in earlier pinhead appearance [24]. Evaluated this characteristic in different casing soil with maximum and minimum days to pinheads appearance being 23.75 and 12.25, respectively. Our results (maximum 16.33 days, minimum 11.5 days) were lower than theirs indicating multilayer cultivation reduced time for pinhead formation. Use of 2 layers of compost could be beneficial in mushroom cultivation.

Conclusion

Multilayer cultivation may result in consuming less energy for heating the cropping room without reduction in yield. In multilayer cultivation as a new idea on mushroom cultivation we put more compost in a mycelium running room and divide ready to fruiting compost in fruiting room. So only a translocation add to our cost while energy consumption for mycelium running room or cooling for shocking specially in warm season at least two fold decreased.

References

1. Gbolagade J, Ajayi A, Oku I, Wankasi D (2006) Nutritive value of common wild edible mushrooms from southern Nigeria. Global Journal Biotechnology Biochemistry 1: 16-21.

2. Toker H, Baysal E, Yigitbasi ON, Colak M, Peker H, et al. (2007) Cultivation of *Agaricus bisporus* on wheat straw and waste tea leaves based composts using poplar leaves as activator material. African Journal of Biotechnology 6: 204-212.

3. Mehta BK, Jain SK, Sharma GP, Doshi A, Jain HK (2011) Cultivation of button mushroom and its processing: An techno-economic feasibility. International Journal of Advanced Biotechnology and Research 2: 201-207.

4. Ebadi A, Alikhani HA, Rashtbari M (2012) Effect of plant growth promoting bacteria (PGPR) on the morphophysiological properties of Button mushroom *Agaricus bisporus* in two different culturing beds. International Research Journal of Applied and Basic Sciences 3: 203-212.

5. Saiqa S, Nawaz BH, Asif HM (2008) Studies on chemical composition and nutritive evaluation of wild edible mushrooms. Iran Journal Chemichal Engineering 27: 151-156.

6. Abah SE, Abah G (2010) Antimicrobial and antioxidant potentials of *Agaricus bisporus*. Advances in Biological Research 4: 277-282.

7. Farsi M, Pouriyanfar HR (2011) Cultivation and breeding of the white button mushroom. Mashhad Jihad University Press, Mashhad, Iran.

8. Iran Statistic Center (2012) Statistical pocketbook of the Islamic republic of Iran.

9. Noble R, Dobrovin-Pennington A, Hobbs P, Rodger A, Pederby J (2009) Volatile C8 compounds and pseudomonas influence primordium formation of *Agaricus bisporus*. Mycologia 101: 583-591.

10. Jafarniya S, Khosroshahi M, Karami SM (2010) Mushroom cultivation: appropriate technology for mushroom growers. Misagh Press, Tehran, Iran.

11. Alam MS, Alam MR, Islam KK (2005) Energy flow in agriculture: Bangladesh. American Journal Environmental Science 1: 213-220.

12. Pahlavan R, Omod M, Akram A (2012) Application of data envelopment analysis for performance assessment and energy efficiency improvement opportunities in greenhouse cucumber production. Journal of Agricultural Science and Technology 14: 1465-1475.

13. Mehrabi H, Esmaili A (2011) Analysis energy input-output in Agriculture in Iran. Agriculture Economic and Development Journal 74: 12-18.

14. Kazemi A, Mehregan MR, Shakouri H, Hosseinzadeh M (2012) Energy resource allocation in Iran: A fuzzy multi-objective analysis. Procedia-Social and Behavioral Sciences 41: 334-341.

15. Faizi F, Nourani M, Ghaedi AK, Mahdavinejad MJ (2011) Design an optimum pattern of orientation in residential complexes by analyzing the level of energy consumption. Proceeding Engineering 21: 1179-1187.

16. Moya A, Mehlitz T, Yildiz I, Kelly SF, Hardin C (2008) Simulated effects of dynamic row spacing on energy and water conservation in semi-arid central California greenhouses. Department of bio-resource and agricultural and mechanical engineering. Polytechnic State University, Marietta, CA.

17. Femor TR, Randle PE, Smith JF (1985) Compost as a substrate and its preparation. In: FLEGG PB, SPENCER DM, WOO DA: The biology and technology of the cultivated mushroom. Wiley, Chichester, UK, pp: 81-109.

18. De Andrade M, Kopytowski J, Minhoni M, Coutinho L, Figueiredo M (2007) Productivity, biological efficiency and number of *Agaricus blazei* mushrooms grown in compost in the presence of *Trichoderma* sp. and *Chaetominum olivacearum* contaminants. Brazilian Journal of Microbiology 38: 243-247.

19. Kashi A (2005) Edible mushroom cultivation. Agricultural Education Press, Karaj, Iran.

20. Peyvast GH, Olfati JA (2005) Improved cultivation of edible mushrooms. Daneshpazir Press, Tehran, Iran.

21. Wakchaure GC, Meena KK, Choudhary RI, Singh M, Yandigeri MS (2013) An improvedrapid composting procedure enhance the substrate quality and yield of *Agaricus bisporus*. African Journal of Agricultural Research 8: 4523-4536.

22. Hussain S, Ali MA, Ahsan A, Ali H, Siddique M (2004) Temperature requirement of button mushroom Agaricus bitorquis for mycelia growth on three different composts. Journal of agricultural research 42: 3-4.

23. Yilmaz F, Baysal E, Toker H, Colak M, Yigitbasi O, Simsek H (2007) An investigation on pinhead formation time of *Agaricus bisporus* on wheat straw and waste tea leaves based composts using some locally available peat materials and secondary casing materials. African Journal of Biotechnology 6: 1655-1664.

30

Identification of an Indigenous Atrazine Herbicide Tolerant Microbial Consortium in Beans (*Phaseolus vulgaris* L.) as a Potential Soil Bioremediator

Margarita Islas-Pelcastre[1], Jose Roberto Villagómez-Ibarra[2], Blanca Rosa Rodríguez-Pastrana[3], Gregory Perry[4] and Alfredo Madariaga-Navarrete[5]*

[1]Agronomy and Forestry Area, Institute of Agricultural Sciences, Universidad Autónoma del Estado de Hidalgo Av, Universidad Km 1, Ex-Hacienda de Aquetzalpa, Tulancingo, Hidalgo, Mexico

[2]Cademic Area of Chemistry. Institute of Basic Science and Engineering, Universidad Autónoma del Estado de Hidalgo, Ciudad del Conocimiento, Carretera Pachuca Tulancingo, Mineral de la Reforma, Hidalgo, Mexico

[3]Agrobusiness and Food Engineering Area. Institute of Agricultural Sciences, Universidad Autónoma del Estado de Hidalgo Av, Universidad Km 1. Ex-Hacienda de Aquetzalpa, Tulancingo, Hidalgo, Mexico

[4]Department of Plant Agriculture, Ontario Agricultural College, University of Guelph, Guelph, Ontario, Canada

[5]Agronomy and Forestry Area, Institute of Agricultural Sciences, Universidad Autónoma del Estado de Hidalgo, Av, Universidad Km 1, Ex-Hacienda de Aquetzalpa, Tulancingo, Hidalgo, Mexico

Abstract

The present article reports the isolation and identification of atrazine-tolerant strains of indigenous microorganisms recovered from three representative agricultural sites representing agronomic characteristics of the Tulancingo Valley, Central part of México (disturbed and undisturbed). Biochemical and morphological tests were performed for microorganism's identification and the minimum inhibitory concentration assay was followed to assess atrazine tolerance. Results showed the microorganism populations varied from 10^{-5} to 10^{-6} UFC g^{-1} of soil for bacteria and 10^4 - 10^5 conidia g^{-1} of soil for fungi. The bacterial genera isolated and identified were: *Agrobacterium* sp., *Bacillus* sp., *Erwinia* sp., *Micrococcus* sp., *Pediococcus* sp., *Rhizobium* sp., *Serrantia* sp. and *Sphingomonas* sp. Identified fungal genera were: *Alternaria* sp., *Aspergillus* sp., *Mucor* sp., *Cladosporium* sp., *Penicillium* sp., *Fusarium* sp. and *Trichoderma* sp. Tests for herbicide tolerance indicate the isolated microorganisms do not show inhibitory growth at 500 to 2,500 ppm of atrazine concentrations under laboratory conditions. The strains of the fungal genera and *Rhizobium* sp. showed greater tolerance rates to atrazine, based on their growth without inhibition in the presence of 5,000 to 10,000 ppm of the agrochemical. Results suggest the isolated microorganisms may be useful as a viable inoculum for bioremediation purposes in agricultural atrazine-contaminated soils.

Keywords: Agricultural sustainability; Soil recuperation; Agrochemicals

Introduction

Soil is a complex and sensitive biomaterial strongly affected by plant-human management, especially agriculture activities. Agriculture requirements include the use of herbicides which results in negative impact to the environment due to contamination by chemicals. Atrazine (2-chloro-4-ethylamino-6-isopropylamino-1,3,5-triazine), one of the most used herbicides worldwide, causes disruption to biogeochemical cycles and biodiversity as a result of the permanency of its soil residue [1]. Therefore, it is necessary to apply friendly technologies for remediation and / or recovery of soils. A biological low cost alternative is the in situ bio augmentation of native microorganisms. This technology consists of inoculating microorganisms with physiological, biochemical and molecular ability to absorb, retain, degrade or transform contaminants into less harmful substances. These microorganisms must be recovered from the same soil that is expected to be restored, in order for these microorganisms to be adapted to the agro ecosystem, and to enhance the efficiency of their degradative ability. Atrazine is classified by its biological action as herbicide and is widely used to kill or inhibit the growth of plants considered to be weeds. In 2006, the application of the herbicide atrazine ranged from 29 to 34 million of kilograms of active ingredient per year in agricultural soils [2]; while between 85% and 100% of agricultural soils in industrialized countries are treated with herbicides.

Atrazine is a systemic, selective and residual herbicide, with

persistent effects for a period of time. It is transported to environmental processes spheres by volatilization or leaching diffusion. It is considered as one of the pollutants responsible for negatively affecting mammalian reproductive systems, carcinogenicity, teratogenicity and ecotoxicity. Evy and Nilanjana [3] report atrazine effects including changes in aquatic species, plants and mammals. Due to its organizational structure, atrazine can experience natural or induced degradation processes such as: (i) hydrolysing and (ii) oxidative - hydrolytic, two very well studied degradation mechanisms. Reports [4-8] showed that atrazine degradation can be achieved by isolated microbial consortia species in combination with other bioremediation techniques.

The primary limiting factor for atrazine mineralization is the absence of specific microorganisms. However, Sene et al. [6];

*Corresponding author: Alfredo Madariaga-Navarrete, Agronomy and Forestry Area, Institute of Agricultural Sciences, Universidad Autónoma del Estado de Hidalgo Av, Universidad Km 1, Ex-Hacienda de Aquetzalpa, Tulancingo, Hidalgo, Mexico, E-mail: alfredomadariaga60@gmail.com

Spaczynski et al. [7]; and Dehghani et al. [8] suggest that rhizosphere bioremediation by native microorganisms has been efficiently achieved by atrazine elimination in soil, when these organisms were isolated from contaminated sites where they grow. And because these native microorganisms grow in contaminated conditions, they may adapt, and develop tolerance, even resistance to such environments, resulting in the generation of metabolic processes that transform the chemical structure of the contaminant into simpler compounds.

This research aims to isolate and identify indigenous microorganisms in the atrazine-contaminated agricultural soils from the Tulancingo Valley, Hidalgo, Mexico and to quantify their herbicide tolerance.

Materials and Methods

Study area

The study area (agricultural land) belongs to the Tulancingo Valley, Hidalgo, Mexico (central), with polygon coordinates: 20° 10' 29" north latitude, 98° 16' 52" west longitude, 19° 57' 20" north latitude and 98° 15' 59" west longitude. Dominated by a humid temperate climate (Cw), the average annual temperature is 14.5°C, with a maximum 30°C from March to July, and a minimum temperature of 3°C from August to February. Precipitation, in the form of rain, is recorded during the months of March through to September, with a minimum of 33.3 mm h^{-1} and a maximum of 190.7 mm h^{-1} with average between 500-553 mm per year [9].

Sampling

Conditions for selection of the sampling site were as follows: (i) Soils with a record of continuous pesticide application and (ii) soils with a record of no pesticide application. Sampling of soils was performed three times during the annual cycle of the legume crop. The sampling depth was 0 cm - 25 cm using a tubular auger (7 cm diameter). Total sample weight was 1 kg.

For sampling collection, the procedures described in the Mexican Official Standard Norm AS-01 method [10] for transport, storage and methods of microbiological, physical and chemical soil analysis were followed. Soil samples for microbiological analysis was transported in labeled sterile plastic bags (Nasco - Whirl - Pak'), to prevent cross contamination. Each of the samples was thoroughly mixed and sieved (2 mm), and then stored at 4°C for physical chemical analysis.

Physical and chemical analytical methods in experimental soils

The physical properties determined in the experimental soils were: available moisture, density, field capacity, moisture, texture and wilting point. Chemical properties determined were: cation exchange capacity, total nitrogen, organic carbon, organic matter and pH. All physical and chemical analytical methods were performed according to the procedures established in the Mexican Official Standard Norm [10].

Diversity of microorganisms in the rhizosphere of legume *Phaseolus vulgaris* L.

In order to determine biodiversity and viability of bacteria and fungi in the representative soils, the following were determined: (i) total score in bacteria (as described in the Mexican Official Standard Norm [11] and (ii) fungi presence, as described by the Waksman method [12]. Six working dilutions (10^{-1} to 10^{-6}) were used in each case. Surface plating with three replications was performed. In order to determine a

statistical record of measurement in time for soil conditions (with and without atrazine application), quantification of microorganisms was performed consecutively for three weeks from the soil sampling date.

Selection and identification of microorganism strains

The used selection criteria and identification of native strains of rhizosphere bacteria and fungi were the comparison of the microorganism growth by soil sample. After the incubation period, a count of viable cells was performed, and the time of growth recorded. Macroscopic evidence for selection of isolates on soil extract agar for bacteria included: (i) colony appearance, (ii) elevation, (iii) shape, (iv) turnover margin or (v) color. Microscopic identification was performed by differential Gram stain.

In fungi, the morphological identification of strains included: (i) thalli color, (ii) thalli texture (iii) growth form (length and radius), (iv) appreciation of the hyphae and (v) the appearance, shape and color of conidia.

The fungal strains were examined by microscopy micro culture (Riddell method) [13] after using the staining technique for micro fixed cultivation of lacto phenol blue with the use of photomicrographs taken through a Motic BA300 microscope, objectives contrast bright field and coupled to a docked digital 480 Moticam camera (software Motic Images Plus 2.0).

Gross observation of bacteria was conducted at 24, 48, 72, 92 and 132 hour intervals after isolation; while for fungi, changes along the fungal morphology life cycle were recorded.

The strains of candidate microorganisms (fungi and bacteria) were purified by consecutive plating (spline surface). Bacteria AES were purified in solid medium and fungal strains by a hyphal needling technique in potato dextrose agar (PDA) (Merk Millipore') and then incubated. This procedure was repeated four times until purification of the isolated strains was obtained; and finally, the inbred strains were stored in inclined test tube for subsequent identification.

To identify the genera of the strains, macroscopic and microscopic description of each one were completed after each biochemical tests; these tests were performed using the Analytical Profile Index System (API tests, bioMérieux'). API tests include: motility, catalase enzyme production, oxidase (cytochrome c oxidase production), NO_3 (nitrate reductase and nitrite reductase enzymes production), tryptophan deaminase enzyme prodcution, glucose fermentation (hexose sugar), lactose (lactose fermentation as a carbon source), urease (urease enzyme production), xylose (xylose fermentation as a carbon source), gelatinase enzyme production (which liquefies gelatin), mannose fermentation (hexose sugar), mannitol (mannitol fermentation as a carbon source), maltose (maltose fermentation as a carbon source), citrate (use of citrate as a carbon source) and starch (α-amylase and oligo-1, 6-glucosidase enzymes production).

In fungi, all strains were examined by microscopy microculture with the use of photomicrographs. Images were extrapolated with the results of lacto phenol blue staining, the reported bibliography, as well as morphological criteria [14].

Bioassays of inhibition of growth tolerance

Atrazine herbicide tolerance was determined with the use of the minimum inhibitory concentrations (MIC) assays performed on eight bacterial strains, seven fungal strains and three unidentified fungal genera using the qualitative disc diffusion method [15].

Fungi and bacteria were plated in 9 cm petri dishes containing 20 mL of potato dextrose agar (PDA) solid culture media (Sigma - Aldrich™) while Mannitol agar (MA) media were used for *Rhizobium* sp. Strains of the identified genera were inoculated with the use of the surface extension technique where sterile Whatman paper discs of 0.5 cm diameter were placed on the agar, equidistant and near the border using vernier caliper. Paper discs impregnated with 10 μL of standard solution of atrazine were used. Tested concentrations were: 0, 500, 1000, 1,500, 2,000, 2,500, 5,000, 7,500 and 10,000 mg L⁻¹. A repetition for each treatment was used. The plates were incubated at 25°C for 5 to 7 days until the growth in the control plates reached the edge of the plates; then, after the incubation time, the inhibition growth zone was measured using an electronic micrometer (IP54) and recorded.

The inhibition percentage of fungal and bacterial growth was calculated by using equations 1 and 2:

$$\% \text{ growth} = \left(\frac{\phi atz}{\phi b} \right) \times 100 \ \% \tag{1}$$

$$\% \text{ CI} = 100 - \% \ growth \tag{2}$$

Where Øatz represents the diameter (mm) of growth of the microorganism in the treatments exposed to atrazine at each assessment. Øb is the diameter of the negative control microorganism growth in each evaluation and % CI is the percentage of inhibitory growth.

The growth rate of the microorganism inhibition was calculated with the use of equation number 3:

$$V = \frac{\phi C}{t} \tag{3}$$

Where V is the speed of growth, ØC represents the diameter (mm) of the microorganism growth (mm) and t is the incubation time (days).

Statistical analysis

The experimental design used for the physical and chemical characterization of soils was completely random. Each sample had three replicates. Analysis of variance (ANOVA) and subsequently Tukey test p > 0.05 were performed in order to assess significant differences.

Result and Discussion

Physical and chemical parameters of the soil samples

The values of the physical and chemical parameters found in the soil samples collected at the three study sites are reported in (Table 1). The pH value for the soil samples was classified as moderately acid for two of the sites (Capulin and Santana) and neutral in the Tepantitla sample. At these intervals, nutrients and contaminants in the soil solution are more available [16]. Also, at these pH values, the rhizosphere microorganisms as bacteria and fungi can grow without a significant reduction of their population.

The values for total nitrogen classified samples at intervals of high (0.15 - 0.25%) and very high (> 0.25), according to Mexican Official Standard Norm [10]. For the organic matter (OM), content of organic carbon (OC), the rule states that values between 4.1% - 6.0% of OM indicate low fertility. According to other reports such as Rodriguez and Rodriguez [17] the amount of organic matter and organic carbon is classified as high and very high (> 3.5%).

Finally, based on the texture (proportion of soil particles: Clay, Silt and Sand) samples are classified as loam and sandy loam [10]. The soil

with a clay texture was that of Capulin. The main feature of the sandy loam soils is their low ability to retain nutrients and water, which can be important for soil agrochemical interactions.

Santana and Tepantitla medium texture soils were classified as loam; while Capuline's as clay texture. Medium-textured or loam soil types are ideal for agricultural production, due to their extensive production capacity and water availability. This soil can be considered versatile for its high cation exchange capacity, which varies in the range of 20 to 45 cmol kg⁻¹ (Table 1).

Comparison tests (Tukey 0.05) of the physical and chemical properties showed significant differences with the soil of Santana, as compared to the other two soil sites analyzed. Taking into account that properties are not only important for the soil fertility and genesis characterization, but also because retention capacity can be strongly affected by them, having an effect on the adsorption and degradation of atrazine rates in the soil [18,19].

Based on a references search [20] and the resulting data, it is estimated that the three sites are classified with good agricultural fertility. The soil sample from the Santana site (coming from an intensive agricultural area, with regular pesticides application) could be vulnerable to deterioration and short-term negative environmental impacts.

Diversity of microorganisms in the rhizosphere of *Phaseolus vulgaris* L.

The results and conditions of the total bacterial count at all sites are shown in (Figure 1). An interval of 1.09 × 10⁵ to 2.0 × 10⁶ CFU g⁻¹ soil was obtained. A higher bacterial population was found in the rhizospheric soils, compared to the population found in the non-rhizospheric soils.

The bacterial population in the Santana soil (all three conditions of the site) is in the same range of that found in the undisturbed soils (no atrazine application), and the total bacteria count is similar to that reported by Sangabriel et al. [21]. The soil bacteria population in the rhizosphere of *Phaseolus coccineus* cultured on fuel oil - contaminated and uncontaminated conditions ranged in 1.7 × 10⁶ CFU g⁻¹. Sangabriel et al. [21] also observed a total increase in the bacteria population after a phytoremediation process using *Phaseolus coccineus*.

Other references [22-24] cite locations of rhizospheric soils treated with legume crops for bioremediation processes with a total count of bacteria ranging from 10⁵ to 10⁶ CFU g⁻¹. These results are similar to the ones found in this experiment, regardless of the condition of the site, hydrology, geographic location or textural class. For fungi, it was observed that the population number (total count) in studied soils is

	Capulín	Tepantitla	Santana	Standard Error
% Moisture	14.89ₐ	15.86ₐ	15.20_b	0.2
pH (1:2 in water)	6.35ₐ	6.69ₐ	5.78_b	0.12
% Total Nitrogen	0.18ₐ	0.24_b	0.30_c	0.0094
% Organic Carbon	2.86ₐ	3.36_b	4.08_c	0.012
% Organic Matter	4.94ₐ	5.79_b	4.04_c	0.0086
CEC (Cmol (+) kg⁻¹)	29.34ₐ	59.63_b	25.46_c	3.63
Density (g cm⁻³)	2.38ₐ	2.43ₐ	2.29ₐ	0.11
Texture	Sandy loam	loam	loam	

For each line, different letters indicate significant differences (α = 0.05, Tukey).

Table 1: Comparison of means for physical and chemical properties of the soils studied.

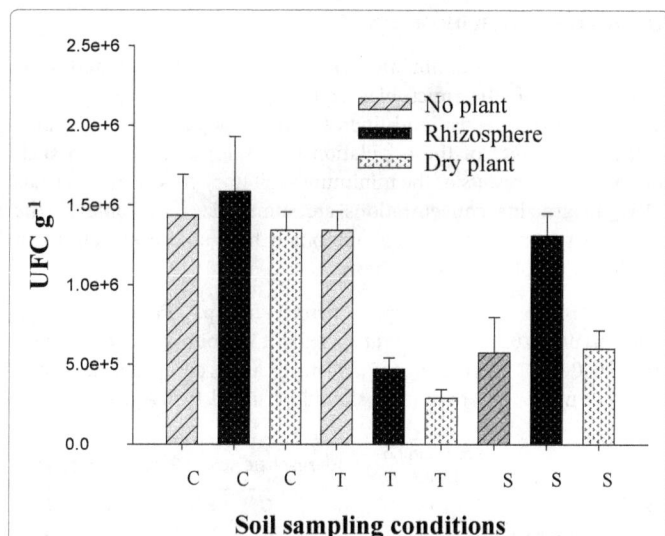

Figure 1: Total count of aerobic bacteria (CFU g⁻¹) of the three agricultural soils of the Tulancingo Valley, Hidalgo, Mexico. C (Capulin soils), T (Tepantitla soils), S (Santana soils). Error bars indicate the standard error of the mean of triplicate replicates.

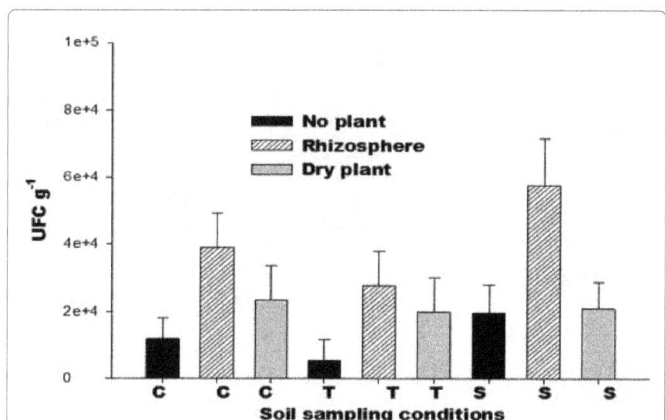

Figure 2: Fungi total account (CFU g⁻¹) of the three agricultural soils from Tulancingo Valley, Hidalgo, Mexico. Where C corresponds to the sample of the Capulin soil, T corresponds to Tepantitla soil and S correspond to Santana soil. Error bars indicate the standard error of the mean of triplicate replicates.

also higher in the rhizosphere area at rates rhizosphere / soil (R / S) of 2:1, in all of the three sites (Figure 2): Capulín (C), Tepantitla (T) and Santana (S).

The proportion of conidia per gram of soil (CFU g⁻¹) reported here was found within the estimated range considered by Maier and Pepper [24]. They reported CFU g⁻¹ ranging from 10^5 to 10^6 in the rhizhosphere. Also, Calvo et al. [22] and Córdova et al. [25] reported ranges of 10^3 CFU g⁻¹ soil, in non- rhizospheric soils in the study area.

Selection and identification of bacterial and fungal strains

The selected fungi strains and bacteria were isolated specifically from the rhizosphere of Santana soil (disturbed) in order to consider the issues addressed by Robinson [26], who relates the genetic resistance of microorganisms in wild systems to their ability to self-regulate and adapt to contaminated (disturbed) environment. Tables 2 and 3 present a summary of the identified organisms to genus level.

As for the identification of *Rhizobium* sp., isolated from nodules of the bean plant, phenological characteristic of the bean plant observed where the nodes extracted to isolate the bacterium *Rhizobium* sp. Results showed an average of 3.2 knots, 3.8 inflorescences and 11.47 cm root length in all the selected plants, the average was 23 peripheral root nodules, pink appearance, different sizes, average diameter of 0.34 cm and a volume of 0.10 mL (in 1 mL conical vial).

The microscopic observation of fresh bean nodules revealed the existence of living organisms in a bacillus form, beige color, translucent and with a mucilaginous consistency. Average Bacillus length was 109.85 μm and Gram negative under a fresh staining. Comparison of morphological and biochemical tests proved that the isolated bean nodule is the bacterial genus *Rhizobium* sp. (Table 3). Similar results were reported by Acosta and Martinez [27] and Richardson et al. [28]. In these reports, Rhizobium strains were isolated from nodules of legumes growing in pesticide contaminated soils. Also, *Agrobacterium* sp., *Bacillus* sp., *Rhizobium* sp., *Sphingomonas* sp., *Micrococcus* sp. and *Serrantia* sp. have been reported for their degradative capacity of organic pollutants, including the herbicide atrazine [4,29,30].

Based in the macro, microscopic and biochemical evidence, eight bacteria genera were identified in the soils, as follows: *Agrobacterium* sp., *Bacillus* sp., *Erwinia* sp., *Micrococcus* sp., *Pediococcus* sp., *Rhizobium*

Genera				
	Bacillus sp.	*Micrococcus* sp.	*Serrantia* sp.	*Sphingomonas* sp.
Macroscopic description				
	Colonia aspect viscous. Circular convex elevation. Entire margin. Color: reddish-brown, slimy	Circular aspect of the colony. Flat. Yellow color. Creamy. Viscose.	Circular form of the colony. Convex elevation. Brown to reddish color. Viscose.	Circular convex colony. Circular orange. Flat. Entire margin. Yellow
				Colonies circular in entire dome margin. Creamy. Opaque beige
Microscopic description				
	Gram negative Bacillus	Gram positive Cocus	Gram negative Bacillus	Gram negative Bacillus
Biochemical tests				
Catalase	+	+	+	+
Citrate	-	-	NA	Negative
Gelatine	+	+	NA	Positive
Glucose	NA	Positivo	+	Negative
Glucose OF	NA	NA	F	Oxidative
Lactose	-	+	-	NA
Maltose	+	+	NA	Negative
Mannitol	NA	+	+	Negative
Manose	+	+	NA	Negative
Mobility	+	++	+	+
NO₃	NA	NA*	+	Negative
Oxidase	NA	-	-	+
Starch	-	NA	NA	NA
Tryptophan	NA	NA		Negative
Urea	+	-	NA	Negative
Xilose	+	+	NA	NA

Note + = positive; - = negative
• NA = Not Applicable.

Table 2: Results of macroscopic, microscopic and biochemical tests of isolated bacterial genera (part I).

Genera				
	Agrobacterium sp.	*Erwinia* sp.	*Pediococcus* sp.	*Rhizobium* sp.
Macroscopic description				
	Circular form of the colony. High grow. White to pale brown color. Viscous consistency.	Sharply colony form. Convex margin. Yellow color.	Circular form of the colony. Lisa. Viscous consistency. Bright yellow color.	Circular colony form. Convex. White to pale brown color. Translucent. Mucilaginous aspect.
Microscopic description				
	Gram negative Bacillus	Gram negative Bacillus	Cocus gram positive	Gram negative Bacillus
Biochemical tests				
Catalase	+	-	-	+
Citrate	NA	NA	+	NA
Gelatine	NA	-	+	NA
Glucosa OF	F	NA	NA	Oxidative
Glucose	Positive	-	NA	-
Hugyleftson test	-	NA	NA	Oxidative
Lactose	NA	+	+	+
Maltose	NA	NA	+	+
Manitol	NA	NA	+	+
Manose	A	NA	+	NA
Mobility	+	+	+	+
NO₃	-	NA	NA	NA
Oxidase	+	NA*	NA	+
Starch	NA	NA	-	NA
Trypthophan		NA	NA	+
Urea	+	NA	+	NA
Xilose	+	NA	+	NA
YMA media (bromo timol blue)	-	NA	NA	+

Note + = positive; - = negative
* NA = Not Applicable

Table 3: Results of macroscopic observations, microscopic and biochemical evidence of isolated bacterial genera (part II).

sp., *Serrantia* sp. and *Sphingomonas* sp. For fungi, seven genera and 23 candidate species were identified and recorded as follows: *Alternaria* (one), *Aspergillus* (one), *Cladosporium* (one), *Fusarium* (three), *Mucor* (two), *Penicillium* (seven) and *Trichoderma* (five). Three unidentified fungal strains were also observed (Table 4).

The frequency of the microorganisms identified by soil condition, highlight the *Micrococcus* sp. and *Bacillus* sp., presence. It can be inferred then, that the microorganism are native and present under different conditions and able to persist over the agricultural cycle of the crop. *Rhizobium* sp. is present in a lower frequency; it was identified only in rhizospheric soil and was isolated from de nodules of the bean plant. Therefore, the condition of the soil (soils without plants) seems to be a limiting factor and responsible for the lower frequency of this bacterium.

Results in terms of frequency and types of fungal genera identified in this study are similar to studies from Ibiene et al., Gopi et al., Martinez et al., and Maldonado et al. [30-33], who report the genus Aspergillus, Fusarium, Mucor, Penicillium and Trichoderma to be present in metals, pesticides and hydrocarbure contaminated soils.

Growth inhibition bioassays

Exposure to contaminants can cause the reduction and / or modification of the microbial soil population. However, some microorganisms persist and increase their population. This may indicate the ability of the population to develop and adapt to such conditions. The results of the minimum inhibitory concentration trials (MIC) to atrazine concentrations are summarized in Table 5. The identified genera grew without inhibition halo at concentrations of 0-2,500 mg L^{-1} of atrazine.

Bacillus sp., *Erwinia* sp., *Micrococcus* sp., *Pediococcus* sp., *Sphingomonas* sp., and *Serrantia* sp. exhibit inhibitory growth at rates up to 5,000 mg L^{-1}, increasing to 7,500 mg L^{-1} and is completely inhibited at 10,000 mg L^{-1} of atrazine. It should be noted that Agrobacterium

Genera	Soils with no plants			Rhizosphere soils			Soils with dry plants		
	T	C	S	T	C	S	T	C	S
Agrobacterium sp.	+	-	+	+	+	+	+	+	+
Alternaria	+	+	+	+	+	+	+	+	+
Aspergillus sp.	+	+	+	+	+	+	+	+	+
Bacillus sp.	+	+	+	+	+	+	+	+	+
Cladosporium	+	+	+	+	+	+	+	+	+
Erwinia sp.	+	-	+	+	+	+	+	+	+
Fusarium sp. 1	+	+	+	+	+	+	+	+	+
Fusarium sp. 2	+	+	+	+	+	+	+	+	+
Fusarium sp. 3	+	+	+	+	+	+	+	+	+
Micrococcus sp.	+	+	+	+	+	+	+	+	+
Mucor sp. 1	+	+	+	+	+	+	+	+	+
Mucor sp. 2	+	+	+	+	+	+	+	+	+
Pediococcus sp	-	+	+	+	+	+	+	+	+
Penicillium sp. 1	+	+	+	+	+	+	+	+	+
Penicillium sp. 2	+	+	+	+	+	+	+	+	+
Penicillium sp. 3	+	+	+	+	+	+	+	+	+
Penicillium sp. 4	+	+	+	+	+	+	+	+	+
Penicillium sp. 5	+	+	+	+	+	+	+	+	+
Penicillium sp. 6	+	+	+	+	+	+	+	+	+
Penicillium sp. 7	+	+	+	+	+	+	+	+	+
Rhizobium sp.	-	-	-	+	+	+	+	+	+
Serrantia sp.	+	+	+	+	+	+	+	-	+
Sphingomonas sp.	+	-	+	+	+	+	+	+	+
Trichoderma 1	+	+	+	+	+	+	+	+	+
Trichoderma 2	+	+	+	+	+	+	+	+	+
Trichoderma 3	+	+	+	+	+	+	+	+	+
Trichoderma 4	+	+	+	+	+	+	+	+	+
Trichoderma 5	+	+	+	+	+	+	+	+	+
Unidentified 1	+	+	+	+	+	+	+	+	+
Unidentified 2	+	+	+	+	+	+	+	+	+
Unidentified 3	+	+	+	+	+	+	+	+	+
Trichoderma 1	+	+	+	+	+	+	+	+	+
Trichoderma 2	+	+	+	+	+	+	+	+	+
Trichoderma 3	+	+	+	+	+	+	+	+	+
Trichoderma 4	+	+	+	+	+	+	+	+	+
Trichoderma 5	+	+	+	+	+	+	+	+	+
Unidentified 1	+	+	+	+	+	+	+	+	+
Unidentified 2	+	+	+	+	+	+	+	+	+
Unidentified 3	+	+	+	+	+	+	+	+	+

T: Tempantitla, C: Capulín, S: Santana

Table 4: Diversity of microorganisms isolated and identified by soil condition.

Minimum Inhibitory Growth Concentration (mg L⁻¹)						
Genera	0-2500	5000	7500	10000	Rise time	Growth rate
	% CI*	% CI	% CI	% CI	(days)	(mm day ⁻¹)
Agrobacterium	0	0	6.25	13.75	6	13.33
Alternaria	0	0	0	0	7	11.42
Aspergillus sp.	0	0	0	0	6	13.33
Bacillus sp.	0	13.75	18.75	100	5	16
Cladosporium	0	0	0	0	5	16
Erwinia sp.	0	12.5	20	100	3	26.66
Fusarium sp. 1	0	0	0	0	5	16
Fusarium sp. 2	0	0	0	0	5	16
Fusarium sp. 3	0	0	0	0	5	16
Micrococcus sp.	0	5	13.75	100	3	26.66
Mucor sp. 1	0	0	0	0	7	11.42
Mucor sp. 2	0	0	0	0	7	11.42
Pediococcus sp.	0	16.25	18.75	100	5	16
Penicillium 1	0	0	0	0	7	11.42
Penicillium 2	0	0	0	0	7	11.42
Penicillium 3	0	0	0	0	7	11.42
Penicillium 4	0	0	0	0	7	11.42
Penicillium sp. 5	0	0	0	0	7	11.42
Penicillium sp. 6	0	0	0	0	7	11.42
Rhizobium sp.	0	0	0	2.5	4	11.42
Serrantia sp	0	6.25	16.25	100	4	20
Sphingomonas	0	8.75	15	100	4	20
Trichoderma 1	0	0	0	0	5	16
Trichoderma 2	0	0	0	0	5	16
Trichoderma 3	0	0	0	0	5	16
Trichoderma 4	0	0	0	0	5	16
Trichoderma 5	0	0	0	0	5	16
Unidentified 1	0	0	0	0	6	13.33
Unidentified 2	0	0	0	0	6	13.33
Unidentified 3	0	0	0	0	6	13.33

Table 5: Minimum inhibitory growth concentrations MIC (%) of atrazine for genera identified.

showed the lowest growth percentage at 7,500 mg L⁻¹, while *Rhizobium* sp. showed the lowest zone of inhibition in the presence of 10, 000 mg L⁻¹ of atrazine.

The seven fungal genera (23 possible species) identified showed no growth inhibition halos of mycelium at 10,000 mgL⁻¹ atrazine concentrations. These results evidence those genera isolates to be tolerant to high concentrations of the chemical. Assays relating minimum inhibitory concentration and type of microorganisms yield good results in terms of the number of species and tolerance to the herbicide. This last is compared to other studies [6,20,34,35]. Bacteria identified are inhibited to a greater extent from 5,000 mg L⁻¹ of atrazine, except *Rhizobium* sp. Moreover, the fungal genera show no growth inhibition at any atrazine concentrations. Of the 15 organisms identified, 74.18% presented no growth inhibition at 10,000 mg L⁻¹ of atrazine concentration. Only 3.22% of bacteria (*Agrobacterium sp.* and *Rhizobium sp.*) grew at this concentration and 22.58% (*Bacillus sp., Erwinia sp., Micrococcus sp., Pediococcus sp., Serrantia sp., Sphingomonas sp.*) showed growth inhibition (Table 5). To the best of our knowledge, there are only a few references describing indigenous microorganisms with high tolerance to atrazine. Regularly, tolerance and / or resistance are induced by genetic modification [3,36].

There are reports of the use of effective bioremediation methods

of atrazine removal. Moreover, the genetic pathways for atrazine detoxification have been well characterized in bacteria, fungi and plants. Along with this, transgenic microbes and plants expressing atrazine degrading enzymes have also been used. A combination of transgenic plants and microbes have been proposed as a combined synergistic method for atrazine bioremediation [37]. However, the gap between bioremediation laboratory conditions, green house bioremediation trials and an effective real soil cleanliness (or large scale soil decontamination) is still under discus [38]. In order to keep research study cost low, no molecular biology techniques for microorganism identification were performed. Further studies will benefit for the use of such techniques.

Conclusion

Organochlorine pesticides, including atrazine, are the most common used herbicide / pesticide in agriculture and forestry for disease and insect control. These compounds have been linked to a various human health hazards such as cancer, reproductive harm and even birth defects. Because of their residual effects, the molecules can remain in the soil and cause water contamination. Many methods for soil bioremediation have been proposed, but some of them seem not to be cost efficient or tend to be difficult to achieve since it is strongly influenced by natural conditions. Then, it is necessary to find ways to reduce variability of results, which can be achieved by adapting the technology to the local native environment. Intensive research needs to be done to discover the biological activity for natural local resources, including microorganisms.

Therefore, this study has resulted in the isolation and identification of indigenous microorganisms with high tolerance to atrazine concentrations, distributed as follows: (i) eight bacterial genera (*Agrobacterium* sp., *Bacillus* sp., *Erwinia* sp., *Micrococcus* sp. *Pediococcus* sp. *Rhizobium* sp., *Sphingomonas* sp., and *Serrantia* sp.), (ii) seven fungal genera (Alternaria, Aspergillus, Cladosporium, Fusarium, Mucor, Penicillium and Trichoderma) and (iii) three unidentified fungal strains. The microorganisms are native, rhizospheric and specific to the bean plant. The consortium identified genera did not show growth inhibition to 2,500 mg L⁻¹ of atrazine concentration assays. In particular, *Rhizobium* sp. and fungal strains showed no growth inhibition to 10,000 mg L⁻¹ atrazine concentration. These microorganisms may be suitable to be used in atrazine agricultural contaminated soils bioremediation. Further studies may include molecular biology techniques for microbe identification and atrazine biodegradation mechanisms.

Acknowledgement

Authors would like to express their gratitude to the National Council for Science and Technology (Consejo Nacional de Ciencia y Tecnología, CONACyT, Mexico) by funding a doctoral scholarship and the funded project call 84961 from Basic Science - 2007. Also, the manuscript revision by Miss Mary Rioux is highly appreciated.

References

1. Udeigwe TK, Teboh JM, Eze PN, Stietiya MH, Kumar V, et al. (2015) Implications of leading crop production practices on environmental quality and human health. Journal of Environmental Management 151: 267-279.

2. Joo H, Choi K, Hodgson H (2010) Human metabolism of atrazine. Pesticide Biochemistry and Physiology 98: 73-79.

3. Evy AAM, Nilanjana D (2012) Microbial Degradation of Atrazine, Commonly Used Herbicide. International Journal of Advanced Biological Research 2: 16-23.

4. Smith D, Alvey S, Crowley ED (2005) Cooperative catabolic pathways with in an atrazine-degrading enrichment culture isolated from soil. Microbiology Ecology 53: 265-273.

5. Lin C, Lerch R, Garrett HG (2008) Bioremediation of Atrazine - Contaminated Soil by Forage Grasses. Transformation, uptake and Detoxification. Journal of Environmental Quality 37: 196-206.

6. Sene L, Converti A, Ribeiro G, Cassia R (2010) New Aspects on Atrazine Biodegradation. Brazilian Archives of Biology and Technology an International Journal 53: 487-496.

7. Spaczynski M, Seta KA, Patrzylas P, Betlej A, Skórzyñska P (2012) Phyto degradation and Biodegradation in rhizosphere as efficient methods of reclamation of soil contaminated by organic chemicals (a Review). Agrophysica Act 19: 155-169.

8. Dehghani M, Nasseri S, Hashemi H (2013) Study of the Bioremediation of Atrazine under Variable Carbon and Nitrogen Sources by Mixed Bacterial Consortium Isolated from Corn Field Soil in Fars Province of Iran. Journal of Environmental and Public Health 1: 1-7.

9. Nacional Weather System (2011).

10. NOM- 021 -2000 - RECNAT (2000) Mexican Official Standard, which sets specifications fertility, salinity and soil classification Studies, sampling and analysis.

11. NOM -112- SSA1- 1994 (1994) Mexican Official Standard, which provides goods and services. Coliform bacteria Determination Most probable number technique.

12. Waksman S (1922) A method for counting the number of fungi in the soil. J Bacteriol 7: 339-341.

13. Riddell RW (1950) Permanent strained mycological preparation obtained by slide culture. Micología 82: 265-270.

14. Barnet HL, Hunter B (2006) General Illustratrated of imperfect fungi, Apss Press. Minnesota, USA.

15. Bauer AW, Kirby W, Sherris JC, Turck M (1966) Antibiotic susceptibility testing by a standardized single disk method. American Journal of Clinical Pathology 45: 493-496.

16. Rinnan R, Michelsen A, Baath E, Jonasson S (2007) Minaralization and carbon turnover in sub-arctic heath soil as affected by warming and additional litter. Soil Biology and Biochemistry 39: 3014-3023.

17. Rodriguez H, Rodriguez J (2002) Methods of soil and plant analysis, interpretation criteria. D.F Mexico.

18. Bridges M, Brien HW, Shaner DL, Khosla R, Westra P, et al. (2008) Spatial Variability of Atrazine and Metolachlor Dissipation on Dry land No-tillage Crop Fields in Colorado. Journal of Environmental Quality 37: 2212-2220.

19. Raymundo E, Nikolskii I, Duwig C, Prado B, Hidalgo C, et al. (2009) Transport of atrazine in andosol and versitol of Mexico. Interscience 34: 330-337.

20. Monard C, Martin LF, Vecchiato C, Francez AJ, Vandenkoornhuyse P, et al. (2008) Combined effect of bioaugmentation and bioturbation on atrazine degradation in soil. Soil Biology & Biochemistry 40: 2253-2259.

21. Sangabriel W, Ferrera CR, Trejo AD, Mendoza L, López, OC, et al. (2006) Tolerance and phytoremediation capacity of fuel oil in the ground for six plant species. International Journal of Environmental Pollution 22: 63-73.

22. Calvo VP, Reymundo ML, Zuniga DD (2008) Study of microbial populations

from the rhizosphere of potato crop (*Solanum tuberosum*) in highlands. Applied Ecology 7: 141-148.

23. Córdova BY, Rivera CM, Ferrera CR, Obrador OJ (2009) Detection of beneficial bacteria in soil with banana cultivar 'Grande naine' and its potential to integrate an biofertilizer. And Science University, Humid Tropics 25: 253-265.

24. Ferrera C, Alarcón A, Mendoza R, Sangabriel W, Trejo A, et al. (2007) Phytoremediation of contaminated fuel oil using Phaseolus coccineus and soil organic and inorganic fertilization. Agrociencia 41: 817-826.

25. Córdova BY, Rivera CM, Ferrera CR, Obrador OJ (2009) Detection of beneficial bacteria in soil with banana cultivar 'Grande naine' and its potential to integrate an biofertilizer. And Science University, Humid Tropics 25: 253-265.

26. Robinson RA (1996) Return to Resistance: Breeding Crops to Reduce Pesticide Dependence. Ag Access. Davis, CA. United States of America.

27. Acosta DC, Martinez RE (2002) Diversity of rhizobia from nodulates of the leguminous trees *Gliricidia sepium*, natural host of Rhizobium tropici. Archives of Microbiology 178: 161-164.

28. Richardson JS, Hynes MF, Oresnik IJ (2004) A genetic locus Necessary for rhamnose uptake and catabolism in *Rhizobium leguminosarum* bv. *trifolii*. Journal of Bacteriology 186: 8433-8442.

29. Vásquez MC, Guerrero J, Quintero A (2010) Bioremediation of contaminated sludge used lubricating oils. Colombian Journal of Biotechnology 12: 141-157.

30. Maldonado CE, Rivera C, Left RF, Palma L (2010) Effects of rhizosphere microorganisms and fertilization in bioremediation and phytoremediation of soils and weathered crude oils again. Humid Tropics 26: 121-136.

31. Ibiene AA, Orji FA, Ezidi CO, Ngwobia CL (2011) Bioremediation of hydrocarbon contaminated soil in the Niger Delta using spent mushroom compost and other organic wastes. Nigerian Journal of Agriculture, Food and Environment 7: 1-7.

32. Gopi V, Upgade A, Soundararajan N (2012) Bioremediation potential of single and consortium non-adapted fungal strains on Azo dye textile container containing effluent. Advances in Applied Science Research 3: 303-311.

33. Martinez TMA, García RM (2012) Applications of immobilized microorganism's environmentalists. Mexican Journal of Chemical Engineering 11: 55-73.

34. Mondragón P, Ruiz ON, Talbya GA, Juárez RC, Curiel QE, et al. (2008) Chemostat selection of a bacterial community able to degrade s - triazinic compounds: continuous simazine biodegradation in a multi -stage packed. Journal of Industrial Microbiology and Biotechnology 35: 767-776.

35. Getenga Z, Dörfler U, Iwobi A, Schmid M, Schroll R (2009) Atrazine and terbuthylazine mineralization by an *Arthrobacter* sp. isolated from a sugarcane - cultivated soil in Kenya. Chemosphere 77: 534-539.

36. Prasad M, Katiyar SC (2010) Drill cuttings and fluids of fossil fuel exploration in North -Eastern India: environmental concern and mitigation options. Current Science 98: 1566-156.

37. Kang JW (2014) Removing environmental organic pollutants with bioremediation and phytoremediation. Biotechnology Letters 36: 1129-1139.

38. Fan XX, Song FQ (2014) Bioremediation of atrazine: recent advances and promises. Journal of Soils and Sediments 14: 1727-1737.

31

Phytoextraction of Cadmium from Contaminated Soil Assisted By Microbial Biopolymers

Jian PU* and Kensuke Fukushi

Integrated Research System for Sustainability Science , The University of Tokyo, Japan

Abstract

Heavy metal contaminated soils remain as a challenging and essential task for environmental engineering. Phytoremediation (plant-based remediation) is effective for the mitigation of large area surface soil contamination, but needs long time and the efficiency is not high. Chemical agents could increase heavy metal bioavailability in soil and bring greater accumulation in plants, but also pose risks to soil, plant growth and ground water environment. In this study, microbial biopolymers, mainly composed of protein and polysaccharide, were obtained from non-induced, copper-induced and cadmium-induced activated sludge culture, and named as ASBP, ASBPCu and ASBPCd, respectively. The influence of microbial biopolymers on phytoextraction of cadmium in contaminated soil was investigated. Microbial biopolymers, compared to other agents, were found to be more effective in improving the phytoextraction of cadmium from soil. In ASBP, ASBPCu and ASBPCd, the cadmium content in plants was found to be 1.52, 1.63 and 1.33 µg (1.9, 2.0 and 1.6 times of the control), respectively. It was also found that in the presence of microbial biopolymers ASBP, ASBPCu and ASBPCd, 10.9%, 26.2% and 13.7% of exchangeable cadmium fraction was extracted from soil matrix to plant or liquid, higher than the control test (4.3%). Microbial biopolymers were more effective in improving cadmium accumulation in plants than other chemical agents. Owing to the benign nature, ease of production, and cadmium binding feasibility, microbial biopolymers may find utility as a new environmentally safe extracting agent for improving phytoextraction of cadmium from contaminated soil.

Keywords: Microbial biopolymers; Phytoextraction; Soil; Heavy metal; Cadmium

Introduction

Heavy metals contamination in soil is of major concern for both developed and developing countries, because of its potential toxicity and high persistence in the environment. The toxic metals, including copper, lead, cadmium, mercury, gold, and so on, are those that will displace essential metal ions in biological processes [1]. They affect crop yields, soil biomass and fertility, contributing to bioaccumulation in the food chain, eventually accumulating in human bodies and generating serious health threat to human and animals [2]. Over the course of recent decades, industrial and agricultural activities have accelerated dramatically metal pollution in different environmental compartments, especially in soil [3]. In soil samples collected up to 20 km in each direction from the Kabwe mine, the cadmium concentration ranged between 0.08 and 28 mg/kg for only the fractions of metals extractable by 0.5 M nitric acid and that could be available for plant uptake in the environment [4]. In 2004, one study in Japan estimated that an arable land area of 7327 hectares (0.16% of the total arable land) was polluted by heavy metals (Cd, Cu and/or As), in which cadmium contamination was observed across 92.6% of land [5]. Various in situ and ex situ remediation methods have been employed for restoration of soils contaminated with heavy metals, including physical/chemical/biological techniques [6,7]. Phytoremediation has attracted much attention because it is environmental friendly and relatively cheap. Phytoremediation uses plants to extract, sequester and/or detoxify heavy metals and other pollutants, and/or accumulate in different parts of them [8,9]. Phytoextraction is the removal of metals from water and soil and concentration into plants parts [10-13]. It usually needs long time and the efficiency is not high. Moreover, a large proportion of metal contaminants are unavailable for root uptake by plants. Thus extracting agents were applied to improve phytoextraction of heavy metals from contaminated soil [14]. When Ethylenediaminetetraacetic acid (EDTA) was added to soil, it forms soluble complexes with metals

from soil, mainly from the exchangeable fraction, organic matter and carbonate-bound fractions [15]. Despite of the stimulation of heavy metal accumulation in plants, the addition of chemicals as extracting agents can also inhibit the uptake of some major elements for plants and pose an additional threat of soil quality and groundwater. Therefore, natural agents seem to be more promising because they are economically acceptable and environmentally benign. Researchers suggested that biomass from biological pollution control processes, especially from activated sludge systems, could be effective in removing heavy metals from polluted waters [16,17]. The activated-sludge bacteria produce extracellular biopolymers to protect themselves from the outer environment with heavy metals [18-20]. Microbial biopolymers produced by bacteria are composed of polysaccharides, proteins, nucleic acids, etc. In many studies, proteins were found to be the main component with a 4 to 5 protein/carbohydrate ratio [21-23]. The proteinaceous biopolymers are considered to be economical and were reported to play an important role in removing heavy metals from aqueous solution [16]. In the other hand, plants release root exudates (eg. amino acid) containing biopolymers with the potential to enhance cadmium uptake, translocation and resistance [24]. However, the concentration of the naturally excreted biopolymers is low. In this paper, high concentration of microbial biopolymers would be produced and used for cadmium removal by plant from contaminated soil. We demonstrate the feasibility of using microbial biopolymers to enhance

*Corresponding author: Jian PU, Integrated Research System for Sustainability Science, The University of Tokyo, Japan, E-mail: pu@ir3s.u tokyo.ac.jp

the removal of cadmium from contaminated soil by phytoextraction. The results presented here may pave the way for the use of microbial biopolymers in phytoextraction of heavy metals in contaminated soils.

Experimental Methods

Soil characterization and preparation

The black soil (pH (H$_2$O)=5.0, pH (KCl)=4.5, TC=146.4 g/kg) is a widely used garden soil in Japan. The black soil was artificially contaminated by soaking autoclaved soil in cadmium nitrate solution at neutral pH condition. Soil samples were then rinsed recovered and dried until the weight was reduced to less than 5%. Dried cadmium-contaminated soil sample was then stored in a closed plastic container for experimental use.

Plant for phytoextraction

In the genus of Crassula, *Crassula helmsii* was reported to be a Cu-accumulator [25], and *Crassula portulacea* was found to have great efficiency to remove benzene from air. Moreover, *Crassula alba* was recorded to be a local indicator and hyperaccumulator for cobalt (>1000 μg/g dry mass). Referring to the accumulation ability of *Crassula* family for pollutants, *Crassula lycopodioides v. pseudolycopodioides* was used in this study, in order to investigate the overall uptake of cadmium by the plant.

Activated sludge cultivation

A sludge seed of MLSS about 4.2 g/l from water treatment plant was used for cultivation. The batch reactors were fed with different synthetic media: Non-selective medium (NSM); Feed 1: NSM with 2 mg/l copper as inducer; Feed 2: NSM with 2 mg/l cadmium as inducer. Seed sludge was introduced into different culture media initially, and cultivated at 120 rpm in 25°C thermostat for 24 h. After that, the 1st generation was harvested and 1 ml mixture was introduced to a series of fresh media for cultivating next generation. To ensure stable composition of the bacterial fauna, this sub-culture process was carried out for several times in the fresh media, generally, for 24 h. Finally, the media were collected as the resource of biopolymer extraction.

Preparation of water-soluble biopolymers

Sonication, detergent and freeze-thaw were used to obtain a full disruption of cells in cultivated sludge. Each culture liquid was centrifuged (10,000 g, 5 min, 4°C) and washed with water prior to extraction. The pellet was re-suspended in 20 mM Tris-HCl buffer (pH=8). Lysozyme was added to reach 0.4 mg ml-1 and incubate for 20 min. Then, the sludge suspension then went through freeze-thaw for 3 cycles at -80°C and 30°C, respectively. The suspension was sonicated for 15 min with 50% burst at 170 W, 20 kHz. The tubes containing the samples were kept in crushed ice during sonication. After that, the cell mass suspensions were centrifugated at 10,000 g for 30 min at 4°C. Supernatants were collected and filtered through 0.45 μm membrane. The filtrates containing water-soluble biopolymers were used for following tests. The water-soluble biopolymers extracted from modified activated sludge were named "Activated Sludge Bio-Polymers" (ASBP), ASBPCu and ASBPCd. ASBP was derived using Non-Selective Media (NSM) as the control, ASBPCu was from NSM with trace copper as biopolymer inducer, and ASBPCd was from NSM with trace cadmium as biopolymer inducer. Lowry method [26] was used for protein quantification, with Bovine Albumin Serum (BSA) as standard.

Measurement analysis of biopolymers

An ultra filtration device was used to separate the biopolymer-bound

cadmium from aqueous solution and determine the concentrations of cadmium complex formed in the aqueous biopolymers. The mixed liquor was introduced to 3 kDa cut-off using Amicon ultra-4 3K device (Millipore) and washed with 1 mM PBS buffer for 3 times. Subsequent determination of the metal content in the filtrate was carried out by FAAS.

Microbial biopolymers-cadmium extraction studies

Batch experiments were carried out using 1 g of soil with a soil-solution (w/v) ratio of 1:5 at different pH condition. Low and high concentrations of extracting agents (including EDTA, citric acid and biopolymer solutions) were added to the contaminated soil and vigorously mixed with a mechanical shaker at 40 rpm for 24 hours. The pH value in the mixture of soil and extracting solutions were adjusted using HNO$_3$ or NaOH. The concentrations of proteins in biopolymers, used for cadmium extraction, were determined before and after its application in soil. Moreover, after the extraction, the biopolymers were separated by ultra filtration with a cut-off at 3 kDa. The Cadmium concentration and protein concentrations in both fractions (with a MW>3 kDa and <3 kDa) were determined. Comparative experiments were carried out using different concentrations of EDTA (0.005 mM, 0.05 mM and 1 mM) and 1 mM of citric acid. Water was used as the control.

Effects of biopolymers in phytoextraction of cadmium from soil

The soil was planted with Crassula *lycopodioides v. pseudolycopodioides* in test tubes. The soil sample (14.9 mg Cd/kg dried soil) was submerged in an aqueous solution of biopolymers (ASBP, ASBPCu and ASBPCd) and other agents (BSA, EDTA, and citric acid) to ensure a uniform contact between agents and soil matrix. EDTA concentrations were 5.0×10^{-3} mM and 0.4 mM. The concentration of citric acid and BSA was 1.7 mM and 3.0×10^{-2} mM, respectively. The biopolymers used were 2493, 2350, and 2768 μg/ml (8.3×10^{-2}, 7.8×10^{-2}, and 9.2×10^{-2} mM) for ASBP, ASBPCu and ASBPCd, respectively. The tubes without chelating agents and those without cadmium were set as control. Five replicates of test tubes, each contained one plant, were used for one batch of experiment. The plants were incubated at 25°C with a 12 h photoperiod in the greenhouse, with pump aeration once a day. Nutrient medium was added regularly to maintain the constant water level. After 3 weeks of incubation, the plant as well as soil samples were collected from the solution. Both protein and cadmium concentrations were determined in the aqueous phase of solutions. The cadmium concentration in plant was analyzed with FAAS after microwave digestion. The soil samples were analyzed for cadmium species, using a modified sequential extraction method.

Results and Discussion

Characteristics of biopolymers

The molecular weight (MW) of activated sludge biopolymers was analyzed by sodium dodecyl sulfate polyacrylamide gel electrophoresis (SDS-PAGE) with Coomassie brilliant blue (CBB) R250 staining, see Figure 1. A broad range of MW was found in ASBP, as summarized in Table 1. About 21.4%, 18.7% and 20.6% had MW over 50 kDa (3.9%, 3.1% and 3.1% over 100 kDa) and 10.5%, 7.9% and 7.6% below 5 kDa, for ASBP, ASBPCu and ASBPCd, respectively. The mean MW was 31 kDa, 30 kDa and 31 kDa for ASBP, ASBPCu and ASBPCd, respectively. The results of MW distribution obtained by this measurement were considerably lower than the 200 kDa reported for extracellular

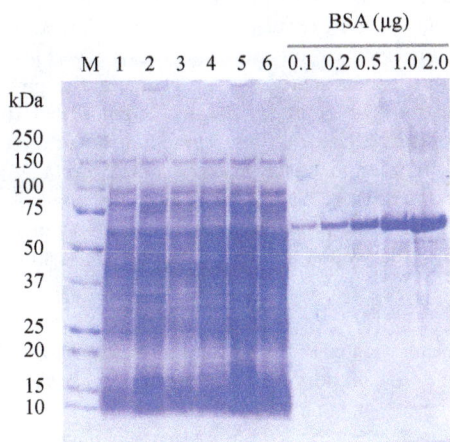

Figure 1: ASBP analysis by sodium dodecyl sulfate polyacrylamide gel electrophoresis (SDS-PAGE) with Coomassie brilliant blue (CBB) R250 staining. (Lane M: precision plus all blue standards; 1: activated sludge biopolymer (ASBP); 2: ASBPCu; 3: ASBPCd; 4: ASBP>3 kDa part; 5: ASBPCu>3 kDa part; 6: ASBPCd>3 kDa part. Standard BAS was used in other lanes).

Molecular weight (kDa)	Fraction (%)		
	ASBP	ASBPCu	ASBPCd
<3	10.0	7.7	7.4
3~5	0.5	0.2	0.2
5~10	8.6	6.8	6.7
10~20	38.4	41.4	38.8
20~30	8.6	11.5	11.4
30~40	6.0	7.2	8.5
40~50	6.4	6.5	6.4
50~60	8.4	6.2	8.2
60~80	3.7	4.0	3.3
80~100	5.4	5.4	6.0
>100	3.9	3.1	3.1
Mean MW	31 kDa	30 kDa	31 kDa

Table 1: Molecular weight distribution of activated sludge biopolymers. MW fraction was calculated based on the band intensity of each protein band of CBB-stained SDS-PAGE gel.

polymeric substances extracted from activated sludge by physical treatment followed by removal of low MW solutes lower than 1 kDa [27] (Figure 1) (Table 1).

Interactions of microbial biopolymers with cadmium in soil solution

Ccadmium uptake of biopolymers in soil solution was shown in Figure 2. For the Cd-biopolymer binding part remained in extracts, ASBPCd showed slightly higher Cd uptake than ASBP and ASBPCu (Figure 2).

The polymer-metal ion interaction might be intra and/or inter-chain. Intra-chain appeared to be the most common for a numerous group of polymer metal binding, which showed comparatively high chemical and thermal stability. Another specific feature of these compounds would be the total saturation of the coordination sphere of the transition metal ion. The polymer interacted with metal ion by binding functional groups of two different macromolecules, which supplied "acidic" functional groups and "basic" groups, respectively. Thus, the process of metal binding with mixed biopolymers would

have promising application with regards to biological reactions [28]. The variables that might affect the polymer-metal ion interactions were intrinsic to the polymer: nature of atoms in the backbone chain, nature of the functional groups attached to the backbone, structure and copolymer composition, molecular weight and polydispersity, distance between functional groups and backbone, degree of branching, etc.; other variables might be extrinsic to the polymer: for example pH, ionic strength, nature and charge of the metal ions, temperature, or nature of the counterion of the metal ion [28]. The sorption of cadmium by microbial biopolymers was significantly lower than cadmium sorption in solution. This difference could be due to the missing of important proteins from solution to solid surface, with or without heavy metal bound with them. In this study, up to 36-60% of protein in ASBP, ASBPCu and ASBPCd was found to be adsorbed onto soil matrix. In the extracts of biopolymers, TOC concentration was reduced to 11-37% of the original.

Effects of microbial biopolymers in phytoextraction of cadmium from soil

Eight sets of experiments were simultaneously assayed: only plant in soil (no agent added), plants in soil supplemented with (i) biopolymers (ASBP, ASBPCu, and ASBPCd), (ii) two different concentrations of EDTA, (iii) citric acid, (iv) BSA. As shown in Figure 3, while biopolymers (ASBP, ASBPCu and ASBPCd) were used, the metal concentrations in the plant biomass were found to be 2.03, 1.70, and 1.74 times as high as the control plants, respectively. The concentration of cadmium in plant biomass was found to be 0.62-1.56 times, 1.22 times, and 1.33 times as high as the control, in the presence of EDTA, citric acid, and BSA, respectively. The plant uptake of cadmium from contaminated soil was reduced only in presence of

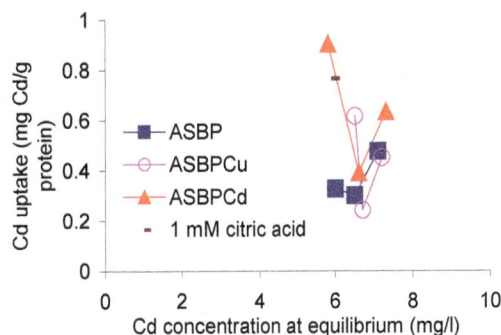

Figure 2: Uptake of cadmium from soil using biopolymers (MW>3 kDa). The x-axis shows cadmium concentration in the soil solution at equilibrium.

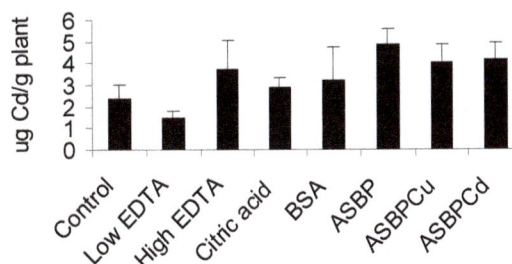

Figure 3: Cadmium concentration in plant biomass. Error bars represented standard deviations for five replicate experiments.

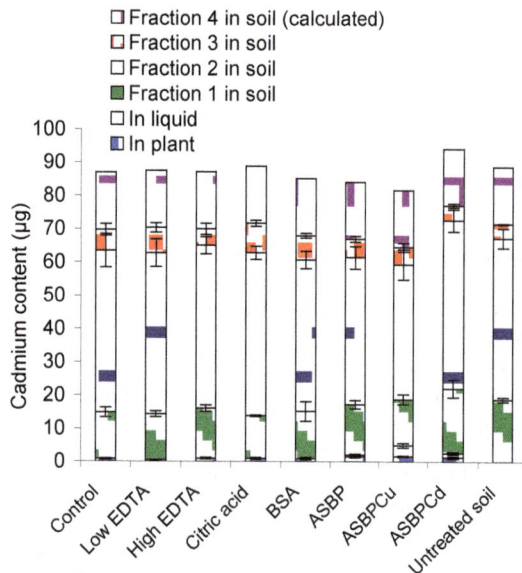

Figure 4: Cadmium distribution in plant, liquid and soil fractions after phytoextraction of cadmium from 6 g contaminated soils using biopolymers and other extracting agents (EDTA, citric acid, BSA). Water was used as the control. For cadmium fractions in soil, fraction 1: exchangeable + water and acid-soluble; fraction 2: Iron and manganese oxides; fraction 3: Organic matter and sulfides; fraction 4: residual. Cadmium contents in fraction 4 were under detection line in this experiment. Control represented the sequential extraction of cadmium from untreated soil (total cadmium concentration: 14.9 mg Cd/kg dried soil). Error bars represented standard deviations for five replicate experiments.

EDTA at concentration of 5×10^{-6} M (Figure-3). The stability constant for EDTA-Cd (16.5) was substantially higher than BSA (logK=3.6) and ASBP-Cd (logK=3.7). One study reported that the EDDS was found more efficient in Pb uptake by *C. sativa*, although the stability constant for EDTA-Pb was significantly higher than EDDS-Pb [29]. LogK value was not deemed enough to represent the potency of specific chelating agents for enhanced phytoextraction. The overall efficiency of agents for aiding phytoextraction was considered to be plant-specific, as well as being controlled by the stability constant and soil conditions.

Mobilization of cadmium from soil using biopolymers during phytoextraction

Sequential extraction of soil cadmium was performed after phytoextraction with biopolymers (ASBP, ASBPCu and ASBPCd) and other agents (water, EDTA, citric acid and BSA), as shown in Figure 4. The existence of biopolymers allowed higher cadmium content accumulated in plant biomass, than other extracting agents. In ASBP, ASBPCu and ASBPCd, the cadmium content in plants was found to be 1.52, 1.63 and 1.33 µg (1.9, 2.0 and 1.6 times of the control), respectively. In phytoextraction of Low EDTA, High EDTA, Citric acid and BSA, the cadmium content in plants was found to be 0.50, 0.97, 0.99 and 0.98 µg (0.6, 1.2, 1.2 and 1.2 times of the control), respectively. In phytoextraction together with biopolymers (ASBP, ASBPCu and ASBPCd), cadmium content was found to be 0.52, 3.31 and 1.24 µg in 9 ml liquid, respectively. Biopolymers solubilized cadmium from soil to liquid. However, cadmium was not detected in condition with other extracting agents. Cadmium in liquid would give an easier uptake by plants, if compared with cadmium fractions in soil (Figure 4).

The exchangeable fraction of heavy metals in the soil is the

fraction that could interact with biological targets and pose a health risk to human via food chain contamination. Thus, this research of phytoextraction focused on removal of exchangeable fractions of cadmium from contaminated soil by plant. In the phytoextraction of cadmium from contaminated soil, it was found that 10.9%, 26.2% and 13.7% of exchangeable cadmium fraction was extracted from soil matrix to plant or liquid, higher than the control test (4.3%). Microbial biopolymers were effective in improving the available cadmium amount in the soil, thus providing a higher and sub sequential potential of cadmium uptake by plants in long term phytoextraction.

Conclusion

This paper elucidated cadmium binding characteristics of microbial biopolymers and a sorption of cadmium was observed by microbial biopolymers both in solution and in soil. With the existence of microbial biopolymers, the cadmium uptake by phytoextraction from the contaminated soil increased. In this paper, 10.9%, 26.2% and 13.7% of exchangeable cadmium fraction was extracted from soil matrix to plant or liquid, higher than the control test (4.3%), when biopolymers were applied to phytoextraction of cadmium from contaminated soil. Microbial biopolymers, acting as a new environmental mobilizing agent, were more effective in improving cadmium accumulation in plants than other chemical agents. Instead of the current chemical agents, microbial biopolymers could significantly reduce the production costs.

References

1. Hughes MN, Poole R K (1991) Metal speciation and microbial growth - the hard (and soft) facts. Journal of General Microbiology 137: 725-734.

2. Lao UL, Chen A, Matsumoto MR, Mulchandani A, Chen W (2007) Cadmium removal from contaminated soil by thermally responsive elastin (ELPEC20) biopolymers. Biotechnol Bioeng 98: 349-355.

3. Gupta SK, Herren T (2000) Wenger K, Krebs R and Hari T In situ gentle remediation measures for heavy metal-polluted soils. In Terry N and Banuelos G (ed.) Phytoremediation of Contaminated Soil and Water. USA: CRC Press LLC. 303-322.

4. Tembo BD, Sichilongo K, Cernak J (2006) Distribution of copper, lead, cadmium and zinc concentrations in soils around Kabwe town in Zambia. Chemosphere 63: 497-501.

5. Kikuchi T, Okazaki M, Toyota K, Motobayashi T, Kato M (2007) The input-output balance of cadmium in a paddy field of Tokyo. Chemosphere 67: 920-927.

6. Makino T, Kamiya T, Takano H, Itou T, Sekiya N, et al. (2007) Remediation of cadmium-contaminated paddy soils by washing with calcium chloride: Verification of on-site washing. Environ Pollut 147: 112-119.

7. Patel MJ, Patel JN, Subramanian RB (2005) Effect of cadmium on growth and the activity of H2O2 scavenging enzymes in Colocassia esculentum. Plant and Soil 273: 183-188.

8. Bizily SP, Rugh CL, Summers AO, Meagher RB (1999) Phytoremediation of methylmercury pollution: merB expression in Arabidopsis thaliana confers resistance to organomercurials. Proceedings of the National Academy of Sciences 96: 6808-6813.

9. Meagher RB, Rugh CL, Kandasamy MK, Gragson G, Wang NJ (2000) Engineered phytoremediation of mercury pollution in soil and water using bacterial genes In Terry N, Banuelos G (Edn.) Phytoremediation of Contaminated Soil and Water. USA: CRC Press LLC. 201-219.

10. Cunningham SD, Berti WR (2000) Phytoextraction and phytostabilization: technical, economic, and regulatory considerations of the soil-lead issue. In Terry N, Banuelos G (Edn.) Phytoremediation of Contaminated Soil and Water. USA: CRC Press LLC, 359-376.

11. Pulford ID, Watson C (2003) Phytoremediation of heavy metal-contaminated land by trees-a review. Environment International 29: 529-540.

12. Ghosh M, Singh AP (2005) A review on phytoremediation of heavy metals and utilization of its byproducts. Applied Ecology and Environmental Research 3: 1-18.

13. Chen TB, Liao XY, Huang ZC, Lei M, Li WX, et al. (2007) Phytoremediation of arsenic-contaminated soil in China. In Willey N (Edn.) Phytoremediation: methods and reviews. Humana Press USA: 393-404.

14. Wu LH, Luo LM, Xing XR, Christie P (2004) EDTA-enhanced phytoremediation of heavy metal contaminated soil with Indian mustard and associated potential leaching risk. Agriculture, Ecosystems and Environment 102: 307-318.

15. Wu LH, Luo YM, Song J (2007) In: Willey N Manipulating soil metal availability using EDTA and low-molecular weight organic acids. (Edn.) Phytoremediation: methods and reviews. Humana Press USA: 291-303.

16. Fukushi K, Chang D and Ghosh S (1996) Enhanced heavy metal uptake by activated sludge cultures grown in the presence of biopolymer stimulators. Water Science and Technology 34: 267-272.

17. Arican B, Gokcay CF, Yetis U (2002) Mechanistics of nickel sorption by activated sludge. Process Biochemistry 37: 1307-1315.

18. Kellems BL, Lion LW (1989) Effect of bacterial exopolymer on lead (II) adsorption by γAl2O3 in seawater. Estuarine, Coastal and Shelf Science 28: 443-457.

19. Chen JH, Czajka DR, Lion LW, Shuler ML, Ghiorse WC (1995) Trace metal mobilization in soil by bacterial polymers. Environ Health Perspectives 103: 53-58.

20. Chen JH, Lion LW, Ghiorse WC, Shuler ML (1995) Mobilization of adsorbed cadmium and lead in aquifer material by bacterial extracellular polymers. Water Research 29: 421-430.

21. Goodwin JAS, Forster CF (1985) A further examination into the composition of activated sludge surfaces in relation to their settlement characteristics. Water Resource 19: 527-533.

22. Fang HHP, Jia XS (1996) Extraction of extracellular polymer from anaerobic sludges. Biotechnology Techniques 10: 803-808.

23. Liao BQ, Allen DG, Droppo IG, Leppard GG, Liss SN (2001) Surface properties of sludge and their role in bioflocculation and settleability. Water Res 35: 339-350.

24. Clemens S, Palmgren MG, Kramer U (2002) A long way ahead: understanding and engineering plant metal accumulation. Trends in Plant Sci 7: 309-315.

25. Küpper H, Götz B, Mijovilovich A, Küpper FC, Meyer-Klaucke W (2009) Complexation and toxicity of copper in higher plants. I. Characterization of copper accumulation, speciation, and toxicity in Crassula helmsii as a new copper accumulator. Plant Physiology 151: 702-714.

26. Lowry OH, Rosebrough NJ, Farr AL, Randall RJ (1951) Protein measurement with the folin phenol reagent. J Biol Chem 193: 265-275.

27. Liu Y, Lam MC, Fang HH (2001) Adsorption of heavy metals by EPS of activated sludge. Water Sci Technol 43: 59-66.

28. Rivas BL, Pereira ED, Villoslada IM (2003) Water-soluble polymer–metal ion interactions. Progress in Polymer Science 28: 173-208.

29. Lestan D (2006) Enhanced heavy metal phytoextraction. In Mackova M, Dowling D, Macek T Phytoremediation and Rhizoremediation (Edn). The Netherlands: Springer 115-132.

32

Information Needs for Climate Change Adaptation among Rural Farmers in Owerri West Local Area of Imo State, Nigeria

Umunakwe PC[1]*, Nnadi FN[1], Chikaire J[1] and Nnadi CD[2]

[1]Department of Agricultural Extension, Federal University of Technology, P.M.B. 1526 Owerri Imo State, Nigeria
[2]Department of Agricultural Economics and Rural Sociology, Niger Delta University, Wilberforce Island, Bayelsa State, Nigeria

Abstract

Information on climate risks communicated timely, in clear and relevant terms and through credible sources is essential for mobilizing decision makers across societies to take actions that will enhance their capacity and willingness to adapt to climate change. An informed public is better able to prepare for a likely occurrence of climate disaster and thus avert or cope with its attendant effects. The study analyzed information needs for climate change adaptation among rural farmers in Imo state, Nigeria. Specifically, it determined the socio-economic characteristics of the farmers, investigated their knowledge of climate change, identified their sources of information on climate change, identified their information needs for climate change adaptation and analyzed the socio-economic determinants of the farmers' needs for climate change adaptation. Data were elicited from 120 farmers using structured questionnaire and interview schedule. These were analyzed using percentages, bar charts and mean statistics. The hypothesis was analyzed using ordinary least square regression model at 0.05%. Results revealed that majority (95.1%) of the farmers described their knowledge of climate change as change in rainfall pattern. It also revealed that the farmers identified radio (61.6%), extension agents (35.8%) and newspaper (27.5%) as their major sources of information on climate change. The result further revealed that the farmers identified effects of climate change (M=4.15), causes of climate change (M=4.06), vulnerable groups to climate change (M=4.03), appropriate socio-cultural practices in climate change adaptation (M=3.99), crops adaptable to climate change (M=3.96), sources of information on climate change (M=3.93), agroforestry practices (M=3.89), flood/erosion control practices (M=3.85), afforestation practices (M=3.75), carbon trading (M=3.68) and adaptation strategies (M=3.34). The study recommended the organization of capacity building programmes relevant to agriculture, the timely generation and dissemination of information on climate change and the reviewing of extension curriculum to accommodate the training of extension personnel on climate change issues as strategies for enhancing adaptation to climate change.

Keywords: Information needs; Climate change; Adaptation; Rural farmers; Nigeria

Introduction

Climate change will have significant impacts on the livelihoods of rural poor in developing countries. The Intergovernmental Panel on Climate Change report 2007 provides an extensive assessment on the expected effects of climate change on agriculture in the African region. It estimates that Africa will be the most vulnerable to climate change globally, due to the increase by between 1.5-4.0°C in temperature in this century. Projections on yield reduction show a drop of up to 50% and crop revenue is forecast to fall by as much as 90% by 2100. The agricultural sector is also expected to experience periods of prolonged droughts and/or flood during the El Nino events. Agricultural losses of between 2-7% of the GDP are expected by 2100 in parts of the Sahara, and 0.4%-1.3% and 2*4% in Western and Central Africa and Northern and Southern Africa respectively. Fisheries will be particularly affected due to changes in sea temperature that could decrease trends in productivity by 50-60% [1].

Furthermore, the Overseas Development Institute (2007) posits that productivity in agriculture will further be undermined by a reduction in fertile agricultural land availability and expansion in the coverage of low potential land [2]. More so, in response to variations in temperature and precipitation, Africa is expected to experience an increase in crop pests and diseases in addition to altered soil fertility. Declining income and rising unemployment are expected to hit agricultural zones in combination to worsening health. A fall in nutrient access is likely to raise susceptibility to diseases such as malaria and HIV/AIDS [3].

In Nigeria, analysis of climatic data collected by the Nigerian Meteorological Institute over several decades reveal that since the 1970s, most parts of Nigeria have experienced some shifts in weather patterns. For instance, in recent years, the average amount of rainfall has decreased by 15 to 20% while rainfall intensity has increased by about 10 to 15% leading to high surface runoff and frequent flooding, and soil erosion in various parts of the country. This has adversely affected crop yields while promoting the development and spread of pests. Complications arising from poor land use and land degradation further compound the problem [4-7].

The overall climate change poses a substantial challenge to Africa's agricultural development. From food security, nutrition to sustainable development, climate change is a significant threat to the welfare of millions of the continent's rural poor. Contending with this challenge therefore requires the development of intensive research and policy frameworks to enable the development of proper adaptation measures which will reduce the continent's vulnerability. This however is

*Corresponding author: Umunakwe, Department of Agricultural Extension, Federal University of Technology, P.M.B. 1526, Owerri Imo State, Nigeria
E-mail: polycarpchika@yahoo.com, limanchiks@gmail.com

dependent largely on the generation, dissemination and adoption of useful information on climate change.

According to the World Meteorological Organization (2011) climate information and prediction services enable better management of climate variability and change and adaptation through the incorporation of science-based practices into planning, policy and practices on the global, regional and national scale. Ilevbaje (1999) corroborates this by stressing that information is crucial in agricultural development. It also enables farmers to take decisions regarding their choice of practices in order to avert or reduce risks related to climate change and promote sustainable development [8-10].

While progress has been made in generating climate change information in sub-Saharan Africa majority of the African countries continue to suffer the full impacts of climate change [11]. Investigations conducted have revealed massive deficits in the transmission of early warning messages in highly vulnerable countries, where climate services where they exist are largely inaccessible to millions of rural poor whose livelihoods are climate-dependent [12]. This may be due to the fact that the largest demographic–the rural farmers–are cut off from the benefits of climate change information despite the designation of such groups as 'key targets' of such projects [13]. Yesuf et al. [14] reported the lack of information as the major adaptation constraint faced by farmers in Ethiopia.

A study conducted by GOZ-UNDP/GEF (2010) in Chiredzi district Zimbabwe revealed that majority of the farmers indicated that the information they received on climate change did not meet their needs because it was too generalized. They however indicated that they required information that would assist them in planting crops that are commonly grown in their area. However, the lack of the required information in climate change information received by the farmers rendered the entire information on climate change unreliable which affected adversely their productivity [15].

According to USAID (2007) appropriate interventions must incorporate disaster planning response and mitigation into governance systems and engage vulnerable groups into participatory fora to address their vulnerability and to identify adaptations to climate change impacts [16]. Sustainable adaptation should also allow for decision making from all stakeholders, including poor men and women and should incorporate site-specific information as the non-consideration of these factors might result to the rejection or slow adoption of such adaptation technologies. Furthermore, involvement of local community members enhances ownership and sustainability [17,12] maintains that transmitting climate information to the grassroots level is one of the major challenges to making climate and weather information relevant to vulnerable communities. It is against this backdrop that the following research questions were asked:

➤ What is the farmers' knowledge on climate change?

➤ What are the farmers' sources of climate change information?

➤ What are the farmers' perceived information needs for climate change adaptation? And

➤ What are the socioeconomic determinants of farmers' needs for information on climate change.

Objectives of the Study

The broad objective of the study is to evaluate information needs for climate change adaptation among rural farmers in Imo State,

Nigeria. Specifically the study seeks to:

I. Determine the socio-economic characteristics of farmers;

II. Investigate the farmers' knowledge about climate change;

III. Identify farmers' sources of information about climate change;

IV. Identify the farmers' information needs for climate change; and

V. Analyse the socio-economic determinants of the farmers' information needs for climate change adaptation.

Research Hypothesis

There is no significant relationship between the socio-economic characteristics of the farmers and their information needs for climate change adaptation.

Materials and Method

The study was conducted in Owerri West local government area. It is in Imo East Senatorial zone and Owerri Agricultural zone of Imo State, Nigeria. It is also among the 27 local government areas in Imo State and has 18 autonomous communities with an estimated population of 101,754 people and a land size of 297, 127 square kilometer (National Population Commissions, 2006). It is located in the rainforest zone about 120 km north of the Atlantic coast and lies on latitude 4°14` and 6°15` and longitude 6°51` and 8°09` East. It is bounded in the east by Owerri North and Owerri Municipal Local Government Areas, in the north by Mbaitoli local government area, in the West by Ohaji/Egbema in the South by Ngor-Okpala Local Government Area.

The average annual rainfall measures up to 2550 mm, the relative mean temperature ranges annually between 24.5° and 25.5°C and the relative humidity varies according to the time of the year (ISADA, 2000). The people are predominantly farmers with cassava, cocoyam, yam, maize and vegetables and goats, sheep, poultry and pigs as major arable crops grown and livestock kept, respectively.

The population for the study comprised all farmers in the study area. Multi-stage sampling technique was used for sample selection. The first stage comprised the selection of six communities out of the 18 communities that make up the local government area using simple random sampling technique. The second stage was the selection of four villages from each of the six communities using simple random sampling technique to give a total of 24 villages. The third and final stage comprised the selection of five farmers from each of the 24 villages using simple random sampling technique giving a total of 120 farmers.

Data were obtained from both primary and secondary sources. Primary data were obtained with the aid of structured questionnaire and interview schedule from literate and illiterate farmers respectively in the study area while secondary data were obtained from annual reports of Imo state Agricultural Development Programme, Imo state Ministry of Petroleum and Environment, text books, journals, internet and previous studies of other researchers.

Descriptive and inferential statistical tools were used to analyze data. Objective 1 was analyzed using frequency and mean statistics. Objectives 2 and 3 were analyzed using bar chart. Objective 4 was analyzed using mean statistics. The hypothesis was analyzed using the Ordinary Least Square Multiple Regression implicitly represented by the equation:

$$Y=f(X_1+X_2+X_3+X_4+X_5+X_6+X_7+X_8)+e$$

Where Y=Sum total of areas of information needs indicated by the farmers

X_1=Age (years)

X_2=Sex (Using dummy, Female=0, Male=1)

X_3=Marital Status (Single=0, Married=1, Separated or Divorced=2, Widowed=3)

X_4=Farming experience (Years)

X_5=Educational level (No formal education=0, Primary School=1, Secondary School=2, Tertiary=3, others=4)

X_6=Household size (Number of people that feed from the same pot)

X_7=Farm size (Hectare)

X_8=Farming status (Part time=0, full time=1)

X_9=Social organization membership (Yes=1, No=0)

The four functional forms of the model were tested and the one with the best fit; highest number of F-value, coefficient of multiple determinations (R2) and highest number of significant variables was chosen for the analysis.

Results and Discussion

Socioeconomic characteristics

Age: Data in Table 1 show that a greater proportion (46.7%) of the farmers were within the age range of 40-49 years while the remaining 35.8%, 15.0% and 2.5% were within the age ranges of 50-59, 30-39 years and 60-69 years respectively. The mean age of the farmers was 47 years. This shows that the study area is dominated by relatively young farmers who are still within the economically active age. According to Agbamu [18] younger farmers are more likely to adopt agricultural innovations. However, this is contrary to the expectation that rural areas are dominated by old farmers as a result of rural-urban migration [19].

Sex: Data in Table 1 show that majority (59.17%) of the respondents were female while the remaining 40.83% were male. This means that the study area was dominated by female farmers. This lays credence that the population of rural Nigeria is dominated by females [20]. This could be attributed to the increasing cases of migration of the male counterparts to cities in search of greener pastures and increasing incidence of female-headed households. As posited by Agarwala (2008) women occupy a significant proportion of agricultural labour force in developing countries and play very crucial roles in agricultural production [21].

Marital status: Data in Table 1 show that majority (80.83%) of the farmers were married while the remaining 9.17%, 5.0% and 5.0% were widowed, single and divorced respectively. Married people are more likely to engage in farming activities to secure households from their requirements and complement their other sources of income. Marriage furnishes more hands for farming as parents and their progenies complement one another in agricultural production. This is in line with the finding of a study by Umunakwe [22] where majority of the farmers in Imo State were married.

Educational level: Data in Table 1 show that majority (98.3%) of the farmers received one form of formal education or the other with 61.0% receiving secondary, 23.3% tertiary and 13.3% primary education. This is in line with the findings of where majority of the farmers interviewed in Delta state, Nigeria received formal education [19].

Socioeconomic characteristic	Frequency	%	M
Age (Years)			
<30	0	0.0	
30-39	18	15.0	
40-49	56	46.7	47
50-59	43	35.8	
60-69	3	2.5	
>69	0	0.0	
Sex			
Male	49	40.83	
Female	71	59.17	
Marital Status			
Single	6	5.0	
Married	97	80.83	
Divorced	6	5.0	
Widowed	11	9.17	
Educational level			
No formal education	2	1.7	
Primary education	16	13.3	
Secondary education	74	61.7	
Tertiary education	28	23.3	
Social organization membership			
Ordinary member	51	42.5	
Regular member	38	31.7	
Financial member	3	2.5	
Committee member	21	17.5	
Executive member	7	5.8	
Household size (Number of persons)			
1-3	36	30.0	
4-6	73	60.8	
7-9	10	8.4	2
>10	1	0.8	
Farm size (Hectare)			
<1.0	48	40.90	
1.0-3.0	60	50.0	
3.0-5.0	8	6.7	
>5.1	4	3.3	
Farming experience (Years)			
<10	1	0.8	
11-20	39	31.8	
21-30	70	58.3	12.9
31-40	10	8.3	
>40	1	0.8	
Farming status			
Full time	104	86.7	
Part time	16	13.3	

Table 1: Distribution of respondents according to socioeconomic characteristics.

Acquisition of education influences the adoption of agricultural technologies by farmers. According to Agbamu [23], formal education enables farmers to obtain useful information from bulletins, agricultural newsletters and other sources. Ogunfiditimi [24] found that the level of education of farmers in Oyo and Ondo States, Nigeria yielded positive significant relationship to the adoption of improved varieties of cassava, maize and cocoa.

Social organization membership: Data in Table 1 show that all the respondents (100%) are members of social organizations. Out of these 42.5% were ordinary members, 31.7% were regular members, 17.5% were committee members, 5.8% were executive members, while 2.5% were financial members. Membership of social organizations satisfied the farmers' social needs and furnished agricultural information which consequently led to adoption and diffusion.

The various statuses are reflective of their levels of commitment. According to Agbamu [18,23] Nigerian farmers belong to a number of social organizations and social or group meetings ranked third among the information sources to farmers in Ogun state, Nigeria.

Household size: Data in Table 1 reveal that majority (60.8%) of the farmers had a household size of 4-6 people while the remaining 30%, 8.4% and 0.8% had a household size of 1-3, 7-9 and more than 10 people, respectively. The mean household size was 2 persons. The consistent population campaign by governmental and non-governmental organizations, increasing economic crunch and influence of religion, especially Christianity may have been responsible for the low household size. Agriculture in developing countries is largely rudimentary and hence is characterized by low or no mechanization and this encourages the reliance on human power for farm operations especially the cheap human labour [20]

Farm size: Data in Table 1 show that majority (50.0%) of the respondents have a farm size of 1-3 hectares while the remaining 40.00%, 6.7% and 3.3% have farm sizes of <1.0, 3-5 and >5.0 hectares respectively. The mean farm size was 1.3 hectares. Farm lands in the traditional Igbo society are communally owned and this leads to fragmentation, leaving farmers with small farm sizes. Resultantly, most of the farmers in Southeastern Nigeria own farm lands that are hardly more than 1.0 hectare [20]. According to Williams et al. [25] the larger the farm business in terms of acreage the earlier the farmer tends to adopt those new technologies which are applicable to his farm business.

Farming experience: Data in Table 1 show that majority (58.3%) of the farmers had been in farming business for 21-30 years while the remaining 31.8%, 8.3%, 0.8% and 0.8% have been in farming business for 11-20, 31-40, <10 and >40 years, respectively. The mean farming experience was 12.9 years. This means that the farmers have been into farming for relatively long periods. Many years of farming experience could entail much knowledge of the climatic, ecological and edaphic conditions of the area which are assets in climate change adaptation. The result corroborates the finding of a study by Nnadi et al. (2012) where majority of the farmers in Imo State had a long farming experience with a mean of.

Farming status: Data in Table 1 show that majority (86.7%) of the farmers were into full time farming while the remaining 13.3% were into part time farming. This result means that majority of the farmers engage in farming as a major occupation and would therefore crave for much information to adapt to climate change to avert food insecurity. Ekong [20] described farming as a dominant activity in rural areas in Nigeria.

Knowledge of climate change: Data in Figure 1 show that majority (95.1%) of the farmers describe climate change as a change in rainfall pattern while the remaining 24.16%, 17.5%, 11.6% and 9.16% know that climate change causes excessive rainfall, prolonged drought, over flooding and erosion. Analyses of climatic data collected by NIMET over several decades reveal that most parts of Nigeria have experienced some shift in weather patterns. For example, some parts of the country now have delay in the onset of rainfall. Also, many parts of the country recently experienced flooding as a result of excessive rainfall [26]. Knowledge and description of climate change is location-specific. However, it will enable the farmer to point out the interventions he requires to remedy the effects/impacts of climate change.

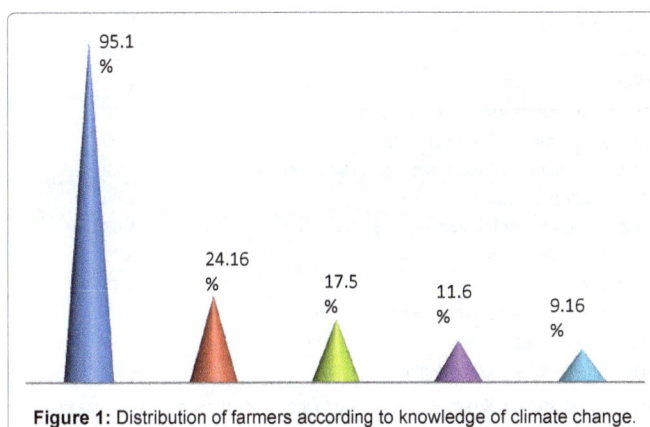

Figure 1: Distribution of farmers according to knowledge of climate change.

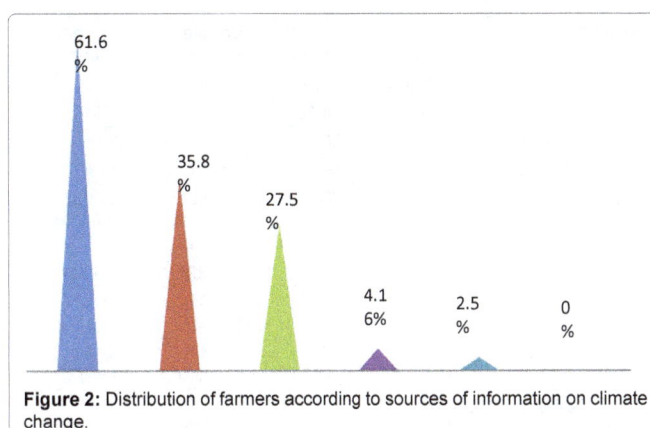

Figure 2: Distribution of farmers according to sources of information on climate change.

Sources of information on climate change: Data in Figure 2 reveal that majority (61.6%) of the farmers received information on climate change through radio while the remaining 35.8%, 27.5%, 4.16% and 2.5% received information on climate change through extension agents, newspapers, research institutions and cooperative societies. According to GoZ-UNDP/GEF (2010) farmers in Zimbabwe relied mainly on radio, newspaper and extension agents for information on climate change. They however lamented that the high cost of newspapers and the fewness of extension agents constrained their access to climate change adaptation information [15].

Timely reception of climate change information will enable farmers take necessary measures to avert or cope with impending impacts/effects of climate change. According to United Nations (2009) early warning system is crucial for preparedness for extreme weather events [27].

Areas of information need for climate change adaptation: Data in Table 2 below show that farmers indicated that they needed information on the effects of climate change (M=4.15), causes of climate change (M=4.06), vulnerable groups to climate change (M=4.03), appropriate socio-cultural practices in climate change (M=3.99), crops adaptable to climate change (M=3.96), sources of information on climate change (M=3.93), agroforestry practices (M=3.89), flood/erosion control practices (M=3.85), afforestation practices (M=3.75), carbon trading (M=3.68) and adaptation strategies (M=3.34). A study conducted in the Chiredzi District, Zimbabwe revealed that farmers interviewed indicated that they needed information to assist them in planting the different crops that are commonly grown, information on rainfall pattern, time of occurrence of the mid-season drought or dry spells and flooding and storm warnings [15].

Areas of information need	M
Causes of climate change	4.06
Effects of climate change	4.15
Sources of information on climate change	3.93
Crops adaptable to climate change	3.96
Appropriate socio-cultural practices in climate change	3.99
Agroforestry practices	3.89
Flood/erosion control practices	3.85
Afforestation practices	3.75
Carbon trading	3.68
Alternative/complementary livelihood activities	2.93
Adaptation strategies	3.34
Vulnerable groups to climate change	4.03

Table 2: Distribution of farmers according to information needs for climate change adaptation.

Explanatory variables	Linear function	Semi-log function	Double-log function	Exponential function
Constant	188.3992	143.0832	129.8117	118.201
R2	0.4993	0.4028	0.8013	0.5618
No. of observation	120	120	120	120
Degrees of freedom	119	119	119	119
F-Value	12.6004	8.2423	49.4629	16.0057 (-0.0079)
X_1 (Age)	-15.0821 (-1.0684)	8.2423 (-1.1212)	-0.0665 (-2.7824)*	-0.0079 (-1.2344)
X_2 (Sex)	16.3913 (1.0845)	1.4107 (1.0136)	0.0638 (1.0687)	0.0053 (1.1277)
X_3 (Marital status)	18.2271 (1.0706)	1.5525 (1.0624)	0.0544 (1.3432)	0.0091 (3.9565)*
X_4 (Farming experience)	7.8827 (3.7684)*	1.5912 (3.9494)*	0.0717 (3.1726)*	0.0054 (2.8421)*
X_5 (Educational level)	-73116 (-1.1647)	-3.7754 (-1.2591)	-1.0843 (-4.0725)*	-0.0087 (-3.7826)*
X_6 (Household size)	8.2107 (3.2951)*	2.1793 (1.0690)	0.0729 (3.344)*	0.0083 (1.1526)
X_7 (Farm size)	5.9278 (1.2238)	1.9602 (1.1142)	0.0527 (2.6616)*	0.0068 (1.1526)
X_8 (Farming status)	10.0806 (1.1061)	2.1843 (1.0420)	0.0692 (3.2488)*	0.0082 (2.6452)*
X_9 (Social organization membership)	13.1314 (3.2143)*	3.0749 (3.0499)*	0.9140 (3.0671)*	0.0792 (2.5631)*

*significant
Table 3: Regression result of the relationship between the farmers' socio-economic characteristics and their information needs for climate change adaptation.

The timely availability of this information to the farmers will enable them make certain decisions pertaining their farming business such as the types of crops to grow, when to grow them, when to weed and apply fertilizer and harvest, when to apply herbicides and when to cull livestock. The knowledge of the type of information required by the farmers will guide researchers in the development of suitable adaptation strategies [28-30].

Hypothesis

There is no significant relationship between the socio-economic characteristics of the farmers and their information needs for climate change adaptation.

Socioeconomic determinants of farmers' information needs for climate change adaptation

The result in Table 3 shows that Double Log Function had the highest number of significant variables (six) and coefficient of multiple

determination R2 (0.08013) and F-value (49.4629). Thus, it serves as the lead equation for the explanation of the relationship.

The result shows that there was a significant relationship between socioeconomic characteristics of the farmers and the areas of information needs for climate change adaptation. The result implies that the combined effects of the variables accounted for 80% variation in the number of areas of information need by the farmers for climate change adaptation. The significant variables were age, educational level, household size, social organization membership, farm size, farming status and farming experience [31].

The variables, age (X_1) and educational level (X_5) were significant negatively related to the farmers' information needs for climate change adaptation with t-values of -2.7824 and -4.0725, respectively. These imply that increasing the age and educational level reduced the farmers' need for information for climate change adaptation. Older farmers have been associated with fatalism, conservation and maintenance of status quo [19] and these attributes do not support innovativeness and quest for information. Education on the other hand furnishes facts and may equip the farmers with the right cognitive domain to analyze and understand climate change. The result agrees with the finding of Agbamu et al. [23] that negative but significant relationship existed between educational level of farmers and their adoption off TMS cassava and TZSR maize varieties.

Farming experience (X_4), household size (X_6), farm size (X_7), farming status (X_8) and membership of social organizations (X_9) were positively and significantly related to the farmers' information needs for climate change adaptation. Several years of farming experience would broaden farmers' knowledge base and consequently inspire their desire for more information on climate change. Large household size entails more pressure on the household for food security and this could incite the need for more information. Again, each member of the farm family is a stakeholder in the farming business and as such is a potential source of information on climate change. The aggregate information need may not be comparable to that of a small household. A large farm size means a big farm investment and this is real business. Increasing the information need consequently sharpens decision making for profitability.

Increasing farming status by part time farmers becoming full time and increase farm investible fund through increased commitment and interest in the business. This could arouse the desire for more information for profitable farm decisions. In the same vein, social organizations produce fora for exchange of agricultural information and some of the information could motivate the search for more especially for a critical issue like climate change adaptation.

However, the variables sex (X_2) and marital status (X_3) were not significantly related to the information needs of the farmers for climate change adaptation. They are inconsequential and as such do not exert any significant effects [32,33].

Conclusion

Adaptation to climate change has remained a viable option for dealing with the impacts and effects of climate change. However, this will remain ineffective and unsustainable without the timely generation and dissemination of useful information on climate change to the people whose livelihoods are mostly affected. This as it has been pointed out will enhance the making the necessary decisions aimed at the reducing the impacts.

The findings from the study indicated that farmers in Nigeria needed information on mitigation issues for easy adaptation to climate change, obtained from diverse information sources. These are determined by the socio-economic characteristics of age, farming experience, educational level, farm size, farming status, household size and membership of social organizations.

From the foregoing, the study makes the following recommendations:

- Capacity building programmes relevant to climate change and agriculture should be organized by the government at all levels through the appropriate ministries, agencies and departments to train extension agents and educate farmers to boost their capacities for effective and sustainable adaptation to climate change;

- Takeholders in the generation of climate change information such as the Nigerian Meteorological Institute (NIMET) should ensure the timely generation and dissemination of relevant information on climate change (e.g. weather forecast) which should cover all the areas of need indicated by the farmers to enable them make decisions which will enhance their adaptation to climate change;

- Curriculum for training prospective extension personnel should be reviewed to incorporate the identified areas of information need for adaptation to climate change; and

- Interventions and advocacies in climate change should be guided by the identified socio-economic characteristics that determine the information needs for proper adaptation.

References

1. Parry ML, Canziani OF, Palutikof JP, van der Linden PJ, Hanson CE (2007) IPCC, Climate change 2007: Impacts, adaptation andvulnerability. Contributions of Working Group II to the Fourth Assessment Report of the Intergovernmental Panel on Climate Change. Cambridge: Cambridge University Press.

2. Rachel S, Leo P, Eva L, David B (2007)Climate change, agricultural policy and poverty reduction: How much do we know? Overseas Development Institute.

3. Food and Agricultural Organization (2009) Food security and agricultural mitigation in developing countries: Options for capturing synergies. FAO, Rome.

4. Apata TG(2010) Effects of global climate change on Nigerian agriculture: An empirical analysis.CBN Journal of Applied Statistics 2: 29-31.

5. Adejuwon JO (2006) Food crop production in Nigeria II: Potential effects of climate change. Climate Research 32: 229-245.

6. Adger WN, Huq S, Brown K, Conway D and Hulme M (2003) Adaptation to climate change in the developing world. Progress in Development Studies 3: 11-17.

7. Agbamu JU (2000) Agricultural Research-Extension Linkage Systems: An International Perspective, Agricultural and Extension and Research Network (Agren). Network Paper 3: 1-24.

8. Smith RE (2003) Land tenure reform in Africa: A shift to the defensive. Progress in Development Studies 3: 210-222.

9. World Meteorological Institute (2007) Climate information for adaptation and development needs. WMO1025.

10. Lyon F (2000) Trust, networks and norms: The creation of social capital in agricultural economies in Ghana. World Development 28: 663-681.

11. Tarhule A (2005) Climate information for development: An integrated dissemination model 11th General Assembly of the Council for the Development of Social Research in Africa (CODESRIA) in Maputo. Mozambique.

12. Lugon R (2010) Climate information for decision-making: Lessons learned from effective user-provider communication schemes. The Graduate Institute Centre for International Governance, Geneva.

13. Ogallo L (2007) CLIPS: RCOFs, Regional networking and consensus building

14. Yesuf M, Di Falco S, Deressa T, Ringler C and kohlin G (2008) The impact of climate change and adaptation on food production in low-income countries: Evidence from the Nile Basin, Ethiopia. International Food Policy Research Institute, IFPRI.

15. GoZ-UNDP/GEF (2010) Capacity needs assessment and strategy for enhanced use of seasonal climate forecasts for small-holder farmers in Chiredzi District, Zimbabwe. Environmental Management Agency, Harare, Zimbabwe.

16. United States Agency for International Development (2007) Adapting to climate variability and change: A guidance manual for development planning. USAID and Stratus Consulting, Washington.

17. Holmes P (1996) Building capacity for environmental management in Hong Kong. Water Resources Development 12: 461-472.

18. Agbamu JU (2006) Essentials of agricultural communication in Nigeria. Lagos: Malthouse Press Limited 203.

19. Ozor, Nicholas, Madukwe MC, Enete AA, Amaechian EC, et al. (2010) Barriers to climate change adaptation among farming households of Southern Nigeria. Journal of Agricultural Extension 14.

20. Ekong EE (2003) Introduction to rural sociology, Uyo: Dove Publications ltd.

21. Agarwala B (2008) Food security, productivity and gender inequality. IEG Working Paper No 314.

22. Umunakwe PC (2011) Strategies for climate change adaptation among rural households in Imo State, Nigeria.

23. Agbamu JU, Fujita Y, Idowu I, Lawal A (1996) Effects of socio-economic factors on adoption of new varieties of maize and cassava: A case study of Ogun State in Nigeria. Journal of Agricultural Development Studies 6.

24. Ogunfiditimi TO (1981) Adoption of improved farm practices: A choice under uncertainty. Indian Journal of Agricultural Extension Education 17.

25. William SKT, Fenley JM, Willaims CE (1984) A manual for agricultural extension workers in Nigeria. Ibadan: Less Shyraden Publishers, Nigeria.

26. Anuforom A (2012) Climate change and the challenge of environmental management and technological development. Lecture delivered at the 25th convocation of the Federal University of Technology, Owerri, Nigeria.

27. United Nations (2009) Guidance on water and adaptation to climate change. United Nations, New York and Geneva.

28. Benhin JKA (2006) Climate change and South African Agriculture: Impacts and adaptation options. CEEPA Discussion Paper No 21.

29. Bryceson DF (2004) Agrarian vista or vortex: African rural livelihood policies. Review of African Political Economy 31: 617-629.

30. Jotoafrika (2009) Managing Africa's water in a changing climate. Issue 2.

31. Osman-Elasha B, Goubti N, Spanger-Siegfried E, Dougherty W, Hanafi A, et al. (2006) Adaptation strategies to increase human resilience against climate variability and change: Lessons from the Arid Regions of Sudan. AIACC, Working Paper 42: 44.

32. Roncoli C, Okoba B, Gathaara V, Ngugi J and Nganga T (2010) Adaptation to climate change for smallholder agriculture in Kenya: Community-Based perspectives from five districts.

33. United States Department for International Development (2007) Adapting to climate variability and change. A guidance manual for development planning, USAID Washington DC.

and use liaison for targeted climate service delivery presented at the Public Weather Services Symposium. Geneva, Switzerland.

33

Study on Degeneration of Potato Seed in Terai Region of Nepal

Ghimire S*, Pandey S and Gautam S

Technical officer, National Potato Research Program, NARC, Khumaltar Lalitpur, Nepal

Abstract

Field experiments were conducted in the experimental field of Regional Agriculture Research Station (RARS), Parwanipur, and Bara, Nepal. The objective of study was to evaluate the rate of degeneration due to viral diseases in Cardinal and Kufri Jyoti. The experimental plot design was Randomized Complete Block Design with 3 replication and 5 treatments considering each farmer as a replication. There were 10 treatment combinations consisting 2 varieties. DAS-ELISA was done to find out the degree of virus infection. Result showed a significant (P<0.01) effect of virus in yield loss of potato. Data of three different years were compared to find out the percentage loss due to damage cause by virus. Field observation was done for other insects and pest. The DAS ELISA results revealed that during third year the presence of PVM and PVY was highest in Cardinal under and Kufri Jyoti under control treatment condition respectively. Yield data of three different year showed that there is serious loss (27–46%) on an average in the subsequent year in the productivity of potato. To decrease the rate of degeneration insect proof net can be used which is cheaper and environment friendly resulting the satisfactory yield in the subsequent generation.

Keywords: Virus; Degeneration; Potato viruses; Small scale farmers

Introduction

Potato (*Solanum tuberosum*) is an important crop both in the hills and the terai of Nepal. Potato, being vegetative propagated crop, is very prune to seed degeneration as several potato viruses accumulate to the seed tubers overtimes resulting into its reduced yield potential [1]. It needs large quantity of healthy seed for its successful cultivation without losing productivity [2]. Seed potato degeneration, the reduction in yield or quality caused by an accumulation of pathogens and pests in planting material due to successive cycles of vegetative propagation, has been a long-standing production challenge for potato growers around the world [3]. In developing countries like Nepal, small scale farmers do not have easy access to the quality seed leading to significant reductions in yield. Studies on the cause of degeneration of seed potatoes in the country showed that the aphid is responsible for the spread of virus diseases in the fields. Vegetative propagation of the same stocks used year after year results in cumulative infiltration of pathogens, particularly the prevalent viruses which spread both through contact and aphid/ vectors [4]. Major potato viruses, namely PLRV, PVS, PVX, PVY, PVA and PVM had been reported to infect potato crops in Nepal [1]. Infection by any one alone or some of them jointly would retard plant growth and reduce tuber yield [5]. Infection of viruses has devastating effect bringing down the yield potential of the infected plants. The varieties react with different degrees of loss in tuber yield depending on the virus stage of infection, and period of field exposure of the seed stocks [6,7].

Materials and Methods

Seed tubers of Cardinal and Kufri jyoti, a commonly cultivated potato variety in terai of Nepal, were used. The experiment was conducted in three consecutive crop seasons of 2070-71, 2071-72, 2072-73 at the experimental field of Regional Agriculture Research Station (RARS), Parwanipur, and Bara. The seed potato obtained from 2070-71 were planted in the subsequent year of 2071-72 and those obtained from that year were planted in 2072-73 [8,9].

Fertilizers were applied @ 100:100:60 kg/ha Urea, DAP, MOP as recommended by NARC (Nepal Agricultural Research Council). One half of urea and full dose of DAP and MOP were applied at the time of planting. Other half of urea was applied as side dressing

after 35 days of planting when first earthling up was done. During land preparation, compost was applied at 20 t/ha. The seeds were planted in the field on second week of Mangsir and harvested in first week of Chaitra. During all the crop seasons of all three years, seed tubers were preserved in the cold storage. Whole tubers were planted maintaining 60 cm row to row and 25 cm seed to seed distances. The experiment was laid out following randomized complete block design (RCBD) with four replications. The unit plot size was 3 m×2.5 m. Intercultural operations, such as irrigation, weeding, mulching, and earthling up were done as and when necessary (Tables 1 and 2). The data were collected with following observation:

a. Emergence at 30 and 60 days after sowing

b. No. of stem per plant

c. Plant Height

d. Tuber yield per plot: For this data the total yield from a single plot was divided to three grades as under seed size; Seed size and Over seed size and the weight and number of each grade were recorded.

The samples were collected after 45 days of emergence for performing virus test and DAS-ELISA test was done. Virus incidences in the potato foliage from different treatments were detected through ELISA and presented in Figure 1.

Results and Discussion

The field trials were conducted at RARS Parwanipur, representing major potato growing area, using predominant varieties (Cardinal

---

*Corresponding author: Ghimire S, Technical officer, National Potato Research Program, NARC, Khumaltar, Lalitpur, Nepal
E-mail: shantwana@narc.gov.np

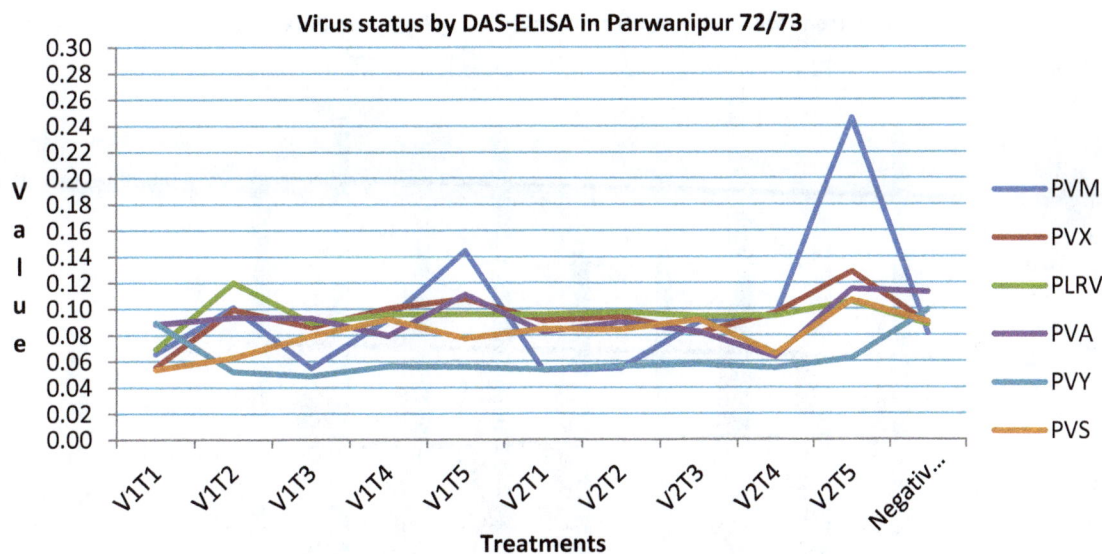

Figure 1: Graph showing the result by DAS-ELISA on status of virus.

SN	Treatments	Variety	Combination
1	Covered by insect proof net	V1: Kufri Jyoti	T1V1
		V2: Cardinal	T1V2
2	Only spraying of appropriate insecticides when aphid population reaches critical	V1: Kufri Jyoti	T2V1
		V2: Cardinal	T2V2
3	Only roughing of infected plant (negative selection)	V1: Kufri Jyoti	T3V1
		V2: Cardinal	T3V2
4	Spraying of appropriate insecticides and roughing of infected plant (2+3)	V1: Kufri Jyoti	T4V1
		V2: Cardinal	T4V2
5	Control	V1: Kufri Jyoti	T5V1
		V2: Cardinal	T5V5

Table 1: Treatment Combination.

Design	RCBD
Total Treatment combination	10
Number of replication	3
Area of individual plot	3 m × 2.5 m
Total plot area	3 × 2.5 × 10
Net experimental area	3 m × 2.5 m × 10 × 3

Table 2: Experimental Setup.

and Kufri Jyoti) of the region to study the rate of degeneration and to find out the optimum period up to which the seed stocks may be used without replacement and reduction in yield when potato seed is planted in the subsequent generations. The data on emergence after 30 and 60 days of sowing, uniformity, vigor, stem/plant, plant height, number of plant harvested, number and weight of under, seed and over size seed, incidence of disease and insects were observed.

The DAS ELISA results revealed that during third year the presence of PVM and PVY was higher in Cardinal under control treatment and in Kufri Jyoti under control condition respectively.

Regarding the percentage yield loss, result showed that the highest percentage loss in second year compared with the first year was highest in Kufri Jyoti (48.2%) in control condition followed by Cardinal (43.5%) in T2. Yield loss in the third year compared with second year was

highest in Kufri Jyoti (43.85% & 43.5%) in control and T3 respectively. In both the years, the yield loss of Kufri Jyoti in the control condition is highest this might be because of high infestation by aphid and presence of viral disease along with other environmental factor. PLRV, PVX and PVY virus was detected in Kufri Jyoti with control treatment in both the year. Surprisingly the highest yield loss was noticed in T2 and T3 is subsequent year in case of Cardinal and PVX, PVY and PLRV virus was detected in DAS ELISA though insecticides were sprayed in T2 and roughing was done in T3. Yield loss in the control treatment was more in every subsequent year compared to other treatment; this might be because of controlled aphid and other insects.

The observation was done on emergence, stem/plant and plant yield. In first and second year the emergence was highest in V1T2 and in the third year in V2T4. And lowest emergence was on V1T1, V2T2 and V2T1 in first, second and third year, respectively. Result on number of stems per plant was highest in V1T4 in the first year and V1T1 in the second and third year. Less number of stem per plant was in V2T4 in first and second year and V2T3 in the third year. Plant height was found higher in V1T1 in first and second year and V1T4 in the third year. Plant with lowest height was observed in V1T2 in first and second year and V2T5 in the third year (Figure 2).

Yield Loss vs. Virus Incidence

While performing ANOVA for finding the effect of virus on yield loss, the result revealed that it is highly significant (p>0.01). The result of this experiment showed that yield and the degree of virus infection have the linear relationship and fits the function (y=26.424+72.23x) with value of r^2 to be 0.1143. The result showed the less percentage yield loss in the plot having insect proof net that might be because of fewer aphids inside the very plot however some loss may be due to other environmental and genetic factors. Loss in production of tuber yield varies from 27% to 46% so it emphasizes the need for farmers to use the clean seed which is the main input for increasing productivity (Figure 3) (Table 3).

Conclusion

Results indicate that the yield reduction was 26–46% in treatment T4 and T5, respectively in Kufri Jyoti and 27.5 to 42.4%

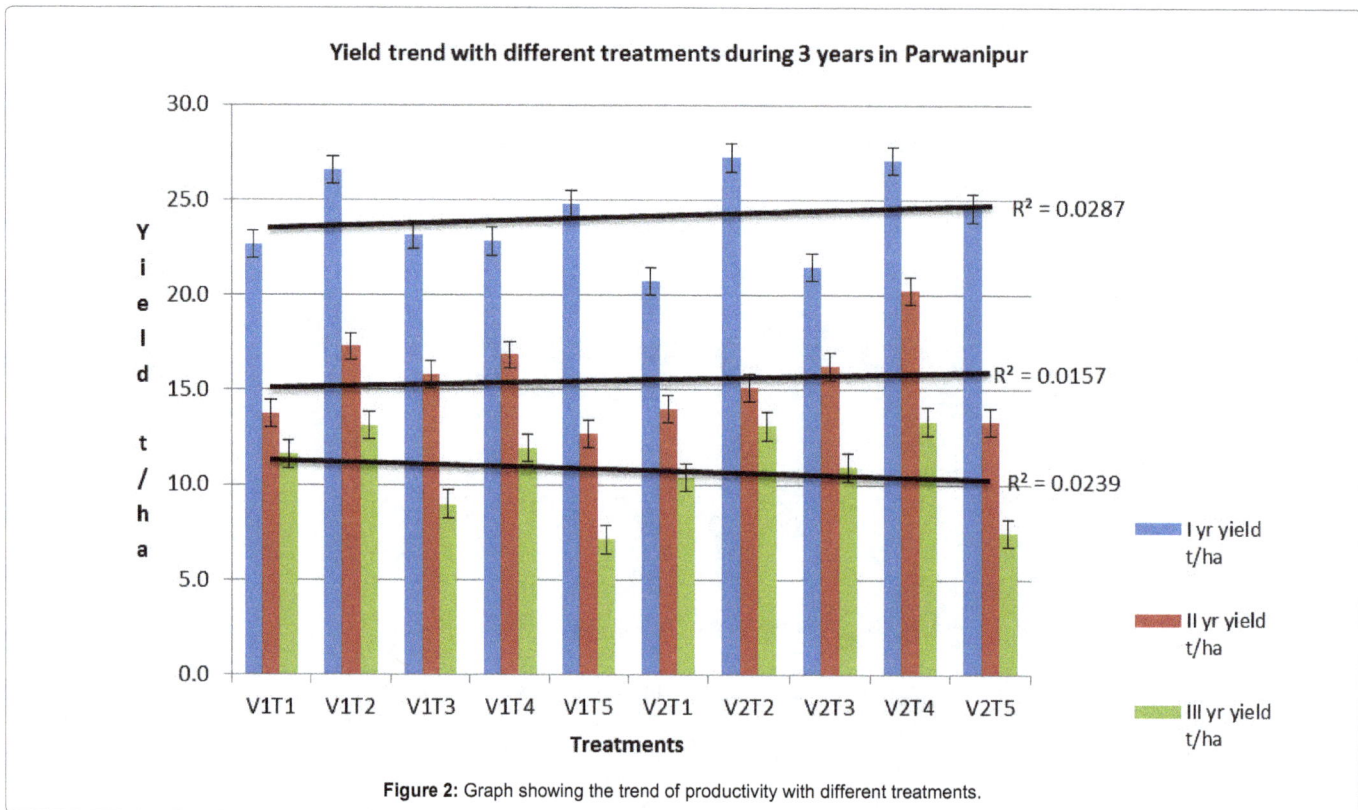

Figure 2: Graph showing the trend of productivity with different treatments.

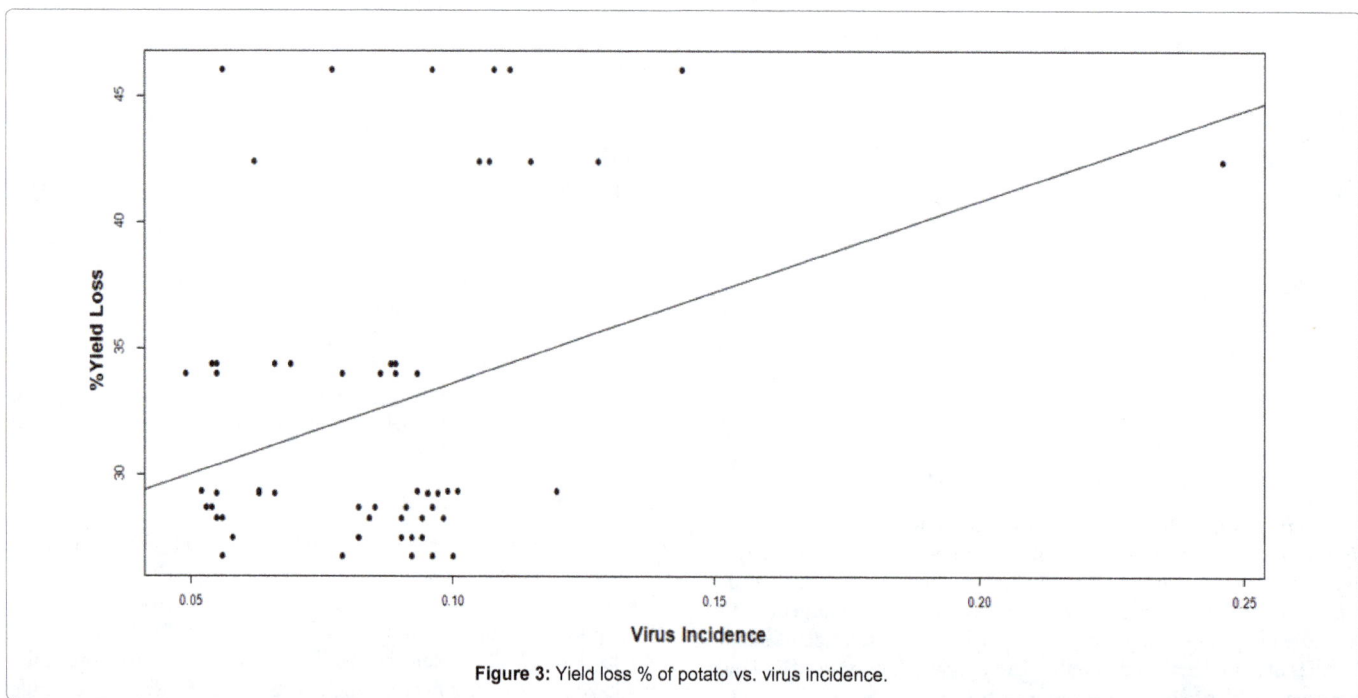

Figure 3: Yield loss % of potato vs. virus incidence.

in treatment T3 and T5 in Cardinal, respectively. PLRV, PVX and PVA virus was detected in Kufri Jyoti with control treatment in both the year and PVX and PLRV virus was detected in Cardinal with treatment T2 and T3 while testing through DAS ELISA. Crop yield was higher in treated plot compared to control in both variety giving the highest yield by Kufri Jyoti 16.6 t/ha under T4 and 13.1 t/ha under T2 in second and 3rd year, respectively. And in case of

Cardinal yield was higher in 20.2 under T4 and 13.1 under T2 in second and 3rd year, respectively. It gave an indication that the crop productivity can be raised 2-3 times higher than present average, if good quality seed socks are made available to the farmers at their door step. To decrease the rate of degeneration insect proof net can be used which is cheaper and environment friendly resulting the satisfactory yield in the subsequent generation.

```
Response: YL
                 Df  Sum Sq  Mean Sq   F value   Pr(>F)
VI                1  272.94  272.939   7.4863  0.008237 **
Residuals        58 2114.58   36.458
---
Signif. codes:  0 '***' 0.001 '**' 0.01 '*' 0.05 '.' 0.1 ' ' 1


Coefficients:
                 Estimate Std. Error  t value  Pr(>|t|)
(Intercept)        26.424      2.408   10.975  8.82e-16 ***
Virus Incidence    72.236     26.401    2.736   0.00824 **
---
Signif. codes:  0 '***' 0.001 '**' 0.01 '*' 0.05 '.' 0.1 ' ' 1

Residual standard error: 6.038 on 58 degrees of freedom
Multiple R-squared: 0.1143,      Adjusted R-squared: 0.09905
F-statistic: 7.486 on 1 and 58 DF,  p-value: 0.008237
```

Table 3: Analysis of Variance.

Acknowledgement

I would like to express my special thanks to the entire National Potato Research Program family.

References

1. Sakha BM, Rai GP, Dhital SP, Nepal RB (2007) Disease-free pre-basic seed potato production through tissue culture in Nepal. NARJ 8: 7-13.

2. Ali S, Kadian M, Ortiz O, Singh BP, Chandla VK (2013) Degeneration of Potato Seed in Meghalaya and Nagaland States in North Eastern Hills of India. Potato J 40: 122-127.

3. Sharma ST, Abdurahman A, Ali, Andrade-Piedra SL, Bao S et al., (2016) Seed degeneration in potato: the need for an integrated seed health strategy to mitigate the problem in developing countries. BSPP 65: 3-16.

4. NPRP (2013) Annual Report, s.l.: National Potato Research Program.

5. Jane M, Hussein S, Rob M (2013) Alleviating potato seed tuber shortage in developing countries: Potential of true potato seeds. Aus Jou Crop Sci 7: 1946-54.

6. Khurana SP et al., (1998) Degeneration of potato varieties in northern and central India. Indian Journal Virol 2: 111-119.

7. Rahman MS, Akanda AM, Mian IH, Bhuian MKA, Karim MR (2010) Growth and yield performance of different generations of seed potato as affected by PVY and PLRV. Banglad J Agric Res 37-50.

8. Bhandari A (1993) Sustainalbility measures of rice-wheat system across agro-ecological regions in Nepal. Doctoral Dissertation. Munoz, Philipinnes: Central Luzon State University.

9. Reddi TRaG (2002) Principle of agronomy. 2nd ed. Delhi, India: Kalayani Publisher.

34

The Effects of Insect Rearing Waste Compost on *Helianthus annuus* and *Tithonia rotundifolia*

Nall I Moonilall[1], Reed S[2] and Jayachandran K[1]*

[1]*Department of Earth and Environment, Florida International University, Miami, FL, USA*
[2]*USDA/ARS Subtropical Horticulture Research Station, Miami, FL, USA*

Abstract

In recent years, there has been a greater demand for growing substrates for ornamental plants. However, as cost rises and quantities of these materials become more limited, alternative forms of growing media are now being sought. A study was conducted to test the efficacy of using insect rearing waste as an alternative growing media for plants. Common sunflower (*Helianthus annuus* L.) and Mexican sunflower (*Tithonia rotundifolia* (Mill) S.F. Blake) were grown in different ratios of insect colony waste compost (ICW) combined with cardboard (Cb) (ICW+Cb) and nursery mix (NM) mixtures. The purpose of this experiment was to determine whether insect colony waste (ICW) from fruit fly rearing would sustain plant growth. Selective characteristics of the potting substrates revealed that the ratio of 100:0 ICW+Cb:NM had a 7.6 pH, 0.86 dS m^{-1} EC (salinity), 0.46 g cm^{-3} bulk density, and 50.1 percent water holding capacity at saturation. For common sunflower, there was a significant difference between the 100:0 and 0:100 ICW+Cb:NM blends for plant height, with the 100:0 ICW+Cb:NM mixture having the greatest height. For the Mexican sunflower, the 100:0 ICW+Cb:NM produced significantly more leaves and had a greater stem diameter than some of the other mixtures of potting substrate. There was no indication that the insect colony waste combined with cardboard (ICW+Cb) would inhibit plant growth. ICW+Cb have the potential to be used as an alternative substrate for growing plants.

Keywords: Soilless substrate; Potting media; Flowering plants; Ornamental plants; Container production

Introduction

Within the agricultural and horticultural industries, there has been an increased demand for substrates in which plants are grown and propagated. Much of these growing substrates require the use of harvesting techniques that are detrimental to the environment. Aspects of the surrounding environment are altered when harvesting these materials. This can result in damaging effects to a species' habitat or any other ecological process or service that is obtained from that area. An example of this can be seen during the harvesting of sphagnum peat moss and the effect there is on this wetland habitat [1,2]. Furthermore, peat and other organic soils, found in cold regions, are currently classified as net carbon (C) sinks [3]. Harvesting from these systems combined with projected climatic change in the future may transform these systems into a net carbon source. Soil organic carbon (SOC) will be lost from these systems as a result [3]. It is estimated that organic peat soils make up about one-third of the total SOC pool in the world [4]. Therefore, these systems assist greatly with the sequestration of carbon dioxide (CO_2). With the rising costs and limited supplies of some of the substrate materials used within the agricultural and horticultural industries, growers are looking for alternative substrates, which are economically and environmentally sound, to cultivate plants [1].

An ideal growing substrate should allow for proper plant growth and development, seed germination, and adequate nutrient and watering holding capacity. Chemical and physical properties also should be taken into consideration when selecting a growing media [5]. An understanding of physical characteristics such as bulk density, particle size distribution, porosity, and pore distribution along with chemical properties such as pH, electrical conductivity, and cation exchange capacity are needed when choosing a growing media [7,8]. Both organic and inorganic materials could be used in developing an alternative-growing medium. Materials such as peat, sawdust, wood fiber, coconut fiber, coir, compost, pumice, vermiculite, and perlite are often times mixed to form a soilless growing media [5,8]. Ideally, it is unwise to use a single particular material to grow plants because this may limit certain desirable characteristics. For example, vermiculite may provide adequate water and nutrient holding capacity but may not provide sufficient drainage. Similarly sawdust or coir could improve water holding capacity, however, nitrogen (N) may be immobilized due to a high C:N ratio. A combination of multiple materials would help to correct some limitations reached by only using one particular material. Furthermore, compost has the potential to sequester carbon and soil structure and aggregation within the soil [4]. Ultimately, the selection of materials used in a growing media includes cost, availability, and experience in using the material [5].

There are a variety of waste types that have been successfully used to cultivate plants. Table 1 summarizes various studies that have been done and shows results of using various waste streams. With regards to insect colony waste (ICW), the state of Florida rears sterile fruit flies for pest management purposes [9]. Some of the rearing facilities in Florida can produce up to a million flies on a weekly basis, while other facilities can produce in the hundred millions. As a result, a large amount of insect waste and waste material used to rear the insects are generated. The spent vermiculite with insect droppings and solid remains from an agar-based media used to rear flies is usually discarded. Cardboard is

*****Corresponding author:** Jayachandran K, Department of Earth and Environment, Florida International University, SW 8th Street, Miami, FL, USA E-mail: jayachan@fiu.edu

Author(s)	Type of waste stream	Plant used
[13]	Composted pig manure and perlite	*Cucumis sativus* (cucumber)
[14]	Pruning waste compost	*Lolium perenne* L. (perennial ryegrass) *Cupressus sempervirens* L. (cypress)
[15]	Grape (*Vitis vinifera*) stalk and grapevine marc	*Cucumis melo* (melon) *Solanum lycopersicum* (tomato) *Lactuca sativa* (lettuce) *Capsicum* (pepper)
[16]	Combinations of yard compost, raw coir, composted manure, forest compost, composted bark, and cattle manure compost	*Viburium tinus* L.
[17]	Various waste materials (pine bark, coconut (*Cocos nucifera*) fiber, and sewage sludge compost)	Coniferous plants: *Pinus pinea*, *Cupressus arizonica*, and *Cupressus sempervirens*
[18]	Municipal solid waste compost	*Solanum lycopersicum* (tomato)
[19]	Sewage sludge sugarcane trash compost and synthetic aggregates	*Lactuca sativa* L. (lettuce)
[20]	Biodegradable urban resources (biosolids and greenwaste) biochar	
[8]	Wood fiber, coconut fiber, and rockwool	*Solanum lycopersicum* (tomato)
[21]	Ground reed canary grass (*Phalaris arundinacea*) straw	*Fragaria x ananassa* (strawberry)
[22]	Spent mushroom substrates	*Solanum lycopersicum* (tomato) *Cucurbita pepo* L. (courgette) *Capsicum annum* L. (pepper)
[23]	Grass and pruning waste compost, vermicompost, and slumgum compost	*Rosmarinus officinalis*, *Cupressocypceris leylandii*, *Lactuca sativa*, *Allium cepa* *Petunia x hybrid Viola tricolor*
[24]	Municipal solid waste, sewage sludge composts, and bark	*Pistacia lentiscus* L.
[25]	Pine bark	*Fragaria x ananassa* (strawberry)
[26]	Palm waste	*Cucumis sativus* (cucumber)
[27]	Sawdust, powder of coconut coir, powder of maize (*Zea mays*) core, powder of soybean (*Glycine max*) stalk, and powder to peanut (*Arachis hypogaea*) hull	*Impatiens hawkeri* (impatiens)
[28]	Sawdust, powder of coconut coir, powder of maize (*Zea mays*) core, powder of soybean (*Glycine max*) stalk, and powder to peanut (*Arachis hypogaea*) hull	*Cyclamen persicum*
[29]	Almond (*Prunus dulcis*) shell waste	*Cucumis melo* (melon) *Solanum lycopersicum* (tomato)
[2]	Sweet corn tassels	*Solanum lycopersicum* (tomato)
[30]	Wood pellet biochar and pelletized wheat (*Triticum aeestivum*) straw biochar	*Solanum lycopersicum* (tomato) *Tagetes patula* (marigolds)
[31]	Potato (*Solanum tuberosum*) anaerobic digest and wood pellet biochar, wheat straw biochar, and pennycress presscake	*Solanum lycopersicum* (tomato) *Tagetes patula* (marigolds)
[32]	Spent mushroom substrate, perlite, and vermiculite	*Cucumis sativus* (cucumber) *Solanum lycopersicum* (tomato)
[33]	Biochar combined with humic acid	*Calathea insignis*

Table 1: Summary of studies on various waste streams as alternative growing media.

a material that is most times discarded after use but can serve a variety of purposes. A study conducted by Chong [10] looked at incorporating cardboard with compost to be used for growing plants in containers.

The objective of this research study was to test the efficacy of insect colony waste plus cardboard (ICW+Cb) as an alternative growing media for successful growth of *Helianthus annuus* and *Tithonia rotundifolia* and to determine characteristics of the ICW+Cb used. It was hypothesized that amendments of ICW+Cb would increase growth and development of both species of plants being grown in this media.

Materials and Methods

Insect colony waste plus cardboard

The United States Department of Agriculture-Agriculture Research Service Subtropical Horticulture Research Station (USDA-ARS-SHRS) in Miami, Florida provided materials for the compost, which included cardboard (Cb) from greenhouse cooling pads and insect colony waste (ICW) from fruit fly rearing. Insect colony waste consisted of a semi-solid agar-based media used in fruit fly larval rearing and spent vermiculite bedding used for pupation. A detailed explanation of the creation and process of obtaining the ICW can be found in Reed et al. [9].

The ICW was a result of media (or diet) used to rear fruit fly (*Anastrepha suspensa*) larvae and the vermiculite used during the pupation of the mature fruit fly larvae. The media that was created for the fruit fly was done by adding 1,296.43 g wheat germ, 1,296.43

g torula yeast, and 1,296 g sugar, 91.6 mL HCl, 65.36 g agar, 21.61 g sodium benzoate, 21.61 g methyl 4-hydroxybenzoate, and 10.80 g cholesterol into 20 L of water. Eggs of *A. suspensa* were then added to trays (58.42 cm × 27.94 cm × 5.08 cm) containing two liters (L) of media. The trays were stored in the fruit fly colony room at the USDA-ARS-SHRS at 26°C. At the end of eight days, the mature larvae were removed from the diet by washing with tap water. One kilogram (kg) of mature larvae was then added to four liters of vermiculite and allowed to pupate. At the end of 12 days, the pupated larvae were removed from the vermiculite and the spent vermiculite was stored in plastic barrel drums. To prepare the colony waste, two trays containing the larvae diet was washed through a 2 mm sieve into a bucket (19 L) containing 2/3 volume of spent vermiculite. The bottom of the bucket was removed and replaced with a 0.25 mm screen mesh. Mature larvae were thoroughly washed to remove any diet that was left. The water used to do the washing was allowed to drain freely in the bucket with the spent vermiculite. The resulting mixture of vermiculite and diet media wash was composted for duration of six weeks and allowed to air dry. The compost was then stored until needed [9].

Cardboard used for this study was dried and chipped into dimensions of 2 cm by ½ cm and combined with the ICW. The ICW+Cb were then composted for six months at the USDA-ARS-SHRS. The compost was turned every two weeks. Prior to potting, the material was steamed for 24 h at 100°C, allowed to cool, then steamed again for an additional 24 h at 100°C.

Potting substrate	pH	Salinity (dS m⁻¹)	B_d (g cm⁻³)	PW_Sat
0% ICW+Cb: 100% NM†	5.8	0.07	0.137	42.9
30% ICW+Cb: 70% NM†	7.5	0.39	0.258	44.7
70% ICW+Cb: 30% NM†	7.4	0.71	0.384	48.4
100% ICW+Cb: 0% NM†	7.6	0.86	0.463	50.1

Potting substrate consisted of composted insect colony waste with cardboard (ICW+Cb) plus different blends with a nursery mix (NM) containing 50% pine bark, 10% sand, and 40% coir pith.

Table 2: Potting substrate mean values for pH, salinity, bulk density (B_d) and percent water holding capacity at saturation (PW_{Sat}) for different treatment groups of potting substrates to be used in plant study.

Potting substrate	Plant height	Number of leaves	Stem diameter	Number of buds
0% ICW+Cb: 100% NM	27.5 b*	21	9.5	1
30% ICW+Cb: 70% NM	30.2 ab	21.8	9.7	0.7
70% ICW+Cb: 30% NM	31.1 ab	20.5	10.9	0.9
100% ICW+Cb: 0% NM	33.2 a	19.7	10	0.9

*Mean values in columns followed by a common letter are not significantly different at $p=0.05$. ns: Not significantly different at $p=0.05$.

Table 3: Common Sunflower (*Helianthus annuus*) mean values for height (cm), number of leaves, stem diameter (mm), number of buds, produced in different insect colony waste plus cardboard compost (ICW+Cb): nursery mix (NM) ratios.

Experimental design

A greenhouse plant growth study was conducted at one of the greenhouses at the USDA-ARS Subtropical Horticulture Research Station. The two plant species tested were common sunflower (*Helianthus annuus* L.) and tithonia (*Tithonia rotundifolia* (Mill.) S.F. Blake). The two species were grown in a mixture of composted insect colony waste plus cardboard (ICW+Cb) and a nursery mix (NM) in the following ratios: 0:100, 30:70, 70:30, and 100:0 ICW+Cb to NM. The nursery mix consisted of 50% pinebark, 10% sand, and 40% coir pith.

Potting material was analyzed for pH, Electrical Conductivity (EC) (salinity), bulk density (B_d), and percent Water Holding Capacity (WHC) at saturation. To measure pH and salinity, a dual channel pH/ion/ conductivity meter was utilized. For each of the four ratios of soil mixture, 90 mL of soil was measured out and combined with 90 mL of reverse osmosis water (ROH_2O). This created a 1 to 1 soil to water ratio by volume. This mixture was then shook for 30 min and then placed in a centrifuge. After being centrifuged, the liquid was decanted. Measurements of pH were taken first followed by electrical conductivity measurements for each of the potting material. Bulk density was obtained by gathering 135.37 cm³ (cylinder with dimensions 5.98 cm for height and 5.37 cm for diameter) of each potting mixture and oven drying it at 110°C. Percent WHC at saturation was determined by obtaining 50 mL of each potting mixture respectively in a beaker and dripping water, from a burette, into the mixture until it became completely saturated. Each of the tests conducted on the potting mixes were completed in duplicate.

For the greenhouse study, seeds for the two flower species were planted in two inch uncovered liners and allowed to germinate. When plants were 5 cm in height, they were transplanted into 2.48 L pots (15.6 cm × 15.9 cm) with the appropriate potting mix. Approximately 8.036 g of fertilizer was added to each pot in each treatment group. The fertilizer used was a 13 N -13 P_2O_5 -13 K_2O blend of Nutricote® Total Controlled Release Fertilizer (Florikan E.S.A. Corp. Sarasota, Florida, U.S.A.). The components of the fertilizer were as follows: 13% total N (6.5% NO_3, 6.5% NH_4), 13% P_2SO_5, 13% K_2O, 1.2% magnesium (water soluble), 0.02% boron, 0.05% copper (water soluble), 0.20% iron (chelated iron), 0.06% manganese (water soluble), 0.02% molybdenum.

Watering was accomplished utilizing drip irrigation three days a week for a total of 5 min each day. There was an initial amount of 157 mL of water being dispensed into each pot with an initial 10% of applied water leaching through the pot. Although not measured, it was observed that there was still adequate water draining from the pots. The design of the experiment was set up so that there was two plant species (*Helianthus annuus* L. and *Tithonia rotundifolia* (Mill.) S.F. Blake), four treatment groups (0:100, 30:70, 70:30, and 100:0 ICW+Cb to NM), and 10 replications per treatment. Once plants were transplanted into their respective pots, this signified the start of the experiment. Each pot was randomly sequenced at the end of each week. Plant data was taken at the 30 days and 60 days. Measurements that were recorded include plant height, number of leaves, stem diameter, bud count, and root mass. Stem diameter was measured using a caliper. The root mass was obtained after plants were harvested and washed to remove potting media. The roots were then excised at the base of the stem. Roots were oven dried at 45°C before weight was recorded. Plant height was measured from the base of the stem, right above the soil surface, to the top most leaf of the plant. Number of leaves and buds were counted respectively for those portions of the plant. The experiment lasted a period of 60 days until each plant had flowered.

Statistical analysis

Statistical analysis of the data was performed using SAS, version 9.4. The data collected for all plant parameters were subjected to analysis of variance using the general linear model (GLM) of SAS. The data that was collected were categorized into four treatment groups-0:100, 30:70, 70:30, and 100:0 ICW+Cb to NM. When there was an indication of a statistically significant difference at p<0.05, a post-hoc comparison utilizing the Tukey-Honest Significant Differences (HSD) test was carried out.

Results

Table 1 shows selective characteristics of potting substrates used for the different ICW+Cb:NM mixtures. The 100:0 ICW+Cb:NM mixture was the highest in pH, electrical conductivity, bulk density, and percent WHC at saturation followed by 70:30, 30:70, and 0:100 ICW+Cb:NM. The 0:100 ICW+Cb:NM yielded results that were lower than 100:0 ICW+Cb:NM. A pH of slightly alkaline was maintained in all mixtures above 30% ICW+Cb:NM. EC was low, with the 100% ICW+Cb:NM being suitable for salt sensitive plants. WHC increased as the percent ICW+Cb increased; however an increase to 42% ICW+Cb doubled B_d of the substrate. The amount of ICW+Cb used in a mixture should depend on the pH of the media and the drainage requirement of the plant.

A statistically significant difference was indicated for plant height for the common sunflower: $F_{(3,36)}=3.19$, p<0.035. The results followed the order of 100:0 ≥ 70:30=30:70 ≥ 0:100 (Table 2). There was a significant difference between the 100:0 and 0:100 ICW+Cb:NM blend for plant height. Stem diameter, number of leaves, and number of buds were each similar in all treatments. No characteristic of ICW+Cb reduced growth compared to the commercial nursery mix control. Visual observation showed that plants grown in the 100:0 ICW+Cb had larger leaves overall than those grown in 0:100 ICW+Cb.

There were no treatment differences in plant height, number of buds and root weight with tithonia (Tables 3 and 4). Plant height for tithonia followed the order of 100:0>70:30>30:70>0:100. A statistically significant difference was found for stem diameter for tithonia: $F_{(3,28)}=9.67$, p<0.0002. An ICW+Cb:NM content of 100:0 produced plants with a greater stem diameter than the 0:100, 30:70, and 70:30

Potting substrate	Plant height	Number of leaves	Stem diameter	Number of buds	Root dry weight
0% ICW+Cb: 100% NM	32.6	88.0 ab	8.0 b	3.4	1.8
30% ICW+Cb: 70% NM	39.8	71.0 b	9.0 b	4.2	2.5
70% ICW+Cb: 30% NM	39.3	81.0 ab	8.9 b	3.1	2.8
100% ICW+Cb: 0% NM	38.8	97.3 a	11.0 a	4	4.2

*Mean values in columns followed by a common letter are not significantly different at *p*=0.05. ns: Not significantly different at p=0.05.

Table 4: Mexican Sunflower (*Tithonia rotundifolia*) mean values for height (cm), number of leaves, stem diameter (mm), number of buds, and root dry weight (g) produced in different insect colony waste plus cardboard compost (ICW+Cb): nursery mix (NM) ratios.

ICW+Cb:NM compost blends. A statistically significant difference existed for number of leaves for tithonia plants: $F_{(3,28)}$=4.70, p<0.01. The 100:0 mixture produced significantly more leaves than the 30:70 ICW+Cb:NM.

Discussion

There is no indication that ICW+Cb would inhibit plant growth. It was hypothesized that the amendments of ICW+Cb would increase growth and development of the two plants being grown. All plants grown in this study exhibited growth that would have been greater than or equal to growth using the Nursery Mix (NM). This is an indication that ICW+Cb can be an amendment with a tradition nursery to cultivate floral plants.

Composted ICW+Cb can be used in a combination with other materials to create a potting mix that is high in nutrients. Reed [9] reported that ICW contained moderate to very high levels of N, K, Mg, Fe, Zn, and Cu plus low to moderate levels of P. Fertilizer use would be minimized by utilizing ICW due to the high nutrients that are available for the plant for the duration of one growing season. Addition of composted ICW+Cb increased the bulk density and drainage while maintaining its ability to hold a higher content of water than the commercial nursery mix. A higher bulk density provides adequate structural support for large plants. Too much of an increase in bulk density can create a disadvantage where there is a reduction of porosity and thus air capacity for the potting media in which the plant is growing [11]. Furthermore, from a nursery management perspective, an increase in bulk density would result in higher transportation costs of the media due to increased weight [11]. Abad [12] provided evidence that the optimal range for bulk density for container media should be close to or less than 0.4 g cm^{-3}. All of the treatment groups within this study fell within that range except for 100:0 ICW+Cb:NM. Optimal pH range for both plant species is between pH 6.0-7.5. The soil pH values of the growing mixes with 30%, 70%, and 100% ICW+Cb all had values that fall within this range. The 0% ICW+Cb treatment group had a pH value lower than the optimal range for this species. The addition of the ICW+Cb allowed for the pH to increase with a minimum of 30% addition. Optimal availability for plant nutrient uptake for both macronutrients and micronutrients are available at pH levels of 6.0 to 7.5, the optimal pH at which sunflowers grow. Growing in a medium with the appropriate pH levels would ensure that nutrients are being obtained efficiently and that the plant is growing without limitations. Addition of up to 30% ICW+Cb to a potting substrate can reduce the amount of fertilizer and irrigation water applied to plants. This will have the potential to save growers money and at the same time conserve natural resources within the environment. Composted ICW+Cb could be a great component of soilless growing media and could, as a considerable waste product, be utilized as an amendment to boost productivity of plant growth.

Acknowledgements

We would like to give special thanks to the Subtropical Horticulture Research Station (USDA) and the Agroecology Program at Florida International University for their continued support of our academic endeavors. Funding for this project was provided by USDA NIFA Multicultural Scholars Program (2009-38413-05236). Mention of a trademark, proprietary product, or vendor does not constitute a guarantee or warranty of the product by the U.S. Department of Agriculture and does not imply its approval to the exclusion of other products or vendors.

References

1. Moral R, Paredes C, Bustamante MA, Marhuenda-Egea F, Bernal MP (2009) Utilisation of manure composts by high-value crops: Safety and environmental challenges. Bior Tech 100: 5454-5460.

2. Vaughn SF, Deppe NA, Palmquist DE, Berhow MA (2011) Extracted sweet corn tassels as a renewable alternative to peat in greenhouse substrates. Ind Crops Prod 33: 514-517.

3. Lal R (2004) Soil carbon sequestration to mitigate climate change. Geoderma 123: 1-22.

4. Lal R (2001) World cropland soils as a source or sink for atmospheric carbon. Advances in Agronomy 71: 145-191.

5. Ashraf S (2011) The effect of different substrates on the vegetative, productivity characters and relative absorption of some nutrient elements by the tomato plant. Adv Environ Biol 5: 3091-3096.

6. Wallach R (2008) Physical characteristics of soilless media. Soilless Culture: Theory and Practice, pp: 41-116.

7. Maher M, Prasad M, Raviv M (2008) Organic soilless media components. Soilless Culture: Theory and Practice, pp: 459-504.

8. Kowalczyk K, Gajc-Wolska J, Marcinkowska M (2011) The influence of growing medium and harvest time on the biological value of cherry fruit and standard tomato cultivars. Veg Crops Res Bul 74: 51-59.

9. Reed S, Epsky ND, Heath RR, Joseph R (2012) Evaluation of composted insect rearing waste as a potting substrate for radish, squash and green bean. Compost Sci Utilizat 20: 87-91.

10. Chong C (1999) Experiences with the utilization of wastes in nursery potting mixes and as field soil amendments. Canadian J Plant Sci 79: 139-148.

11. Atiyeh RM, Edwards CA, Subler MJ (2001) Pig manure vermicompost as a component of a hortcultural bedding plant medium: Effects on physicochemical properties and plant growth. Bioresource Technol 78: 11-20.

12. Abad M, Noguera P, Burés S (2001) National inventory of organic wastes for use as growing media for ornamental potted plant production: Case study in Spain. Bioresource Technol 77: 197-200.

13. Al Naddaf O, Livieratos I, Stamatakis A, Tsirogiannis I, Gizas G (2011) Hydraulic characteristics of composted pig manure, perlite, and mixtures of them, and their impact on cucumber grown on bags. Scientia Hortic 129: 135-141.

14. Benito M, Masaguer A, De Antonio R, Moliner A (2005) Use of pruning waste compost as a component in soilless growing media. Bioresource Technol 96: 597-603.

15. Carmona E, Moreno MT, Avilés M, Ordovás J (2012) Use of grape marc compost as substrate for vegetable seedlings. Scientia Hortic 137: 69-74.

16. Guérin V, Lemaire F, Marfà O, Caceres R, Giuffrida F (2001) Growth of Viburnum tinus in peat-based and peat-substitute growing media. Scientia Hortic 89: 129-142.

17. Hernández-Apaolaza L, Gascó AM, Gascó JM, Guerrero F (2005) Reuse of waste materials as growing media for ornamental plants. Bioresource Technol 96: 125-131.

18. Herrera F, Castillo JE, Chica AF, López Bellido L (2008) Use of municipal solid waste compost (MSWC) as a growing medium in the nursery production of tomato plants. Bioresource Technol 99: 287-296.

19. Jayasinghe GY, Tokashiki Y, Arachchi IDL, Arakaki M (2010) Sewage sludge sugarcane trash based compost and synthetic aggregates as peat substitutes in containerized media for crop production. J Hazard Mater 174: 700-706.

20. Kaudal BB, Chen D, Madhavan DB, Downie A, Weatherley A (2015) Pyrolysis of urban waste streams: Their potential use as horticultural media. Journal of Analytical and Applied Pyrolysis 112: 105-112.

21. Kuisma E, Palonen P, Yli-Halla M (2014) Reed canary grass straw as a substrate in soilless cultivation of strawberry. Scientia Hortic 178: 217-223.

22. Medina E, Paredes C, Pérez-Murcia MD, Bustamante MA, Moral R (2009) Spent mushroom substrates as component of growing media for germination and growth of horticultural plants. Bior Tech 100: 4227-4232.

23. Morales-Corts MR, Gómez-Sánchez MA, Pérez-Sánchez R (2014) Evaluation of green/pruning wastes compost and vermicompost, slumgum compost and their mixes as growing media for horticultural production. Scientia Hortic 172: 155-160.

24. Ostos JC, López-Garrido R, Murillo JM, López R (2008) Substitution of peat for municipal solid waste- and sewage sludge-based composts in nursery growing media: Effects on growth and nutrition of the native shrub Pistacia lentiscus L. Bior Tech 99: 1793-1800.

25. Paranjpe AV, Cantliffe DJ, Stoffella PJ, Lamb EM, Powell CA (2008) Relationship of plant density to fruit yield of "Sweet Charlie" strawberry grown in a pine bark soilless medium in a high-roof passively ventilated greenhouse. J Scienta 115: 117-123.

26. Pooyeh F, Peyvast G, Olfati JA (2012) Growing media including palm waste in soilless culture of cucumber. Int J Veg Sci 18: 20-28.

27. Qing-chao L, Kui-ling W, Qing-hua L, Hui-tang P, Qi-xiang Z (2013) The effect of the substitute media on the development of the potted New Guinea Impatiens (Impatiens hawkeri). Acta Ecologica Sinica 33: 293-300.

28. Qing-chao L, Kui-ling W, Qing-hua L, Hui-tang P, Qi-xiang Z (2014) Effects of Substitute Media on Development of Potted Cyclamen percicum Mill. Journal of Northeast Agricultural University (English Edition) 21: 28-37.

29. Urrestarazu M, Martínez GA, Salas MDC (2005) Almond shell waste: Possible local rockwool substitute in soilless crop culture. Scientia Hortic 103: 453-460.

30. Vaughn SF, Eller FJ, Evangelista RL, Moser BR, Lee E, et al. (2015) Evaluation of biochar-anaerobic potato digestate mixtures as renewable components of horticultural potting media. J Indcrop 65: 467-471.

31. Vaughn SF, Kenar JA, Thompson AR, Peterson SC (2013) Comparison of biochars derived from wood pellets and pelletized wheat straw as replacements for peat in potting substrates. Industrial Crops and Products 51: 437-443.

32. Zhang L, Sun XY, Tian Y, Gong XQ (2014) Biochar and humic acid amendments improve the quality of composted green waste as a growth medium for the ornamental plant Calathea insignis. Scientia Hortic 176: 70-78.

33. Zhang RH, Duan ZQ, Li ZG (2012) Use of Spent Mushroom Substrate as Growing Media for Tomato and Cucumber Seedlings. Pedosphere 22: 333-342.

35

The Impact of Phosphorus Fertilizers on Heavy Metals Content of Soils and Vegetables Grown on Selected Farms in Jordan

Asad M F AlKhader*

Water, Soil and Environment Department, National Center for Agricultural Research and Extension, Jordan

Abstract

A survey was conducted to investigate the levels of Cd, Pb and As heavy metals in soils, leafy vegetables (lettuce plant), and irrigation water in areas characterized by intensive agricultural activities in Jordan. Thirteen farms from three locations (Jordan Valley, Alyadoda, and Jarash) were selected for this purpose. Ten P fertilizers that are most widely used by farmers were also collected and analyzed for heavy metals content. The results indicated that the lettuce, used as an indicator plant for possible vegetables contamination with heavy metals, was within allowable levels of Cd and Pb of 0.2 and 0.3 mg kg^{-1} of fresh weight for leafy vegetables, respectively. The plant was, also, safe with respect to As as the level of this metalloid was much less than the established acceptable concentration of 1 mg kg^{-1} fresh weight. The results suggested that the most probable sources of the heavy metals (Cd and Pb) and metalloid (As) in the collected samples of soils and plant from the selected farms were soil parent materials and pesticides application. Long term P fertilizers additions are, also, likely sources of heavy metals in agricultural soils and crops. This implies a risk to the human health and environment in the future is expected.

Keywords: Heavy metals; Lettuce; Phosphorous; Fertilizers; Contamination; Fresh weight

Introduction

Phosphorous (P), which is supplied mainly through chemical fertilizers application, is considered an essential nutrient element for growth and development of plant crops. Its deficiency constitutes a major limiting factor in the crop production of the world [1]. Heavy metals like cadmium (Cd), lead (Pb) and arsenic metalloid (As) have been found in P fertilizers and are considered the most important of health concern [2]. These elements are regarded toxic [2,3] and classified as carcinogenic [2,4,5]. For example, Galadima and Garba [6] and Agwaramgbo et al. [7] reported that poisoning by Pb in Nigeria killed more than 500 children, and left thousands in severe health conditions in 2010. Recent studies, also, have demonstrated that As and other toxic heavy metals like Cd and Pb were responsible for causing a chronic kidney disease, known as toxic nephropathy, in contaminated areas in Siri Lanka [8]. Moreover, it was pointed out that P fertilizers (triple super phosphate) and pesticides were the main source of the heavy metals. Cadmium is a highly mobile metal and found to accumulate in plants in large amounts without showing phytotoxic symptoms. It is, therefore, considered as one of the most serious heavy metals to human health [9-11]. Moreover, Cd tends to accumulate in vegetables more than other heavy metals; for this reason Cd can enter the food chain by ingestion of vegetables [12]. In order to protect human health from food contaminants, JECFA (1989) [13] set provisional tolerable weekly intakes (PTWI) for inorganic As and Cd at 15 and 7 µg kg^{-1} body mass, respectively, while PTWI for Pb was established at 25 µg kg^{-1} body mass [14]. Lettuce (*Lactuca sativa* L.) showed a high capability to absorb Cd from the soil and considered an accumulator for heavy metals in its leave tissues [15,16]. Moreover, Wrobel [17] reported that lettuce plant (cv. Loreto) grown on a contaminated soil from a copper smelting plant showed Cd and Pb levels of up to 0.14 and 0.8 ppm (on fresh mass basis), respectively, compared with 0.02 and 0.2 ppm for vegetation grown on uncontaminated soil. Similarly, Mausi et al. [18] indicated that the permissible level of Pb (0.3 mg/kg) in oranges and mangos fruits was exceeded as they recorded mean values of 0.65 and 0.61 mg/kg, respectively. These high levels were attributed to the use of pesticides, fertilizers and wastewater. However, the levels of Cd in both fruits were within the recommended level of 0.2 mg/kg, where the recorded concentrations were 0.089 mg/kg in mangos and 0.057 mg/kg in oranges. Maalem, et al. [19] also found that P fertilization of *Atriplexes* induced higher levels of Cd in their leaves than control plants and even exceeded the standards. They suggested that phosphates raise the levels of Cd in the soil and, thus, its bioavailability for the plant is increased. However, the rate of transfer of the Cd from soil to the plant varied from 3 to 6, depending on the plant species. Actually, *Atriplexes*, used as fodder, are regarded as hyper-bioaccumulaters for toxic heavy metals like Cd and constitute a potential risk of contamination of the food chain [19]. Moreover, Wagesho and Chandravanshi [20] confirmed the reliance of the metals levels (Cd among them) in the plant (ginger) on their relevant levels in the soil where it has been grown. On the other hand, local research works have indicated that heavy metals like Cd (9.2-10.9 ppm) and Pb (1.2-32.5 ppm) are found in phosphate rock of Jordan which is used primarily in the production of P fertilizers [21,22]. In Addition, Alkhader and Abu Rayyan [22] investigated some P fertilizers like di ammonium phosphate (DAP), mono ammonium phosphate (MAP) and single super phosphate (SSP) and reported that Cd (0.5-7.9 ppm), Pb (1.8-2.2 ppm) and As (2.8-43.0 ppm) were contained as contaminants. Moreover, Ghrefat et al. [23] pointed out that fertilizers application induced high levels of Cd (4.6 ppm) and Pb (58.4 ppm) in soils located beside the Zerqa River. Therefore, the objective of this study is to investigate the possible contamination of soils, vegetable plants, and irrigation water in intensively cultivated

***Corresponding author:** Asad M F AlKhader, Water, Soil and Environment Department, National Center for Agricultural Research and Extension (NCARE), P. O. Box (639)-Baqa 19381 Jordan, E-mail: asad_fathi@yahoo.com

areas in Jordan, with heavy metals (Cd and Pb) and As metalloid. Some widely used P fertilizers will also be investigated as prolonged fertilizers application might be one of the most probable reasons for the contamination. Lettuce was used as an indicator plant for potential heavy metals contamination of vegetables.

Materials and Methods

Farms selection

Thirteen farms from three locations characterized by intensive agricultural activities in Jordan (Jordan Valley, Alyadoda, and Jarash) were selected for soil, plant, fertilizers and irrigation water sampling during the spring/summer period of the 2010 year.

Soil

Three composite soil samples at 0-20 cm depth were collected from each selected farm for some chemical and physical analysis. The samples were air dried, crushed and passed through a 2 mm sieve. Soil pH and electrical conductivity (EC) for the paste extract were determined according to Bower and Wilcox [24], cation exchange capacity (CEC) according to Chapman [25], organic matter according to Allison [26], calcium carbonate (calcimeter method) according to Allison and Moodie [27], total N (Kjeldhal method) according to Bremner [28], available P (using spectrophotometer) according to Olsen and Dean [29] and available K (using flame photometer) according to Pratt [30]. Soil 0.005 M diethylenetriaminepentaacetic acid (DTPA)-extractable Cd and Pb, and 0.5 M $NaHCO_3$- extractable As were determined according to Lindsay and Norvell [31], and Shiowatana et al. [32], respectively. Atomic absorption spectrophotometer (AAS) (Model Varian, Spectr. AA-200, Australia) was used in these determinations, with instrument detection limits for Cd, Pb and As as 0.002 ppm, 0.01 ppm and 0.2 ppb, respectively. Soil texture (hydrometer method) was determined according to Day [33].

Fertilizer

Levels of nutrients (N, P and K), heavy metals (Cd and Pb) and metalloid (As) in ten P fertilizers which are widely used in the investigated farms were determined according to Horwitz and Latimer [34]. Cadmium, Pb and As were, also, determined using AAS.

Irrigation water

Chemical analysis for the irrigation water samples collected from the investigated farms was conducted to determine pH, EC, major cations and anions according to Chapman and Pratt [35]. Cadmium, Pb and As concentrations were measured using AAS.

Plant

Lettuce plant (iceberg type) samples were collected from only three farms (farms no. 11, 12 and 13) which were cultivated with this crop out of the 13 selected farms. Three plants from each farm were used to make representative samples. Plants were weighed firstly to have fresh weight and then rinsed with tap water followed with distilled water and dried in an oven at 65^0 C for 72 h and their dry matters were determined. After that they were ground by stainless steel grinder to pass a 1 mm stainless steel sieve for chemical analysis. Each plant sample of 1.0 g weight was transferred into a silica crucible and placed in a muffle furnace at 500^0C for 4 h in dry-ashing process. The crucible was left to cool and then 5 ml of 6 N HCl was added. After that, the crucible was placed on a hot plate and digested to obtain a clear solution. The residue was dissolved in 0.1 N HNO_3 and transferred to a 50 ml volumetric flask and completed to the mark with deionized water. Standard solutions of the elements (Cd, Pb, and As) were prepared from stock solutions (1000 ppm) by dilution with 0.1 N HNO_3 for linearity inspection. The plant contents of these elements were determined using AAS. Measurements were taken in triplicate and averaged. Total N, P and K in the plant samples were determined according to Chapman and Pratt [35].

Results and Discussion

Chemical and physical analysis for the soils

The results of some chemical and physical properties of the collected soil samples (0-20 cm depth) from the 13 selected farms are presented in Table 1. The investigated soils showed properties that ranged in their values as follows: pH (7.9-8.7), salinity (0.65-32.8 dS/m), total N (0.02-0.22%), available P (4.9-130.4 ppm), and K (275.4-730.9 ppm), DTPA-extractable Cd (0.01-0.22 ppm), and Pb (0.40-1.90 ppm), and $NaHCO_3$–extractable As (0.69-17.77 ppm). The values were averages of three composite soil samples collected from each farm. The soil samples had different textural classes, as shown in the table. The results indicated that uncultivated virgin soil (Farm no. 10) had relatively

Table 1: Average values for some chemical and physical properties of the soils (0-20 cm depth) from the selected farms in the survey.

Farm number	Location	pH	Salinity	Total N	Available		Extractable			Texture
					P	K	Cd	Pb	As	
			dS/m	%	(ppm)					
1	Middle Jordan Valley	8.1	2.76	0.08	23.1	572.9	0.028	0.62	9.04	Clay
2		8.0	1.26	0.22	130.4	730.9	0.066	0.78	7.45	
3		8.2	24.2	0.12	63.7	461.3	0.158	0.84	16.41	Clay
4		8.1	3.47	0.14	98.4	535.7	0.058	0.92	0.69	
5		8.2	2.05	0.07	87	284.7	0.024	0.56	12.56	Clay loam
6		8.3	1.96	0.12	97.6	479.9	0.044	0.74	4.99	
7		8.2	2.28	0.08	70.6	294	0.014	0.46	2.69	Sandy clay loam
8		8.1	5.72	0.06	42.9	396.2	0.136	0.72	7.92	Clay loam
9		8.2	1.52	0.08	68.4	275.4	0.022	1.9	17.77	Sandy clay loam
10	Southern Jordan Valley/cultivated	8.7	27.3	0.1	84.9	684.4	0.028	0.4	16.24	Sandy clay loam
	Southern Jordan Valley/uncultivated	8.5	32.8	0.02	4.9	331.1	0.01	0.64	0.85	Sandy loam
11	Al-Yadoda	7.9	1.13	0.14	81.4	572.9	0.216	0.84	2.46	Clay
12		8.3	0.65	0.09	33.1	377.6	0.11	0.9	1.08	
13	Jarash	8.0	1.95	0.08	17.9	563.6	0.024	0.52	0.75	Clay

lower values of available P(4.9 ppm), K(331.1 ppm), total N(0.02%) and extractable Cd (0.01 ppm), Pb (0.64 ppm) and As (0.85 ppm) than those of most of the cultivated soils of the same farm and other farms. The source of the heavy metals in this virgin soil is mainly from natural resources like soil parent material [36]. However, fertilizer addition is the most probable reason for the high contents of the nutrients in the agricultural soils [37], whereas, the potential source for the heavy metals might be, primarily, from long-term using of pesticides [18,23,36,38,39]. Addition of P fertilizers might, also, contribute to the heavy metals contents of these soils [11,18,23,39-42]. This can be inferred from the poorly positive correlations between the soil available P and soil DTPA-extractable Cd and Pb, and NaHCO$_3$-extractable As (r=0.109, 0.111 and 0.249, respectively) as depicted in Figure 1. However, the cultivated soil in the Farm no. 10 can be considered as contaminated soil, especially, with respect to Cd (0.03 ppm) and As (16.24 ppm), as compared with the virgin uncultivated soil of the same farm. This is because the levels of these heavy metals exceeded those in the unfertilized soil more than 2-3 times [11]. On the other hand, the relatively high salinity level for the uncultivated soil might be due to the inadequate rainfall to leach downward the natural accumulated soluble salts from surface where evaporation exceeds precipitation [43]. Also, the high content of the available K in this soil was attributed to the presence of high amounts of soluble salts, whereas long-term additions of fertilizers might, also, be responsible for the high K contents in the other agricultural soils [44].

Chemical analysis for fertilizers

As shown in Table 2, the maximum levels of Cd, Pb and As in the ten most commonly forms of P fertilizers used by farmers in the selected farms were 7.9 ppm (DAP), 8.2 ppm (NPK, 16:8:24) and 43.0 ppm (MAP), respectively. Generally, these levels are considered below the critical limits of Cd (20 ppm), Pb (500 ppm) and As (75 ppm) in fertilizers according to the Canadian Standards [45]. However, the investigated P fertilizers can be regarded as one of the potential sources of these heavy metals in the agricultural soils and crops in Jordan under long-term, heavy, and continuous application [11,42,46,47].

Chemical analysis for the irrigation water

Results of the chemical analysis of the irrigation water samples collected from the selected farms showed that the pH values ranged from 7.1 to 8.4, while, the water salinity (EC) varied from 0.7 to 3.4 dS/m. On the other hand, the levels of Cd, Pb and As were below the instrumental detection limits, as shown in Table 3.

Chemical analysis for lettuce plant

The results suggested that lettuce plants collected from the three farms were within allowable levels of Cd and Pb of 0.2 and 0.3 mg kg^{-1} of fresh weight for leafy vegetables, respectively [48], as shown in Table 4. The high pH values of the investigated soils may be responsible for the lower levels of the positively charged heavy metals of Cd and Pb [49]. This could be related to immobilization of the heavy metals which limits their bioavailability to the plant [49,50]. The formation and precipitation of metal hydroxides are enhanced under such environments. This means that the concentration of the heavy metals in the immobile fraction, consecutively, increases [51]. The plant lettuce samples, also, were safe with respect to As as their contents of this heavy metal were much less than the established permissible concentration of 1 mg kg^{-1} fresh weight [38]. The nutrients content (N, P and K) of the plant samples are also presented in the table. It was ranged from 2.56-3.50% for N, 0.61-0.65% for P and 7.59-12.09% for K. According to the previously established nutrient leaf sufficiency ranges [52]; the N and P

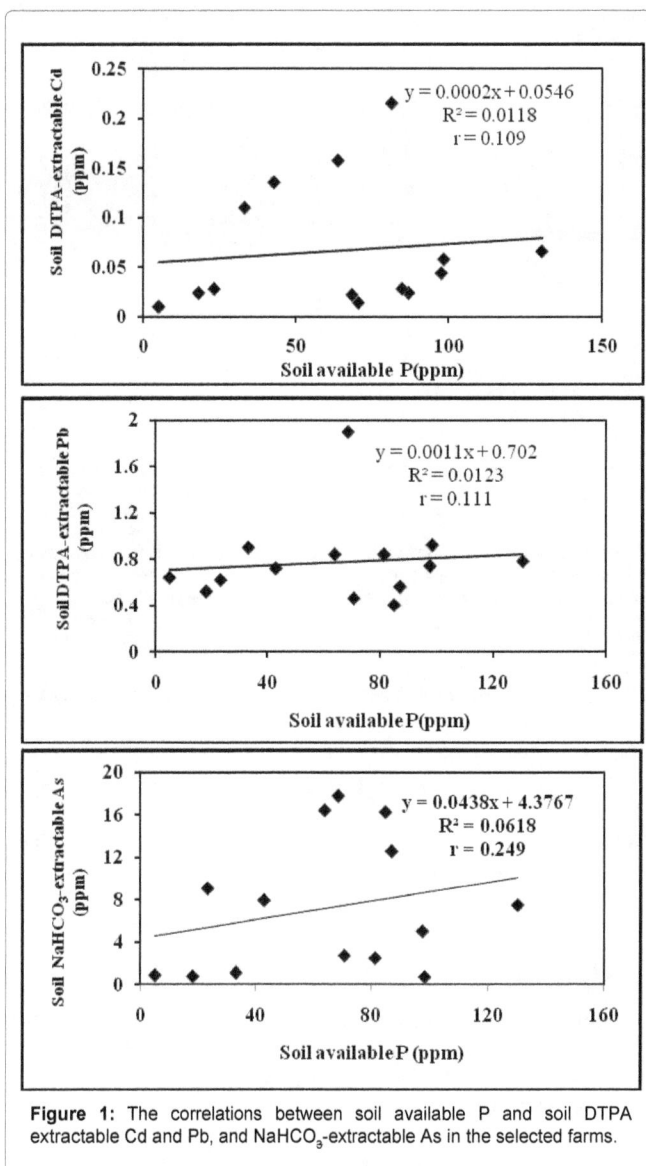

Figure 1: The correlations between soil available P and soil DTPA extractable Cd and Pb, and NaHCO$_3$-extractable As in the selected farms.

Table 2: Average values of some nutrients and heavy metals contents for some selected chemical fertilizers usually used by farmers in Jordan.

Fertilizer		Nutrients			Heavy metals*		
		N	P$_2$O$_5$	K$_2$O	Cd	Pb	As
		%			(ppm)		
1.	Urea Phosphate	17.4	46.7	0	2.76	0.4	13.74
2.	DAP	18.2	44.0	0	7.9	2.1	2.8
3.	MAP	12.3	61.1	0	0.5	1.8	43.0
4.	SSP	0	17.4	0	6.1	2.2	5.5
5.	NPK	13	40	13	1.02	5.8	0.26
6.	NPK	15	15	30	0.7	6	3.77
7.	NPK	30	10	10	0.42	3.4	7.85
8.	NPK	20	5	10	0.6	5	1.85
9.	NPK	19	19	19	0.86	5.6	16.36
10.	NPK	16	8	24	0.8	8.2	0.70

*The critical limits of Cd, Pb and AS in chemical fertilizers are 20, 500, and 75 ppm, respectively, according to the Canadian Standards (Heckman 2006).

Table 3: Results of chemical analysis for the irrigation water samples from the selected farms.

Farm number	Location	pH	EC	Cd	Pb	As
			dS/m	(ppm)		(ppb)
1	Middle Jordan Valley	7.1	2.2	<0.002	<0.01	< 0.2
2		7.3	1.4	<0.002	<0.01	< 0.2
3		7.2	3.4	<0.002	<0.01	< 0.2
4		8.3	2.3	<0.002	<0.01	< 0.2
5		8.4	1.7	<0.002	<0.01	< 0.2
6		8.4	1.7	<0.002	<0.01	< 0.2
7		8.3	1.7	<0.002	<0.01	< 0.2
8		8.4	1.7	<0.002	<0.01	< 0.2
9		8.3	1.8	<0.002	<0.01	< 0.2
10	Southern Jordan Valley	8.4	1.8	<0.002	<0.01	< 0.2
11	Al-Yadoda	7.6	0.9	<0.002	<0.01	< 0.2
12		7.7	0.7	<0.002	<0.01	< 0.2
13	Jarash	7.7	0.8	<0.002	<0.01	< 0.2

Table 4: Average values of the heavy metals (Cd, Pb), metalloid (As) and nutrients (N, P and K) contents of lettuce plant (iceberg type) from three investigated farms in the conducted survey during spring-summer period of the year 2010.

Farm no.	Cd	Pb	As	N	P	K
	(ppm)		(ppb)	%		
	Fresh weight basis				Dry weight basis	
11	0.05	0.2	10.76	3.5	0.65	12.09
12	0.04	0.25	12.76	3.31	0.62	11.5
13	0.03	0.12	12.78	2.56	0.61	7.59

contents were generally in close agreement, meanwhile the K content was higher. Hartz et al. [52] suggested that the leaf optimum ranges for lettuce (iceberg and romaine types) were 3.3-4.8% for N, 0.35-0.75% for P and 2.9-7.8% for K.

Conclusions

Lettuce which was considered as an indicator plant for potential heavy metals contamination of vegetables was within the allowable levels of Cd and Pb of 0.2 and 0.3 mg kg^{-1} of fresh weight for leafy vegetables, respectively. The plant was, also, safe with respect to As as the level of this metalloid was much less than the established acceptable concentration of 1 mg kg^{-1} fresh weight. Long term applications of P fertilizers and pesticides are likely sources of heavy metals in agricultural soils and crops in Jordan. This, essentially, may constitute a threat to the human health and surrounding environment.

Recommendations

A national strategy should be developed and adopted in Jordan to monitor and minimize the concentration of the heavy metals and inputs into agricultural soils and their transfer to the plant crops. This could help protect the environment from pollution and, thus, jeopardy to the human health could be reduced.

Acknowledgements

The author would like to thank farmers of the investigated farms for their collaboration during samples collection of plant, water, soil and fertilizers. Special gratitude and appreciation for the team work of The National Center for Agricultural Research and Extension (NCARE) Laboratories (Soil, Water and Fertilizer Section) for their assist in samples preparation and analysis. Technical assistance and support of NCARE during the survey are, also, highly valued.

References

1. George TS, Richardson AE (2008) In: White PJ, Hammond JP (eds) Potential and Limitations to Improving Crops For Enhanced Phosphorus Utilization. The Ecophysiology of Plant-Phosphorus Interactions, Springer Science+Business Media B.V. 247-270.

2. Minnesota Department of Health (1999) Screening Evaluation of Arsenic, Cadmium, and Lead Levels in Minnesota Fertilizer Products.

3. Wolnik KA, Fricke FL, Capar SG, Braude GL, Meyer MW, et al. (1983) Elements in Major Raw Agricultural Crops in the US. 1. Cadmium and Lead in Lettuce, Peanuts, Potatoes, Soybeans, Sweet Corn, and Wheat. J Agric Food Chem 31: 1240-1244.

4. Mensah E, Kyei Baffour N, Ofori E, Obeng G (2009) In: Yanful EK (Ed) Influence of Human Activities and Land Use on Heavy Metal Concentrations in Irrigated Vegetables in Ghana and Their Health Implications. Appropriate Technologies for Environmental Protection in the Developing World, Springer Science+Business Media B.V. 9-14.

5. Oymen HH, Oymen IM, Usese AI (2015) Iron, Manganese, Cadmium, Chromium, Zinc and Arsenic Groundwater Contents of Agbor and Owa Communities of Nigeria. SpringerPlus 4: 104.

6. Galadima A, Garba ZN (2012) Heavy Metals Pollution in Nigeria: Causes and Consequences. Elixir Pollution. 45: 7917-7922.

7. Agwaramgbo L, Iwuagwu A, Alinnor J (2014) Lead Removal from Contaminated Water by Corn and Palm Nut Husks. British Journal of Applied Science and Technology 4: 4992-4999.

8. Jayasumara C, Fonseka S, Fernando A, Jayalath K, Amarasingle M, Siribadana S, Gunatilake S, Paranagama P (2015) Phosphate Fertilizer is a Main Source of Arsenic in Areas Affected with Chronic Kidney Disease of Unknown Etiology in Sri Lanka. Springer Plus. 4: 90.

9. Moustakas NK, Akoumianakis KA, Passam HC (2001) Cadmium Accumulation and its Effects on yield of Lettuce, Radish, and Cucumber. Commun. Soil Sci. Plant Anal 32: 1793-1802.

10. Kirkham MB (2006) Cadmium in Plants on Polluted Soils: Effects of Soil Factors, Hyperaccumulation, and Amendments. Geoderma 137: 19-32.

11. Al-faiyz YS, El-Garaway MM, Assubaie FN, Al-Eed MA (2007) Impact of Phosphate Fertilizer on Cadmium Accumulation in Soil and Vegetable Crops. Bull Environ Conatm Toxicol 78: 358-362.

12. Podar D, Ramsey MH (2005) Effect of alkaline pH and associated Zn on the concentration and total uptake of Cd by lettuce: comparison with predictions from the CLEA model. Science of the Total Environment 347: 53-63.

13. Joint FAO/WHO Expert Committee on Food Additives (JECFA) (1989) Evaluation of Certain Food Additives and Contaminants. Twenty-Third Report, World Health Organization Technical Report Series, No. 776, Geneva.

14. Galal Gorchev H (1993) Dietary Intake, Levels in Food and Estimated Intake of Lead, Cadmium, and Mercury. Food Additives and Contaminants 10: 115-128.

15. Smical AI, Hotea V, Oros V, Juhasz J, Pop E (2008) Studies on Transfer and Bioaccumulation of Heavy Metals from Soil into Lettuce. Environmental Engineering and Management Journal 7: 609-615.

16. Yargholi B, Azimi AA, Baghvand A, Liaghat AM, Fardi GA (2008) Investigation of Cadmium Absorption and Accumulation in Different Parts of Some Vegetables. American Eurasian J. Agric. & Environ. Sci 3: 357-364.

17. Wrobel S (2012) Lettuce Yields as an Indicator of Remediation Efficacy of Soils Contaminated with Trace Metals Emitted by Copper Smelting. Journal of Food, Agriculture and Environment 10: 828-832.

18. Mausi G, Simiyu G, Lutta S (2014) Assessment of Selected Heavy Metal Concentrations in Selected Fresh Fruits in Eldoret Town, Kenya. Journal of Environment and Earth Science 4: 1-8.

19. Maalem S, Dellaa Y, Rahmoune C (2014) Absorption and Accumulation of Heavy Metals by Atriplex under Phosphoric Fertilizers (Phosphate rock, Organophosphate and TSP). Advances in Environmental Biology 8: 429-435.

20. Wagesho Y, Chandravanshi BS (2015) Level of Essential and Non-essential Metals in Ginger (Zingiber officivale) Cultivated in Ethiopia. Springer Plus 4:107.

21. Javied S, Mehmood T, Chaudhry MM, Tufail M, Irfan N (2009) Heavy Metal Pollution from Phosphate Rock Used for the Production of Fertilizer in Pakistan. Microchemical Journal 91: 94-99.

22. Alkhader AMF, Abu Rayyen AM (2014) Effects of Phosphorus Fertilizer Type and Rate on Plant Growth and Heavy Metal Content in Lettuce (Lactuca sative

L) Gown on Calcareous Soil. Jordan Journal of Agricultural Sciences 10: 796-810.

23. Ghrefat HA, Yusuf N, Jamarh A (2012) Fractionation and Risk assessment of Heavy Metals in Soil Samples Collected along Zerga River, Jordan. Environ Earth Science 66: 199-208.

24. Bower CA, Wilcox LV (1965) Soluble salts. Methods of Soil Analysis. Part 2. Chemical and Microbiological Properties. Agronomy 9. In: Black, C. (Ed.) Amer Soc. of Agronomy. Madison, Wisconsin, USA. 933-951.

25. Chapman HD (1965) Caution-exchange capacity. Methods of Soil Analysis. Part 2. Chemical and Microbiological Properties. Agronomy 9. In: Black, C. (Ed.) Amer Soc. of Agronomy. Madison, Wisconsin, USA. 891-900.

26. Allison LE (1965) Organic Carbon. Methods of Soil Analysis. Part 2. Chemical and Microbiological Properties. Agronomy 9. In: Black, C. (Ed.) Amer Soc. of Agronomy. Madison, Wisconsin, USA. 1376-1378.

27. Allison LE, Moodie CD (1965) Carbonates. Methods of Soil Analysis, Part 2. Chemical and Microbiological Properties, Agronomy 9. In: Black, C. (Ed.), Amer Soc of Agronomy Madison, Wisconsin, USA, 1379-1396.

28. Bremner IM (1965) Total Nitrogen. Methods of Soil Analysis. Part 2. Chemical and Microbiological Properties. Agronomy 9. In: Black, C. (Ed.) Amer Soc. of Agronomy. Madison, Wisconsin, USA. 1149-1178.

29. Olsen SR, Dean LA (1965) Phosphorus. Methods of Soil Analysis. Part 2. Chemical and Microbiological Properties. Agronomy 9 In: Black, C. (Ed.) Amer. Soc. of Agronomy. Madison, Wisconsin, USA. 1035-1048.

30. Pratt PF (1965) Potassium. Methods of Soil Analysis. Part 2. Chemical and Microbiological Properties. Agronomy 9. In: Black, C. (Ed.) Amer. Soc. of Agronomy. Madison, Wisconsin, USA. 1022-1030.

31. Lindsay WL, Norvell WA (1978) Development of a DTPA Soil Test for Zinc, Iron, Manganese, and Copper. Soil Sci. Soc. Am. J. 42: 421-428.

32. Shiowatana J, McLaren RG, Chanmekha N, Samphao A (2001) Fractionation of Arsenic in Soil by a Continuous-Flow sequential Extraction Method. Journal of Environ. Qual 30: 1940-1949.

33. Day TR (1965) Particle size analysis. Methods of soil analysis. Part 1. Agronomy 9. In: Black, C. (Ed.) Amer Soc. of Agronomy. Madison, Wisconsin, USA. 562-566.

34. Horwitz W, Latimer G (2005) Fertilizer analysis: Official Methods of Analysis of AOAC International, (18thedn), Gaithersburg, Maryland, USA. 20877-2417.

35. Chapman HD, Pratt PF (1962) Methods of Analysis for Soils, Plants and Waters. University of California.

36. Intawongse M, Dean JR (2006) Uptake of Heavy Metals by Vegetable Plants Grown on Contaminated Soil and their Bioavailability in the Human Gastrointestinal Tract. Food Additives and Contaminants 23: 36-48.

37. Al-Zu bi Y (2007) Effect of Irrigation Water on Agricultural Soil in Jordan Valley. An Example from Arid Area Conditions. Journal of Arid Environments.70: 63-79.

38. Warren GP, Alloway BJ, Lepp NW, Singh B, Bochereau FJM, Penny C (2003) Field Trials to Assess the Uptake of Arsenic by Vegetables from Contaminated Soils and Soil Remediation with Iron Oxides. The Science of the Total Environment 311: 19-33.

39. Yap DW, Adezrian J, Khairiah J, Ismail BS, Ahmad Mahir R (2009) The Uptake of Heavy Metals by Paddy Plants (Oryza sativa) in Kota Maruda, Sabah, Malysia. American-Eurasian J. Agric. and Environ. Sci 6: 16-19.

40. Zarcinas BA, Pongsakul P, McLaughlin MJ, Cozens G (2004) Heavy Metals in Soils and Crops in Southeast Asia. 2. Thailand. Environmental Geochemistry and Health 26: 359-371.

41. Sheng MY, Sukhdev S, Malhi, FM, Suo MG, et al. (2007) Long term Effects of Manure and Fertilization on Soil Organic Matter and Quality Parameter of a Calcareous Soil in NW China. J. Plant Nutr. Soil Sci 170: 234-243.

42. Khoshgoftarmanesh AH, Aghili F, Sanaeiostovar A (2009) Daily Intake of Heavy Metals and Nitrate through Greenhouse Cucumber and Bell Pepper Consumption and Potential Health Risks for Human. International Journal of Food Sciences and Nutrition 60: 199-208.

43. Al-Abed N, Amayreh, J, Al-Afifi A, Al-Hiyari G (2004) Bioremediation of a Jordanian Saline Soil a Laboratory Study. Communications in Soil Science and Plant Analysis 35: 1457-1467.

44. Shadfan H (1983) Clay minerals and Potassium Status in Some Soils of Jordan. Geodermas 31: 41-56.

45. Heckman JR (2006) The Soil Profile, A Newsletter Providing Information on Issues Relating to Soils and Plant Nutrition in New Jersey. 16: 1-4.

46. Roberts AHC, Longhurst RD, Brown MW (1994) Cadmium Status of Soils, Plant and Grazing Animals in New Zealand. New Zealand Journal Agricultural Research 37: 119-129.

47. Osztoics E, Csatho P, Nemeth T, Baczo G, Magyar M, et al. (2005) Influence of Phosphate Fertilizer Sources and Soil Properties on Trace Element Concentrations of Red Clover. Communications in Soil Science and Plant Analysis 36: 557-570.

48. European Community (EC) (2006) Commission Regulation (EC) 1881/2006 Setting Maximum Levels for Certain Contaminants in Foodstuffs. Official Journal of the European Union. 364: 5-24.

49. Bolan NS, Adriano DC, Mani PA, Duraisamy A (2003a) Immobilization and Phytoavailability of Cadmium in Variable Charge Soils, II. Effect of Lime Addition. Plant and Soil. 251: 187-198.

50. Castro E, Manas P, Heras JDL (2009) A comparison of the Application of Different Waste Products to a Lettuce Crop, Effects on Plant and Soil Properties. Scientia Horticulturea. 123: 148-155.

51. Bolan NS, Adriano DC, Duraisamy P, Mani A (2003b) Immobilization and Phytoavailability of Cadmium in Variable Charge Soils. III. Effect of Biosolid Compost Addition. Plant and Soil. 256: 231-241.

52. Hartz TK, Johnstone PR, Williams E, Smith RF (2007) Establishing Lettuce Leaf Nutrient Optimum Ranges through DRIS Analysis. Hort Science 42: 143-146.

36

The Extraction Technology of Flavonoids from Buckwheat

Wang L[1]* and Bai X[2]

[1]The College of Life Science, Yangtze University, Jingzhou, Hubei, China

[2]The First People's Hospital of Jingzhou, Jingzhou, Hubei, China

Abstract

Buckwheat (*Fagopyrum esculentum*) is a kind of medicinal and edible crops with high Flavonoid content. Flavonoids from tartary buckwheat have significant therapeutic effects on vascular diseases, diabetes and obesity. In this paper, the extracting technologies of flavonoids from buckwheat were investigated. The results showed that the optimum parameters for the extraction is temperature 60℃, alcohol concentration 60%, solid to liquid ratio 1:20, pH=2, duration 120 min.

Keywords: Buckwheat; Flavonoid; Extraction; Orthogonal design

Introduction

Flavonoids have anti-inflammatory, antiallergic, diuretic, antispasmodic, antitussive and hypolipidemic effects, and have significant therapeutic effects on vascular diseases, diabetes and obesity [1,2]. Buckwheat (*Fagopyrum esculentum*) is a kind of medicinal and edible crops. The content of flavonoids in Tartary buckwheat is particularly rich, and the effect is the most remarkable [3,4]. We analyzed the effects of the extraction conditions to tartary buckwheat flavonoids and got the optimal preparing conditions for Tartary buckwheat flavonoids. The results of this work will lay the foundation of theory and application for the further study of Tartary buckwheat flavonoids.

Materials and Methods

Preparation of buckwheat flour

Tartary buckwheat (Chuanqiao No. 1) was purchased from Liangshan Yi Autonomous Prefecture. Buckwheat was grinded into flour using flour mill, then filtered using 200 mesh sieve.

Determination of flavonoid content

Accurately prepare 0.1 mg/mL rutin methanol solution (rutin standard solution). Add 0.1 mL, 0.2 mL, 0.4 mL, 1.0 mL, 0.6 mL, 0.8 mL of rutin standard liquid in the calibration tubes and add the 100% methanol solution volume up to 1.0 mL, and then add the 2 mL 0.1 mol/L and 3 mL 1 mol/L acetic acid potassium chloride. Finally, add 30% ethanol up to 10 mL, after resting 30 min, the absorbance was measured at 420 nm. The standard curve was made with rutin concentration X as abscissa and absorbance difference (Y) as ordinate. The regression equation was y=8.0068x, and the correlation coefficient was r=0.9996 (Figure 1) [5,6].

The optimization of the preparation process of buckwheat flavonoids

To optimize the preparation process of buckwheat flavonoids, the major factors and their levels were determined according the effects of various factors (such as solid to liquid ratio (S/L), ethanol concentration, extraction temperature, extraction time, pH value) on buckwheat flavonoid content. The optimum preparation conditions of buckwheat flavonoids were further determined using orthogonal test.

Results and Discussion

The effects of solid to liquid ratio on buckwheat flavonoid

The buckwheat flavonoids were extracted at different solid to liquid ratio for 60 min with ethanol concentration is 50% and temperature is 30℃. The buckwheat flavonoids content was analyzed. The optimum solid to liquid ratio is 1:20 (Figure 2).

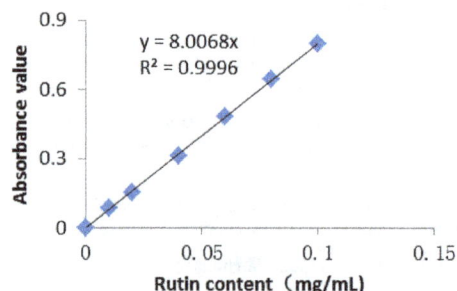

Figure 1: Standard curve of rutin.

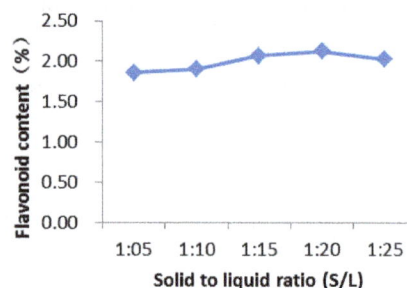

Figure 2: Effects of solid to liquid ratio on flavonoid extraction.

*Corresponding author: Wang L, The College of Life Science, Yangtze University, Jingzhou, Hubei, China, E-mail: ljwang516@126.com

Effects of ethanol concentration on buckwheat flavonoid extraction

The buckwheat flavonoids were extracted at different ethanol concentration for 60 min with solid to liquid ratio is 1:10 and temperature is 25℃. The buckwheat flavonoids content was analyzed. The optimum ethanol concentration is 60% (Figure 3).

Effects of extracting time on buckwheat flavonoid extraction

The buckwheat flavonoids were extracted at different extracting time with solid to liquid ratio is 1:20, pH 5, ethanol concentration is 50% and temperature is 50℃. The buckwheat flavonoids content was analyzed. The optimum extracting time is 160 min (Figure 4).

Effects of temperature on buckwheat flavonoid extraction

The buckwheat flavonoids were extracted at different temperature with solid to liquid ratio is 1:20, pH 5, ethanol concentration is 50% and extracting time is 60 min. The buckwheat flavonoids content was analyzed. The optimum extracting temperature is 70℃ (Figure 5).

Effects of pH values on buckwheat flavonoid extraction

The buckwheat flavonoids were extracted at different pH values with solid to liquid ratio is 1:20, ethanol concentration is 50%, temperature is 50℃ and extracting time is 60 min. The buckwheat flavonoids content was analyzed. The optimum pH is 2 (Figure 6).

Orthogonal experiment of buckwheat flavonoid extraction

According the effects of individual factors on buckwheat flavonoid extraction, orthogonal experiments were conducted using extracting time, temperature, solid-to-liquid ratio (S/L) and pH as factors and flavonoid content as index (Tables 1 and 2).

As the results shown in the Table 2, solid-to-liquid ratio had the largest effect on flavonoid content. The pH Value had the second largest effect on flavonoid content. Temperature had the third largest effect on flavonoid content. Exacting time had the fourth largest effect on flavonoid content. The optimum parameters for producing technology of flavonoids from buckwheat using ethanol solvent are A1B2C2D1, that is exacting time 120 min, temperature at 60℃, solid-liquid ratio 1:20, pH 2. The sequence of effects on flavonoid content: C>D>B>A.

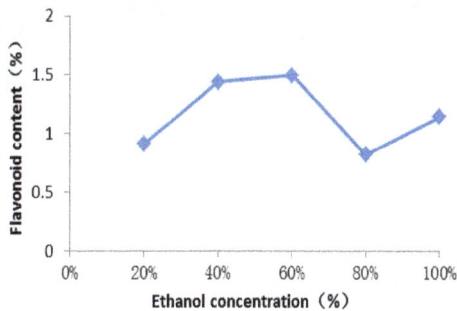

Figure 3: Effects of ethanol concentration on flavonoid extraction.

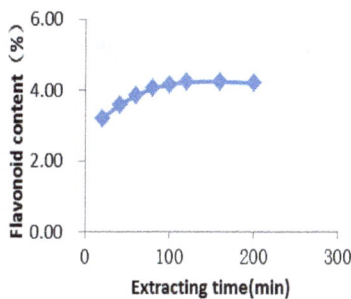

Figure 4: Effects of extracting time on flavonoid extraction.

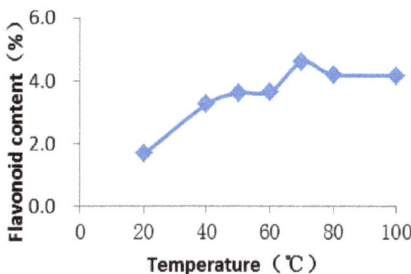

Figure 5: Effects of temperature on flavonoid extraction.

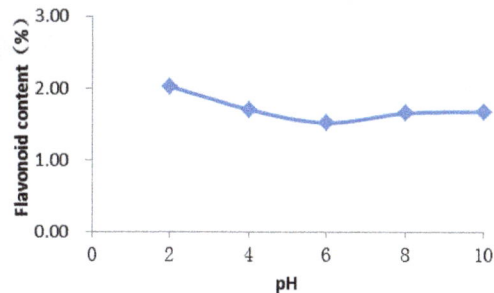

Figure 6: Effects of pH on flavonoid extraction.

Level	A (Extracting time/min)	B (Temperature/°C)	C (Solid-liquid ratio)	D (pH)
1	120	50	01:15	2
2	160	60	01:20	6
3	180	70	01:25	10

Table 1: Factor level table.

S. No.	A (Extracting time/min)	B (Temperature/℃)	C (Solid-liquid ratio)	D (pH)	Flavonoid Content (%)
1	1 (120 min)	1 (50℃)	1 (1:15)	1 (2)	3.895
2	1 (120 min)	2 (60℃)	2 (1:20)	2 (6)	4.017
3	1 (120 min)	3 (70℃)	3 (1:25)	3 (10)	2.26
4	2 (160 min)	1 (50℃)	2 (1:20)	3 (10)	3.59
5	2 (160 min)	2 (60℃)	3 (1:25)	1 (2)	2.576
6	2 (160 min)	3 (70℃)	1 (1:15)	2 (6)	3.973
7	3 (180 min)	1 (50℃)	3 (1:25)	2 (6)	2.027
8	3 (180 min)	2 (60℃)	1 (1:15)	3 (10)	3.872
9	3 (180 min)	3 (70℃)	2 (1:20)	1 (2)	4.214
K1	10.172	9.512	11.74	10.685	-
K2	10.139	10.465	11.821	10.017	-
K3	10.113	10.447	6.863	9.722	-
R	0.02	0.317667	1.652667	0.321	-

Table 2: $L_9(3^4)$ flavonoid extracting orthogonal experiment design and results.

Conclusion

The main factors affecting the extraction of flavonoids were solid-liquid ratio, pH value, and extraction temperature and extraction time. Through the analysis of single factor gradient experiment and orthogonal experiment, the optimum extraction conditions were obtained: the extraction temperature was 60℃, the concentration of ethanol was 60%, the ratio of material to liquid was 1:20, and pH was 2.

References

1. Li D, Li X, Ding X (2010) Composition and antioxidative properties of the flavonoid-rich fractions from tartary buckwheat grains. Food Sci Biotechnol 19: 711-716.

2. Sun T, Ho CT (2005) Antioxidant activities of buckwheat extracts. Food Chem 90: 743-749.

3. Yao H, Li C, Zhao H, Zhao J, Chen H, et al. (2017) Deep sequencing of the transcriptome reveals distinct flavonoid metabolism features of black tartary buckwheat (*Fagopyrum tataricum* Garetn.). Prog Biophys Mol Bio 124: 49-60.

4. Li B, Li Y, Hu Q (2016) Antioxidant activity of flavonoids from tartary buckwheat bran. Toxicol Environ Chem 98: 429-438.

5. Liu B, Zhu Y (2007) Extraction of flavonoids from flavonoid-rich parts in tartary buckwheat and identification of the main flavonoids. J Food Eng 78: 584-587.

6. Sathishkumar T, Baskar R, Shanmugam S, Rajasekaran P, Sadasivam S, et al. (2008) Optimization of flavonoids extraction from the leaves of *Tabernaemontana heyneana* wall. using L16 orthogonal design. Nature Sci 6: 10-21.

37

The Biology, Utilization and Phytochemical Composition of the fruits and leaves of *Gongronema latifolium* Benth

Osuagwu AN*, Ekpo IA, Okpako EC, Otu P and Ottoho E

Department of Genetics and Biotechnology, University of Calabar, Nigeria

Abstract

Basic information on plant species is important for the improvement of the species. This study was carried out to investigate and understand the biology, utilization and phytochemical composition of *Gongronema latifolium* which is a spice plant growing in the humid forest vegetation of South- Eastern Nigeria. Results showed that the species had culinary and medicinal properties. *G. latifolium* has simple and opposite leaves, dehiscent seed pod (follicle), that opens along a single seam. The seeds are flat with white hairy pappus, The flowers are bisexual, regular with pale yellow coloured petals and superior ovary. Phytochemical analysis of the tender fruits and mature leaves of *Gongronema latifolium* revealed the presence of alkaloids, tannins, saponins, flavonoids, phenols, phytic acid and hydrocyanic acid. The phytochemicals were higher in the fruitscompared to the leaves ($P<0.001$) except the flavonoids which were higher in the leaves than in the fruits ($P<0.001$). The presence of these phytochemicals account for the nutriceutical /medicinal properties of *Gongronema latifolium*.

Keywords: *Gongronema latifolium*; biology; utilization; phytochemical analysis; plant morphology

Introduction

The tropical rainforest is the most biologically diverse ecosystem on the earth and it is the predominant natural forest in Nigeria. It is favorable for the production of a wide range of useful plants and spices including *Gongronema latifolium* [1,2]. Plants and plant-products are good sources of medications and provide raw materials for modern pharmaceuticals used for various ailments [3]. A great number of the world's population particularly Nigerians, rely on traditional medicines for their primary health care needs.

The medicinal value of plants lies in some chemical substances that produce a definite physiologic action on the human body. The most active of these bioactive compounds (phytochemicals) of plants are alkaloids, flavonoids, tannins and phenolic compounds [4]. Phytochemicals are chemical compounds that occur naturally in plants and which may affect health and are not yet established as essential nutrients [5]. Phytochemicals give plants their color, flavour, smell and texture. Kushi et al., Setchell and Cassidy, and Wang et al. [6-8] reported that phytochemicals work to affect antioxidant activity, hormonal action, stimulation of enzymes, interference with DNA replication, and antibacterial effect among others. As time goes by, our typical diets seem to increasingly contain more fatty processed foods and less natural plant based foods. The result of this disturbing trend can be seen in the alarming statistics on cancer, heart diseases, stroke and many other degenerative diseases. Aside from dietary problems, there is a problem of inadequate intake of plant based foods and all the benefits they bring with them. On the other hand, with increase in population and socio-economic changes leading to high rates of urbanization, need for more houses as well as increased commercialization of agricultural productions, natural forests that are rich sources of useful plants are being destroyed in a large scale [9]. *Gongronema latifolium* is of huge importance in food and medicine, but there is a gross lack of knowledge in the biology of this plant species and its utilization viz a viz its phytochemical composition. This consideration has made it important to investigate and understand the biology, phytochemical composition and utilization of the species. This information will be invaluable for any subsequent improvement and conservation of the plant.

Materials and Methods

The *Gongronema latifolium* samples used in this study were collected from four States in the South- Eastern zone of Nigeria and planted in the experimental fields of the Department of Genetics and Biotechnology, University of Calabar. The four States are Abia State, Akwa Ibom State, Cross River State and Imo State (Table 1). Stem cuttings of the species were raised in the nursery for one month before transplanting in the field. Cultural practices such as weeding and staking were carried out. The plants were monitored for three years (Feb, 2009–Feb, 2012) to determine the time of maturity (flowering and fruiting) and the consistency in the biological processes such as germination of its seeds. Taxonomic identification and classification of different parts of the plant (leaves, flowers, fruits, and seeds) were carried out in the Department of Botany, University of Calabar and through literature review.

Studies on the utilization of *Gongronema latifolium* were based on oral interviews with various respondents at each State where collections were made. A total of 80 respondents (20 from each State), were interviewed. This number included farmers, traditional doctors,

State	LGA	Village	Latitude °N	Longitude °E	Altitude (m)
Cross River	Akamkpa	Iko-Ekperem	05.6068	08. 2170	10.32
Imo	Ngor Okpala	Umuogba Ntu	04. 9975	08. 3330	45.3
Abia	Ikwuano	Umudike-Uku	04. 8604	7.7843	121
Akwa Ibom	Etinan	Ikot Udobia	04. 8863	7.8299	67

Table 1: Collection sites of *Gongronema latifolium*.

*Corresponding author: Osuagwu AN, Department of Genetics and Biotechnology, University of Calabar, P.M.B 1115,Calabar, Nigeria
E-mail: anniosuagwu@gmail.com

housewives, traders and other users of *G. latifolium*. Phytochemical screening of the fruits and leaves of *G. latifolium* was carried out in the laboratory of the Department of Pure and Applied Chemistry, University of Calabar, Calabar. The mature leaves and tender green fruits samples were oven dried at 50°C for three days, ground to fine powder and stored in dry containers for further analysis. Quantitative phytochemical analyses were carried out to determine alkaloids, tannins, saponins, flavonoids, phenols, phytic acid, oxalate and hydrocyanic acid, Using standard methods described by Sofowora [10], Edeoga et al. [11], Trease et al. [12], Andrews [13], with three replications and the results were recorded in percentages.

Determination of Phytochemicals

Alkaloids: 2 g of sample were weighed and added to 1 ml of concentrated acetic acid and ethanol (1:2), covered and kept to stand for 4 hrs then filtered and concentrated in a water bath to one-quarter (1/4) of the original volume. Concentrated ammonium hydroxide was added drop-wise to the extract until the precipitate formation was completed. It was then allowed to settle and washed and filtered with dilute ammonium hydroxide solution. The residue was dried in an oven and taken as crude alkaloid. It was weighed and recorded.

Tannins: Tannin was extracted from 0.5 g of the sample with methanol then purified with Whatman filter paper. Colour was developed using Vanillin hydrochloric acid reagent and the concentration was quantitatively measured using a spectrophotometer at 500 nm.

Saponins: 2 g of sample were extracted with ethanol which was subsequently removed using a rotary evaporator. The solution was washed with diethyl ether until colorless. The pH was adjusted to 5.0 with sodium chloride. Finally, it was extracted with n-butanol and washed with sodium chloride and evaporated to dryness to give saponins which were weighed and recorded.

Flavonoids: 2 g of samples were extracted with 100 ml of 80% aqueous methanol at room temperature until the supernatant became colorless. The solution was filtered through Whatman filter paper. The filtrate was transferred into a beaker and evaporated to dryness over a hot plate to give flavonoids which was weighed and recorded.

Phenols: 2 g of samples were weighed and 50 ml of ether was added. The slurry was agitated for 15 minutes. 5 ml amyl alcohol was added and allowed for 30 minutes for the development of the color. The concentration of the solution was determined using uv-vis spectrophotometer at 505 nm.

Phytic acid: 2 g of sample were weighed and added to 25 ml of 0.5 N NaCl then was shaken for 30 minutes. 2 ml of ferric chloride was added to the extract. The precipitate (ferric phytate) was converted to sodium phytate by adding 3 ml of sodium hydroxide. The precipitate was digested with acid mixture of equal portions of concentrated tetraoxosulphate (VI) acid and perchloric acid in a digestion set. The liberated phosphorus was quantified calorimetrically at 620 nm after color development with molybdate reagent.

Oxalate: 2 g of sample were weighed and extracted with dilute hydrochloric acid. The oxalate in the extract was precipitated with calcium chloride as salts. The precipitated extract was washed with 50 ml of 25% H_2SO_4 and dissolved in hot water then was titrated with 0.05 N $KMnO_4$.

1 ml of 0.05 N $KMNO_4$=2.2 mg oxalate.

Hydrocyanic acid: 2 g of sample was weighed and soaked in 50 ml of distilled water for about 4 hrs and then steam distilled into 50 cm^3 of 2.5% NaOH in a beaker. 8 ml of 6 N NH_4OH and 2 ml of 5% KI was added to the 25 cm^3 portion of the distillate and 0.05 N silver nitrate was titrated against it to a faint and permanent turbidity.

1 ml of 0.02 N $AgNO_3$=1.08 mg HCN

Data analysis

Data collected were analyzed using the *t*-test and descriptively using the means and standard error.

Results and Discussion

Biology of *Gongronema latifolium*

Gongronema latifolium is a flowering plant of the order Gentiales and the family Apocynaceae, Subfamily Asclepiadaceae, and genus Calotropis. It is a tropical climbing plant (lianas woody) distributed mainly in the tropical and sub tropical regions of Africa, Asia and Oceania [14-16]. It is propagated by stem cuttings as well as by seeds. The species takes up to one year to fully mature and flower.

The study showed that the leaves of *Gongronema latifolium* are simple, opposite, decussate and whorled, peltate with entire margin and long petiole (about 5.2 cm long) (Plates A and B). The leaves area measured approximately 50.2 cm^2 ± 0.35 cm^2. The inter node length varied according to the habitat, and reached up to 30 cm in length in shady environments such as forest floors or less than 5 cm in exposed regions. The roots are adventitious arising wherever the soft woody stem makes contact with soil.

Flowers of *G. latifolium* are bisexual and actinomorphic. Inflorescences are extra-axillary cymes. The calyx has five basal glands; the corolla is urceolate with five lobes. There are five lobes of scale-like corona inserted at the base of gynostegium. The stamens are five while the filaments are connected into a tube. Anthers are erect with membranous apical appendages. There are two pollinia per pollinarium, and the styles are short. The stigma heads are vertically conical in shape. Ovary is superior. The flowers are pale yellow in colour (Plate C).

The fruit of *Gongronema latifolium* is a dehiscent seed pod called a follicle which is oblong- lanceolate. The colour of the fruits varied from green in small fruits to dark brown to black at maturity. During maturity stage, the fruit splits open length wise, along the seam releasing flat seeds, the seeds are attached to a white silky tuft (pappus) which aids dispersal .The seeds are strongly compressed, coma shaped and measure about 0.5 cm in length (Plates D, E and F (Figure 1)).

G. latifolium established from stem cuttings grow and mature in twelve months. Flowering was initiated in late January and went on through March. Day temperatures at this period in the region ranged between 32°C-37.5°C. The flowers are insect pollinated. Fruit development in the species is a slow process lasting from April through November, when the fruits begin to mature and change color from green to black and eventually dehisce in the later part of dry season (December to February). New flowers often meet old fruits on the plant. Mature seeds of *G. latifolium* planted within two weeks after harvest germinated in seven to fourteen days at temperatures of about 27°C, with a germination rate of 67%.

Utilization of *Gongronema latifolium*

Gongronema latifolium is an important plant that is utilized for

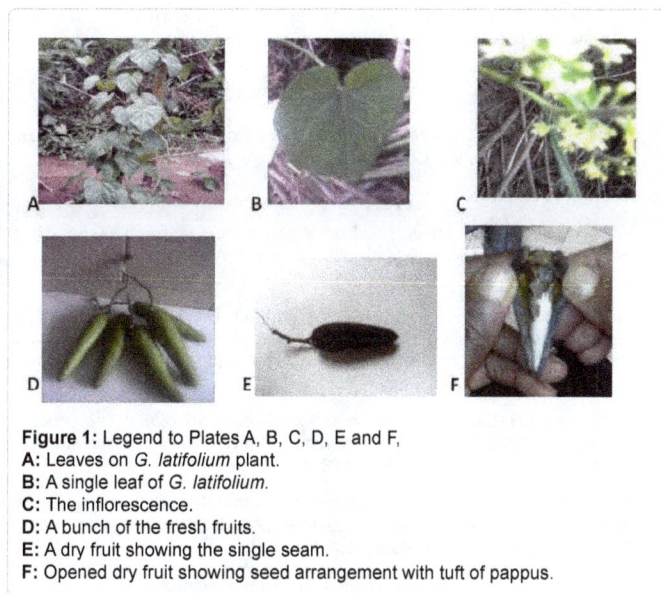

Figure 1: Legend to Plates A, B, C, D, E and F,
A: Leaves on *G. latifolium* plant.
B: A single leaf of *G. latifolium*.
C: The inflorescence.
D: A bunch of the fresh fruits.
E: A dry fruit showing the single seam.
F: Opened dry fruit showing seed arrangement with tuft of pappus.

Phytochemicals (%)	df	Fruits (x ± S.E)	Leaves (x ± S.E)	tcalculated	5%	1%	0.1%
Alkaloids	2	50.0 ± 0.58	10.0 ± 0.02	70***	2.78	4.60	8.61
Tannins	2	35.5 ± 0.43	30.0 ± 0.01	13.1***	2.78	4.60	8.61
Saponins	2	150.0 ± 0.05	38.0 ± 0.01	5600***	2.78	4.60	8.61
Flavonoids	2	21.2 ± 0.02	23.4 ± 0.01	157.1***	2.78	4.60	8.61
Phenols	2	10.4 ± 0.01	7.1 ± 0.02	194.1***	2.78	4.60	8.61
Phytic acid	2	25.0 ± 0.10	20.0 ± 0.01	16.1***	2.78	4.60	8.61
Oxalate	2	ND	ND	-	-	-	-
Hydrocyanic acid	2	70.0 ± 0.29	48.5 ± 0.01	76.8***	2.78	4.60	8.61

ND–not detected   ***-highly significant

5%, 1%, 0.1% represent tabulated significant levels

Table 2: Phytochemicals analysis of the leaves and fruits of *Gongronema latifolium*.

its medicinal and culinary properties. In Nigeria, the species is utilized mostly by people from the South-Eastern region where it is popularly known as 'Utazi' by the Ibos and 'Utasi' by the Efiks and Ibibios. The parts used are the leaves/vines and the tender fruits.

Leaves: A few of the leaves are chopped and added to foods such as porridges, stews and pepper soups. Two very popular Ibo delicacies known as *Nkwobi* (Cow leg pepper soup) and *Isi ewu* (Goat head pepper soup) are prepared with *G. latifolium* leaves. According to findings from respondents, the leaves impart a sharp bitter taste and sweet aroma to food and it increases appetite and this has also been reported by Adelaja and Fasidi, FAO [9,17].

The crude extract from the leaves and vines are mixed with lime juice and drunk to expel worms and dispel stomach upsets and crams. The crude extract is mixed with extracts of bitter leaf (*Vernonia amygdalina*) and scent leaf (*Ocimum gratissimum*) and taken to treat malaria and typhoid fever [18].

The extract is taken alone to maintain healthy blood glucose level and to check excesses of diabetes and hypertension. Ugochukwu et al. and Ogundipe et al. [19,20] have established the hypoglycemic, hypolipidemic and antioxidant properties of aqueous and ethanolic extracts of *G. latifolium* leaves. The extract is also used as enema, for treatment of malaria and stomach disorders.

Fruits: One or two young follicles are eaten daily with or without the seeds to treat stomach ache, check diabetes, treat malaria and tone the blood.

Phytochemical differences in fruits and leaves of *Gongronema latifolium*

The results showed that there is high significant difference, (P<0.001) in the quantity of phytochemicals (alkaloids, tannins, saponins, flavonoids, phytic acid and hydrocyanic acid) in the fruits and leaves. Oxalate was not detected in either the leaf or fruit samples. However, alkaloids, tannins, saponins, phenols, phytic acid and hydrocyanic acid have relatively higher percentage quantities in the fruits compared to the leaves while flavonoid content was however higher in the leaves than in the fruits (Table 2).

The presence of phytochemicals confers nutritional, industrial, therapeutic, as well as economic potentials on *Gongronema latifolium*. Saponins, flavonoids, tannins and alkaloids have chemo-preventive properties and the concept of chemo-prevention has assumed a global significance as a result of its acceptance in the management, prevention and treatment of a wide range of life threatening diseases such as cancer, diabetes and coronary diseases and in the maintenance of good health [21]. Alkaloids also have antipyretic effects [22]. Phenols are considered as antimicrobial (bacteriostatic and fungistatic) agents and play an active role in disease resistance and prevention [23]. Many reports suggest a positive correlation between total phenolic content and antioxidant activity [24]. A higher phenolic content recorded in the fruit of *Gongronema latifolium* as compared to the leaves indicate good antioxidant property. Alkaloids and phenols are more in the fruits than in the leaves of *G. latifolium*, this may be due also, to the need to preserve the seeds from microbial attack [25]. Flavonoids provide antioxidant and anti-inflammatory actions while saponins are used to recover homeostasis and are also anti-fungal [26,27]. The presence of saponins and tannins may be responsible for the bitter and astringent taste of the plant. The presence of saponins in *G. latifolium* also indicate that intake of the plant can clear fatty compounds from the body, lowering the blood cholesterol [28]. Tannins have traditionally been considered as anti-nutritional but it is now known that their beneficial or anti-nutritional properties depend on their dosage. Recent studies have demonstrated that low dosages are beneficial while high dosages especially in sensitive individuals may cause bowel irritation, kidney and stomach irritation, liver damage and gastro-intestinal pain. Excess intake of tannins is therefore not recommended, and *G. latifolium* leaves or fruits should be taken in small amounts. Tannins are effective in protecting the kidney and deactivating the effect of poliovirus, herpes simplex virus and various enteric viruses. They also show anti-bacterial and anti-parasitic effects and are used in the treatment of hereditary hemochromatosis [29]. Phytic acid has also been considered asanti-nutritional substance but however, recent research has shown its many health benefits such as antioxidant, anticancer, hypocholesterol and hypolipidemic effects. Phytic acid releases inositol that might help reduce depression and inflammation [27]. Kushi et al. [6], reported that hydrocyanic acid is a successful remedy for whooping cough, employed to alloy cerebral disorders, relieve itching, curbing various kinds of nervous disorders such as vomiting, gastralgia, and pain caused by indigestion, irritative dyspepsia, and enteralgia. It is also an expellant for worm infection.

Conclusion

This work has X-rayed the biology and utilization of *Gongronema latifolium* and has confirmed the observation by locals that the fruits of *G. latifolium* are even more potent than its leaves due to the higher concentration of most of its phytochemical constituents. It is advisable therefore that lesser quantities of fruits should be consumed to avoid the effect of over dosage. The results from this work suggest that *Gongronema latifolium* may find its use in food/feed formulation/supplementation as well as nutriceutical/medicinal and industrial uses.

Acknowledgement

This research is an aspect of the research project in the University Of Calabar funded by STEP-B in conjunction with the World Bank. We acknowledge and thank the donors.

References

1. Akinsanmi FA, Akindele SO (2002) Timber yield assessment in the natural forest area of Oluwa forest reserve, Nigeria. Nigerian Journal of Forestry 32: 16-22.

2. Gillespie TW (2004) Prospects for quantifying structure, floristic composition and species richness of tropical forest. International Journal of Remote Science 25: 707-715.

3. Abraham Z (1981) Glimpse of Indian Ethnobotany. Oxford & Publishing Co., New Delhi, India. 308-320.

4. Chhetri HP, Vogol NS, Sherchan J, Anupa KC, Mansoor S, et al. (2008) Phytochemical and antimicrobial evaluations of some medicinal plants of Nepal.Kathmandu University Journal of Science, Engineering and Technology. 1: 49-54.

5. Liu RH (2004) Potential synergy of phytochemicals in cancer prevention: mechanism of action. J Nutr 134: 3479S-3485S.

6. Kushi LH, Byers T, Doyle C, Bandera EV, McCullough M, et al. (2006) American Cancer Society Guidelines on Nutrition and Physical Activity for cancer prevention: reducing the risk of cancer with healthy food choices and physical activity. CA Cancer J Clin 56: 254-281.

7. Setchell KD, Cassidy A (2003) Dietry isoflavones: Biological effects and relevance to human health. Journal of Nutrition 129: 758-767.

8. Wang YH, Chao PD, Hsiu SL, Wen KC, Hou YC (2004) Lethal quercetin-digoxin interaction in pigs. Life Sci 74: 1191-1197.

9. Adelaja BA, Fasidi IO (2009) Survey and collection of indigenous spice germplasm for conservation and genetic improvement of Nigeria. Bioversity International 153: 67-71.

10. Sofowora A (1993) Screening of plants for bioactive agents in medicinal plants and traditional medicine in Africa. (2nd ed) Spectrum Books Ltd, Sunshine House, Ibadan. 81-93, 135-156.

11. Edeoga HO, Okwu DE, Mbaebie BO (2005) Phytochemical constituents of some Nigerian medical plants. African Journal of Biotechnology 4: 685-688.

12. Trease GE, Evans WC(1978) A textbook of Pharmacognosy. (11th Edition). Bailliese, Tindall, London. 397-53.

13. Andrews WH (1994) Update on validation of microbiological methods by AOAC International. J AOAC Int 77: 925-931.

14. Tsiang Ving, Li Ping-tao (1977) Asclepiadaceae. Fl. Republic Popularis Sin 63: 249-575.

15. Burkhill HM (1985) The useful plants of West Tropical Africa. (2nd Ed) Kew, Royal Botanical Gardens, Great Britain. 960

16. Global Biodiversity Information Facility (GBIF).

17. FAO/Food and Agriculture Organization of the United Nations (1983) Food and fruit bearing forest species: Examples from Eastern Africa. FAO Forestry paper 44: 172.

18. Atangwho IJ,Ebong PE, Eyong EU, Williams IO, Eteng MU, et al. (2009). Comparative Chemical composition of leaves of some antidiabetic medicinal plants: Azadirachta indica, Vernonia amygdalina and Gongronema latifolium. African Journal of Biotechnology 8: 4685-4689.

19. Ugochukwu NH, Babady NE, Cobourne M, Gasset SR (2003) The effect of Gongronema latifolium extracts on serum lipid profile and oxidative stress in hepatocytes of diabetic rats. J Biosci 28: 1-5.

20. Ogundipe OO, Moody JO, Akinyemi TO, Raman A (2003) Hypoglycemic potentials of methanolic extracts of selected plant foods in alloxanized mice. Plant Foods and Human Nutrition 58: 1-7.

21. Greenwald P, Kelloff GJ (1996) The role of chemoprevention in cancer control. IARC Sci Publ: 13-22.

22. Okwu DE (2005) Phytochemicals, vitamins and mineral content of two Nigeria medicinal plants. International Journal of Molecular Medicine and Advance Sciences. 1: 375-381.

23. Matern U, Kneusel RE (1980) Phenolic compounds in plant disease resistance. Phytoparasitica 16: 153-170.

24. Salah N, Miller NJ, Paganga G, Tijburg L, Bolwell GP, et al. (1995) Polyphenolic flavanols as scavengers of aqueous phase radicals and as chain-breaking antioxidants. Arch Biochem Biophys 322: 339-346.

25. Okwu DE, Emenike IN (2006) Evaluation of the phytonutrients and vitamin content of citrus fruits. International Journal of Molecular Medicine and Advance Sciences 2: 1-6.

26. Galeotti F, Barlie E, Curir R, Doki M, Lanzotti V (2008) Flavonoids from Carnation (Dianthus caryophyllus) and their anti fungal activity. Phytochemistry letters 1-44.

27. Russel AD , Chopra I (1990) Understanding anti bacterial action and resistance. Ellis Horwood Limited, New York, USA. 131.

28. Vaghasiya Y, Dave R, Chanda S (2011) Phytochemical analysis of some medicinal plants from Western region of India. Research Journal of Medicinal Plant 5: 567-576.

29. Burkhill HM (1977) The useful plants of West Tropical Africa. (3rd ed). Royal Botanic Gardens Kew, Great Britain. 55-56.

38

Study of Delay Cultivation on Seed Yield and Seed Quality of Canola (*Brassica Napus* L.) Genotypes

Asadollah Gholamian[1]* and Mahdi Bayat[2]

[1]*Department of Biology, Mashhad Branch, Islamic Azad University, Mashhad, Iran*
[2]*Department of Agriculture, Mashhad Branch, Islamic Azad University, Mashhad, Iran*

Abstract

In order to study the effects of delay cultivation on seed yield and germination parameters in canola, an experiment was conducted in Torbat-Jam region during 2010-2011. The experimental design was a split plot arranged in RCBD with three replications. Three sowing dates (6 September, 7 October and 6 November) were assigned to main plots and three canola genotypes (Hyola 401, Zarfam and Mudena) were randomized to subplots. The results of variance analysis about yield components showed genotypes, sowing dates and their interactions had significant effects on all agronomical traits. Also delaying in culture leads to decrease yield components and seed yield subsequently. However, the remarkable point was that genotypes could affect seed yield more than sowing dates; so seed yield, to change genotype from Hyola 401 to Modena, decreased 40%; whereas to change sowing date from 6 September to 6 November, decreased 10%. The results of variance analysis about germination parameters showed that genotypes, sowing dates and their interaction had significant effects on germination parameters. In other hands, sowing dates were more effective on qualification and seed vigor than genotypes. As a conclusion, genotypes and sowing dates affected seed yield and seed quality significantly, therefore, produced seed of cultivated crops in appropriate date will were more vigorous which will increase canopy and growth rate at next year cultivation.

Keywords: Agronomical traits; Germination parameters; Germination test; Sowing date

Introduction

Canola is one of the most important oil seed that developing in two recent decades extremely. Canola after soybeen is the second oil plant in the world [1]. Reaching maximum yield needs to select strong seeds with high quality. Selection of high quality seeds is essential for more stability and growth. It is clear that seed vigour is dependent on some factors such as sowing date, temperature, humidity, drought, diseases, pathogens and genetic factors [2]. So selection of appropriate sowing date is important for cold resistant in winter and scape from high temperature and unfavorite conditions at the end of growth period. In other hands, the successful passing of winter in canola is due to formation of advanced root system; which is dependent on sowing dates [3]. Elias et al. [4] reported that factors such as sowing date, genotype, ripening stage, gathering stages, method of gathering, seed desiccation and storage conditions are effective on seed quality. At the present study is tried to determine the best sowing date and genotype in Torbat-jam.

Materials and Methods

In order to study the effects of delay cultivation on seed yield and germination parameters in canola, an experiment was conducted in Torbat-jam during 2010-2011. Torbat-Jam region locate in 35'.15" Latitude, 60'.35" longitude and 950.4 meters above sea level. The experimental design was a split plot arranged in RCBD with three replications. Three sowing dates (6 September, 7 October and 6 November) were assigned to main plots and three canola genotypes (Hyola 401, Zarfam and Mudena) were randomized to subplots (Hyola 401 is a spring and hybrid genotype (single cross) and cultivate in warm and moist regions commonly. However, Mudena and Zarfam are winter genotypes and cultivate in cold and moderate regions commonly). Each plot has eight rows (5 m length). Distance between each row was 25 cm and seeds implanted on rows by 8 cm distances from each other. To evaluate agronomy traits, 10 plants were selected randomly. Then the averages of the traits were used to analysis. To study germination parmeters, the seeds are provided from three genotypes in three sowing dates separately. The experimental design was a factorial arranged in CRD with three replications in laboratory. Results analysis is done by SAS ver. 9.12 software.

Results

The result of variance analysis in agronomy traits indicated significant differences between sowing dates, canola genotypes and their interactions (Table 1). Also, the results of means comparisons of interaction between sowing dates and canola genotypes (Table 2) indicated that Hyola401 has cultivated at 6 September had the best performance in many traits including: number of sub-branches, number of pod, 1000 seed weight, oil content and especially seed yield; whereas Mudena has cultivated at 6 November had the worst performance in many agronomy traits (Table 2). The result of correlation coefficients indicated that there are significant and positive relations between seed yield with number of sub-branches, number of pod, pod diameter, number of seed in pod, 1000 seed weight and oil content, whereas there are significant and negative relations between seed yield with pod length and protein content (Table 3). The results of variance analysis for germination parameters indicated that there is significant difference between sowing dates and genotypes in all traits. Also, there is significant difference between interaction of genotypes and sowing dates regard to all traits except root and stem lenght (Table 4). The result

*Corresponding author: Asadollah Gholamian, Department of Biology, Mashhad Branch, Islamic Azad University, Iran, E-mail: Gholamian.academia@yahoo.com

S.O.V	Df	MS									
		Seed yield (kg/ha)	Plant height	Number of subbranches	Number of pod	Pod length	Pod diameter	Number of seed	1000 Seed	Oil percentage	Protein percentage
Row	2	634345.3**	314.5 ns	0.25 ns	389.1**	0.55*	1.01**	0.72**	0.04 ns	3.19**	3.07**
Sowing date	2	167435.1*	271.6 ns	0.96*	356.7**	1.2*	2.03**	0.49*	0.18*	0.03 ns	0.25*
MSEa	4	18087.7	94.4	0.06	18.6	0.08	0.02	0.03	0.03	0.53	0.12
genotype	2	2663417.3*	1912.7**	7.25**	3438.3**	1.70**	9.84**	50.95**	50.95**	14.56**	17.58**
genotype* Date	4	184731.3*	397.4**	0.14*	480.1**	0.15*	3.22**	0.92**	0.92**	1.16**	1.24**
MSE	12	38599.3	55.9	0.03	80.7	0.07	0.25	0.20	0.20	0.12	0.20
CV%	---	18.8	16.3	13.8	16.4	14.9	19.9	13.6	13.6	4.86	9.8

Ns: not significant, *: P<0.05 and **: P<0.01

Table 1: Variance analysis of yield and yield components in canola genotypes.

genotypes	Sowing dates	Seed yield (kg/ha)	Plant height (cm)	Number of sub-branches	Number of pod	Pod length (cm)	Pod diameter (mm)	Number of seed in pod	1000 Seed weight	Oil percentage	Protein percentage
Mudena	6 September	1847.2 cde	141 a	4.1 cd	128.3 de	5.7 b	4.3 de	15.5 a	3.2 bc	37.9 cd	25.8 ab
	15 October	1630.4 de	130.3 ab	4.3 c	118 e	6.5 a	4.0 de	14.3 b	3.3 bc	37.5 d	26.2 a
	6 November	1432.4 e	123.6 abc	3.7 d	110.3 e	5.7 b	3.3 e	15.1 ab	2.6 c	38.2 cd	26.6 a
Hyola401	6 September	2786.7 a	131.6 ab	6.1 a	170.3 a	4.7 c	5.9 ab	9.8 e	4.2 a	40.5 a	23.2 e
	15 October	2570.6 ab	122.3 abc	6.0 a	162.3 ab	5.5 b	6.5 a	10.3 e	4.0 ab	40.0 a	23.8 ed
	6 November	2760.5 ab	105.3 cd	5.3 b	140.3 cd	5.2 bc	5.5 abc	10.5 e	3.6 ab	40.5 a	23.4 ed
Zarfam	6 September	2297.1 bc	100.3 d	5.2 b	130.3 cde	5.2 bc	5.5 abc	11.6 d	3.4 b	39.5 ab	25.0 bc
	15 October	2716.2 ab	94.0 d	4.6 c	150.6 abc	5.7 b	5.1 bcd	12.5 cd	3.3 bc	40.4 a	24.4 cd
	6 November	2022.6 cd	113.6 bcd	4.6 c	146.6 bcd	5.4 b	4.4 cd	12.7 c	3.8 ab	38.9 bc	23.4 de

Table 2: Means comparisons of interactions between sowing dates and canola cultivars in agronomy traits.

traits	Plant height	Number of sub-branches	Number of pod	Pod length	Pod diameter	Number of seed in pod	1000 Seed weight	Oil percentage	
Seed Yield	-0.47 ns	0.83**	0.84**	-0.66*	0.88**	0.88**	0.7*	0.95**	-0.84**
Plant height	1	-0.13 ns	-0.13 ns	0.22 ns	-0.25 ns	0.39 ns	0.03 ns	-0.56 ns	0.33 ns
Number of sub-branches		1	0.84**	-0.67*	0.96**	-0.95**	0.88**	0.78*	-0.78*
Number of pod			1	-0.60 ns	0.83**	-0.81**	0.88**	0.78*	-0.88**
Pod length				1	-0.60 ns	0.72*	-0.55 ns	-0.75*	-0.70*
Pod diameter					1	-0.90**	0.80**	0.81**	-0.76*
Number of seed in pod						1	-0.81**	-0.90**	0.86**
1000 Seed weight							1	0.59 ns	-0.86**
Oil percentage								1	-0.81**

Ns: not significant, *: P<0.05 and **: P<0.01

Table 3: Phenotypic correlation coefficient among agronomy traits in canola.

of meas comparison for interaction between genotypes and sowing dates indicated that Hyola 401 cultivated at 6 September and 7 October has the best quality at all germination parameters, whereas Mudena has cultivated at 6 November has the least quality (Table 5). The result of correlation coefficient for germination parameters indicated that there were significant and positive relations between all germination parameters (Table 6).

Discussion

Yield and yield components

According to the present study, it is determined that canola genotypes have contradictory reaction to sowing date, as that in early sowing dates some genotypes were tall and another was short; this is due to genetic variations and adaptation differences among genotypes. In other hands, it was observed that delay cultivation led to reducing in number of sub-branches and number of pod in all genotypes; it is probably due to shortening of vegetative period and lack of enough time to growth. Niknam et al. [5] stated that delay cultivation led to shortening of vegetative period and unfavorable conditions at the end of growth season such as heat and drought; it caused, at last, would

reduce number of pod and seed yield significantly. In this study, it was showed negative relation between number of seed in pod and 1000 seed weight, so more seeds in pod led to less 1000 seed weight. This result seems to be logical, because since sources of photosynthetic producer are constant, more seed received less photosynthetic material whereas less seed received more photosynthetic materials. Mendham et al. [6] stated that letar cultivation caused to coinciding seed growth period with high temperature of season that lead to decrease in photosynthetic productions, shortening of seed growth period, accelerating of ripening period and subsequent lead to reduction in 1000 seed weight and seed yield. Finally, it revealed that dilatary cultivation reduced yield components such as number of sub-branchs, pod diameter, 1000 seed weight and oil content, so overall seed yield surely reduced. So these traits regard as the most important yield components that their improvements lead to improvement of seed yield. Anyway, it should rememberd that different genotypes effect on seed yield is rather than different sowing dates effect, because changing of genotypes from Hyola 401 to Mudena, seed yield reduced from 2705.93 to 1636.66 Kg/ha (about 40%), whereas changing sowing dates from 6 Septmber to 6 November, seed yield reduced from 2310 to 2071 Kg/ha (about 10%); these results approved the results of Farre et al. [7].

S.O.V	MS					
	Germination percentage	Germination rate	Vigor of germination	root lenght	stem lenght	Seedling lenght
genotypes	182.3**	177.6**	232.6**	1.0**	0.16**	1.9**
Sowing time	235.7**	246.9**	354.5**	1.1**	0.19**	2.1**
Sowing time × genotypes	15.6**	26.9**	25.0**	0.2	0.02	0.3**
MSE	3.1	5.0	3.5	0.1	0.01	0.1
CV (%)	5.0	10.1	9.6	7.4	5.3	5.1

Ns: not significant, *: P<0.05 and **: P<0.01

Table 4: Variance analysis of germination parameters in three canola genotypes.

genotypes	Sowing date	Germination percentage	Germination rate	Vigor of germination	root lengh	stem lengh	Seedling lengh
Mudena	6 September	86.33 b	27.53 b	23.78 b	4.54 a	3.36 ab	7.91 a
	15 October	83.33 bc	21.48 c	17.91 c	3.89 bc	3.14 bc	7.03 ab
	6 November	80.67 c	20.66 c	16.71 c	3.7 c	3.09 bc	6.78 b
Hyola401	6 September	97.00 a	35.52 a	34.81 a	4.74 a	3.49 a	8.22 a
	15 October	93.67 a	35.91 a	31.8 a	4.73 a	3.47 a	8.24 a
	6 November	86.00 b	24.9 bc	21.39 bc	4.57 a	3.36 ab	7.92 a
Zarfam	6 September	93.33 a	34.47 a	33.63 a	4.78 a	3.51 a	8.25 a
	15 October	85.67 b	27.7 b	23.74 b	4.34 ab	3.29 abc	7.63 a
	6 November	79.33 c	20.47 c	16.47 c	3.76 bc	3.04 c	6.79 b

Table 5: Meas comparison for interaction between genotypes and sowing dates in germination parameters.

Traits	Germination rate	Vigor of germination	Root lenght	stem lenght	Seedling lenght
Germination percentage	0.97**	0.98**	0.87**	0.94**	0.91**
Germination rate	1	0.99**	0.89**	0.93**	0.93**
Vigor of germination		1	0.85**	0.96**	0.90**
Root lenght			1	0.99**	0.99**
stem lenght				1	0.99**

*: P<0.05 and **: P<0.01

Table 6: Phenotpice correlation coefficient of germination parameters in canola.

Germination parameters

We observed that produced seeds of different sowing dates had significant differences in germination parameters; on the late sowing date the seed quality was low. While cultivated seed in suit sowing date due to has more favorite growth conditions has more seed quality and vigour. Elias and Copland [8] stated that percent of germination is affected severly by seed quality, and it is necessary to use standard germination test to determine seed quality and vigour in canola. At the present paper is observed seeds had higher germination percentage, had fast germination rate and longer seedling length. In this way, Hyola 401 and Zarfam cultivated at 6 September due to enough time to growth could produce higher quality seeds with longer seedling length. Hampton and Tekrony [9] showed there is significant positive correlation between seedling length and seed quality. They, at last, reported seedling length is an index to evaluate quality of seedling. The vigour of germination is a very important parameter in germination test, because it calculated by multiplying length of seedling to germination percent. So the seeds had higher seedling length and more germination percent the seeds had higher vigour of germination. Also it is distinguished, at this study, vigour of germination is effected more sowing date rather than genotype.

Conclusion

The results of this study indicated that 6 September is the best sowing date for canola in Torbat-jam region, so the cultivation is done later the traits such as number of sub-branches, number of pod, oil content,

1000 seed weight and seed yield are worse. Also the results showed that delay cultivation lead to decreasing of seed qualities (germination percent, seedling length, vigour of germination, etc). So the seeds with high quality are producted only in genotypes cultivated in suit sowing date. At last, we recommended Hyola401 and 6 September as desirable genotype and perfect sowing date in Torbat-jam region.

References

1. FAO (2007) FAO Statistic Service.

2. Gusta LV, Johnson EV, Nesbitt NT, Klikland KJ (2003) Effect of seeding date on canola seed quality and seed vigour. Can J Plant Sci 84: 463-471.

3. Laaniste P, Joudu J, Eremeev V, Maeorg E (2007) institute of Agricultural and environmental science, stonian university of life science, tarto estonia.acta agricultural scandinavia section b-soil and plant science, 57: 342-348.

4. Elias S, Garary A, Schweitzer L, Henning S (2006) Seed quality testing of native species. Native Plants Journal 7: 15-19.

5. Niknam S, Ma RO, turner DW (2003) Osmatic adjustment seed yield of Brassica napus B. juneed genotypes in a water limited environment in sout Western Australia. Aus J of Experimental agriculture 43:1127-1135.

6. Mendham NJ, Shipway PA, Scott RK (1981) The effects of delayed sowing and weather on growth, development and yield of winter oil-seed rape (*Brassica napus*). J Agric Sci 96: 389-416.

7. Farre MJ, Robertson G, Walton H, Asseng S (2002) The effect of delayed sowing and weather on growth, development and yield of winter oilseed rape (*Brassica napus* L.). Aust J Agric Res 53: 1155-1164.

8. Elias SG, Copleland Lo (1994) The effect of storage condition on canola (*Brassica Napus* L.) seed quality. Journal of Seed Technology 18: 21-29.

9. Hampton JG, Tekrony DM (1995) Handbook of vigour test methods. (3rdedn) International Seed Testing Assoction (ISTA). Zurich, Switzerland.

39

The Producing Technology of Resistant Starch from Buckwheat Using Ultrasonic Treatment

Wang L[1]* and Bai X[2]

[1]The College of Life Science, Yangtze University, Jingzhou, Hubei, China
[2]The First People's Hospital of Jingzhou, Jingzhou, Hubei, China

Abstract

Resistant Starch (RS) has various functions in controlling the Glycemic Index (GI), lowering concentration of cholesterol and triglycerides, inhibiting fat accumulation, preventing colonic cancer, reducing gall stone formation, maintaining intestinal tract healthy and enhancing the absorption of minerals. Elevated RS in food is an important and effective approach for public health. RS is also an important material for industries. In this paper, the producing technologies of resistant starch from buckwheat were investigated. The results showed that the optimum parameters for producing technology of resistant starch from buckwheat using ultrasonic treatment are ultrasonic treatment time is 20 min, ultrasonic power is 300 W, and ultrasonic frequency is 63 KHz, Solid-to-liquid ratio 1:8.

Keywords: Buckwheat; Resistant Starch (RS); Orthogonal design

Introduction

Resistant Starch (RS) is also called enzyme resistant starch, definited as the starch and starch degradation products which cannot be digested and absorbed in the healthy small intestine of human [1]. RS provides functional properties in controlling GI [2], lowering concentration of cholesterol and triglycerides [3,4], inhibiting fat accumulation [5], preventing colonic cancer [6], reducing gall stone formation [7], maintaining intestinal tract healthy [8,9] and enhancing the absorption of minerals [10]. RS a novel insulin receptor sensitizer is benefited to diabetes, which can enhance insulin function and regulate blood glucose [11]. Elevated RS in food is an important and effective approach for public health. RS is also an important material for industries. Buckwheat (*Fagopyrum esculentum*) belonging to plants of the genus Polygonaceae Buckwheat, is edible biologic medicine with relative high starch content, with various values of nutritional therapy health care [12].

The mechanism of RS formation is largely unknown. There are several factors affect the RS formation. It's reported that RS content is positive related to AC [13,14]. Starch granule size and structure are related the RS content. Starch granule in potato is larger than that in cereals, the potato starch digested more slowly than that of cereals [15]. Starch Crystalline structure can be classified into A type, B type and C type, according X-ray scattering pattern. The digestibility of the starch with B types less than A type, C type in the middle [16]. The chain length of amylose and amylopectin is another major factor affect the RS formation. RS increase according Degree of Polymerization (DP) of amylose (from 10 DP to 610 DP) by hydrothermal treatment with retention [17]. The effect of the chain length amylopectin on RS formation is unclear in detail. It reported that amylopectin starch debranched by Pullulanase followed by heat-processing can increase RS content [18]. It's due to long unbranched chains of amylopectin involve into RS formation [19]. Other components in cell, such as protein, lipid, cellulose can also effect RS content [20,21]. Among them, Lipids is most important effect on RS formation. Lipids can decrease RS content significantly [19]. Food additives and food processing technologies are another factors can affect RS content [22,23]. We analyzed the effects of the preparing conditions to buckwheat RS content and got the optimal preparing conditions for buckwheat RS

content. The results of this work will lay the foundation of theory and application for the further study of buckwheat RS.

Materials and Methods

Preparation of Buckwheat flour

Buckwheat was purchased from Jilin City. Buckwheat was grinded into flour using flour mill, then filtered using 200 mesh sieve.

Determination of RS content

RS content was measured according to AOAC method (2002.02) with a slight modification [24]. 100 ± 1 mg milled maize flour (only endosperm) were accurately weighed and placed directly into screw-cap tubes (16×125 mm). 500 µL water was added into each tube, then boiled in electric cooker for 20 min and at warm keeping status at 50°C for 10 min. Tubes were taken out and cooled to room temperature. KCl-HCl buffer (pH=1.5) containing 6 IU/mg pepsin was added into each tube and the rice floury was ground and dispersed by a stirring rod, mimicking the chewing in mouth and warmed at 37°C for 1 h. Other procedures were carried out as described in the method AOAC (2002.02) [24].

The optimization of the preparation process of buckwheat RS

To optimize the preparation process of buckwheat RS, the major factors and their levels were determined according the effects of various factors (such as ultrasonic power, ultrasonic frequency, ultrasonic treatment time, heating temperature after ultrasonic treatment, Solid-to-Liquid ratio (S/L) using ultrasonic treatment) on RS content using ultrasonic treatments. The optimum preparation conditions of buckwheat RS were further determined using orthogonal test.

*Corresponding author: Wang L, The College of Life Science, Yangtze University, Jingzhou, Hubei, China, E-mail: ljwang516@126.com

Results and Discussion

The effects of ultrasonic frequency on buckwheat RS content

The buckwheat starch was heated at 100℃ for 20 min after using different ultrasonic frequency for 30 s with 1:2 solid-to-liquid ratio, then after storage at 4℃ for 24 h, dried at 50℃ for 18 h. The RS content of the dried buckwheat was analyzed. The optimum ultrasonic frequency is 40 KHz (Figure 1).

The effects of ultrasonic power on buckwheat RS content

The buckwheat starch was heated at 100℃ after using different ultrasonic power for 30 s with 1:2 solid-to-liquid ratio, then after storage at 4℃ for 24 h, dried at 50℃ for 18 h. The RS content of the dried buckwheat was analyzed. The optimum ultrasonic power is 300 W (Figure 2).

The effects of ultrasonic treatment time on buckwheat RS content

The buckwheat starch was heated at 100℃ after using 28 KHz ultrasonic power for different time with 1:2 solid-to-liquid ratios, then after storage at 4℃ for 24 h, dried at 50℃ for 18 h. The RS content of the dried buckwheat was analyzed. The optimum ultrasonic treatment time is 20 min (Figure 3).

Effects of heating temperature after ultrasonic treatment on buckwheat RS

The buckwheat starch was heated at different temperature after using 28 KHz ultrasonic power for 30s with 1:2 solid-to-liquid ratio, and then heated at different temperature after storage at 4℃ for 24 h, dried at 50℃ for 18 h. The RS content of the dried buckwheat was analyzed. The optimum heating temperature after ultrasonic treatment is 120℃ (Figure 4).

The effects of solid-to-liquid ratio using ultrasonic treatment on buckwheat RS content

The buckwheat starch was heated at 100℃ after using 28 KHz ultrasonic power for 30s with different solid-to-liquid ratio, then after storage at 4℃ for 24 h, dried at 50℃ for 18 h. The RS content of the dried buckwheat was analyzed. The optimum solid-to-liquid ratio using ultrasonic treatment is 1:6 (Figure 5).

RS processing orthogonal experiment

According the effects of individual factors on the RS contents, orthogonal experiments were conducted using microwave power, treatment time using microwave power, solid-to-liquid ratio and annealing time after microwave treatment as factors and RS content as index (Tables 1 and 2).

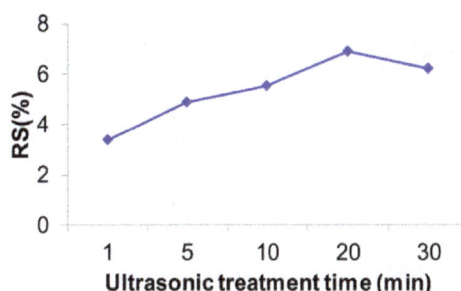

Figure 3: Effects of ultrasonic treatment time on buckwheat RS.

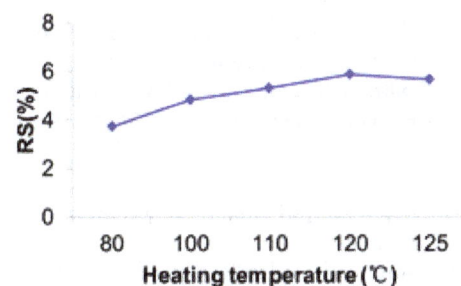

Figure 4: Effects of heating temperature after ultrasonic treatment on buckwheat RS.

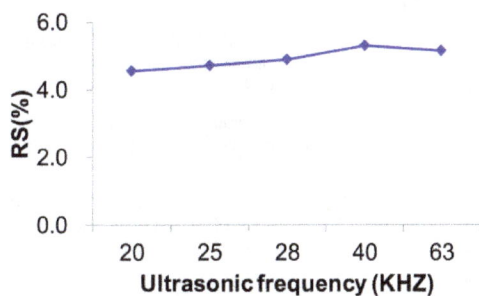

Figure 1: Effects of ultrasonic frequency on buckwheat RS.

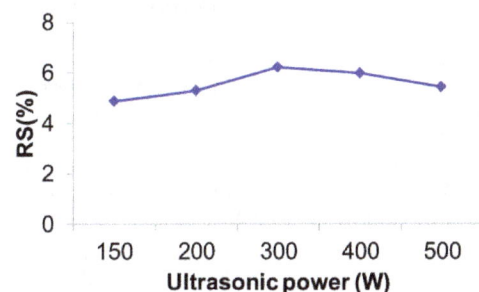

Figure 2: Effects of ultrasonic power on buckwheat RS.

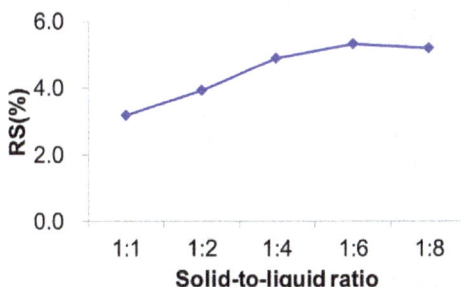

Figure 5: Effects of solid-to-liquid ratio using ultrasonic treatment on buckwheat RS content.

Level	A	B	C	D
	Power (W)	Ultrasonic frequency (KHz)	Solid-liquid ratio	Treatment time (min)
1	200	28	01:04	10
2	300	40	01:06	20
3	400	63	01:08	30

Table 1: Factor level table.

S. No.	A	B	C	D	RS
	Power (W)	Ultrasonic frequency (KHz)	Solid-liquid ratio	Treatment time (min)	(%)
1	1 (200)	1 (28)	1 (1:4)	1 (10)	5.93
2	1 (200)	2 (40)	2 (1:6)	2 (20)	6.95
3	1 (200)	3 (63)	3 (1:8)	3 (30)	6.84
4	2 (300)	1 (28)	2 (1:6)	3 (30)	7.14
5	2 (300)	2 (40)	3 (1:8)	1 (10)	7.07
6	2 (300)	3 (63)	1 (1:4)	2 (20)	7.73
7	3 (400)	1 (28)	3 (1:8)	2 (20)	7.32
8	3 (400)	2 (40)	1 (1:4)	3 (30)	6.89
9	3 (400)	3 (63)	2 (1:6)	1 (10)	6.67
K1	19.72	20.39	20.55	19.67	-
K2	21.94	20.91	20.76	22	-
K3	20.88	21.24	21.23	20.87	-
R	2.22	0.85	0.68	2.33	-

Table 2: L_9 (3^4) RS processing orthogonal experiment design and results.

As the results shown in the Table 2, ultrasonic treatment time had the largest effect on RS content. Ultrasonic Power had the second largest effect on RS content. Ultrasonic frequency had the third largest effect on RS content. Solid-liquid ratio had the fourth largest effect on RS content. The optimum parameters for producing technology of resistant starch from buckwheat using ultrasonic treatment are D2A2B3C3, that is ultrasonic treatment time is 20 min, ultrasonic Power is 300 W, ultrasonic frequency is 63 KHz, Solid-to-liquid ratio 1:8. The sequence of effects on RS content: D>A>B>C.

Conclusion

The major factors on RS content using ultrasonic treatment are ultrasonic treatment time, microwave power, ultrasonic frequency, and solid-to-liquid ratio. The optimum parameters for producing technology of resistant starch from buckwheat using ultrasonic treatment are ultrasonic treatment time is 20 min, ultrasonic Power is 300 W, and ultrasonic frequency is 63 KHz, Solid-to-liquid ratio 1:8.

Acknowledgement

This work was supported by the PhD Start-up Fund of Natural Science Foundation under Grant 801100010121.

References

1. Escarpa A, Gonzalez MC, Morales MD, Saura-Calixto F (1997) An approach to the influence of nutrients and other food constituents on resistant starch formation. Food Chem 60: 527-532.

2. Hasjim J, Lee SO, Hendrich S, Setiawan S, Ai Y, et al. (2010) Characterization of a novel resistant-starch and its effects on postprandial plasma-glucose and insulin responses. Cereal Chem 87: 257-262.

3. Han KH, Iijuka M, Shimada KI, Sekikawa M, Kuramochi K, et al. (2005) Adzuki resistant starch lowered serum cholesterol and hepatic 3-hydroxy-3-methylglutaryl-CoA mRNA levels and increased hepatic LDL-receptor and cholesterol 7α-hydroxylase mRNA levels in rats fed a cholesterol diet. Brit J Nutr 94: 902-908.

4. Martinez-Puig D, Mourot J, Ferchaud-Roucher V, Anguita M, Garcia F, et al. (2006) Consumption of resistant starch decreases lipogenesis in adipose tissues but not in muscular tissues of growing pigs. Livest Sci 99: 237-247.

5. Higgins JA, Jackman MR, Brown IL, Johnson GC, Steig A, et al. (2011) Resistant starch and exercise independently attenuate weight regain on a high fat diet in a rat model of obesity. Nutr Metab 8: 49.

6. Burn J, Bishop DT, Chapman PD, Elliott F, Bertario L, et al. (2011) A randomized placebo-controlled prevention trial of aspirin and/or resistant starch in young people with familial adenomatous polyposis. Cancer Prev Res 4: 655-665.

7. Birkett AM, Mathers JC, Jones GP, Walker KZ, Roth MJ, et al. (2000) Changes to the quantity and processing of starchy foods in a Western diet can increase polysaccharides escaping digestion and improve in vitro fermentation variables. Br J Nutr 84: 63-72.

8. Lesmes U, Beards EJ, Gibson GR, Tuohy KM, Shimoni E (2008) Effects of resistant starch type III polymorphs on human colon microbiota and short chain fatty acids in human gut models. J Agr Food Chem 56: 5415-5421.

9. Phillips J, Muir JG, Birkett A, Lu ZX, Jones GP, et al. (1995) Effect of resistant starch on fecal bulk and fermentation-dependent events in humans. American J Clin Nutr 62: 121-130.

10. Yonekura L, Suzuki H (2005) Effects of dietary zinc levels, phytic acid and resistant starch on zinc bioavailability in rats. Eur J Nutr 44: 384-391.

11. Robertson MD, Bickerton AS, Dennis AL, Vidal H (2005) Insulin-sensitizing effects of dietary resistant starch and effects on skeletal muscle and adipose tissue metabolism. Am J Clin Nutr 82: 559-567.

12. Zhang ZL, Zhou ML, Tang Y, Li FL, Tang YX, et al. (2012) Bioactive compounds in functional buckwheat food. Food Res Int 49: 389-395.

13. Yadav BS, Sharma A, Yadav RB (2009) Studies on effect of multiple heating/cooling cycles on the resistant starch formation in cereals, legumes and tubers. Int J Food Sci Nutr 60: 258-272.

14. Leeman AM, Karlsson ME, Eliasson AC, Bjorck IM (2006) Resistant starch formation in temperature treated potato starches varying in amylose/amylopectin ratio. Carbohyd Polym 65: 306-313.

15. Ring SG, Gee JM, Whittam M, Orford P, Johnson IT (1988) Resistant starch: its chemical form in foodstuffs and effect on digestibility in vitro. Food Chem 28: 97-109.

16. Englyst HN, Veenstra J, Hudson GJ (1996) Measurement of rapidly available glucose (RAG) in plant foods: a potential in vitro predictor of the glycaemic response. Brit J Nutr 75: 327-337.

17. Eerlingen RC, Crombez M, Delcour JA (1993) Enzyme-resistant starch .1. quantitative and qualitative influence of incubation-time and temperature of autoclaved starch on resistant starch formation. Cereal Chem 70: 339-344.

18. Berry CS (1986) Resistant starch-formation and measurement of starch that survives exhaustive digestion with amylolytic enzymes during the determination of dietary fiber. J Cereal Sci 4: 301-314.

19. Mangala SL, Udayasankar K, Tharanathan RN (1999) Resistant starch from processed cereals: the influence of amylopectin and non-carbohydrate constituents in its formation. Food Chem 64: 391-396.

20. Escarpa A, Gonzalez MC (1997) Technology of resistant starch. Food Sci Technol Int 3: 149-161.

21. Torre M, Rodriguez AR, Sauracalixto F (1992) Study of the interactions of calcium-ions with lignin, cellulose, and pectin. J Agr Food Chem 40: 1762-1766.

22. Gelencser T, Gal V, Salgo A (2008) Effects of applied process on the in vitro digestibility and resistant starch content of pasta products. Food Bioprocess Tech 3: 491-497.

23. Mulinacci N, Ieri F, Giaccherini C, Innocenti M, Andrenelli L, et al. (2008) Effect of cooking on the anthocyanins, phenolic acids, glycoalkaloids, and resistant starch content in two pigmented cultivars of Solanum tuberosum L. J Agr Food Chem 56: 11830-11837.

24. McCleary BV, McNally M, Rossiter P, Aman P, Amrein T, et al. (2002) Measurement of resistant starch by enzymatic digestion in starch and selected plant materials: collaborative study. J Aoac Int 85: 1103-1111.

Permissions

List of Contributors

Fangbin Cao, Wasim Ibrahim, Yue Cai and Feibo Wu
Department of Agronomy, College of Agriculture and Biotechnology, Zhejiang University, Hangzhou, China

Li Liu
Department of Agronomy, College of Agriculture and Biotechnology, Zhejiang University, Hangzhou, China Hangzhou Wanxiang Vocational and Technical College, Hangzhou, China

V.G. Stanley, P. Shanklyn, M. Daley, C. Gray and V. Vaughan
Prairie View A&M University, Prairie View, Texas 77446, USA

A. Hinton Jr
Poultry Processing and Swine Physiology Unit, Agricultural Research Service, United States Department of Agriculture, 950 College Station Road, Russell Research Center, Athens, GA 30604, USA

M. Hume
United States Department of Agriculture, Agricultural Research Service, College Station, Texas 77845, USA

Adhi Shankar and M Pratap
Department of Horticulture, College of Horticulture, Rajendranagar, Dr. Y.S.R. Horticultural University, Hyderabad, India

RVSK Reddy
Principal Scientist (Hort.), Vegetable Research Station, Rajendranagar, Dr. Y.S.R. Horticultural, Hyderabad, India

M Sujatha
Department of Genetics and Plant Breeding, College of Agriculture, Hyderabad, India

Kisan B, Shruthi H, Sharanagouda H, Revanappa SB and Pramod NK
Department of Biotechnology, College of Agriculture, University of Agricultural Sciences, Raichur-584104, India

Vanilarasu K
Ph. D. Scholar (Horticulture) in Fruit Science, Department of Fruit Crops, Tamil Nadu Agricultural University, Coimbatore – 641 003, Tamil Nadu, India

Balakrishnamurthy G
Professor (Horticulture), in Fruit Science, Department of Fruit Crops, Tamil Nadu Agricultural University, Coimbatore – 641 003, Tamil Nadu, India

Uday Chand Basak, Ajay K Mahapatra and Satarupa Mishra
Regional Plant Resource Centre, R and D Institute of Forest and Environment Department, Govt. of Odisha, Bhubaneswar, India

Chung U
Climate Application Department, APEC Climate Center, Busan 48059, Republic of Korea

Kim YU and Seo BS
Department of Plant Science, College of Agriculture and Life Science, Seoul National University, Seoul 08826, Republic of Korea

Seo MC
Crop Production and Physiology Research Division, National Institute of Crop Science, Rural Development Administration, Jeonju 55365, Republic of Korea

Subudhi R
Department of Soil and Water Conservation Engineering, College of Agricultural Engineering and Technology, Orissa University of Agriculture and Technology, Bhubaneswar, Odisha, India

Franklin E Nlerum
Department of Agricultural and Applied Economics/Extension, Rivers State University of Science and Technology, Nkpolu-Oroworukwo, Nigeria

Talukder A and Islam MA
Statistics Discipline, Khulna University, Khulna-9208, Bangladesh

Sakib MS
Department of Statistics, Jagannath University, Dhaka-1100, Bangladesh

Mathew MK
National Centre for Biological Sciences (TIFR), Bellary Road, GKVK Campus, Bangalore, Karnataka 560065, India

Chowdery RA
National Centre for Biological Sciences (TIFR), Bellary Road, GKVK Campus, Bangalore, Karnataka 560065, India Manipal University, Madhav Nagar, Manipal, Karnataka 576104, India

Shashidhar HE
Department of Plant Biotechnology, University of Agricultural Sciences, Bellary Road, GKVK Campus, Bangalore, Karnataka 560065, India

Tanmay K
Jawaharlal Nehru Technological University, Hyderabad India

Umakanth AV, Madhu P and Bhat V
Indian Institute of Millets Research, Hyderabad India

Vijaya K. Varanasi
Department of Plant Sciences, North Dakota State University, Fargo, ND 58102, USA

Wun S. Chao, James V. Anderson and David P. Horvath
Department of Plant Sciences, North Dakota State University, Fargo, ND 58102, USA
United States Department of Agriculture, Agricultural Research Service, Biosciences Research Laboratory, P.O. Box 5674, State University Station, Fargo, ND 58105- 5674, USA

Mousa Khani
Department of Cultivation and Development of medicinal plants, Iranian Institute of Medicinal Plants, [ACECR], P.O. Box: 13145-1446, Tehran, Iran Department of Plant Protection, Faculty of Agriculture, University Putra Malaysia, 43400 UPM Serdang, Selangor, Malaysia

Rita Muhamad Awang and Dzolkhifli Omar
Department of Plant Protection, Faculty of Agriculture, University Putra Malaysia, 43400 UPM Serdang, Selangor, Malaysia

Mawardi Rahmani
Department of Chemistry, Faculty of Science, University Putra Malaysia, 43400 UPM Serdang, Selangor, Malaysia

Ghimire S
Technical officer, Nepal Agricultural Research council, Nepal

Sherchan DP
ARTC Manager, Cereal System Initiative for South Asia, Nepal

Andersen P
Associate Professor, Departrment of Geography, University of Bergen, Norway

Pokhrel C
Associate Professor, Central Departrment of Botany, Tribhuvan University, Nepal

Ghimire S
Assistant Manager, Rastriya Banijya Bank, Nepal

Khanal D
Senior Program Officer, Karuna Foundation Nepal

Harb OM, Abd El-Hay GH, Hager MA and Abou El-Enin MM
Agronomy Department, Faculty of Agriculture, Al-Azhar University, Cairo, Egypt

Gautam AK, Gupta N, Bhadkariya R and Bhagyawant SS
School of Studies in Biotechnology, Jiwaji University, Gwalior, 474011, India

Srivastava N
Department of Bioscience and Biotechnology, Banasthali University, Banasthali, 304022, India

Gurmu F and Mekonen S
South Agricultural Research Institute, Hawassa Research Center, Hawassa, Ethiopia

Karma Landup Bhutia, NG Tombisana Meetei and VK Khanna
School of Crop Improvement, College of Post Graduate Studies, CAU, Umiam, Meghalaya, India

Rajeshwar Malavath, Ravinder Naik, Pradeep T and Sreedhar Chuhan
Acharya N.G Ranga Agricultural University, District Agricultural Advisory and Transfer of Technology Center, KVK, ARS, Adilabad-504 001, RARS, Jagitial, India

Simon MK and Jegede CO
Department of Veterinary Parasitology and Entomology, Faculty of Veterinary Medicine, University of Abuja, Nigeria

Rediet Girma and Awdenegest Moges
School of Biosystems and Environmental Engineering, Hawassa University, Ethiopia

Shoeb Quraishi
School of Natural Resource and Environmental Engineering, Haramaya University, Ethiopia

Marcelo Zolin Lorenzoni, Roberto Rezende, Álvaro Henrique Cândido De Souza, Cássio De Castro Seron, Tiago Luan Hachmann and Paulo Sérgio Lourenço De Freitas
State University of Maringá, Maringá, Paraná, Brazil

C Devendra
Consulting Tropical Animal Production Systems Specialist, 130A Jalan Awan Jawa, Kuala Lumpur, Malaysia

V.G. Stanley, K. Hickerson and M.B. Daley
Prairie View A & M University, Prairie View, Texas, USA

M. Hume
Food and Feed Safety Research Unit, Agricultural Research Service, United States Department of Agriculture, College Station, Texas, USA

A. Hinton
Poultry Processing and Swine Physiology Unit, Agricultural Research Service, United States Department of Agriculture, 950 College Station Road, Russell Research Center, Athens, Georgia

Zahra Hosseini Cici S
School of Crop Science, University of Guelph, Ontario, Canada
School of Sustainable Agriculture, University of Payame-Noor, Tehran, Iran

Adewumi IO, Oluwatoyinbo FI, Omoyajowo AO, Ajisegiri GO and Akinsete AE
Department of Agricultural Engineering, Federal College of Agriculture, P.M.B. 5029, Moor Plantation, Ibadan, Oyo State Nigeria

Parwada C
Department of Agronomy, University of Fort Hare, Alice, South Africa
Department of Crop Science, Bindura University of Science Education, Bindura, Zimbabwe

van Tol J
Department of Agronomy, University of Fort Hare, Alice, South Africa
Department of Soil-and Crop-and Climate Sciences, University of the Free State, Bloemfontein, South Africa

Elham Hassanpour and Jamal-Ali Olfati
Faculty of Agriculture, Horticultural Department, University of Guilan, Rasht, Iran

Mohammad Naqashzadegan
Faculty of Engineering, Mechanic Department, University of Guilan, Rasht, Iran

Margarita Islas-Pelcastre
Agronomy and Forestry Area, Institute of Agricultural Sciences, Universidad Autónoma del Estado de Hidalgo Av, Universidad Km 1, Ex-Hacienda de Aquetzalpa, Tulancingo, Hidalgo, Mexico

Jose Roberto Villagómez-Ibarra
Cademic Area of Chemistry. Institute of Basic Science and Engineering, Universidad Autónoma del Estado de Hidalgo, Ciudad del Conocimiento, Carretera Pachuca Tulancingo, Mineral de la Reforma, Hidalgo, Mexico

Blanca Rosa Rodríguez-Pastrana
Agrobusiness and Food Engineering Area. Institute of Agricultural Sciences, Universidad Autónoma del Estado de Hidalgo Av, Universidad Km 1. Ex-Hacienda de Aquetzalpa, Tulancingo, Hidalgo, Mexico

Gregory Perry
Department of Plant Agriculture, Ontario Agricultural College, University of Guelph, Guelph, Ontario, Canada

Alfredo Madariaga-Navarrete
Agronomy and Forestry Area, Institute of Agricultural Sciences, Universidad Autónoma del Estado de Hidalgo, Av, Universidad Km 1, Ex-Hacienda de Aquetzalpa, Tulancingo, Hidalgo, Mexico

Jian PU and Kensuke Fukushi
Integrated Research System for Sustainability Science, The University of Tokyo, Japan

Umunakwe PC, Nnadi FN and Chikaire J
Department of Agricultural Extension, Federal University of Technology, P.M.B. 1526 Owerri Imo State, Nigeria

Nnadi CD
Department of Agricultural Economics and Rural Sociology, Niger Delta University, Wilberforce Island, Bayelsa State, Nigeria

Ghimire S, Pandey S and Gautam S
Technical officer, National Potato Research Program, NARC, Khumaltar Lalitpur, Nepal

Nall I Moonilall and Jayachandran K
Department of Earth and Environment, Florida International University, Miami, FL, USA

Reed S
USDA/ARS Subtropical Horticulture Research Station, Miami, FL, USA

Asad M F AlKhader
Water, Soil and Environment Department, National Center for Agricultural Research and Extension, Jordan

Wang L
The College of Life Science, Yangtze University, Jingzhou, Hubei, China

Bai X
The First People's Hospital of Jingzhou, Jingzhou, Hubei, China

Osuagwu AN, Ekpo IA, Okpako EC, Otu P and Ottoho E
Department of Genetics and Biotechnology, University of Calabar, Nigeria

Asadollah Gholamian
Department of Biology, Mashhad Branch, Islamic Azad University, Mashhad, Iran

Mahdi Bayat
Department of Agriculture, Mashhad Branch, Islamic Azad University, Mashhad, Iran

Wang L
The College of Life Science, Yangtze University, Jingzhou, Hubei, China

Bai X
The First People's Hospital of Jingzhou, Jingzhou, Hubei, China

Index